MW00760962

Texts and Monographs in Physics

Series Editors: R. Balian   W. Beiglböck   H. Grosse   E. H. Lieb
N. Reshetikhin   H. Spohn   W. Thirring

Springer
Berlin
Heidelberg
New York
Barcelona
Hong Kong
London
Milan
Paris
Singapore
Tokyo

Texts and Monographs in Physics

Series Editors: R. Balian   W. Beiglböck   H. Grosse   E. H. Lieb
N. Reshetikhin   H. Spohn   W. Thirring

Francisco J. Ynduráin

The Theory
of Quark and Gluon
Interactions

Third Revised and Enlarged Edition
With 79 Figures

 Springer

Professor Dr. Francisco J. Ynduráin

Departamento de Física Teórica, C-XI
Universidad Autónoma de Madrid, Canto Blanco
E-28049 Madrid, Spain
e-mail: fjy@delta.ft.uam.es

Editors

Roger Balian

CEA
Service de Physique Théorique de Saclay
F-91191 Gif-sur-Yvette, France

Nicolai Reshetikhin

Department of Mathematics
University of California
Berkeley, CA 94720-3840, USA

Wolf Beiglböck

Institut für Angewandte Mathematik
Universität Heidelberg, INF 294
D-69120 Heidelberg, Germany

Herbert Spohn

Zentrum Mathematik
Technische Universität München
D-80290 München, Germany

Harald Grosse

Institut für Theoretische Physik
Universität Wien
Boltzmanngasse 5
A-1090 Wien, Austria

Walter Thirring

Institut für Theoretische Physik
Universität Wien
Boltzmanngasse 5
A-1090 Wien, Austria

Elliott H. Lieb

Jadwin Hall
Princeton University, P.O. Box 708
Princeton, NJ 08544-0708, USA

ISSN 0172-5998
ISBN 3-540-64881-X 3rd Ed. Springer-Verlag Berlin Heidelberg New York

ISBN 3-540-55803-9 2nd Ed. Springer-Verlag Berlin Heidelberg New York

Library of Congress Cataloging-in-Publication Data. Ynduráin, F. J. The theory of quark and gluon inter-actions / F. J. Ynduráin – 3rd rev. and enl. ed. p. cm. – (Texts and monographs in physics, ISSN 0172-5998) Includes bibliographical references and index. ISBN 3-540-64881-X (alk. paper) 1. Quantum chromody-namics. I. Title. II. Series. QC793.5.Q2528Y58 1999 539.7'548–dc21 99-14725

Typesetting: Data conversion by Fa. Imsieke & Schmidt, Leipzig
Cover design: *design & production*, Heidelberg
SPIN 10687252   55/3144/ba - 5 4 3 2 1 0 – Printed on acid-free paper

To Elsa, Marcos and Elena

Preface to the Third Edition

QCD is an ever growing area of physics; when writing a new edition of a book dealing with it, it is thus impossible not to take into account some at least of the new developments in the field. In selecting those to be included, I have followed the principle of incorporating developments pertaining to topics already treated in the former editions. Thus I have *not* included sections on chiral dynamics or effective theories.* What *has* been included are expansions of a number of sections, in particular of those dealing with deep inelastic scattering. Here new material is included both on higher order calculations, quite a number of which have become available in the last few years and, especially, on the small x limit of structure functions where, triggered by the results of HERA, there has been considerable activity. In this chapter dealing with perturbative QCD we have also added a section on τ and Z decays, at present the more reliable processes from which to extract the QCD coupling constant, α_s.

Two more topics have also received special attention. One is the matter of bound states of heavy quarks, where inclusion of higher order perturbative and nonperturbative evaluations has led to a clarification of the QCD description, particularly of lowest states of heavy quarkonia: Chapter 6 has been almost entirely rewritten. The other concerns Chapter 10, expanded to include the results of high order (four loop) calculations of β, γ_m; an updated determination of the parameters of the theory (quark masses, Λ, condensates); and a few considerations on the character of the perturbative QCD series, including discussion of saturation and renormalons.

Besides this, I have profited to improve some of the features of the book: a general polishing, including updating of Sect. 7.4 and relocating of the sections on instantons and lattice QCD from Chapter 8 to two separate chapters (plus addition of a small Subsection 9.5iii); and technical improvements such as replacement of the old fashioned figures by computer generated graphs.

For this new edition, I would like to record my gratitude to, besides people already quoted in former ones, A. Pich, J. Vermaseren and Yu. Simonov, who read some of the new material in the manuscript; and to K. Adel for his program "Kdraw", used for the computer generation of the figures. Finally,

---

* With respect to the first, see bibliography in text. For the second, the interested reader may consult the original paper of Caswell and Lepage (1986) and the reviews of Lepage and Thacker (1988) and Grinstein (1991).

the continued support of Springer–Verlag in all editorial matters is warmly acknowledged.

Madrid, December 1998   *F. J. Ynduráin*

Preface to the Second Edition

The present book is not merely an elaboration of the 1982 text *Quantum Chromodynamics: An Introduction to the Theory of Quarks and Gluons*. In fact, a lot of material has been added: some of which is entirely new, and some of which is an extension of topics in the older text. Among the latter there are two sections dealing with the background field method, and the expansion of the section devoted to processes describable by perturbative QCD (other than deep inelastic scattering) into a whole chapter, containing a detailed description of Drell–Yan scattering, jet physics, exclusive processes, QCD sum rules, etc. Besides this, I have added a completely new chapter on constituent quark models of hadrons, including a derivation of the quark–quark potential and, also entirely new, a half chapter containing a detailed introduction to lattice QCD. An extra short chapter on the parameters of QCD and an appendix on group integration contribute to making the present book a really new text, sufficiently so to justify the change of title to *The Theory of Quark and Gluon Interactions*: a change that also emphasizes the consolidation of QCD as *the* theory of strong interactions.

Of course, even with the inclusion of new material there are whole areas of quark and gluon physics not covered at all. Among these, let me mention the large N_c limit of QCD (the interested reader may consult 't Hooft, 1974a,b, Witten, 1979b, 1980); the infrared problems in QCD, very poorly understood (see, e.g., the classic paper of Lee and Nauenberg, 1964; Muller, 1978 and Zachariasen, 1980), and, especially, QCD at finite temperature, a fashionable subject at present, which the reader may follow from the review of Gross, Pisarski and Yaffe (1981).

Besides the people quoted in the preface to the 1982 edition, I would like to acknowledge the valuable influence of my scientific involvement with R. Akhoury, F. Barreiro, G. López Castro and M. Veltman, and thank again A. González-Arroyo, who kindly undertook a most useful critical reading of the part concerning lattice QCD.

Madrid, 1992   *F. J. Ynduráin*

Preface to the First Edition

Quantum Chromodynamics – An Introduction to the Theory of Quark and Gluon Interactions

It is almost thirty years since Yang and Mills (1954) performed their pioneering work on gauge theories, and it is probably safe to say that we have in our hands a good candidate for a theory of the strong interactions based, precisely, on a non-Abelian gauge theory. While our understanding of quantum chromodynamics (QCD) is still incomplete, there have been sufficient theoretical developments, many of them enjoying a degree of support from experimental evidence, to justify a reasonably systematic treatise on the subject.

Of course, no presentation of QCD can claim to be complete, since the theory is still in the process of elaboration. The selection of topics reflects this: I have tried to discuss those parts of the theory that are more likely to endure, and particularly those developments that can, with a minimum of rigour, be derived from "first principles". To be sure, prejudice has also influenced my choice: one necessarily tends to give more attention to subjects with which one is familiar, and to eschew unfamiliar ones. I will not pause here to point out topics which perhaps should have been included; the list of references should fill the gaps.

This work grew out of graduate courses I have been teaching for the last few years: the book is intended to reflect the pedagogical and introductory nature of those lectures. With this aim in mind, I have tried to write a self-contained text which avoids as far as possible the maddening circumventions of sentences like "it can be shown" or "as is well known". However, I have assumed the reader to have a basic knowledge of field theory and particle phenomenology, and have no doubt that occasional recourse to the literature will be necessary.

What this book owes to the standard reviews and articles on the subject should be apparent and is recorded in the references. I have directly benefited from collaboration with my colleagues: A. González-Arroyo, C. Becchi, S. Narison, J. Bernabeu, E. de Rafael, R. Tarrach and, particularly, C. López and P. Pascual (who also spotted several mistakes in a preliminary version of this work), to name only a few. I also wish to acknowledge the invaluable secretarial help of Antoinette Malene.

Madrid, 1982

F. J. Ynduráin

Contents

Appendices

1 Generalities

1.1 The Rationale for QCD

Historically, quantum chromodynamics (QCD) originated as a development
of the quark model. In the early sixties it was established that hadrons could
be classified according to the representations of what today we would call
flavour $SU_F(3)$ (Gell-Mann, 1961; Ne'eman, 1961). This classification presented a number of features that are worth noting. First of all, only a few,
very specific representations occurred; they were such that they built representations of a group $SU(6)$ (Gürsey and Radicati, 1964; Pais, 1964) obtained
by adjoining the group of spin rotations $SU(2)$ to the internal symmetry
group, $SU_F(3)$. However, neither for $SU_F(3)$, or $SU(6)$ did the fundamental
representations (3 and $\bar{3}$ for $SU_F(3)$) appear to be realized in nature. This led
Gell-Mann (1964a) and Zweig (1964) to postulate that physical hadrons are
composite objects, made up of three *quarks* (baryons) or a quark–antiquark
pair (mesons). These three quarks are now widely known as the three *flavours*,
u (up), d (down) and s (strange); the first two carry the quantum numbers
of isospin, and the third strangeness. It has been found that precisely those
representations of $SU_F(3)$ occur that may be obtained by reducing the prod-
ucts $3 \times 3 \times 3$ (baryons) or $3 \times \bar{3}$ (mesons); when the spin $1/2$ of the quarks
is taken into account, the $SU(6)$ scheme is obtained. In addition, the mass
differences of the hadrons may be understood by assuming

$$m_d - m_u \approx 4 \text{ MeV}, \quad m_s - m_d \approx 150 \text{ MeV}, \qquad (1.1.1)$$

together with eventual electromagnetic radiative corrections. The electric
charges of the quarks, in units of the proton charge, are

$$Q_u = \tfrac{2}{3}, \quad Q_d = Q_s = -\tfrac{1}{3}. \qquad (1.1.2)$$

That hadrons are composite objects was a welcome hypothesis on other
grounds, too. For example, it is known that the magnetic moment of the
proton is $\mu_p = 2.79 \times e\hbar/2m_p$, instead of the value $\mu_p = e\hbar/2m_p$ expected if
it were elementary. The values of the magnetic moments calculated with the

quark model are, on the other hand, in reasonable agreement with experimental results.

These successes stimulated a massive search for quarks that still goes on. None of the candidates found to this date has been confirmed, but at least we have a *lower* bound (of the order of dozens of GeV) for the mass of free quarks, which seems to imply that hadrons are very tightly bound states of quarks. This picture, however, can be challenged on at least two grounds. First, the fundamental state of a composite system, in the $SU(6)$ scheme, is one in which all relative angular momenta vanish. Thus the Δ^{++} resonance had to be interpreted as being made up of

$$u_\uparrow, u_\uparrow, u_\uparrow, \tag{1.1.3}$$

(where the arrows stand for spin components) at relative rest. However, this is preposterous: being spin one-half objects, quarks should obey Fermi–Dirac statistics and their states should be antisymmetric, which is certainly not the case in (1.1.3). Second, one can use current algebra techniques (Gell-Mann, Oakes and Renner, 1968; Glashow and Weinberg, 1968; Leutwyler, 1974) to calculate m_s/m_d with the result

$$m_s/m_d \simeq 20, \tag{1.1.4}$$

which is a flat contradiction of (1.1.1) for quarks of a few GeV of mass.

With respect to the first objection, a possible solution was proposed by Greenberg (1964), who assumed that quarks obey parastatistics of rank three. It is known that such parastatistics can be disposed of by taking ordinary Fermi–Dirac statistics and introducing a new internal quantum number,[1] which Gell-Mann and his collaborators[2] called *"colour"*, so that each species of quark may come in any of the three colours $i = r,\ y,\ v$ (red, yellow, violet). Then, one can reinterpret the Δ^{++} as

$$\sum \epsilon^{ikl}(u_\uparrow^i, u_\uparrow^k, u_\uparrow^l),$$

i.e., perfectly antisymmetric. In addition, the absence of states with, say, two or four quarks (so-called "exotics") could be explained by postulating that all physical hadrons are colourless; that is to say, that they are singlets under rotations in colour space:

$$U_c : q^i \to \sum_k U_c^{ik} q^k, \quad U_c U_c^\dagger = 1. \tag{1.1.5}$$

If we take these transformations of determinant unity so as to eliminate a trivial overall phase, they build a new invariance group, namely colour

---

[1] In fact, a colour quantum number was first introduced by Han and Nambu (1965).

[2] See Bardeen, Fritzsch and Gell-Mann (1972); Fritzsch and Gell-Mann (1972); Fritzsch, Gell-Mann and Leutwyler (1973).

$SU_c(3)$. Now the singlet representation of this group only appears in the products $3_c \times 3_c \times 3_c$ (baryons) or $\bar{3}_c \times 3_c$ (mesons), and this explains why we have these particles, and no exotics, which we do not find in nature.

We will not yet discuss a solution to the second difficulty, but rather make it worse by digressing to current algebra. If quarks are elementary, one must build *currents* out of quarks. Thus, the electromagnetic (e.m.) current is

$$J_{\text{em}}^\mu = \tfrac{2}{3}\bar{u}\gamma^\mu u - \tfrac{1}{3}\bar{d}\gamma^\mu d - \tfrac{1}{3}\bar{s}\gamma^\mu s + \tfrac{2}{3}\bar{c}\gamma^\mu c; \tag{1.1.6a}$$

and the charged weak current

$$J_{\text{W}}^\mu = \bar{u}\gamma^\mu \frac{1-\gamma_5}{2} d_\theta + \bar{c}\gamma^\mu \frac{1-\gamma_5}{2} s_\theta,$$

$$d_\theta = d\cos\theta_C + s\sin\theta_C; \quad s_\theta = -d\sin\theta_C + s\cos\theta_C \tag{1.1.6b}$$

(θ_C is the Cabibbo angle). Summing over omitted colour indices is understood, and we have included the contribution of the c charmed quark. Gell-Mann (1962, 1964b) then postulated that, at short distances, the commutation relations of these currents appear as if the quark fields entering into them were free:

$$\mathcal{L}_{\text{quarks}} \approx \mathcal{L}_0 = \sum_{q=u,d,\dots} \sum_j \bar{q}^j(x)(i\slashed{\partial} - m_q)q^j(x). \tag{1.1.7}$$

It was difficult to understand how this could be so, but the hypothesis met with spectacular success in the Adler–Weissberger sum rule, the Cabibbo–Radicatti sum rule, and the calculations by Sirlin and others of radiative corrections to β decay in nuclei.

Another view of the quark model came from deep inelastic scattering experiments. Here a virtual photon, or W, with large invariant mass, $-Q^2$, and high energy, ν, is scattered off some target (a proton, for example). One found the surprising result, which had been anticipated by Bjorken (1969), that the cross section was of the form (for the kinematics, see Sect. 4.3)

$$\frac{\partial\sigma}{\partial\Omega\partial k_0'} = \frac{\alpha}{4m_p k_0^2 \sin^4\theta/2}\left\{W_2\cos^2\frac{\theta}{2} + 2W_1\sin^2\frac{\theta}{2}\right\}, \tag{1.1.8a}$$

where, if we write

$$F_1(x,Q^2) = W_1, \quad F_2(x,Q^2) = \frac{\nu}{m_p^2}W_2, \quad x = Q^2/\nu;$$

$$\int d^4 z\, e^{iq\cdot z}\langle p|[J^\mu(z), J^\nu(0)]|p\rangle \approx -g^{\mu\nu}W_1 + \frac{1}{m_p^2}p^\mu p^\nu W_2, \tag{1.1.8b}$$

then the F_i are approximately independent of Q^2 for $Q^2 \to \infty$, when x has a fixed value (Bjorken scaling). Feynman showed how this could be interpreted if we consider that as Q^2, $\nu \to \infty$ (which, in view of (1.1.8b) means short distances), we consider the proton to be made up of parts, the "partons", that do not interact among themselves. It took only one step to identify these partons with quarks, which appear again free at short distances, thus creating another puzzle.

Clearly, all these difficulties are dynamical and can therefore only be solved by building a theory of strong interactions, so we come to the crux of the matter: Which are the interactions among hadrons? A remarkable fact of hadron physics is that in spite of the variety of hadrons (compare, for example, the π, K masses), interactions among them (coupling constants and high energy cross-sections, where one can neglect mass differences) are flavour independent. This means that whatever agency causes quarks to interact, it must act equally on u or d, s or c.

In the meantime, renormalizable theories of weak and electromagnetic interactions had been constructed by Glashow, Weinberg and Salam and Ward and others. Weinberg (1973a) and Nanopoulos (1973) have shown that, to avoid catastrophic violations of parity to order α, one needs that strong interactions act on quantum numbers other than flavour. These were among the reasons that led physicists to consider the possibility that whatever glued the quarks (the *gluons*) interacted precisely with colour to which weak and electromagnetic interactions are blind, cf. (1.1.6). One takes eight vector gluons, with fields B_a^μ, $a = 1$ to 8, in the adjoint representation of $SU_c(3)$ interacting universally with all quark flavours:

$$\mathcal{L}_1 = \mathcal{L}_0 + g \sum_q \sum_{ika} \bar{q}^i(x)\gamma_\mu t_{ik}^a q^k(x) B_a^\mu(x), \qquad (1.1.9)$$

where $\mathcal{L}_0$ is still given by (1.1.7) and the t_{ik}^a matrices are $t_{ik}^a = \frac{1}{2}\lambda^a$, with λ^a the Gell-Mann matrices; they generate the fundamental representation of $SU_c(3)$, and satisfy the commutation relations[3]

$$[t^a, t^b] = \mathrm{i} \sum f^{abc} t^c. \qquad (1.1.10)$$

The colour and vector character of gluons has the extra virtue of explaining the split between the masses of the Δ_{33} and the nucleons, and even the mass difference between the Λ and Σ^0 particles (De Rújula, Georgi and Glashow, 1975).

A further step is taken if it is realized that, for *massless* vector fields, a non-Abelian gauge theory (first introduced by Yang and Mills, 1954) presents hideous infrared singularities that could prevent the liberation of individual quarks and gluons. Thus, we can at least be reconciled to (1.1.1) and (1.1.4): one cannot see isolated quarks because they cannot escape from hadrons due

---

[3] For group theoretic relations, see Appendix C. We will write colour indices as subscripts or superscripts indifferently: $f^{abc} = f_{abc}$, $t_{ik}^a = t_a^{ik}$, etc.

to their interactions, not because they are heavy. This is the hypothesis of confinement. Now, we modify (1.1.9) to

$$\mathcal{L}_{\text{QCD}} = \mathcal{L}_1 - \tfrac{1}{4} \sum_a G_a^{\mu\nu}(x) G_{a\mu\nu}(x),$$

$$G_a^{\mu\nu}(x) = \partial^\mu B_a^\nu - \partial^\nu B_a^\mu + g \sum f_{abc} B_b^\mu B_c^\nu.$$

(1.1.11)

We get a bonus: as occurs in all non-Abelian gauge theories, the coupling constant is automatically universal. Eq. (1.1.11) shows the standard QCD Lagrangian, which will be our starting point in the following sections.

Until now the entire construction has been rather fragile. It consists of a set of assumptions, culminating in Eq. (1.1.11), in which each hypothesis takes us further away from the real world (pions, protons, etc.) into a fictitious realm (quarks and gluons) with a set of predictions that hardly outnumber the assumptions. However, the situation changed radically in the early seventies. At that time 't Hooft (unpublished), Politzer (1973) and, independently, Gross and Wilczek (1973a,b, 1974), proved that in a theory such as that described by the Lagrangian (1.1.11), the effective coupling constant vanishes at short distances (*asymptotic freedom*) and increases at long distances. Thus at one stroke they explained the success of current algebra and the parton model, and made confinement probable. What is more, the corrections to the free-field behaviour of quarks are calculable; when calculated, they are found in systematic agreement with experiment – as much, indeed, as the accuracy of calculations (and of the experimental data!) allows. By and large, there is impressive evidence that QCD is *the* theory of strong interactions.

Another important property of QCD, which is perhaps not sufficiently emphasized in most presentations, is its character as a *local field theory*, which leads (at least if the ideas about confinement really work) to local observables. To be precise, the expected pattern is the following. The fields in the Lagrangian (1.1.11) are defined in a Hilbert space, $\mathfrak{H}_{\text{QCD}}$, made up of quark and gluon states, and built from a perturbation theoretic treatment of Eq. (1.1.11), for example. The quarks and gluons are described there by local fields, $q(x)$, $B(x)$. If confinement ideas are correct, however, it is the space of bound states $\mathfrak{H}_{\text{phys}}$ that contains physical states. That is, if we solved the theory exactly, only colour singlet operators would survive. These include *currents*, such as

$$\sum \bar{q}^i \gamma^\mu (1 \pm \gamma_5) q'^i,$$

or other composite operators: the operator for a π or a proton,

$$\sum \bar{u}^i \gamma_5 d^i, \quad \sum \epsilon^{ikl} u^i u^k d^l,$$

etc. The point is that these operators, tough composite, are still local; so if the picture is right, it follows that operators in the physical Hilbert space,

$\mathfrak{H}_{\text{phys}}$, are local. This is sufficient[4] to derive all the standard results of old-fashioned hadron physics – fixed t dispersion relations, Froissart-like bounds, etc. – whose success, when tested experimentally, is very impressive.

One more property of QCD which, although at a speculative level is still worth noting, is that of allowing naturally grand unifications. Because $SU(3)$ of colour is a larger group than the standard $SU(2) \times U(1)$ of electroweak interactions, it follows that an energy scale may exist where all couplings are equal. Although the more naive grand unification schemes have been proven in disagreement with experiment, the possibility for more sophisticated ones remains.

1.2 Perturbative Expansions;
S-Matrix and Green's Functions; Wick's Theorem

In this section we review very briefly a few basic topics of relativistic field theory. Of course this is by no means intended to cover the subject. The topics in this section are presented merely to establish the notation and to outline the prerequisites for understanding what will follow; the details will have to be sought for elsewhere.[5]

A field theory may be specified by giving the relevant Lagrangian. If $\Phi_i(x)$ are the fields in the theory, the Lagrangian is a function of the $\Phi_i(x)$ and their space–time derivatives, $\partial\Phi_i(x)$. It is customary and convenient to split the Lagrangian, $\mathcal{L}$ ($\mathcal{L}$ is actually the Lagrangian density) into $\mathcal{L}_0$ and $\mathcal{L}_{\text{int}}$, with $\mathcal{L}_0$ obtained from $\mathcal{L}$ by setting all interactions to zero, and $\mathcal{L}_{\text{int}}$ is defined as $\mathcal{L}_{\text{int}} \equiv \mathcal{L} - \mathcal{L}_0$. For example, in QCD, the Lagrangian is (1.1.11) and

$$\mathcal{L}_0 = \sum_q \bar{q}(x)(\mathrm{i}\partial\!\!\!/ - m_q)q(x)$$

$$- \tfrac{1}{4}\sum_a \left(\partial^\mu B_a^\nu(x) - \partial^\nu B_a^\mu(x)\right)\left(\partial_\mu B_{a\nu}(x) - \partial_\nu B_{a\mu}(x)\right).$$

Besides the basic, or elementary, fields $\Phi_i(x)$ that enter into the theory (the q, B for QCD), we often require *composite operators*, usually local combinations of the $\Phi_i(x)$, i.e., combinations involving *finite* products of the $\Phi_i(x)$ and its derivatives at the same point. For example, in QCD we will use the currents, $\bar{q}(x)\gamma^\mu q'(x)$. Of course, $\mathcal{L}$ itself is a composite local operator.

With local fields or, more generally, local operators (whether elementary or composite), we can form new local operators. The simplest method is by ordinary multiplication; but there are other types of product that we will consider repeatedly. These are the *Wick product* and the *time ordered*

---

[4] See Epstein, Glaser and Martin (1969) and Bogolubov, Logunov and Todorov (1975), from where one can trace the relevant literature.

[5] For example, in standard textbooks such as Bogoliubov and Shirkov (1959); Bjorken and Drell (1965) or Itzykson and Zuber (1980).

product. The *Wick,* or *normal product,* is defined primarily for free basic fields, as follows. Expand the Φ_i in creation–annihilation operators:

$$\Phi_i(x) = \sum_k C_i^{(k)}(x)a_k + \sum_k \bar{C}_i^{(k)}(x)\bar{a}_k^{\dagger},$$

where a, $\bar{a}$ may, or may not coincide. For example, if Φ is q,

$$q(x) = \frac{1}{(2\pi)^{\frac{3}{2}}} \sum_\sigma \int \frac{\mathrm{d}^3\mathbf{p}}{2p_0} \left\{ \mathrm{e}^{-\mathrm{i}p\cdot x}u(p,\sigma)a(p,\sigma) + \mathrm{e}^{\mathrm{i}p\cdot x}v(p,\sigma)\bar{a}^{\dagger}(p,\sigma)\right\},$$

with u, v standard Dirac spinors and a ($\bar{a}^{\dagger}$) creates quarks (annihilates anti-quarks). Then, the Wick product, $: \Phi_1(x_1)\Phi_2(x_2) :$ is obtained by placing all creators to the left of all annihilators as if they commuted/anticommuted if the fields Φ_i corresponded to bosons/fermions:

$$: \Phi_1(x_1)\Phi_2(x_2) : \equiv \sum_{k,k'} \left\{ C_1^{(k)}(x_1)C_2^{(k')}(x_2)a_ka_{k'} + \bar{C}_1^{(k)}(x_1)\bar{C}_2^{(k')}(x_2)\bar{a}_k^{\dagger}\bar{a}_{k'}^{\dagger} \right.$$
$$\left. \bar{C}_1^{(k)}(x_1)C_2^{(k')}(x_2)\bar{a}_k^{\dagger}a_{k'} + (-1)^{\delta}C_1^{(k)}(x_1)\bar{C}_2^{(k')}(x_2)\bar{a}_{k'}^{\dagger}a_k \right\},$$

$\delta = 1$ for fermions, $\delta = 0$ for bosons.

The extension to Wick products of more factors,

$$: \Phi_1(x_1) \ldots \Phi_N(x_N) :,$$

or to Wick products of Wick products like

$$: \left(: \Phi_1(x_1)\Phi_2(x_2) : \right)\left(: \Phi_3(x_3)\Phi_4(x_4) : \right) :$$

is straightforward – one always expands in creators and annihilators, and writes creators to the left of annihilators as if they commuted/anticommuted.

The proof is not totally trivial, yet it is not too difficult to verify that the Wick product of local operators at the same point is itself local,[6] i.e., if $O_1, \ldots, O_n$ are local, so is $: O_1(x) \ldots O_n(x) :$.

Another important property of the Wick product is that it is regular; that is to say, for any states a, b, the matrix elements of a Wick product, $\langle a | : O_1(x_1) \ldots O_n(x_n) : | b \rangle$ are regular (i.e., differentiable) functions of the $x_1^{\mu}, \ldots, x_n^{\rho}$.

The *time-ordered product,* or *T-product* of local (elementary or composite) operators, $O_1(x) \ldots O_n(x)$ is defined as follows:

$$\mathrm{T}O_1(x) \ldots O_n(x) \equiv \mathrm{T}\{O_1(x) \ldots O_n(x)\} = (-1)^{\delta}O_{i_1}(x) \ldots O_{i_n}(x).$$

---

[6] For our purposes a local operator $O_\alpha(x)$ is one transforming locally under Poincaré transformations: $U(a,\Lambda)O_\alpha(x)U^{-1}(a,\Lambda) = \sum P_{\alpha\alpha'}(\Lambda)O_{\alpha'}(\Lambda x + a)$.

Here the permutation $i_1, \ldots, i_n$ is such that, in the right hand side, the times are ordered: $x_{i_1}^0 \geq x_{i_2}^0 \ldots \geq x_{i_n}^0$, and δ is the number of transpositions of the indices corresponding to *fermion* operators necessary to bring $1, \ldots, n$ to $i_1, \ldots, i_n$. Otherwise stated, the time ordered product is obtained by rearranging the operators in the natural sequence of times as if they commuted/anticommuted for boson/fermion operators. For example, for two factors,

$$Tq_1(x)q_2(y) = \theta(x^0 - y^0)q_1(x)q_2(y) - \theta(y^0 - x^0)q_2(y)q_1(x)$$

or

$$Tq(x)B(y) = \theta(x^0 - y^0)q(x)B(y) + \theta(y^0 - x^0)B(y)q(x).$$

Note that boson–fermion operators are always taken to be commuting. The time ordered product of local fields is relativistically invariant.

The S matrix is the operator that transforms free states at time $-\infty$ into free states at time $+\infty$. S may be obtained in terms of the interaction Lagrangian using Matthews' formula:

$$S = \mathrm{T} \exp \mathrm{i} \int \mathrm{d}^4 x \, \mathcal{L}_{\mathrm{int}}^0(x). \tag{1.2.1a}$$

Here $\mathcal{L}_{\mathrm{int}}^0(x)$ is the interaction Lagrangian with all fields in it taken as if they were free, and in normal order. The time-ordered exponential is a formal device; it actually is defined by its series expansion,

$$\begin{aligned}
S = \mathrm{T} \exp \mathrm{i} \int \mathrm{d}^4 x \, \mathcal{L}_{\mathrm{int}}^0(x) &\equiv 1 + \mathrm{i} \int \mathrm{d}^4 x \, \mathcal{L}_{\mathrm{int}}^0(x) \\
+ \frac{\mathrm{i}^n}{n!} \int \mathrm{d}^4 x_1 \ldots \mathrm{d}^4 x_n \, &\mathrm{T} \mathcal{L}_{\mathrm{int}}^0(x_1) \ldots \mathcal{L}_{\mathrm{int}}^0(x_n) + \cdots .
\end{aligned} \tag{1.2.1b}$$

Many times, instead of matrix elements of S, we will require the matrix elements of currents, or products of currents, or more general composite operators. These may be obtained by adding a fictitious extra term to the interaction. For example, suppose that we require

$$\langle a | \mathrm{T} J_1^\mu(x) J_2^\nu(y) | b \rangle, \tag{1.2.2}$$

where the J are weak or electromagnetic currents; see (1.1.6). We then change $\mathcal{L}_{\mathrm{int}}$ according to

$$\mathcal{L}_{\mathrm{int}} \to \mathcal{L}_{\mathrm{int}}^\phi = \mathcal{L}_{\mathrm{int}} + J_{1\mu}(x)\phi_1^\mu(x) + J_{2\mu}(x)\phi_2^\mu(x), \tag{1.2.3}$$

where the ϕ are c-number auxiliary fields. We expand again:

$$\langle a|\mathrm{T}\exp\int \mathrm{d}^4x\,\mathcal{L}_{\mathrm{int}}^{\phi}(x)|b\rangle = \langle a|b\rangle$$

$$+\mathrm{i}\langle a|\int \mathrm{d}^4x\left\{\mathcal{L}_{\mathrm{int}}^0(x) + \sum_i J_{i\mu}^0(x)\phi_i^{\mu}(x)\right\}|b\rangle + \cdots$$

$$+\frac{\mathrm{i}^n}{n!}\langle a|\int \mathrm{d}^4x_1\ldots\mathrm{d}^4x_n\mathrm{T}\left\{\mathcal{L}_{\mathrm{int}}^0(x_1) + \sum_i J_{i\mu}^0(x_1)\phi_i^{\mu}(x_1)\right\}\ldots$$

$$\times\left\{\mathcal{L}_{\mathrm{int}}^0(x_n) + \sum_i J_{i\mu}^0(x_n)\phi_i^{\mu}(x_n)\right\}|b\rangle + \cdots.$$

We let ϕ be infinitesimal and keep only terms $O(\phi)$, $O(\phi^2)$. The last are of the form

$$\frac{\mathrm{i}^n}{n!}\langle a|\int \mathrm{d}^4x_1\ldots\mathrm{d}^4x_n\sum_{ij}\mathrm{T}\mathcal{L}_{\mathrm{int}}^0(x_1)\ldots[\mathcal{L}_{\mathrm{int}}^0(x_i)]\ldots$$

$$\times[\mathcal{L}_{\mathrm{int}}^0(x_j)]\ldots\mathcal{L}_{\mathrm{int}}^0(x_n)J_{1\mu}^0(x_i)J_{2\nu}^0(x_j)|b\rangle\phi_1^{\mu}(x_i)\phi_2^{\nu}(x_j),$$

where $[\mathcal{L}]$ means that we have dropped the bracketed term. Letting $\phi_{i\mu}(x) = \epsilon_{i\mu}\delta(x-y_i)$, differentiating with respect to ϵ_1, ϵ_2 and setting $\epsilon_1 = \epsilon_2 = 0$ we get the Gell-Mann–Low equation,[7]

$$\langle a|\mathrm{T}J_1^{\mu}(x)J_2^{\nu}(y)|b\rangle$$

$$=\frac{\delta^2}{\delta\phi_{1\mu}(x)\delta\phi_{2\nu}(y)}\;\langle a|\mathrm{T}\exp\mathrm{i}\int \mathrm{d}^4z\left\{\mathcal{L}_{\mathrm{int}}^0(z) + \sum_i J_{i\lambda}^0(z)\phi_i^{\lambda}(z)\right\}|b\rangle \qquad (1.2.4)$$

$$=\sum_{n=0}^{\infty}\frac{\mathrm{i}^n}{n!}\langle a|\int \mathrm{d}^4x_1\ldots\mathrm{d}^4x_n\mathrm{T}\mathcal{L}_{\mathrm{int}}^0(x_1)\ldots\mathcal{L}_{\mathrm{int}}^0(x_n)J_1^{0\mu}(x)J_2^{0\nu}(y)|b\rangle.$$

To identify the right hand side with (1.2.2) we have used the formula, proved e.g. in Bogoliubov and Shirkov (1959) (see also Sects. 1.3, 2.6 below)

$$\frac{\delta^2 S_{\phi}}{\delta\phi_{1\mu}(x)\delta\phi_{2\nu}(y)} = \mathrm{T}J_1^{\mu}(x)J_2^{\nu}(y). \qquad (1.2.5)$$

We then turn to the requirements of relativistic invariance and unitarity. If (a, Λ) is a transformation in the Poincaré group, then

$$U(a,\Lambda)SU^{-1}(a,\Lambda) = S: \qquad (1.2.6)$$

S is relativistically invariant. It is also unitary,

$$S^{\dagger}S = SS^{\dagger} = 1. \qquad (1.2.7)$$

---

[7] Functional derivatives are defined in Appendix H.

If we write
$$S = 1 + \mathrm{i}\mathcal{T},$$
where $\langle a|\mathcal{T}|b\rangle$ is the so-called transition amplitude, then (1.2.7) may be rewritten in terms of $\mathcal{T}$,

$$\mathrm{Im}\langle a|\mathcal{T}|b\rangle = \tfrac{1}{2} \sum_{\mathrm{all}\,c} \langle c|\mathcal{T}|b\rangle\langle c|\mathcal{T}|a\rangle^*, \tag{1.2.8}$$

and we have assumed time reversal invariance, which holds for QCD, to derive (1.2.8). When expanding in powers of g, Eqs. (1.2.6) and (1.2.8) imply relations order by order in perturbation theory. As (1.2.6) is linear, it must hold for each order; but, because of its nonlinearity, (1.2.8) mixes different orders. For example, if we let

$$\mathcal{T} = g \sum_{n=0}^{\infty} g^n \mathcal{T}_n,$$

then the second order constraint is

$$\mathrm{Im}\langle a|\mathcal{T}_2|b\rangle = \tfrac{1}{2} \sum_{\mathrm{all}\,c} \Big\{ \langle c|\mathcal{T}_0|b\rangle\langle c|\mathcal{T}_2|a\rangle^* \\ + \langle c|\mathcal{T}_2|b\rangle\langle c|\mathcal{T}_0|a\rangle^* + \langle c|\mathcal{T}_1|b\rangle\langle c|\mathcal{T}_1|a\rangle^* \Big\}. \tag{1.2.9}$$

We will finish by introducing reduction formulas. Consider a scattering amplitude, say $a + b \to a' + b'$, with a, a' bosons with fields ϕ_a, $\phi_{a'}$. We may write the scattering amplitude as

$$\langle a', b'|S|a, b\rangle = \lim_{\substack{t' \to +\infty \\ t \to -\infty}} \langle a', b'; t'|a, b; t\rangle.$$

Now, if p_i is the momentum of particle i, we may use the expression for the creation operator in terms of the field,

$$a^\dagger(p_a) = \lim_{t \to -\infty} \frac{\mathrm{i}}{2(2\pi)^{\frac{3}{2}}} \int \mathrm{d}^3\mathbf{x}\, \mathrm{e}^{-\mathrm{i}p_a \cdot x}\, \overset{\leftrightarrow}{\partial}_0 \phi^\dagger(x),$$

to write, after some manipulations, *reduction formulas*; for example,

$$\langle a', b'|S|a, b\rangle = \frac{\mathrm{i}}{(2\pi)^{\frac{3}{2}}} \int \mathrm{d}^4 x\, \mathrm{e}^{-\mathrm{i}p_a \cdot x} \left(\Box + m_a^2\right) \langle a', b'|\phi_a^\dagger(x)|b\rangle.$$

We will not prove this, or give a full set of reduction formulas, which may be found in Bjorken and Drell (1965); but we will at least present a few typical cases. If we also "reduce" a' we obtain

$$\langle a', b'|S|a, b\rangle = \frac{\mathrm{i}}{(2\pi)^{\frac{3}{2}}} \frac{-\mathrm{i}}{(2\pi)^{\frac{3}{2}}}$$
$$\times \int \mathrm{d}^4 x \int \mathrm{d}^4 y\, \mathrm{e}^{-\mathrm{i}p_a \cdot x} \mathrm{e}^{\mathrm{i}p_{a'} \cdot y} \left(\Box_x + m_a^2\right) \left(\Box_y + m_{a'}^2\right) \langle b'|\mathrm{T}\phi_{a'}(y)\phi_a^\dagger(x)|b\rangle.$$

If we continue reducing we ultimately obtain the Fourier transform of the *vacuum expectation value* (VEV) of the T-product of four fields,

$$\langle 0|T\phi_{a'}(y)\phi_{b'}(z)\phi_a^\dagger(x)\phi_b^\dagger(w)|0\rangle.$$

The extension to spinor fields is easy. If we reduce a fermion with momentum p_a and spin σ, and denote its corresponding field by ψ, we obtain

$$\langle a',b'|S|(p_a,\sigma),b\rangle = \frac{i}{(2\pi)^{\frac{3}{2}}}\int d^4x\,\langle a',b'|\bar\psi(x)|b\rangle(i\overleftarrow{\partial}+m_a)u(p,\sigma)e^{-ip_a\cdot x},$$

etc.

Finally, we turn to Wick's theorem. An expression such as (1.2.1b) permits us to calculate, order by order in perturbation theory, the S matrix elements (or current matrix elements, or Green's functions). The tool that allows us to do this is Wick's theorem. Consider the time-ordered product of two free fields, $T\Phi_1^0(x_1)\Phi_2^0(x_2)$. We expand the Φ_i in creation-annihilation operators:

$$\Phi_i(x) = \frac{1}{(2\pi)^{\frac{3}{2}}}\int\frac{d^3\mathbf{k}}{2k^0}\sum_\sigma\left\{e^{-ik\cdot x}\xi_+(k,\sigma)a_+(k,\sigma) + e^{ik\cdot x}\xi_-(k,\sigma)a_-^\dagger(k,\sigma)\right\}.$$

Here σ is the spin, $\xi_\pm$ the wave functions and $a_\pm$, $a_\pm^\dagger$ the annihilation, creation operators for particles $(+)$ and antiparticles $(-)$. Their commutation relations

$$\left[a_\pm(k,\sigma), a_\pm^\dagger(k',\sigma')\right] = 2\delta_{\sigma\sigma'}k^0\delta(\mathbf{k}-\mathbf{k}'),$$

$$[a_+, a_-^\dagger] = 0$$

(where the symbol $[\ ,\ ]$ is to be interpreted as the anticommutator for fermions) can be applied to check that the difference

$$T\Phi_1^0(x_1)\Phi_2^0(x_2) - :\Phi_1^0(x_1)\Phi_2^0(x_2): \equiv \underline{\Phi_1^0(x_1)\Phi_2^0(x_2)}$$

is a *c*-number, called the *contraction*. Thus it coincides with its vacuum expectation value (the *propagator*):

$$\underline{\Phi_1^0(x_1)\Phi_2^0(x_2)} = \langle 0|T\Phi_1^0(x_1)\Phi_2^0(x_2)|0\rangle \equiv \langle T\Phi_1^0(x_1)\Phi_2^0(x_2)\rangle_0.$$

Applying this repeatedly to, say, (1.2.1) we find that $T\mathcal{L}_{\text{int}}^0\ldots\mathcal{L}_{\text{int}}^0$ may be written as a combination of contractions times fully normal ordered products of operators. As matrix elements of these may be easily calculated, the full result for each term in the perturbation expansion may be so evaluated. The Feynman rules are such that they summarize the manipulations, allowing us to write the final result directly. For QCD, they are as shown in Appendix D (see also Sect. 2.6, where some of the Feynman rules are explicitly deduced).

1.3 Path Integral Formulation of Field Theory

In many applications we are concerned with the perturbative aspects of QCD. For these, the use of either a canonical or path integral formalism is largely a matter of taste. Nonperturbative aspects of QCD, however, can be formulated with greater clarity in functional language. In this section we review briefly the Feynman path integral formalism, in particular as applied to field theory. Of course, this is no substitute for a detailed treatment, for which the interested reader may consult the lectures of Fadeyev (1976) and Lee (1976) or the textbooks of Itzykson and Zuber (1980) and Fadeyev and Slavnov (1980).

Let us start with nonrelativistic quantum mechanics, in one dimension (Feynman and Hibbs, 1965). Here we have a Hamiltonian, $\hat{H}$, which is a function of momentum and position[8] $\hat{P}$, $\hat{Q}$; we assume that it has been written in "normal form", with all $\hat{P}$'s to the left of all $\hat{Q}$'s. The classical Hamiltonian may be obtained as

$$\langle p|\hat{H}|q\rangle = (2\pi)^{-\frac{1}{2}}e^{-ipq}H(p,q), \qquad (1.3.1)$$

where $\hat{P}|p\rangle = p|p\rangle$, $\hat{Q}|q\rangle = q|q\rangle$, $\langle p|q\rangle = (2\pi)^{-\frac{1}{2}}e^{-ipq}$. We then evaluate the matrix elements of the evolution operator,

$$\langle q''|e^{-i(t''-t')\hat{H}}|q'\rangle. \qquad (1.3.2)$$

To do so we write

$$e^{-it\hat{H}} = \lim_{N\to\infty}\left(1 - \frac{it}{N}\hat{H}\right)^N, \quad t = t'' - t',$$

and insert sums over complete sets of states:

$$\langle q''|e^{-i(t''-t')\hat{H}}|q'\rangle = \lim_{N\to\infty}\int \prod dp_n \prod dq_n \, \langle q''|p_N\rangle\langle p_N|1 - \frac{it}{N}\hat{H}|q_N\rangle$$
$$\times \langle q_N|p_{N-1}\rangle\langle p_{N-1}|1 - \frac{it}{N}\hat{H}|q_{N-1}\rangle \ldots \langle p_1|1 - \frac{it}{N}\hat{H}|q'\rangle.$$

Now, using (1.3.1) we find

$$\langle p_n|\left(1 - \frac{it}{N}\hat{H}\right)|q_n\rangle = \frac{\exp\{-ip_nq_n - (it/N)H(p_n,q_n)\}}{\sqrt{2\pi}} + O\left(\frac{1}{N^2}\right),$$

so

$$\langle q''|e^{-i(t''-t')\hat{H}}|q'\rangle =$$
$$\lim_{N\to\infty}\int \prod \frac{dp_n}{2\pi} \prod dq_n \, \exp i\Big\{p_N(q'' - q_N) + \cdots \qquad (1.3.3)$$
$$+ p_1(q_1 - q') - \frac{t}{N}\big(H(p_N,q_N)\ldots H(p_1,q')\big)\Big\}.$$

---

[8] We place carets over operators temporarily.

Feynman's procedure consists of defining two functions, $p(t)$, $q(t)$, with $p(t_n) = p_n$, $q(t_n) = q_n$ so we may replace the integrals

$$\prod_n \frac{\mathrm{d}p_n}{2\pi} \to \prod_t \frac{\mathrm{d}p(t)}{2\pi}, \quad \prod_n \frac{\mathrm{d}q_n}{2\pi} \to \prod_t \frac{\mathrm{d}q(t)}{2\pi}, \qquad (1.3.4)$$

i.e., we now integrate over all *functions*, and the term in brackets in (1.3.3) gives

$$\int_{t'}^{t''} \mathrm{d}t\{p(t)\dot{q}(t) - H(p(t), q(t))\}, \quad \dot{f} \equiv \frac{\mathrm{d}f}{\mathrm{d}t}.$$

The entire (1.3.3) thus becomes

$$\langle q''|e^{-\mathrm{i}(t''-t')\hat{H}}|q'\rangle = \int \prod_t \frac{\mathrm{d}q(t)\mathrm{d}p(t)}{2\pi} \exp \mathrm{i} \int_{t',q'}^{t'',q''} \mathrm{d}t\,(p\dot{q} - H). \qquad (1.3.5)$$

Of course this expression is formal,[9] and it only makes sense as a limit of (1.3.3), but in this it is not so very different from the usual Riemann definition of an ordinary integral. The important thing about (1.3.5) is that only classical c-number functions enter: we have traded the complexities of operator calculus for those of functional integrations.

Equation (1.3.5) may be simplified in some circumstances. If $H = p^2/2m + V(q)$, then the integration over $\mathrm{d}p$ is Gaussian and can be explicitly evaluated. Shifting the integration variables by $p \to p - m\dot{q}$,

$$\int \prod_t \frac{\mathrm{d}p(t)}{2\pi} \exp \mathrm{i} \int \mathrm{d}t \left(p\dot{q} - \frac{p^2}{2m}\right)$$

$$= \int \prod_t \frac{\mathrm{d}p(t)}{2\pi} \exp \left(-\mathrm{i} \int \mathrm{d}t \frac{p^2}{2m}\right) \exp \left(\mathrm{i} \int \mathrm{d}t \frac{m\dot{q}^2(t)}{2}\right);$$

therefore,

$$\langle q''|e^{-\mathrm{i}(t''-t')\hat{H}}|q'\rangle = F \int \prod_t \mathrm{d}q(t) \exp \mathrm{i} \int_{q',t'}^{q'',t''} \mathrm{d}t\, L(q(t), \dot{q}(t)). \qquad (1.3.6)$$

Here we have identified $\frac{1}{2}m\dot{q}^2 - V$ with the (integrated) Lagrangian, and extracted the normalization factor, independent of the dynamics,

$$F = \int \prod_t \frac{\mathrm{d}p(t)}{2\pi} \exp \left\{-\mathrm{i} \int \mathrm{d}t \frac{p^2(t)}{2m}\right\}.$$

The generalization of (1.3.6) to several degrees of freedom is obvious. Let us write the coordinates as $q(t, k)$ instead of $q_k(t)$, $k = 1, \ldots, N$, to

---

[9] See, however, Wiener (1923) for a rigorous treatment of functional integrals similar to (1.3.5).

facilitate the transition to the field theoretic case; and let us also introduce the Lagrangian density, writing $L = \sum_k \mathcal{L}$. We find

$$\langle q''|e^{-\mathrm{i}(t''-t')\hat{H}}|q'\rangle = F \int \prod_{t,k} \mathrm{d}q(t,k) \exp\left\{ \mathrm{i} \int_{q',t'}^{q'',t''} \mathrm{d}t \sum_k \mathcal{L}\big(q(t,k),\dot{q}(t,k)\big) \right\}.$$
(1.3.7)

The product in (1.3.7) *excludes* the endpoints q', t'; q'', t''. For an arbitrary state $|\Psi\rangle$ we have,

$$\langle \Psi|e^{-\mathrm{i}(t''-t')\hat{H}}|\Psi\rangle = \int \mathrm{d}q'\mathrm{d}q'' \langle \Psi|q''\rangle\langle q''|e^{-\mathrm{i}(t''-t')\hat{H}}|q'\rangle\langle q'|\Psi\rangle$$

$$= \int \mathrm{d}q'\mathrm{d}q'' \, \Psi^*(q'')\langle q''|e^{-\mathrm{i}(t''-t')\hat{H}}|q'\rangle\Psi(q').$$

The state with minimum energy, which will correspond to the vacuum in field theory, is one with zero momentum. Thus its wave function is constant, $\Psi_0(q) = $ const. Therefore, for it,

$$\langle \Psi_0|e^{-\mathrm{i}(t''-t')\hat{H}}|\Psi_0\rangle = F \int \prod_{k,t} \mathrm{d}q(t,k) \exp\mathrm{i} \int_{q',t'}^{q'',t''} \mathrm{d}t \sum_k \mathcal{L}\big(q(t,k),\dot{q}(t,k)\big),$$

and the product now *includes* the endpoints q', t'; q'', t''.

We consider now the case of field theory. Here k is replaced by $\mathbf{x}$, $\sum_k$ by $\int \mathrm{d}^3\mathbf{x}$; $q(t,k)$ is replaced by a field (or fields, if there is more than one) $\phi(t,\mathbf{x}) = \phi(x)$ and $|\Psi_0\rangle$ by the vacuum state $|0\rangle$. Then we find

$$\langle 0|e^{-\mathrm{i}(t''-t')\hat{H}}|0\rangle = N \int \prod_x \mathrm{d}\phi(x) \exp\left\{ \mathrm{i} \int_{t'}^{t''} \mathrm{d}^4 x \, \mathcal{L}(\phi,\partial\phi) \right\},$$
(1.3.8)

N a constant factor. Of course, just as in ordinary quantum mechanics, the functional integral has to be interpreted via a limiting procedure. Consider a large volume of space, V, and divide the four-dimensional volume $(t''-t', V)$ into a finite number n of cells. Let x_j, $j = 1,\ldots,n$, be points inside each cell, and let δ be the four-dimensional volume of each cell. Then the right hand side of (1.3.8) is defined as

$$\lim_{\substack{V\to\infty \\ n\to\infty \\ \delta\to 0}} \int \mathrm{d}\phi(x_1)\ldots\mathrm{d}\phi(x_n)\, e^{\mathrm{i}\delta \sum_j \mathcal{L}(\phi(x_j),\partial\phi(x_j))}.$$
(1.3.9)

Later on we will see that the normalization factor N in e.g. (1.3.8) can be disposed of when considering transition amplitudes.

To evaluate S matrix elements or Green's functions, we require VEVs $\langle T\phi(x)\ldots\phi(z)\rangle_0$. For this we consider the vacuum-to-vacuum amplitude,

$$\langle 0|\hat{S}|0\rangle = \lim_{\substack{t'\to-\infty \\ t''\to+\infty}} \langle 0|e^{-\mathrm{i}(t''-t')\hat{H}}|0\rangle,$$

and obtain the Green's functions through the introduction of sources. According to (1.3.7),

$$\langle 0|\hat{S}|0\rangle = N \int \prod_x \mathrm{d}\phi(x) \exp \mathrm{i}\mathcal{A}, \quad \mathcal{A} = \int \mathrm{d}^4x\, \mathcal{L}; \qquad (1.3.10)$$

$\mathcal{A}$ is the action. We add a source term to $\mathcal{L}$,

$$\mathcal{L}_\eta = \mathcal{L} + \eta(x)\phi(x), \quad \mathcal{A}_\eta = \int \mathrm{d}^4x\, \mathcal{L}_\eta,$$

and define the *generating functional*

$$Z[\eta] = N \int \prod_x \mathrm{d}\phi(x) \exp \mathrm{i}\mathcal{A}_\eta. \qquad (1.3.11)$$

We will see that from this it follows that

$$\left.\frac{\delta^n \log[Z\eta]}{\delta\eta(x_1)\ldots\delta\eta(x_n)}\right|_{\eta=0} = \frac{\mathrm{i}^n \langle \mathrm{T}\hat{\phi}(x_1)\ldots\hat{\phi}(x_n)\rangle_0}{\langle S\rangle_0}, \qquad (1.3.12)$$

where the right hand side is the *connected Green's function*, which we have until now denoted simply by

$$\langle \mathrm{T}\hat{\phi}(x_1)\ldots\hat{\phi}(x_n)\rangle_0,$$

absorbing the phase $\langle S\rangle_0$ in the definition of the physical $\hat{S}$. We will prove (1.3.12) in the free field case (we will consider interactions later). The Lagrangian is thus

$$\mathcal{L} = \tfrac{1}{2}\partial_\mu\phi\partial^\mu\phi - \tfrac{1}{2}m^2\phi^2 = -\tfrac{1}{2}\phi(\Box+m^2)\phi + \text{four-divergence}.$$

The trick lies in reducing the integral to a Gaussian. For this, define ϕ' so that

$$\phi'(x) = (\Box+m^2)^{1/2}\phi(x),$$

which is accomplished with

$$\phi'(x) = \int \mathrm{d}^4y\, K^{-1/2}(x-y)\phi(y),$$

$$K(z) = \frac{-1}{(2\pi)^4} \int \mathrm{d}^4k\, \frac{e^{\mathrm{i}k\cdot z}}{k^2 - m^2 + \mathrm{i}0} = \mathrm{i}\Delta(z). \qquad (1.3.13)$$

The $+\mathrm{i}0$ prescription guarantees that we will obtain time ordered products. Then,

$$Z[\eta] = N \int \prod_x \mathrm{d}\phi'(x)\, \det(\partial\phi/\partial\phi')$$

$$\times \exp \mathrm{i} \int \mathrm{d}^4x \left\{ -\tfrac{1}{2}\phi'(x)\phi'(x) + \int \mathrm{d}^4y\, \eta(x)K^{1/2}(x-y)\phi'(y) \right\};$$

$\det(\partial\phi/\phi')$ is the (infinite dimensional) Jacobian of the change of variables. The final step is a shift of the integration variable

$$\phi'(x) = \phi''(x) + \int d^4y\, K^{1/2}(x-y)\eta(y)$$

so that

$$Z[\eta] = \left\{ N \int \prod_x d\phi''(x) \det\left(\frac{\partial\phi}{\partial\phi''}\right) e^{-\frac{i}{2}\int d^4x\,\phi''^2} \right\} e^{\frac{i^2}{2}\int d^4x\,d^4y\,\eta(x)\Delta(x-y)\eta(y)}.$$

$$(1.3.14)$$

where $\Delta(x-y)$ is the propagator

$$\Delta(x) = \frac{i}{(2\pi)^4} \int d^4k\, \frac{e^{-ik\cdot x}}{k^2 - m^2 + i0} = \langle T\phi(x)\phi(0)\rangle_0.$$

The term in braces in the right hand side of (1.3.14) is independent of η; hence it will cancel when taking the logarithmic derivative. So we may write

$$Z[\eta] = \bar{N} \exp\left\{ \frac{i^2}{2}\int d^4x\,d^4y\,\eta(x)\Delta(x-y)\eta(y) \right\},\qquad (1.3.15)$$

from which (1.3.12) follows directly.

The treatment of vector fields presents no problems, and we will describe it in Sect. 2.5. Operator insertions are dealt with by the introduction of extra sources (an example will be found in Sect. 2.6). Only fermion fields require some elaboration. We have to introduce, at the classical level, *anticommuting c-numbers*,[10] defined by the relations

$$\psi(x)\psi(y) = -\psi(y)\psi(x), \quad [\psi(x)]^2 = 0.$$

A functional of (classical) fermion fields will be of the general form

$$F[\psi] = K_0 + \int dx_1\, K_1(x_1)\psi(x_1) + \cdots$$

$$+ \int dx_1\ldots dx_n\, K_n(x_1,\ldots,x_n)\psi(x_1)\ldots\psi(x_n) + \cdots,$$

where K_1 is an anticommuting function and the K_n, $n \geq 2$, may be taken as fully antisymmetric in their arguments. The extension of the definition

$$\frac{\delta F[\psi]}{\delta\psi} = \lim_{\epsilon\to 0} \frac{F[\psi + \epsilon\delta_x] - F[\psi]}{\epsilon},$$

---

[10]The corresponding structure is known as a Grassmann algebra in the standard mathematical literature. More details may be found in the treatise of Berezin (1966).

where ϵ is an anticommuting number,

$$\epsilon\psi = -\psi\epsilon, \quad \epsilon^2 = 0,$$

yields the derivatives

$$\left.\frac{\delta^n F[\psi]}{\delta\psi(x_n)\dots\delta\psi(x_1)}\right|_{\psi=0} = n!K_n(x_1,\dots,x_n).$$

Note the reversed order of the x; this is so because

$$\frac{\delta^2}{\delta\psi_1\delta\psi_2} = -\frac{\delta^2}{\delta\psi_2\delta\psi_1}.$$

The integration over anticommuting functions also presents peculiarities because, in order to be consistent, we have to define

$$\int d\psi(x) = 0, \quad \int d\psi(x)\,\psi(y) = \delta(x-y).$$

Finally, if we want to generate *one particle irreducible* (1PI) Green's functions, i.e., Green's functions that remain connected when cutting one internal line, we do so by functional differentiation with respect not to η, but to the new field $\bar\phi$, a functional $\Gamma[\bar\phi]$:

$$\Gamma[\bar\phi] = \frac{1}{i}\log Z[\eta] - \int d^4x\,\eta(x)\bar\phi(x), \tag{1.3.16a}$$

$$\bar\phi(x) \equiv \frac{-i\delta\log Z[\eta]}{\delta\eta(x)}. \tag{1.3.16b}$$

Note that $\bar\phi$ is the VEV of $\hat\phi$.

The proof that Γ generates 1PI Green's functions is apparent from an identity that we now prove. Differentiating Γ twice,

$$\frac{\delta^2\Gamma}{\delta\bar\phi(x)\delta\bar\phi(y)} = -\frac{\delta\eta(x)}{\delta\bar\phi(y)} = \left[-\frac{\delta\bar\phi(y)}{\delta\eta(x)}\right] = -i\Delta^{-1}(x-y),$$

so that, in particular, $\Delta\{\delta^2\Gamma/\delta\bar\phi(x)\delta\bar\phi(y)\}\Delta = i\Delta$: up to an i, the propagator is obtained by dressing the 1PI Green's function with propagators. More generally,

$$\frac{\delta}{\delta\bar\phi} = \left[\frac{\delta\eta}{\delta\bar\phi}\right]\frac{\delta}{\delta\eta} = -i\Delta^{-1}(x-y)\frac{\delta}{\delta\eta}, \tag{1.3.17}$$

which is the required equation.

2 QCD as a Field Theory

Puisque ces mystères nous depassent, feignons de les avoir organisés.

J. COCTEAU

2.1 Gauge Invariance

Let us consider the set of fields that we have postulated for QCD: three $q^j(x)$ for each quark flavour and the eight $B_a(x)$. The first set builds up the fundamental representation of $SU(3)$: if U is a 3×3 unitary matrix of determinant unity, then the q^j transform as

$$U : q^j(x) \to \sum_k U_{jk} q^k(x).$$

It is possible to write any matrix in $SU(3)$, U, in terms of the eight generators of its Lie algebra, t^a. (The explicit form of these 3×3 matrices is given in Appendix C.) We have

$$U = \exp\left\{ -ig \sum_a \theta_a t^a \right\}.$$

The θ_a are the parameters of the group, and the factor g is introduced for future convenience. Representing q^j by a vertical matrix, it transforms according to

$$q(x) \to e^{-ig \sum \theta_a t^a} q(x).$$

For B, we consider the *adjoint* (dimension 8) representation of $SU(3)$. We let C^a be the corresponding matrices, with elements $C^a_{bc} = -if_{abc}$ (see again Appendix C for the explicit values). Then

$$B^\mu(x) \to e^{-ig \sum \theta_a C^a} B^\mu(x).$$

If the θ_a are constant independent of the space-time point, the analysis is complete; we have a *global* $SU(3)$ invariance. However, as we know from quantum electrodynamics (QED), we have an interest in extending the transformations to transformations with parameters $\theta_a(x)$ which depend upon the space–time point. We thus define the (local) *gauge* transformations, considering for the moment only *classical* fields,

$$q(x) \to e^{-ig \sum \theta_a(x) t^a} q(x). \tag{2.1.1a}$$

Similarly, we generalize the usual QED transformations and define

$$B^\mu(x) \to e^{-ig \sum \theta_a(x) C^a} B^\mu(x) - \partial^\mu \theta(x), \qquad (2.1.1b)$$

or, for infinitesimal θ,

$$q^j(x) \to q^j(x) - ig \sum_{a,k} \theta_a(x) t^a_{jk} q^k(x),$$

$$B^\mu_a(x) \to B^\mu_a(x) + g \sum_{b,c} f_{abc} \theta_b(x) B^\mu_c(x) - \partial^\mu \theta_a(x). \qquad (2.1.1c)$$

We will assume invariance under the transformations (2.1.1): in fact, the Lagrangian (1.1.11) has this property built in. As we shall see, this invariance forces the fields to appear in very precise combinations, and it will be clear at the end of the present section that indeed (1.1.11) is the most general Lagrangian invariant under (2.1.1) *and* involving no couplings with dimensions of a *negative* power of the mass (cf., however, Sect. 7.7 and Chap. 8).

Let us consider the transformation properties of the derivative of a field, say $\partial^\mu q(x)$. From (2.1.1c),

$$\partial^\mu q^j(x) \to \partial^\mu q^j(x) - ig \sum t^a_{jk} \theta_a(x) \partial^\mu q^k(x) - ig \sum t^a_{jk} \big(\partial^\mu \theta_a(x)\big) q^k(x).$$

We see that it transforms differently from the field itself. To obtain an invariant Lagrangian, all derivatives must appear in covariant combinations:

$$D^\mu q^j(x) \equiv \sum_k \left\{ \delta_{jk} \partial^\mu - ig \sum_a B^\mu_a(x) t^a_{jk} \right\} q^k(x), \qquad (2.1.2)$$

where D^μ is called the (gauge) *covariant derivative*. The proof that D^μ is covariant is straightforward. Using matrix notation,

$$D^\mu q(x) \to \partial^\mu q(x) - ig \sum t^a \theta_a(x) \partial^\mu q(x) - ig \sum t^a \big(\partial^\mu \theta_a(x)\big) q(x)$$

$$- g^2 \sum B^\mu_a(x) t^a t^b \theta_b(x) q(x) - ig \sum B^\mu_a t^a q(x)$$

$$- ig^2 \sum f_{abc} t^a \theta_b(x) B^\mu_c(x) q(x) + ig \sum \big(\partial^\mu \theta_a(x)\big) t^a q(x). \qquad (2.1.3a)$$

Because

$$t^a t^b = t^b t^a + \big[t^a, t^b\big], \quad \big[t^a, t^b\big] = i \sum f^{abc} t^c,$$

the right hand side of (2.1.3a) is

$$D^\mu q(x) - ig \sum t^a \theta_a(x) D^\mu q(x), \qquad (2.1.3b)$$

as we wished to prove. Similarly, the *covariant curl* of the field B is

$$(D^\mu \times B^\nu)_a \equiv G^{\mu\nu}_a = \partial^\mu B^\nu_a - \partial^\nu B^\mu_a + g \sum f_{abc} B^\mu_b B^\nu_c, \qquad (2.1.4)$$

and the analogy of $G_a^{\mu\nu}$ with the electromagnetic field strength tensor, $F^{\mu\nu} = \partial^\mu A^\nu - \partial^\nu A^\mu$, is apparent. In terms of these, we can write (1.1.11) in a manner that is manifestly gauge invariant. Dropping the QCD index, we have

$$\mathcal{L} = \sum_q \left\{ i\bar{q}(x)\not{D}q(x) - m_q\bar{q}(x)q(x) \right\} - \tfrac{1}{4}\left(D \times B\right)^2. \tag{2.1.5a}$$

Here $(D \times B)^2$ is short for the *pure Yang–Mills* component,

$$\left(D \times B\right)^2 \equiv G^2 = \sum_a G_a^{\mu\nu}G_{a\mu\nu}; \quad \mathcal{L}_{\mathrm{YM}} \equiv \tfrac{1}{4}\left(D \times B\right)^2. \tag{2.1.5b}$$

The importance of (non-Abelian) gauge invariance is threefold. First, as is clear from the proof of (2.1.3), it requires universality of the coupling; i.e., one single constant g characterizes the coupling of quarks to gluons, or of gluons among themselves. Second, 't Hooft (1971) has proved that a non-Abelian theory is renormalizable, but only if it is gauge invariant. Third, it has been shown by Coleman and Gross (1973) that *only* a non-Abelian theory can be asymptotically free.

At first sight it looks as if Eq. (2.1.5) could be carried over to the quantum theory directly by simply reinterpreting the fields as quantum fields. However, and as we already know from quantum electrodynamics (QED), this is not so. It is clear from gauge invariance that the fields B are undefined, since we may effect transformations such as (2.1.1) that will alter the commutation relations. Of course, this is related to the fact that the particles corresponding to the fields B, being massless, have only two degrees of freedom; whereas the fields B^μ have four independent components. To effect the quantization, we will be forced to select definite representatives of each gauge class (*gauge fixing*), which breaks manifest gauge invariance. Because of the presence of gluon self-interactions, we expect this to cause more trouble than in the Abelian case and, indeed, we will see that the Lorentz covariant gauges require the introduction of extra, nonphysical fields[1] (*ghosts*) to restore gauge invariance and unitarity. Alternatively, we may choose ghost-free gauges (like the so-called axial gauges) which, however, break manifest Lorentz invariance.

To complete this, and before considering the quantized theory, we write the equations of motion that follow from (2.1.5), at the classical level. The Euler–Lagrange equations for a generic field Φ are obtained by requiring stationary *action*, $\mathcal{A} = \int \mathrm{d}^4x\, \mathcal{L}$: they are

$$\partial_\mu \frac{\partial \mathcal{L}}{\partial(\partial_\mu \Phi)} = \frac{\partial \mathcal{L}}{\partial \Phi};$$

---

[1] Peculiar gauges with ghosts may also be constructed for Abelian theories.

so for QCD we find

$$\bar{q}(\mathrm{i}\overleftarrow{D} - m) = 0, \quad (\mathrm{i}D\!\!\!/ - m)q(x) = 0,$$
$$D_\mu G_a^{\mu\nu}(x) \equiv \partial_\mu G_a^{\mu\nu}(x) + g\sum f_{abc}B_{b\mu}(x)G_c^{\mu\nu}(x) = 0. \tag{2.1.6}$$

2.2 Canonical Quantization; Gauge Fixing; Covariant Gauges

Let us start by trying to quantize the free gluon fields. The free gluon (Yang–Mills) Lagrangian is

$$\mathcal{L}_{\mathrm{YM}}^0 \equiv -\tfrac{1}{4}\sum G_a^{0\mu\nu}G_{a\mu\nu}^0,$$
$$G_a^{0\mu\nu} = \partial^\mu B_a^{0\nu} - \partial^\nu B_a^{0\mu} \tag{2.2.1}$$

and the index 0 denotes free fields ($g \equiv 0$). Eq. (2.2.1) is similar to the Lagrangian for eight uncoupled electromagnetic fields; as such it is invariant under the *free* gauge transformations. Dropping the labels 0 that denote free fields, these are,

$$B_a^\mu(x) \to B_a^\mu(x) - \partial^\mu\theta_a(x). \tag{2.2.2}$$

We expect all the problems and benefits associated with gauge invariance. In particular, since B is undefined, it will be impossible to quantize (2.2.1) directly. In fact, suppose we want to implement the standard canonical quantization procedure. We define momenta conjugate to the B_a^0:

$$\pi_a^\mu(x) = \frac{\partial\mathcal{L}}{\partial(\partial_0 B_{a\mu})} = G_a^{\mu 0}, \tag{2.2.3}$$

and we see that $\pi_a^0(x)$ vanishes identically. The *canonical commutation relations* are

$$[\pi_a^\mu(x), B_b^\nu(y)]\,\delta(x^0 - y^0) = -\mathrm{i}\delta_{ab}g^{\mu\nu}\delta(x - y), \tag{2.2.4}$$

so that the B_a^0 would commute with all operators and should thus be c-numbers.

At this point, two paths lie open to us. We may choose a gauge in which the nonphysical degrees of freedom are absent. It is quite clear that this violates manifest Lorentz invariance. Or we may treat all the B^μ in the same manner. Since this introduces nonphysical degrees of freedom, we will be forced to work in a space with an indefinite metric. We shall discuss physical gauges later on and for the moment consider covariant ones.

As is known from the case of the electromagnetic field (and at the level at which we are working now there is no difference), we cannot have the Lorentz condition, $\partial \cdot B_a = 0$, and at the same time keep covariant commutation relations: therefore, we have to give up $\partial \cdot B = 0$ as an operator statement. We then introduce the *Gupta–Bleuler* space $\mathfrak{H}_{\mathrm{GB}}$ where Eq. (2.2.4) is realized.

We shall see that this implies an indefinite metric for $\mathfrak{H}_{\text{GB}}$. Physical vectors will be those for which

$$\langle \Phi_{\text{Ph}}|\partial_\mu B_a^\mu(x)|\Phi_{\text{Ph}}\rangle = 0. \tag{2.2.5}$$

If we identify vectors that differ by a vector of zero norm, i.e.,

$$|\Phi_{\text{Ph}}\rangle \sim |\Phi'_{\text{Ph}}\rangle = |\Phi_{\text{Ph}}\rangle + |\Phi^{(0)}\rangle, \tag{2.2.6}$$

when $\langle \Phi^{(0)}|\Phi^{(0)}\rangle = 0$, we finally obtain the space of physical vectors $\mathfrak{L}$.

To maintain Eq. (2.2.4) we have to modify the Lagrangian (2.2.1). We do this by adding a term $-(\lambda/2)\sum_a(\partial \cdot B_a)^2$ (*gauge fixing*):

$$\mathcal{L}_{\lambda\text{YM}} = -\frac{1}{4}\sum_a G_a^{\mu\nu}G_{a\mu\nu} - \frac{\lambda}{2}\sum_a (\partial_\mu B_a^\mu)^2. \tag{2.2.7}$$

This should have no physical consequences, at least in the free field case, because the term added vanished between physical vectors, as in Eq. (2.2.5). The momenta conjugate to the B are now

$$\pi_{\lambda a}^\mu(x) = G_a^{\mu 0}(x) - \lambda g^{\mu 0}\partial_\nu B_a^\nu(x), \tag{2.2.8}$$

which is not zero, and thus we may keep (2.2.4). However, there arises an indefinite metric. For example, consider (2.2.4) with $\mu = 0$:

$$\lambda\big[\partial_\mu B_a^\mu(x), B_b^\nu(y)\big]\delta(x^0 - y^0) = i\delta_{ab}g_{0\nu}\delta_4(x - y). \tag{2.2.9}$$

The sign is undefined. To see this more clearly, we consider momentum space. Let us take the case $\lambda = 1$ and introduce a canonical tetrad $\epsilon^{(\rho)}(k)$ associated to the lightlike vector k:

$$\epsilon_\mu^{(0)} = \delta_{\mu 0}, \quad \epsilon_0^{(i)} = 0, \quad \epsilon_\mu^{(3)} = \frac{1}{k^0}k_\mu - \delta_{\mu 0};$$
$$\mathbf{k}\epsilon^{(i)} = 0, \quad i = 1,2; \quad \epsilon_\mu^{(i)}\epsilon^{(j)\mu} = -\delta_{ij}, \quad i, j = 1, 2, 3. \tag{2.2.10}$$

Of these, only $\epsilon^{(i)}$, $i = 1$, 2, are associated to physical zero mass particles; $\epsilon^{(3)}$ is longitudinal, and $\epsilon^{(0)}$ corresponds to a spin-zero object. We may expand B into creation and annihilation operators:

$$B_b^\mu(x) = \frac{1}{(2\pi)^{\frac{3}{2}}}\int \frac{d^3\mathbf{k}}{2k^0}\sum_\rho \big\{e^{-ik\cdot x}\epsilon^{(\rho)\mu}(k)a_\rho(b, k) + e^{ik\cdot x}\epsilon^{(\rho)\mu}(k)^* a_\rho^\dagger(b, k)\big\}. \tag{2.2.11}$$

From (2.2.4) we then find

$$[a_\mu(b, k), a_\nu^\dagger(b', k')] = -g_{\mu\nu}\delta_{bb'}2k^0\delta(\mathbf{k} - \mathbf{k}') : \tag{2.2.12}$$

thus $\langle 0|a_0(k)a_0^\dagger|0\rangle$ is negative in the gauge we are considering.

Using (2.2.12) we may calculate the propagator. If we let

$$\langle T B_a^\mu(x) B_b^\nu(0) \rangle_0 = D_{ab}^{\mu\nu}(x),$$

then, writing the answer for an arbitrary value of the *gauge parameter*, λ, we have

$$D_{ab}^{(0)\mu\nu}(x) = \delta_{ab} \frac{i}{(2\pi)^4} \int d^4k \, e^{-ik\cdot x} \frac{-g^{\mu\nu} + (1 - \lambda^{-1})k^\mu k^\nu/(k^2 + i0)}{k^2 + i0},$$

(2.2.13a)

the superindex 0 in $D_{ab}^{(0)\mu\nu}$ indicating free fields. We have used the notation, that will be consistently employed in this text,

$$\langle fg \ldots h \rangle_0 \equiv \langle 0 | fg \ldots h | 0 \rangle.$$

It is convenient to write $1 - 1/\lambda = \xi$; this simplifies the expression for the propagator. In momentum space,

$$D_{ab}^{(0)\mu\nu}(k) = i\delta_{ab} \frac{-g^{\mu\nu} + \xi k^\mu k^\nu/(k^2 + i0)}{k^2 + i0}.$$

(2.2.13b)

An especially simple case is the *Fermi–Feynman gauge*, $\xi = 0$; also useful is the *Landau* or *transverse gauge*, $\xi = 1$.

Actually, and for $\lambda \neq 1$, Eqs. (2.2.13) have to be obtained somewhat indirectly because, for physical, massless gluons, the term $k^\mu k^\nu / k^2$ is infinite. The solution is obtained by introducing a fictitious mass, M. With it we obtain, in momentum space,

$$D_{ab}^{(0)\mu\nu}(k, M) = i\delta_{ab} \frac{-g^{\mu\nu} + (1 - \lambda^{-1})k^\mu k^\nu/(k^2 - \lambda^{-1}M^2 + i0)}{k^2 - M^2 + i0};$$

taking the limit $M \to 0$, Eqs. (2.2.13) follow.

In QED, because photons do not possess self-interactions, one can work with covariant gauges without additional considerations. In QCD, self-interactions cause further complications. This will be seen in next section.

2.3 Unitarity; Lorentz Gauges; Ghosts; Physical Gauges

i Covariant Gauges

The fact that not all states in the space where fields are defined correspond
to physical vectors means that we have to be careful with unitarity. The
unitarity condition, (1.2.7) or (1.2.8), is valid only for physical states. To
extend it to our case, we introduce the projector into physical states, P:

$$P\mathfrak{H}_{\mathrm{GB}} = \mathfrak{L}, \quad P^2 = P^\dagger = P. \tag{2.3.1}$$

The unitarity condition then reads

$$(PSP)(PSP)^\dagger = P. \tag{2.3.2}$$

If the Lagrangian is Hermitian, the S matrix is unitary in $\mathfrak{H}_{\mathrm{GB}}$, so we find
that (2.3.2) will be satisfied if S commutes with P. In QED this is automatic
for the gauges defined previously. In QCD such is not the case because, except
for $g = 0$, gauge transformations involve interactions. This means that the
Lagrangian

$$\mathcal{L}^\xi = \sum_q \left\{ i\bar{q}\slashed{D}q - m_q\bar{q}q \right\} - \tfrac{1}{4}\left(D \times B \right)^2 - \frac{\lambda}{2}(\partial \cdot B)^2, \quad \xi = 1 - 1/\lambda, \tag{2.3.3}$$

obtained by adjoining the gauge-fixing term to (2.1.5) is not complete as
it stands and will have to be modified. To see how this modification comes
about, we will check, in the Fermi–Feynman gauge, how (2.3.2) is violated.

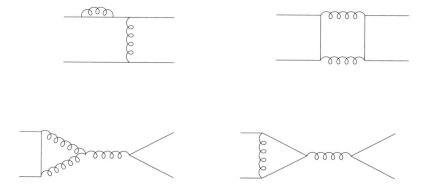

Fig. 2.3.1A. Diagrams contributing to quark–antiquark scattering to second order. Only one diagram of each kind is shown, in particular those containing two gluons as intermediate states.

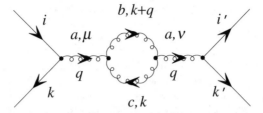

Fig. 2.3.1B. The diagram, contributing to quark–antiquark scattering, where the conflict with unitarity appears. i, k, i', k'; a, b, c are colour indices.

Consider second order quark–antiquark scattering. The Feynman diagrams that describe this process are shown in Figs. 2.3.1A,B,C. It is not difficult to see that only the diagram involving a modification of the gluon operator, depicted in Fig. 2.3.1B, may cause trouble. The diagram in Fig. 2.3.1C contains what is known as a "tadpole". In the regularization scheme known as *dimensional regularization*, it vanishes identically as it contains an integral

$$\int \mathrm{d}^D k \, (k^2 + \mathrm{i}0)^{-1} \equiv 0;$$

see Sect. 3.1.

We will calculate the diagram in Fig. 2.3.1B in dimension D (see below, Sect. 3.1 for more details) and at the end we will take the physical limit, $D \to 4$. The corresponding amplitude, with the routing of momenta of the Fig. 2.3.1B, is

$$\mathcal{T} = \frac{-g^2}{(2\pi)^2} \sum_{aa'} \bar{v}\gamma_{\mu'} u \, t_{ik}^a \, \frac{-\mathrm{i}g^{\mu'\mu}}{q^2} \Pi_{aa'\mu\nu} \frac{-\mathrm{i}g^{\nu'\nu}}{q^2} \bar{u}'\gamma_{\nu'} v' t_{i'k'}^{a'} \delta(P_f - P_i), \quad (2.3.4a)$$

where

$$\Pi_{aa'}^{\mu\nu} = \frac{-\mathrm{i}g^2}{2} \sum f^{abc} f^{a'bc} \int \frac{\mathrm{d}^D k}{(2\pi)^D} \frac{1}{k^2(k+q)^2}$$
$$\times \left\{ \left[-(2k+q)^\mu g_{\alpha\beta} + (k-q)_\beta g_\alpha^\mu + (2q+k)_\alpha g_\beta^\mu \right] \right. \qquad (2.3.4b)$$
$$\left. \times \left[-(2k+q)^\nu g^{\alpha\beta} + (k-q)^\beta g^{\nu\alpha} + (2q+k)^\alpha g^{\nu\beta} \right] \right\}.$$

(we will consistently omit the terms i0 in the denominators, which should be understood). Using the relation $\sum f^{abc} f^{a'bc} = \delta_{aa'} C_A$, $C_A = 3$ (Appendix C)

Fig. 2.3.1C. "Tadpole" diagram for second order $\bar{q}q$ scattering.

and carrying out standard manipulations (that the reader may find in any textbook, or in Appendix B here) we find the expression

$$\Pi_{aa'}^{\mu\nu} = \delta_{aa'} C_A \frac{g^2}{32\pi}$$

$$\times \left\{ \left[\frac{19}{6} N_\epsilon - \frac{1}{2} - \int_0^1 dx \, (11x^2 - 11x + 5) \log\left(-x(1-x)q^2 \right) \right] q^2 g^{\mu\nu} \right.$$

$$\left. - \left[\frac{11}{3} N_\epsilon + \frac{2}{3} - \int_0^1 dx \, (-10x^2 + 10x + 2) \log\left(-x(1-x) \right) \right] q^\mu q^\nu \right\};$$

$$(2.3.5)$$

$N_\epsilon \equiv 2/\epsilon - \gamma_E + \log 4\pi, \quad \epsilon = 4 - D \to 0.$

This is divergent, but that is not the difficulty worrying us at present. Unitarity tells us that $\text{Im}\,\mathcal{T} = \frac{1}{2}\mathcal{T}\mathcal{T}^\dagger$. Now, $\text{Im}\,\mathcal{T}$ is obtained from (2.3.4) by replacing Π by $\text{Im}\,\Pi$ which, according to (2.3.5), is

$$\text{Im}\,\Pi_{aa'}^{\mu\nu}(q) = \delta_{aa'} C_A \frac{g^2}{32\pi}\theta(q^2) \left\{ -\frac{19}{6}q^2 g^{\mu\nu} + \frac{22}{6}q^\mu q^\nu \right\}, \quad (2.3.6)$$

which is finite even for $D = 4$. This should be proportional to

$$\frac{1}{2} \sum_{c,\text{phys.}} \langle \bar{q}q | \mathcal{T} | c, \text{phys.} \rangle \langle c, \text{phys.} | \mathcal{T}^\dagger | \bar{q}q \rangle,$$

i.e., to the square of the amplitude for $\bar{q}q \to BB$ with *physical* gluons BB (Fig. 2.3.2). If we make use of the Feynman rules we see that the expression for this is similar to $\text{Im}\,\mathcal{T}$, with the replacement of $\text{Im}\,\Pi_{aa'}^{\mu\nu}(q)$ by

$$\delta_{aa'} C_A \sum_{\substack{\eta_1,\eta_2 \\ k_1+k_2=q}} \mathcal{A}_\mu(k_1,k_2;\eta_1,\eta_2)\mathcal{A}_\nu^*(k_1,k_2;\eta_1,\eta_2), \quad (2.3.7a)$$

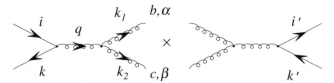

Fig. 2.3.2. Imaginary part of $\mathcal{T}$, $\mathcal{T} \times \mathcal{T}^\dagger$.

where η_1, η_2 are the physical helicities of the gluons and

$$\mathcal{A}^\mu = \left[(k_1 + q)_\beta g_\alpha^\mu - (q + k_2)_\alpha g_\beta^\mu + (k_2 - k_1)^\mu g_{\alpha\beta}\right] \epsilon_{\text{phys}}^\alpha(k_1, \eta_1) \epsilon_{\text{phys}}^\beta(k_2, \eta_2).$$
$$(2.3.7\text{b})$$

Here the ϵ_{phys} are the polarization vectors for physical gluons given by, e.g.,

$$\epsilon_{\text{phys}}^\alpha(k, \eta) = \frac{1}{\sqrt{2}} \left\{ \epsilon^{(1)\alpha}(k) + i\eta \epsilon^{(2)\alpha}(k) \right\},$$

with the $\epsilon^{(i)}$ of (2.2.10). Because the gluons are physical, the ϵ_{phys} verify

$$k_\alpha \epsilon_{\text{phys}}^\alpha(k, \eta) = 0, \quad k^2 = 0,$$

so (2.3.7b) may be written as (note that $q = k_1 + k_2$)

$$\mathcal{A}^\mu = \left[2k_{1\beta} g_\alpha^\mu - 2k_{2\alpha} g_\beta^\mu + (k_2 - k_1)^\mu g_{\alpha\beta}\right] \epsilon_{\text{phys}}^\alpha(k_1, \eta_1) \epsilon_{\text{phys}}^\beta(k_2, \eta_2),$$

and it is then easy to check that

$$q_\mu \mathcal{A}^\mu = 0$$

(transversality). Unitarity cannot be satisfied in the space of physical gluons. Indeed, from (2.3.6) it is clear that $q_\mu \Pi_{aa'}^{\mu\nu}(q) \neq 0$, which contradicts transversality.

Of course, what happens is that $\mathcal{L}_{\text{int}}^\xi$ sends physical states into nonphysical ones. This fact was first noted by DeWitt (1964) and by Feynman; the solution was given by Feynman (1963) in particular cases, and by Fadeyev and Popov (1967) in general. The idea is to introduce extra nonphysical particles (*ghosts*) that will cancel exactly the nonphysical states produced by $\mathcal{L}_{\text{int}}^\xi$. We thus modify $\mathcal{L}^\xi$ by adding the ghost term

$$\mathcal{L}_{\text{all}}^\xi = \mathcal{L}^\xi + \sum \left(\partial_\mu \bar{\omega}_a(x)\right)\left(\delta_{ab}\partial^\mu - gf_{abc} B_c^\mu(x)\right)\omega_b(x), \qquad (2.3.8)$$

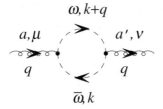

Fig. 2.3.3. Ghost loop contribution to the gluon propagator.

with $\mathcal{L}^\xi$ given by (2.3.3). The fields ω, $\bar{\omega}$ are of spin zero, but they satisfy Fermi–Dirac statistics. Since they will never appear in initial or final states (they are, by hypothesis, nonphysical) this should not worry us. [2]

Let us proceed with our analysis, introducing the ghost contribution. Since ghosts only couple to gluons, they will only modify Fig. 2.3.1B, which suits us. Their contribution to Π is easily evaluated to be, with the conventions of Fig. 2.3.3,

$$\Pi^{\mu\nu}_{(\text{ghost})aa'} = \delta_{aa'} C_A \text{i}g^2 \int \frac{\mathrm{d}^D k}{(2\pi)^D} \frac{k^\mu (k+q)^\nu}{k^2 (k+q)^2} = \frac{\delta_{aa'} g^2 C_A}{32\pi^2}$$

$$\times \left\{ \left[\tfrac{1}{6} N_\epsilon + \tfrac{1}{6} - \int_0^1 \mathrm{d}x\, x(1-x) \log\left(-x(1-x)q^2 \right) \right] q^2 g^{\mu\nu} \right.$$

$$\left. - \left[-\tfrac{1}{3} N_\epsilon + 2 \int_0^1 \mathrm{d}x\, x(1-x) \log\left(-x(1-x)q^2 \right) \right] q^\mu q^\nu \right\}.$$

After some manipulations, using the integration formulas of Appendix B, adding the result to (2.3.5) and integrating $\mathrm{d}x$, we find

$$\Pi^{\mu\nu}_{(\text{all})aa'} = \frac{\delta_{aa'} g^2 C_A}{32\pi^2} \left(-g^{\mu\nu} q^2 + q^\mu q^\nu \right) \left\{ -\tfrac{10}{3} N_\epsilon - \tfrac{62}{9} + \tfrac{10}{3} \log(q^2) \right\}, \quad (2.3.9)$$

which certainly verifies the transversality condition,

$$q_\mu \Pi^{\mu\nu}_{(\text{all})aa'} = \Pi^{\mu\nu}_{(\text{all})aa'} q_\nu = 0. \qquad (2.3.10)$$

---

[2] It is convenient at times, although not necessary, to think of ω, $\bar{\omega}$ as mutually adjoint. The deduction of (2.3.8) to all orders will be given in Sect. 2.5.

We leave it to the reader to check that now $\operatorname{Im} \Pi \sim \sum \mathcal{A}\mathcal{A}^*$. We will henceforth drop the index "all" and consider the QCD Lagrangian in a covariant (*Lorentz*) gauge to be (2.3.8), i.e.,

$$
\begin{aligned}
\mathcal{L}_{\text{QCD}}^{\xi} = \sum_q \left\{ i\bar{q}\not{D}q - m_q\bar{q}q \right\} &- \tfrac{1}{4}\left(D \times B\right)^2 - \frac{\lambda}{2}(\partial \cdot B)^2 \\
&+ \sum_{abc}(\partial_\mu\bar{\omega}_a)\left(\delta_{ab}\partial^\mu - gf_{abc}B_c^\mu\right)\omega_b, \quad \xi = 1 - 1/\lambda.
\end{aligned}
\tag{2.3.11}
$$

We will also drop at times the index QCD from $\mathcal{L}_{\text{QCD}}^{\xi}$ as given by (2.3.11).

ii Physical Gauges

Since the appearance of ghosts was caused by the fact that the projection over physical states P does not commute with the QCD Lagrangian in a Lorentz gauge, it may appear that the problem will not arise if we choose a gauge with only physical gluons, so that the whole Hilbert space is physical. As we already know at the level of QED, we cannot simultaneously have positivity, locality and manifest Lorentz invariance; so we will have to work with a noncovariant gauge. A Coulomb gauge still has ghosts,[3] but ghost-free gauges exist if we require

$$
n \cdot B = 0, \quad n^2 \le 0. \tag{2.3.12}
$$

For $n^2 \le 0$, one talks of an *axial gauge*; $n^2 = 0$ gives a *lightlike gauge*.[4] Since n is an external vector, manifest Lorentz invariance is lost; of course, gauge invariance guarantees that *physical* quantities will be independent of n, hence Lorentz invariant.

Let us begin with an axial gauge. The Lagrangian is

$$
\mathcal{L}_n = \sum_q \left\{ i\bar{q}\not{D}q - m_q\bar{q}q \right\} - \tfrac{1}{4}\left(D \times B\right)^2 - \frac{1}{2\beta}(n \cdot B)^2, \tag{2.3.13}
$$

and the limit $\beta \to 0$ is to be taken so that the condition (2.3.12) holds as an operator statement over the entire Hilbert space. The propagator that corresponds to (2.3.13) is

$$
i\frac{-g^{\mu\nu} - k^\mu k^\nu(n^2 + \beta k^2)/(k \cdot n)^2 + (n^\mu k^\nu + n^\nu k^\mu)/(n \cdot k)}{k^2 + i0}, \tag{2.3.14}
$$

---

[3] And it also presents further complications. The formulation of QCD in a Coulomb gauge may be found in Christ and Lee (1980).

[4] Axial gauges are discussed in Kummer (1975) and work quoted there. For a lightlike gauge, see for example Tomboulis (1973) and references therein.

which in the limit $\beta \to 0$ becomes

$$\text{i}\frac{-g^{\mu\nu} - n^2 k^\mu k^\nu/(k\cdot n)^2 + (n^\mu k^\nu + n^\nu k^\mu)/(n\cdot k)}{k^2 + \text{i}0}. \tag{2.3.15}$$

These propagators may be obtained with some effort from the canonical formulation or, more easily, in the path integral formalism (Sects. 2.5, 2.6).

The extension of the theory to axial gauges is nontrivial due to the singularities of (2.3.14, 15) for $n\cdot k = 0$. Actually, the way we choose to circumvent these singularities is in principle irrelevant, provided that we choose it consistently: because of gauge invariance the terms proportional to k^μ or k^ν in the propagator will end up cancelling out. But this indicates that care has to be exercised in effecting this cancellation *before* removing whatever regulator on has introduced. In this text we will only make one loop calculations, for which the problem is absent.

For lightlike gauges it is convenient to introduce so-called "null" coordinates: for any vector v,

$$v^\pm = \frac{1}{\sqrt{2}}\left(v^0 \pm v^3\right), \quad \underline{v} = \begin{pmatrix} v^1 \\ v^2 \end{pmatrix}; \quad v^\alpha = v^\pm \text{ or } v^i \ (i = 1, 2).$$

We also define the metric

$$g_{+-} = g_{-+} = 1, \quad g_{++} = g_{--} = 0, \quad g_{ij} = -\delta_{ij}, \text{ for } i,j = 1,2.$$

Note that

$$v\cdot w = v_+ w_- + v_- w_+ - \underline{v}\,\underline{w} = v^\alpha v_\alpha.$$

For a lightlike vector, we may choose $n = u$ with $u^2 = 0$, and the specific value for u, $u_+ = 1$, $u_- = 0$, $\underline{u} = 0$. Then the supplementary condition $u\cdot B = 0$ may be written as

$$B_-^a(x) = 0. \tag{2.3.16}$$

The propagator is now

$$\text{i}\frac{P^{\mu\nu}(k,u)}{k^2 + \text{i}0} = \text{i}\frac{-g^{\mu\nu} + (u^\mu k^\nu + u^\nu k^\mu)/(u\cdot k)}{k^2 + \text{i}0}, \tag{2.3.17}$$

which will be recognized as the limit of (2.3.15) for $n = u$, $u^2 \to 0$. In terms of null coordinates (2.3.17) may be rewritten as

$$\text{i}\frac{P^{\alpha\beta}}{k^2} = \text{i}\frac{-g^{\alpha\beta} + (\delta_-^\alpha k^\beta + \delta_-^\beta k^\alpha)/k_-}{k_\alpha k^\alpha + \text{i}0}.$$

As an example of the use of a lightlike gauge, we consider the second-order gluon propagator. In this gauge, and with ordinary Minkowski coordinates,

$$\Pi_{\text{light}.ab}^{\mu\nu} = -\frac{\text{i}g^2 C_A \delta_{ab}}{2} \int \frac{d^D k}{(2\pi)^D} \frac{1}{k^2(k+q)^2}$$
$$\times \left[-(2k+q)^\mu g^{\alpha\beta} + (k-q)^\beta g^{\mu\alpha} + (2q+k)^\alpha g^{\mu\beta}\right] P_{\alpha\rho}(k,u)$$
$$\times \left[-(2k+q)^\nu g^{\rho\sigma} + (k-q)^\sigma g^{\nu\rho} + (2q+k)^\rho g^{\nu\sigma}\right] P_{\sigma\beta}(k+q,u).$$

We will only consider the divergent and logarithmic part. This simplifies the calculation and we find easily,

$$\Pi^{\mu\nu}_{\text{light.}ab}(q) = \frac{11 C_A g^2 \delta_{ab}}{48\pi^2}\left(-q^2 g^{\mu\nu} + q^\mu q^\nu\right)\left\{N_\epsilon - \log(-q^2) + \text{const. terms}\right\}.$$
(2.3.18)

We check that Π is transverse; no ghosts are required. It should also be noted that the propagator is "self-reproducing" under the transverse tensor in the sense that

$$\frac{P^{\mu\alpha}(q,u)}{q^2}\left\{-q^2 g_{\alpha\beta} + q_\alpha q_\beta\right\}\frac{P^{\beta\nu}(q,u)}{q^2} = \frac{P^{\mu\nu}(q,u)}{q^2}.$$
(2.3.19)

2.4 The Becchi–Rouet–Stora Transformations

In the previous section, we showed that the QCD Lagrangian without ghosts violates unitarity in the space of physical states. Since gauge invariance guarantees that this should not occur, it is clear that the phenomenon must be due to the introduction of the gauge-fixing term which, by its very nature, is not gauge invariant. One may wonder whether the ghosts could not be interpreted as an addition that somehow restores something equivalent to gauge invariance. This is indeed so, as will be discussed in the present section.

Let us begin with QED.[5] The Lagrangian in a covariant gauge is

$$\mathcal{L}^\xi = \bar\psi(i\slashed{D} - m)\psi - \tfrac{1}{4}F_{\mu\nu}F^{\mu\nu} - \frac{\lambda}{2}(\partial \cdot A)^2,$$
(2.4.1)

where now

$$F_{\mu\nu} = \partial^\mu A^\nu - \partial^\nu A^\mu, \quad D^\mu = \partial^\mu - ieA^\mu.$$

It is not gauge invariant because of the gauge fixing term $-(\lambda/2)(\partial \cdot A)^2$. Invariance under suitably generalized gauge transformations may, however, be restored by means of the following trick. Add a term

$$\mathcal{L}_\omega = -\tfrac{1}{2}(\partial^\mu\omega)\partial_\mu\omega$$
(2.4.2)

to $\mathcal{L}^\xi$, with ω a massless field without interactions. We could take ω to be fermionic, and then the parameter ϵ would be an anticommuting c-number, but this is not necessary in the Abelian case.

The gauge transformations are then generalized as follows. Consider infinitesimal transformations and set $\theta(x) = \epsilon\omega(x)$; then the extended gauge transformations are defined as

$$\psi(x) \to \psi(x) + ie\omega(x)\psi(x), \quad A^\mu(x) \to A^\mu(x) - \epsilon\partial^\mu\omega(x),$$
$$\omega(x) \to \omega(x) - \epsilon\lambda\partial_\mu A^\mu(x).$$
(2.4.3)

---

[5] We follow the discussion of de Rafael (1977, 1979).

Then, up to a four-divergence, the Lagrangian

$$\mathcal{L}_{\text{QED}} = \mathcal{L}^{\xi} + \mathcal{L}_{\omega} \tag{2.4.4}$$

is invariant under (2.4.3).

The restoration of gauge invariance was easy here: because A has no self-interactions, we could take ω to be real and free. However, the simplicity of $\mathcal{L}_{\omega}$ does not mean that the extended gauge invariance does not have deep consequences. In fact, one may show that the transformations (2.4.3) generate all the Ward identities of QED – which, in particular, ensure that the interaction does not lead from physical to unphysical states. As an example, we will show how the transversality condition for the photon propagator can be deduced from (2.4.3) and (2.4.4). Of course, it can also be verified by direct computation of the vacuum polarization tensor.

Consider the VEV

$$\langle \text{T} A_{\mu}(x)\omega(0)\rangle_0.$$

Effecting a generalized gauge transformation we find, to first order in ϵ,

$$\lambda \langle \text{T} A_{\mu}(x)\,(\partial_{\nu} A^{\nu}(0))\rangle_0 = \langle \text{T}\,(\partial_{\mu}\omega(x))\,\omega(0)\rangle_0.$$

A Fourier transformation gives

$$
\begin{aligned}
\mathrm{i}\, q_{\nu} D^{\mu\nu}(q) &= \mathrm{i} q_{\nu} \int \mathrm{d}^4 x\, \langle \text{T} A^{\mu}(x) A^{\nu}(0)\rangle_0 \\
&= \int \mathrm{d}^4 x\, \mathrm{e}^{\mathrm{i}q\cdot x} \langle \text{T} A^{\mu}(x)\,(\partial_{\nu} A^{\nu}(0))\rangle_0 = -\frac{1}{\lambda}\int \mathrm{d}^4 x\, \mathrm{e}^{\mathrm{i}q\cdot x} \langle \text{T}\,(\partial^{\mu}\omega(x))\,\omega(0)\rangle_0 \\
&= \frac{\mathrm{i}q^{\mu}}{\lambda}\int \mathrm{d}^4 x\, \mathrm{e}^{\mathrm{i}q\cdot x} \langle \text{T}\omega(x)\omega(0)\rangle_0 = \frac{q^{\mu}}{\lambda}\frac{1}{q^2 + \mathrm{i}0}.
\end{aligned}
\tag{2.4.5}
$$

The last equality holds true because the field ω is free, so its propagator is a free-field propagator. We have thus proved in particular that if we write $D^{\mu\nu}$ as a sum of a transverse and a longitudinal part,

$$D^{\mu\nu}(q) = \left(-q^2 g^{\mu\nu} + q^{\mu}q^{\nu}\right) D_{\text{tr}} + \frac{q^{\mu}q^{\nu}}{q^2} D_L(q^2), \tag{2.4.6}$$

then

$$D_L(q^2) = -\frac{1}{\lambda}\frac{\mathrm{i}}{q^2 + \mathrm{i}0}, \tag{2.4.7}$$

i.e., the longitudinal part of D remains as in the free-field case; recall that, for free fields,

$$D^{(0)\mu\nu}(q) = \mathrm{i}\frac{-g^{\mu\nu} + (1 - \lambda^{-1})q^{\mu}q^{\nu}/(q^2 + \mathrm{i}0)}{q^2 + \mathrm{i}0}.$$

Otherwise stated: if, in perturbation theory, we write

$$D^{\mu\nu}(q) = D^{(0)\mu\nu}(q) + \frac{e^2}{16\pi^2} D^{(2)\mu\nu}(q) + \cdots,$$

then all the $D^{(n)\mu\nu}(q)$ with $n > 0$ will satisfy the transversality condition

$$q_\nu D^{(n)\mu\nu}(q) = 0, \quad n = 2, 4, \ldots,$$

so the equivalent of (2.3.10) is automatic here.

For a non-Abelian theory the generalization of (2.4.3) are the Becchi–Rouet–Stora (1974, 1975), or BRST, transformations.[6] They extend gauge invariance to the ghost fields and leave invariant (up to a four-divergence) the full QCD Lagrangian (2.3.11). As for QED, they generate the analogue of the Ward identities, the *Slavnov–Taylor* identities (Taylor, 1971; Slavnov, 1975). The BRST transformations for QCD are, for infinitesimal ϵ, assumed to be an anticommuting, x independent c-number,[7]

$$
\begin{aligned}
B_a^\mu &\to B_a^\mu - \epsilon \sum \{\delta_{ab}\partial^\mu - g f_{abc} B_c^\mu\} \omega_b, \\
q &\to q - i\epsilon g \sum t^a \omega_a q, \\
\omega_a &\to \omega_a - \frac{\epsilon}{2} g \sum f_{abc} \omega_b \omega_c, \\
\bar\omega_a &\to \bar\omega_a + \epsilon\lambda\partial_\mu B_a^\mu.
\end{aligned}
\tag{2.4.8}
$$

Using them it is easy to derive, by the same method as for QED, the result analogous to (2.4.7). If we write

$$D_{ab}^{\mu\nu}(q) = \delta_{ab}\Big\{ \left(-g^{\mu\nu} + q^\mu q^\nu\right) D_{\rm tr} + \frac{q^\mu q^\nu}{q^2} D_L(q^2)\Big\}, \tag{2.4.9}$$

then also here

$$D_L(q^2) = -\frac{1}{\lambda}\frac{i}{q^2 + i0}. \tag{2.4.10}$$

Therefore if we expand,

$$D_{ab}^{\mu\nu} = \sum_{n=0}^{\infty} \left(\frac{g^2}{16\pi^2}\right)^n D_{ab}^{(n)\mu\nu},$$

and recall that, say to second order,

$$D_{ab}^{(2)\mu\nu} = \sum D_{aa'}^{(0)\mu\mu'} \Pi_{a'b';\mu'\nu'}^{(2)} D_{b'b}^{(0)\nu'\nu},$$

---

[6] The name BRST records also the work of Tyutin (1974); see also Iofa and Tyutin (1976).

[7] Hence $\epsilon^2 = 0$, $\epsilon\omega = -\omega\epsilon$, $\epsilon q = -q\epsilon$, $\epsilon B = B\epsilon$, etc. Remember also that the ω are fermions, so $\omega_b\omega_c = -\omega_c\omega_b$.

($\Pi^{(2)}$ is the second order vacuum polarization tensor), then we have

$$q_\mu \Pi_{ab}^{(2)\mu\nu} = 0.$$

This we have already checked in Eqs. (2.3.9, 10).

A last important point is that all of the above derivations are formal; that is, we neglected to consider, when manipulating propagators (for example) that they are singular functions. To actually verify the identities, one has to check that they go through the renormalization program; see Sects. 3.1, 3.2 and 3.3. Indeed, some formal identities do break down: an example will be found in Sect. 7.5. But even those that do not break down may have to be interpreted. This is true for Eq. (2.4.10), because the gauge parameter becomes renormalized in QCD.

2.5 Functional Formalism for QCD. Gauge Invariance

In this and the following sections, we present the path integral formulation of QCD. The formalism of Sect. 1.3 may be applied to QCD provided we first tackle the question of gauge invariance. To do so, we start with a physical gauge,

$$u \cdot B_a(x) = 0, \quad u^2 \le 0, \tag{2.5.1}$$

and we have to integrate over all B compatible with (2.5.1), i.e., we define, with N an arbitrary normalization factor,

$$Z = N \int \mathcal{D}q \, \mathcal{D}\bar{q} \, \mathcal{D}B \prod_{a,x} \delta \left(u \cdot B_a(x) \right) \exp \mathrm{i} \int \mathrm{d}^4 x \, \mathcal{L}_u, \tag{2.5.2}$$

where we have introduced the notation, to be used systematically,

$$\mathcal{D}q \equiv \prod_{x,f,i,\alpha} \mathrm{d}q_{f,\alpha}^i(x), \quad \mathcal{D}B \equiv \prod_{x,\mu,a} \mathrm{d}B_a^\mu(x), \text{ etc.}$$

with α the Dirac index, and f labelling the quark flavour, $f = u, d, s \ldots$. In (2.5.2), $\mathcal{L}_u$ is the QCD Lagrangian without the gauge fixing term, which is unnecessary since the delta function in Eq. (2.5.2) automatically selects the gauge. If we want to work in a physical gauge, this is all we need. However, we will want to extend the formalism to other gauges, particularly covariant ones. We may write the gauge condition as

$$K_a[B(x)] = 0, \tag{2.5.3}$$

where K_a is the gauge-fixing functional. For example, the Lorentz gauges are selected by choosing

$$K_a[B(x)] = \partial \cdot B_a(x) - \varphi_a(x), \tag{2.5.4}$$

where φ is a prescribed function (we could in particular take $\varphi = 0$).

Let $T(\theta)$ be a gauge transformation with parameters $\theta(x)$, and let B_T be the transform of B under T:

$$B^\mu_{Ta}(x) = B^\mu_a(x) + g \sum f_{abc}\theta_b(x)B^\mu_c(x) - \partial^\mu\theta_a(x),$$

cf. Sect. 2.1. The quantity

$$\Delta_K^{-1}[B] = \int \prod_{x,a} d\theta_a(x) \prod_{x,a} \delta\big(K_a\left[B_T(x)\right]\big) \qquad (2.5.5)$$

is independent of the gauge:

$$\Delta_K^{-1}[B_{T'}] = \Delta_K^{-1}[B].$$

The proof only requires the fact that the integration element $\prod_{x,a} d\theta_a(x)$ is a gauge invariant measure. This is indeed obvious for infinitesimal θ, because then

$$T(\theta)T(\theta') = T(\theta + \theta'),$$

and by iteration we get it for any θ.

Let us temporarily neglect quarks, which play no role in gauge shifts. We may rewrite (2.5.2) as

$$Z = N \int \mathcal{D}B\,\mathcal{D}\theta \prod \delta\big(u \cdot B_a(x)\big) \prod \delta\big(K_b[B_T]\big)\Delta_K[B_T]e^{i\mathcal{A}_{YM}}, \qquad (2.5.6)$$

where $\mathcal{A}_{YM}$ is the pure Yang–Mills action,

$$\mathcal{A}_{YM} = -\tfrac{1}{4} \int d^4x \sum_a G_{a\mu\nu}(x)G^{\mu\nu}_a(x).$$

Suppose we change variables in (2.5.6) via a gauge transformation,

$$B(x) \to B_{T_0}(x),$$

choosing $T_0 = T^{-1}$. We find

$$Z = N \int \mathcal{D}B\,\mathcal{D}\theta \Delta_K[B] \prod \delta\big(u \cdot B_{T_0a}(y)\big) \prod \delta\big(K[B(y)]\big)e^{i\mathcal{A}_{YM}}.$$

Let B_u be a gluon field that verifies (2.5.1). We may find B_{T_0} by a gauge transformation $U(\theta)$. Then,

$$\delta(u \cdot B_{T_0}) = \delta(u \cdot B_{uU}),$$

and thus

$$\int \mathcal{D}\theta \prod \delta(u \cdot B_{T_0a}(y)) = \int \mathcal{D}\theta \prod \delta\big(-u \cdot \partial\theta_{ua}(y)\big),$$

which is independent of B and can therefore be absorbed into the normalization. We have obtained the evaluation

$$Z = N' \int \mathcal{D}B \, \Delta[B] \prod \delta\big(K[B]\big) e^{i\mathcal{A}_{\text{YM}}}. \tag{2.5.7}$$

We have to eliminate the δ function and calculate Δ_K. For the former we choose a Lorentz gauge, (2.5.4); integrating (2.5.7) over $d\varphi$ with the weight

$$\exp\left\{-\frac{i\lambda}{2} \int d^4x \, [\varphi_a(x)]^2\right\},$$

we obtain, on the left hand side, Z times a factor independent of B, namely

$$\int \mathcal{D}\varphi \, \exp\left\{-\frac{i\lambda}{2} \int d^4x \, [\varphi_a(x)]^2\right\},$$

which can again be lumped into N', while, on the right hand side, the integration over $\mathcal{D}\varphi$ may be performed trivially with the help of the δ function:

$$Z = N'' \int \mathcal{D}B \Delta_K[B] e^{i(\mathcal{A}_{\text{YM}} + \mathcal{A}_{\text{GF}})}, \tag{2.5.8a}$$

where the gauge-fixing action is

$$\mathcal{A}_{\text{GF}} = -\frac{\lambda}{2} \int d^4x \, [\partial_\mu B_a^\mu(x)]^2. \tag{2.5.8b}$$

Let us then turn to Δ_K. Because of Eq. (2.5.7), we only require B's such that they verify (2.5.3). Thus, for infinitesimal θ, $K[B_T] = K[B] + (\delta K/\delta B)\delta B \sim (\delta K/\delta B)\delta B$, $\delta B = B_T - B$, so that

$$\Delta_K^{-1}[B] = \int \mathcal{D}\theta \prod \delta\left(\frac{\delta(\partial \cdot B_a)}{\delta B_b^\mu}\left(\partial^\mu \theta_b - g\sum f_{bcd}B_d^\mu \theta_c\right)\right).$$

This may be cast into a more convenient form by introducing the Fadeyev–Popov ghost fields, $\omega, \bar{\omega}$ as anticommuting c-number functions for then, with $\bar{N}$ a number independent of B and ω,

$$\Delta_K[B] = \bar{N} \int \mathcal{D}\omega \, \mathcal{D}\bar{\omega} \, \exp\left\{-i \int d^4x \, d^4y \, \bar{\omega}_a(y)\frac{\delta(\partial \cdot B_a)}{\delta B_b^\mu}\right.$$
$$\left. \times \left[\partial^\mu \omega_b(x) - g\sum f_{bcd}B_d^\mu \omega_c(x)\right]\right\}. \tag{2.5.9}$$

The proof is based on the formula

$$\int \prod_i dc_i \prod_j d\bar{c}_j \, e^{\sum \bar{c}_k A_{kk'} c_{k'}} = (\text{const.}) \times \det A,$$

which is valid for anticommuting c-numbers,[8] c_j, and on the fact that, due to the relation

$$\int \mathrm{d}x_1 \ldots \int \mathrm{d}x_k \prod_{i=1}^{k} \delta\big(f_i(x_1, \ldots, x_k)\big) = \frac{1}{|\det(\partial f_i / \partial x_j)|},$$

Δ_K is simply the determinant of the (infinite) matrix

$$\frac{\partial}{\partial\theta} \left\{ \frac{\delta(\partial \cdot B_a)}{\delta B_b^\mu} \left(\partial^\mu \theta_b - g \sum f_{bcd} B_d^\mu \theta_c \right) \right\}.$$

There only remains one step to complete the analysis. The functional derivative entering (2.5.9) is (cf. Appendix H)

$$\frac{\delta(\partial \cdot B_a(x))}{\delta B_b^\mu(y)} = \delta_{ab} \partial_\mu \delta(x - y),$$

so we may transfer the ∂_μ to the left-hand side and partial integrate $\mathrm{d}^4 y$. We finally obtain

$$Z = N \int \mathcal{D}B \, \mathcal{D}\omega \, \mathcal{D}\bar{\omega} \, \mathrm{e}^{\mathrm{i}(\mathcal{A}_{\mathrm{YM}} + \mathcal{A}_{\mathrm{GF}} + \mathcal{A}_{\mathrm{FP}})} \qquad (2.5.10a)$$

where the Fadeyev–Popov ghost action is

$$\mathcal{A}_{\mathrm{FP}} = \int \mathrm{d}^4 x \sum \big(\partial_\mu \bar{\omega}_a(x)\big) \Big[\delta_{ab} \partial^\mu - g f_{abc} B_c^\mu(x)\Big] \omega_b(x), \qquad (2.5.10b)$$

in agreement with the result we found to one loop in Sect. 2.3 using unitarity.

---

[8] PROOF.

$$\int \prod_{i=1}^{N_0} \mathrm{d}c_i \prod_{j=1}^{N_0} \mathrm{d}\bar{c}_j \, \mathrm{e}^{\sum \bar{c}_k A_{kk'} c_{k'}} = \int \prod_{i=1}^{N_0} \mathrm{d}c_i \prod_{j=1}^{N_0} \mathrm{d}\bar{c}_j \sum_{N=0}^{\infty} \left\{ \sum \bar{c}_k c_{k'} A_{kk'} \right\}^N \frac{1}{N!}.$$

Due to the rules for integrating anticommuting variables, only the term with $N = N_0$ will not vanish and there we obtain

$$\frac{(-1)^{N_0}}{N_0!} \sum \mathrm{sign}(k_1, \ldots, k_{N_0}) \, \mathrm{sign}(k'_1, \ldots, k'_{N_0}) A_{k_1 k'_1} \cdots A_{k_{N_0} k'_{N_0}},$$

where the sum is extended over all permutations $k_1, \ldots, k_{N_0}; k'_1, \ldots, k'_{N_0}$ of the sequence $1, 2, \ldots, N_0$. This is nothing but $(-1)^{N_0} \det A / N_0!$. The extra $-\mathrm{i}$ of the exponent in (2.5.9) only contributes an overall factor. In this way we also see that the phase of the Fadeyev–Popov term is arbitrary. We have chosen it to agree with that of conventional scalar fields.

To generate Green's functions, we have to introduce anticommuting sources $\bar{\eta}_a$, η_a; $\bar{\xi}_{if}$, ξ_{if} for the ghost ω_a, $\bar{\omega}_a$ and quark q_f^i, $\bar{q}_f^i$ fields (f being the flavour index), and commuting sources λ_a^μ for the gluons B_a^μ. Thus, our starting point will be the functional

$$Z[\eta, \bar{\eta}; \xi, \bar{\xi}; \lambda] = \int \mathcal{D}q\, \mathcal{D}\bar{q}\, \mathcal{D}\omega\, \mathcal{D}\bar{\omega}\, \mathcal{D}B \exp i \int d^4x \left\{ \mathcal{L}_{\text{QCD}}^\xi + \mathcal{L}_{\text{source}} \right\},$$

(2.5.11a)

where $\mathcal{L}_{\text{QCD}}^\xi$ is given in Eq. (2.3.11) and

$$\mathcal{L}_{\text{source}} = \sum \left\{ \bar{\eta}_a \omega_a + \bar{\omega}_a \eta_a + \bar{\xi}_{if} q_f^i + \bar{q}_f^i \xi_{if} + \lambda_{a\mu} B_a^\mu \right\}.$$

(2.5.11b)

2.6 Feynman Rules in the Path Integral Formalism

In Sect. 1.3 we stated that the expansion of the Green's functions generated by (2.5.11) in powers of g reproduces the usual Feynman rules which were previously obtained with Wick's theorem and the decomposition of field operators in creation–annihilation operators. Alternatively, we could have derived the Feynman rules from Eq. (2.5.11). We will exemplify this with three typical cases: the gluon propagator, the ghost–gluon vertex, and the insertion of certain composite operators which will appear in studies of deep inelastic scattering.

With respect to the first, we consider

$$\langle T\hat{B}_a^\mu(x)\hat{B}_b^\nu(y)\rangle_0 \Big|_{g=0} = (-i)^2 \frac{\delta^2 \log Z}{\delta\lambda_{a\mu}(x)\delta\lambda_{b\nu}(y)} \Big|_{\substack{\text{sources}=0 \\ g=0}}.$$

(2.6.1)

We here use čarets to denote operators. Imitating the discussion of Sect. 1.3, we write, up to a four-divergence, and with $\lambda = a^{-1}$ for the gauge parameter,

$$-\tfrac{1}{4} \sum \left(\partial^\rho B_a^\sigma(x) - \partial^\sigma B_a^\rho(x)\right)\left(\partial_\rho B_{a\sigma}(x) - \partial_\sigma B_{a\rho}(x)\right) - \frac{1}{2a} \sum \left(\partial \cdot B_a(x)\right)^2$$

$$= \tfrac{1}{2} \sum B_a^\sigma(x)\left(\Box B_{a\sigma} - (1-a^{-1})\partial_\sigma \partial^\rho B_{a\rho}(x)\right) + \partial_\mu f^\mu$$

$$= \tfrac{1}{2} \sum B_{a\sigma}(x)\left(K^{-1}\right)_{ab}^{\sigma\rho} B_{b\rho}(x) + \partial_\mu f^\mu,$$

where

$$\left(K^{-1}\right)_{ab}^{\sigma\rho} = \delta_{ab}\left\{g^{\sigma\rho}\frac{\partial^2}{\partial x^2} - (1-a^{-1})\frac{\partial}{\partial x_\sigma}\frac{\partial}{\partial x_\rho}\right\}.$$

(2.6.2)

Thus, setting η, $\bar{\eta}$, ξ, $\bar{\xi}$ and g in (2.5.11) to zero,

$$Z = \int \mathcal{D}q\, \mathcal{D}\bar{q}\, \mathcal{D}\omega\, \mathcal{D}\bar{\omega}\, \mathcal{D}B \exp i \int d^4x \left\{ \sum \left(i\bar{q}(x)\slashed{\partial}\, q(x) - m_q\right) \right.$$

$$\left. + \tfrac{1}{2} \sum B_{a\sigma}(x)\left(K^{-1}\right)_{ab}^{\sigma\rho} B_{b\rho}(x) + \sum \lambda_{a\mu}(x)B_a^\mu(x) \right\}.$$

(2.6.3)

The integrals over q, $\bar{q}$, ω and $\bar{\omega}$ yield a constant that will drop when the logarithmic derivative is taken. If we change variables,

$$B \to B' = K^{-1/2}B,$$

(2.6.3) becomes

$$Z = (\text{const.}) \times \int \mathcal{D}B'\, J(K)$$
$$\times \exp\,\mathrm{i} \int \mathrm{d}^4x \sum \left\{ \tfrac{1}{2} B'_{a\mu}(x) B'^{\mu}_a(x) + \lambda_{a\mu}(x)(K^{1/2}B')^{\mu}_a(x) \right\},$$

where $J(K)$ is the Jacobian. Finally, we displace the integration variable,

$$B' \to B'' = B' + K^{1/2}\lambda,$$

so that, with N a constant,

$$Z = N \int \mathcal{D}B''\, J(K) \exp\,\mathrm{i} \int \mathrm{d}^4x \sum \left\{ \tfrac{1}{2} B''_{a\mu}(x) B''^{\mu}_a(x) \right.$$
$$\left. - \tfrac{1}{2}\lambda_{a\mu}(x)(K\lambda)^{\mu}_a(x) \right\}. \tag{2.6.4a}$$

It is convenient to write K in integral form: for any φ,

$$(K\varphi)^{\mu}_a(x) = -\mathrm{i}\sum \int \mathrm{d}^4y\, D^{\mu\nu}_{ab}(x-y)\varphi_{b\nu}(y). \tag{2.6.4b}$$

We then obtain the desired result:

$$(-\mathrm{i})^2\, \frac{\delta^2 \log Z}{\delta\lambda_{a\mu}(x)\delta\lambda_{b\nu}(y)}\bigg|_{\substack{\text{sources}=0\\ g=0}} = D^{\mu\nu}_{ab}(x-y), \tag{2.6.4c}$$

and we omit the index 0 denoting order zero in g. The value of D is obtained from its definition, Eq. (2.6.4b). It is such that for any function φ

$$(K^{-1}\varphi)^{\mu}_a(x) = \sum \delta_{ab}\left\{ g^{\mu\nu}\Box - (1-a^{-1})\partial^{\mu}\partial^{\nu} \right\}\varphi_{b\nu}(x),$$

so, with the help of a Fourier transform which we denote by a tilde,

$$(\widetilde{K^{-1}\varphi})^{\mu}_a(k) = \sum \delta_{ab}\left\{ -g^{\mu\nu}k^2 + (1-a^{-1})k^{\mu}k^{\nu} \right\}\widetilde{\varphi}_{a\nu}(k).$$

Letting $\widetilde{K}f = \varphi$, and inverting the relation above, this immediately yields

$$(\widetilde{Kf})^{\mu}_a(k) = \sum \delta_{ab}\frac{-g^{\mu\nu} + (1-a)k^{\mu}k^{\nu}/k^2}{k^2}\widetilde{f}_{b\nu}(k),$$

and therefore we have the explicit expression

$$\langle T\hat{B}_a^\mu(x)\hat{B}_b^\nu(y)\rangle_0\Big|_{g=0} = D_{ab}^{\mu\nu}(x-y)$$

$$= \delta_{ab}\frac{i}{(2\pi)^4}\int d^4k\, e^{-ik\cdot(x-y)}\frac{-g^{\mu\nu}+(1-a)k^\mu k^\nu/k^2}{k^2}, \qquad (2.6.5)$$

$$a = \lambda^{-1},$$

as was to be expected. This evaluation identifies the propagators as the inverses of the free Lagrangian differential operators,[9] which very much simplifies their evaluation. The proof that the poles, say in (2.6.5), have to be circumvented with the $+i0$ prescription requires a consideration of asymptotic states or other similar boundary conditions; it may be found in the lectures of Fadeyev (1976).

For the vertex we require

$$\langle T\hat{\omega}_a(x_1)\hat{\omega}_b(x_2)\hat{B}_c^\mu(x_3)\rangle_0\Big|_{\text{order } g} = \frac{i\delta^3\log Z}{\delta\eta_a(x_1)\delta\bar{\eta}_b(x_2)\delta\lambda_{c\mu}(x_3)}\Big|_{\substack{\text{sources}=0 \\ \text{first order } g}}. \qquad (2.6.6)$$

We perform the change of variables

$$B \to B' = K^{-1/2}B,\ \omega \to \omega' = \Box^{-1/2}\omega,\ \bar{\omega} \to \bar{\omega}' = \Box^{-1/2}\omega,$$

and integrate out the quarks, which play no role here. Then Z becomes (with, as usual, N a constant)

$$Z = N\int \mathcal{D}\omega'\,\mathcal{D}\bar{\omega}'\,\mathcal{D}B'\, J(K)J(\Box)\exp i\int d^4x \sum\left\{g\left[\partial_\mu(\Box^{1/2}\bar{\omega}')_a(x)\right]\right.$$

$$\times f_{abc}(K^{1/2}B')_c^\mu(x)(\Box^{1/2}\omega')_b(x) + \tfrac{1}{2}B'^2 - \bar{\omega}'\omega'$$

$$+ \bar{\eta}_a(x)(\Box^{1/2}\omega')_a(x) + (\Box^{1/2}\bar{\omega}')_a(x)\eta_a(x) + \lambda_a^\mu(x)(K^{1/2}B')_{a\mu}(x) + \cdots\Big\}$$

where the dots represent terms that will vanish for $g \approx 0$, sources $= 0$. Next, we translate:

$$B' \to B'' = B' - K^{1/2}\lambda,\quad \omega' \to \omega'' = \omega' + \Box^{1/2}\eta,\quad \bar{\omega}' \to \bar{\omega}'' = \bar{\omega}' + \Box^{1/2}\bar{\eta}.$$

The only term that will yield a contribution is the one containing a product of three sources. This is

$$g\sum\left(\partial_\mu(\Box\bar{\eta})_a(x)\right)f_{abc}(K\lambda)_c^\mu(x)(\Box\eta)_b(x),$$

---

[9] This property may also be deduced in the canonical formalism by the identification of the propagators as Green's functions of the corresponding differential equations.

and therefore

$$\langle T\hat{\tilde{\omega}}_a(x_1)\hat{\omega}_b(x_2)\,\hat{B}_c^\mu(x_3)\rangle_0\Big|_{\text{order }g}$$

$$= \int \frac{d^4p_1}{(2\pi)^4}\,e^{-ix_1\cdot p_1}\,\frac{i}{p_1^2}\int \frac{d^4p_2}{(2\pi)^4}\,e^{-ix_2\cdot p_2}\,\frac{i}{p_2^2}\int \frac{d^4p_3}{(2\pi)^4}\,e^{-ix_3\cdot p_3}$$

$$\times\,i\,\frac{-g^{\mu\nu}+(1-\lambda^{-1})p_3^\mu p_3^\nu/p_3^2}{p_3^2}\,(2\pi)^4\delta(p_1+p_2+p_3)g f_{abc}p_{1\nu},$$

again as expected.

Finally, we consider the vertex

$$\langle T\hat{\tilde{q}}_1(x_1)\hat{N}_{NS}^{\mu_1\ldots\mu_n}(x_2)\hat{q}_2(x_3)\rangle_0 \tag{2.6.7}$$

to order zero in g and where

$$\hat{N}_{NS}^{\mu_1\cdots\mu_n}(x) = \tfrac{1}{2}i^{n-1}\mathcal{S}:\hat{\tilde{q}}_2(x)\gamma^{\mu_1}\hat{D}^{\mu_2}\ldots\hat{D}^{\mu_n}\hat{q}_1(x): -\text{traces} \tag{2.6.8}$$

(its usefulness will appear in Sect. 4.5). Here the indices $f = 1, 2$ for the quarks q_f are flavour indices; $\mathcal{S}$ stands for symmetrization in the indices $\mu_1,\ldots,\mu_n$, and by "traces" we mean terms obtained replacing $D^{\mu_i}D^{\mu_j} \to g^{\mu_i\mu_j}D\cdot D$. To calculate (2.6.7) we introduce into (2.5.11) a new source term,

$$j_{\mu_1\ldots\mu_n}N_{NS}^{\mu_1\ldots\mu_n},$$

so that

$$\langle T\hat{\tilde{q}}_1(x_1)\hat{N}_{NS}^{\mu_1\cdots\mu_n}(x_2)\hat{q}_2(x_3)\rangle_0 = \left.\frac{i\delta^3\log Z}{\delta\xi_1(x_1)\delta\bar{\xi}_2(x_3)\delta j_{\mu_1\ldots\mu_n}(x_2)}\right|_{\substack{g=0\\ \text{sources}=0}}. \tag{2.6.9}$$

To zero order in g, the gluons or ghosts play no role and can be eliminated through integration. Similarly, the covariant derivatives of N may be replaced by ordinary derivatives. The quark fields may be treated in the same way in which we treated the gluon fields before. Using the definitions

$$q'_f = S^{-1/2}q_f, \quad \bar{q}'_f = \bar{q}_f\bar{S}^{-1/2}, \quad f = 1, 2,$$

where

$$S^{-1}q_f(x) = \not{\partial}\,q_f(x), \quad \bar{q}_f\bar{S}^{-1} = \bar{q}_f(x)\overleftarrow{\not{\partial}},$$

we find, to zero order in g,

$$Z = (\text{const.})\int \mathcal{D}q\,\mathcal{D}\bar{q}\,J(S)J(\bar{S})$$

$$\times \exp i\int d^4x\,\{\bar{q}'_1q'_1 + \bar{q}'_2q'_2 + \bar{\xi}_1S^{1/2}q'_1 + \bar{\xi}_2S^{1/2}q'_2 \tag{2.6.10}$$

$$+ (\bar{q}'_1\bar{S}^{1/2})\xi_1 + (\bar{q}'_2\bar{S}^{1/2})\xi_2 + (\bar{S}^{1/2}N'^{\mu_1\cdots\mu_n}_{NS}S^{1/2})j_{\mu_1\ldots\mu_n}\},$$

$$N'^{\mu_1\cdots\mu_n}_{NS} \equiv \tfrac{1}{2}i^{n-1}\mathcal{S}:\bar{q}'_2(x)\gamma^{\mu_1}\partial^{\mu_2}\ldots\partial^{\mu_n}\hat{q}'_1(x): -\text{traces}.$$

Then we shift:

$$q''_f = q'_f + S^{1/2}\xi_f, \quad \bar{q}''_f = \bar{q}'_f + \bar{\xi}_f \bar{S}^{1/2}.$$

The only term that contains all three sources ξ_1, ξ_2 and j is

$$\tfrac{1}{2} i^{n-1} \{ \mathcal{S}(\bar{\xi}_2 \bar{S}^{-1})(x) \gamma^{\mu_1} \partial^{\mu_2} \ldots \partial^{\mu_n} (S^{-1}\xi_1)(x) - \text{traces} \} j_{\mu_1 \ldots \mu_n}(x),$$

so that, using the explicit expression for S, we find

$$\langle \mathrm{T}\hat{\bar{q}}_1(x_1) \hat{N}_{NS}^{\mu_1 \cdots \mu_n}(x_2) \hat{q}_2(x_3) \rangle_0$$

$$= \int \frac{\mathrm{d}^4 p_2}{(2\pi)^4} \, e^{-ip_2 \cdot x_2} \int \frac{\mathrm{d}^4 p_1}{(2\pi)^4} \, e^{-ip_1 \cdot x_1} \, \frac{i}{\not{p}_1} \, \tfrac{1}{2} \{ \mathcal{S} \gamma^{\mu_1} p_3^{\mu_2} \ldots p_3^{\mu_n} - \text{traces} \}$$

$$\times \int \frac{\mathrm{d}^4 p_3}{(2\pi)^4} \, e^{ip_3 \cdot x_3} \, \frac{i}{\not{p}_3} \, (2\pi)^4 \delta(p_1 + p_2 - p_3).$$

$$(2.6.11)$$

We can simplify the formula by introducing a vector Δ^μ with $\Delta^2 = 0$, and contracting (2.6.11) with it:

$$\Delta_{\mu_1} \ldots \Delta_{\mu_n} \langle \mathrm{T}\hat{\bar{q}}_1(x_1) \hat{N}_{NS}^{\mu_1 \cdots \mu_n}(x_2) \hat{q}_2(x_3) \rangle_0 = (2\pi)^4 \delta(p_1 + p_2 - p_3)$$

$$\times \int \frac{\mathrm{d}^4 p_2}{(2\pi)^4} \, e^{-ip_2 \cdot x_2} \int \frac{\mathrm{d}^4 p_1}{(2\pi)^4} \, e^{-ip_1 \cdot x_1} \, \frac{i}{\not{p}_1} \, \not{\Delta} (\Delta \cdot p_3)^{n-1} \int \frac{\mathrm{d}^4 p_3}{(2\pi)^4} \, e^{ip_3 \cdot x_3} \, \frac{i}{\not{p}_3} :$$

$$(2.6.12)$$

the symmetrization is automatic and the traces give zero (they contain terms $g^{\mu\mu'} \Delta_\mu \Delta_{\mu'}$). The vertex may be recovered by differentiation,

$$(\partial/\partial\Delta_{\mu_1}) \ldots (\partial/\partial\Delta_{\mu_n}).$$

Eq. (2.6.12) generates the Feynman rule given in Appendix E and to be used in Sect. 4.6.

2.7 The Background Field Method

The functional formalism allows a simple introduction of the *background field method*, an elegant and powerful formalism whereby gauge invariance of the generating functional (in a sense to be specified) is preserved. The method was first introduced by DeWitt (1967), and was extended by 't Hooft, Boulware and Abbott. In our exposition we will follow the very readable account of the last author (Abbott, 1981), which may also be consulted for more details and references.

Consider the generating functional for pure Yang–Mills fields; fermions play no role in the background field method (i.e., they are treated as in the ordinary formalism) and will be neglected. We write it as

$$Z[\eta, \bar{\eta}, \lambda] = N \int \mathcal{D}\omega \, \mathcal{D}\bar{\omega} \, \mathcal{D}B \, \exp i \int \mathrm{d}^4 x \, (\mathcal{L}_{\text{source}} + \mathcal{L}_{\text{QCD}} + \mathcal{L}_{\text{GF}}),$$

$$(2.7.1a)$$

where (cf. Eq. (2.5.11)),

$$\mathcal{L}_{\text{source}} = \sum \left(\bar{\eta}_a \omega_a + \bar{\omega}_a \eta_a + \lambda_{a\mu} B_a^\mu \right),$$

$$\mathcal{L}_{\text{QCD}} = -\tfrac{1}{4}(D \times B)^2 + \sum (\partial_\mu \bar{\omega}_a)(\delta_{ab}\partial^\mu - g f_{abc} B_c^\mu)\omega_b,$$

(2.7.1b)

and $\mathcal{L}_{\text{GF}}$ is the gauge fixing term, which we do not specify yet.

We now shift the gauge field by an external, background field, $\phi_a^\mu(x)$:

$$B_a^\mu(x) \to B_a^\mu(x) + \phi_a^\mu(x).$$

(2.7.2)

We get a new generating functional,

$$\widetilde{Z}[\eta, \bar{\eta}, \lambda, \phi] = N \int \mathcal{D}\omega \, \mathcal{D}\bar{\omega} \, \mathcal{D}B \, \exp i \int d^4x \left(\widetilde{\mathcal{L}}_{\text{source}} + \widetilde{\mathcal{L}}_{\text{QCD}} + \widetilde{\mathcal{L}}_{\text{GF}} \right),$$

(2.7.3a)

and now

$$\widetilde{\mathcal{L}}_{\text{source}} = \sum \left(\bar{\eta}_a \omega_a + \bar{\omega}_a \eta_a + \lambda_{a\mu} B_a^\mu \right),$$

$$\widetilde{\mathcal{L}}_{\text{QCD}} = -\tfrac{1}{4}(D \times B + D \times \phi)^2$$

$$+ \sum (\partial_\mu \bar{\omega}_a)(\delta_{ab}\partial^\mu - g f_{abc} B_c^\mu - g f_{abc} \phi_c^\mu)\omega_b;$$

(2.7.3b)

we have omitted a term

$$\sum \lambda_{a\mu} \phi_a^\mu$$

in the definition of $\widetilde{\mathcal{L}}_{\text{source}}$ as we want ϕ to be external, and thus we need not attach a source to it.

Now we choose the form of the gauge fixing term. We write

$$\widetilde{\mathcal{L}}_{\text{GF}} = -\frac{1}{2a}\left(\partial_\mu B_a^\mu + g \sum f_{abc}\phi_{b\mu}B_c^\mu \right)^2 = -\frac{1}{2a}(D_\phi \cdot B)^2,$$

(2.7.3c)

where D_ϕ is the covariant derivative with respect to the field ϕ,

$$(D_\phi^\mu)_{ab} = \delta_{ab}\partial^\mu + g \sum f_{acb}\phi_c^\mu.$$

The gauge fixed by (2.7.3) is called the *background gauge (or gauges)*. We remark that (2.7.3c) contains the parameter a: in the background field method one still has the liberty to alter the gauge by changing a, so there is a background Fermi–Feynman gauge, a background Landau gauge, etc.

We next prove the announced gauge invariance: the generating functional (2.7.3) is left unaltered by gauge shifts of the background field ϕ. To verify this, consider infinitesimal transformations:

$$\phi_a^\mu \to \phi'^\mu_a + \sum_{bc} f_{abc}\theta_b\phi'^\mu_c - \partial^\mu \theta_a \equiv \phi'^\mu_a + \delta\phi'^\mu_a.$$

(2.7.4)

We perform the same transformation on the source,

$$\lambda_a^\mu \to \lambda_a'^\mu + \sum_{bc} f_{abc}\theta_b\lambda_c'^\mu - \partial^\mu\theta_a. \qquad (2.7.5)$$

Then we shift the variable of integration

$$B_a^\mu \to B_a'^\mu + \sum_{bc} f_{abc}\theta_b B_c'^\mu - \partial^\mu\theta_a. \qquad (2.7.6)$$

Because (2.7.6) is a unitary transformation, one has $\mathcal{D}B = \mathcal{D}B'$. Applying (2.7.4–6) to (2.7.3) we find that $\widetilde{\mathcal{L}}_{\text{source}}$ does not change as (2.7.4) and (2.7.6) are compensated in (2.7.3): $\mathcal{L}_{\text{QCD}}$ is invariant under gauge transformations, and hence under (2.7.4–6). Finally, and this is the point of using the background fixing (2.7.3c), $\widetilde{\mathcal{L}}_{\text{GF}}$ contains the covariant derivative $D_\phi B$, manifestly invariant under (2.7.4) and (2.7.6) by its very construction.

Using (2.7.3) we can generate the Feynman rules in the background gauge formalism. The propagator of the field ϕ is undefined, which is irrelevant since ϕ, being classical, will not appear in loops. Vertices with only one field B need not be considered if we are interested in 1PI Green's functions. Compared with ordinary Feynman rules, the only difference lies in the appearance of the ϕ field in external legs; we denote this field by a fuzzy blob. The full set of Feynman rules is shown in Appendix D. An example of application of the background field method will be given in Sect. 3.3iii.

2.8 Global Symmetries of the QCD Lagrangian: Conserved Currents

In this section we will discuss the *global* symmetries of the QCD Lagrangian. Since its form is unaltered by renormalization (as we will see), we can neglect the distinction between bare and renormalized $\mathcal{L}$.

Clearly, $\mathcal{L}$ is invariant under Poincaré transformations, $x \to \Lambda x + a$. The currents corresponding to (homogeneous) Lorentz transformations Λ are not of great interest for us here. Space–time translations generate the *energy–momentum tensor*. Its form is fixed by Noether's theorem, which gives

$$\Theta^{\mu\nu} = \sum_i \frac{\partial\mathcal{L}}{\partial(\partial_\mu\Phi_i)}\,\partial_\nu\Phi_i - g^{\mu\nu}\mathcal{L}, \qquad (2.8.1)$$

and the sum over i runs over all the fields in the QCD Lagrangian. These currents are conserved,

$$\partial_\mu\Theta^{\mu\nu} = 0,$$

and the corresponding "charges" are the components of the four-momentum

$$P^\mu = \int \mathrm{d}^3\mathbf{x}\,\Theta^{0\mu}(x).$$

The explicit expression for $\Theta^{\mu\nu}$ in QCD is

$$
\Theta^{\mu\nu} = i \sum_q \bar{q}\gamma^\mu D^\nu q - ig^{\mu\nu} \sum_q \bar{q}\slashed{D}q + g^{\mu\nu} \sum_q m_q \bar{q}q
$$
$$
- g_{\alpha\beta}G^{\mu\alpha}G^{\nu\beta} + \tfrac{1}{4}g^{\mu\nu}G^2 + \text{gauge fixing} + \text{ghost terms}
$$

(2.8.2)

(sum over omitted colour indices understood).

In the quantum version, we understand that products are replaced by Wick ordered products. Θ is not unique and, as a matter of fact, direct application of (2.8.1) does not yield a gauge invariant tensor. To obtain the gauge invariant expression (2.8.2) one may proceed by replacing derivatives by covariant derivatives. A more rigorous procedure would be to reformulate (2.8.1) in a way consistent with gauge invariance by performing gauge transformations simultaneously to the spacetime translation. For an infinitesimal one, $x^\mu \to x^\mu + \epsilon^\mu$, we then define

$$
B_a^\mu \to B_a^\mu + (\epsilon_\alpha \partial^\alpha B_a^\mu \equiv D^\mu \epsilon_\alpha B_a^\alpha + \epsilon_\alpha G_a^{\alpha\mu}).
$$

The term $D^\mu \epsilon_\alpha B_a^\alpha$ may be absorbed by a gauge transformation, so we may write the transformation as $B_a^\mu \to B_a^\mu + \epsilon_\alpha G_a^{\alpha\mu}$. For a discussion of the arbitrariness in the definition of the energy–momentum tensor, see Callan, Coleman and Jackiw (1970) and Collins, Duncan and Joglekar (1977).

Next, we have the currents and charges associated with colour rotations. We leave it to the reader to write them explicitly; they are particular cases of colour gauge transformations (with constant parameters). We now pass over to a different set of currents *not* associated with interactions of quarks and gluons among themselves.

If all the quark masses vanished, we would have invariance of $\mathcal{L}$ under the transformations,

$$
q_f \to \sum_{f'=1}^{n_f} U_{ff'}q_{f'}, \quad q_f \to \sum_{f'=1}^{n_f} U_{ff'}^5 \gamma_5 q_{f'}
$$

(2.8.3)

where f, f' are flavour indices, and U, U^5 unitary matrices. This implies that the currents

$$
V_{qq'}^\mu(x) = \bar{q}(x)\gamma^\mu q'(x),
$$
$$
A_{qq'}^\mu(x) = \bar{q}(x)\gamma^\mu \gamma_5 q'(x)
$$

(2.8.4)

are each separately conserved. When mass terms are taken into account, only the diagonal V_{qq}^μ are conserved; the others are what is called *quasi-conserved* currents, i.e., their divergences are proportional to masses. These divergences are easily calculated: since the transformations in (2.8.3) commute with the interaction part of $\mathcal{L}$, we may evaluate them with free fields, in which case use of the free Dirac equation $i\slashed{\partial}q = m_q q$ gives

$$
\partial_\mu V_{qq'}^\mu = i(m_q - m_{q'})\bar{q}q', \quad \partial_\mu A_{qq'}^\mu = i(m_q + m_{q'})\bar{q}\gamma_5 q'.
$$

(2.8.5)

In fact, there is a subtle point concerning the divergence of axial currents. Eq. (2.8.5) is correct as it stands for the nondiagonal currents, $q \neq q'$; for $q = q'$, however, one has instead

$$\partial_\mu A_{qq}^\mu(x) = \mathrm{i}(m_q + m_q)\bar{q}(x)\gamma_5 q(x) + \frac{T_F g^2}{16\pi^2} \epsilon^{\mu\nu\rho\sigma} G_{\mu\nu}(x) G_{\rho\sigma}(x), \qquad (2.8.6)$$

with $T_F = 1/2$ a colour factor. This is the so-called Adler–Bell–Jackiw anomaly that will be discussed in detail in Sects. 7.5, 7.6 and 7.7.

The *equal time commutation relations* (ETC) of the V, A with the fields are also easily calculated, for free fields. Using (2.8.4) and the ETC of quark fields, one finds,

$$\begin{aligned}
\delta(x^0 - y^0)[V_{qq'}^0(x), q''(y)] &= -\,\delta(x - y)\delta_{qq''}q'(x), \\
\delta(x^0 - y^0)[A_{qq'}^0(x), q''(y)] &= -\,\delta(x - y)\delta_{qq''}\gamma_5 q'(x), \text{ etc.}
\end{aligned} \qquad (2.8.7)$$

The V, A commute with gluon and ghost fields. The equal time commutation relations of the V, A among themselves (again for free fields) are best described for three flavours, $f = 1, 2, 3 = u, d, s$ by introducing the Gell-Mann λ^a matrices in flavour space; cf. Appendix C. So we let

$$V_a^\mu(x) = \sum_{ff'} \bar{q}_f(x)\lambda_{ff'}^a \gamma^\mu q_{f'}(x), \quad A_a^\mu(x) = \sum_{ff'} \bar{q}_f(x)\lambda_{ff'}^a \gamma^\mu \gamma_5 q_{f'}(x),$$

$$(2.8.8)$$

and we then obtain the commutation relations

$$\begin{aligned}
\delta(x^0 - y^0)[V_a^0(x), V_b^\mu(y)] &= 2\mathrm{i}\delta(x - y)\sum f_{abc} V_c^\mu(x), \\
\delta(x^0 - y^0)[V_a^0(x), A_b^\mu(y)] &= 2\mathrm{i}\delta(x - y)\sum f_{abc} A_c^\mu(x), \qquad (2.8.9) \\
\delta(x^0 - y^0)[A_a^0(x), A_b^\mu(y)] &= 2\mathrm{i}\delta(x - y)\sum f_{abc} V_c^\mu(x), \text{ etc.}
\end{aligned}$$

Equations (2.8.7) and (2.8.9) have been derived only for free fields. However, they involve short distances; therefore, in QCD and because of asymptotic freedom, they will hold as they stand even in the presence of interactions.

Equal time commutation relations of conserved or quasi-conserved currents with the Hamiltonian (or Lagrangian) may also be easily obtained. If J^μ is conserved, then the corresponding charge

$$Q_J(t) = \int \mathrm{d}^3\mathbf{x}\, J^0(t, \mathbf{x}), \quad t = x_0,$$

commutes with $\mathcal{H}$:

$$[Q_J(t), \mathcal{H}(t, \mathbf{y})] = 0.$$

Here $\mathcal{H}$ is the *Hamiltonian density*, $\mathcal{H} = \Theta^{00}$. If J is quasi-conserved, let $\mathcal{H}_m$ be the mass term in $\mathcal{H}$,

$$\mathcal{H}_m = \sum_q m_q \bar{q}q.$$

Then,

$$[Q_J(t), \mathcal{H}_m(t, \mathbf{y})] = \mathrm{i}\partial_\mu J^\mu(t, \mathbf{y}). \qquad (2.8.10)$$

Of course, Q_J still commutes with the rest of $\mathcal{H}$.

3 Renormalization in QCD

3.1 Regularization (Dimensional)

As we saw in the example of Sect. 2.3i, some amplitudes are divergent. This is due to the fact that field operators are singular objects: it is easy to trace the divergence of the $\mathrm{d}^4 k$ integral in (2.3.4b) at large values of k to the occurrence, in position space, of products of field operators at the same space–time point. Because of this we must, in order to discuss QCD (or indeed any local relativistic field theory), give a meaning to the integrals that appear when we evaluate Feynman diagrams. This is called *regularization*, and it amounts to altering the Lagrangian $\mathcal{L}$ to $\mathcal{L}_\epsilon$ in such a way that $\mathcal{L}_\epsilon$ produces finite answers (at least in perturbation theory) and, in some sense, we have that $\mathcal{L}_\epsilon \to \mathcal{L}$ for $\epsilon \to 0$. Since the classical work of Bohr and Rosenfeld (1933, 1950), we know that field operators are intrinsically singular; therefore, any regularization must destroy some physical feature of the theory. Thus, Pauli–Villars regularization destroys hermiticity and gauge invariance for non-Abelian theories; lattice regularization destroys Poincaré invariance, etc. In the limit $\epsilon \to 0$ these properties are recovered (if one was careful enough!). Because gauge invariance is essential for QCD we will use now dimensional regularization, that destroys only scale invariance. The method is related to so-called analytical regularization (Bollini, Giambiagi and González-Domínguez, 1964; Speer, 1968) and has been thoroughly developed by 't Hooft and Veltman[1] (1972). It amounts to working in an arbitrary dimension, $D = 4 - \epsilon$; the physical limit is of course $\epsilon \to 0$. Divergences appear as poles in $1/\epsilon$. A mathematical treatment of dimensional regularization has been given by Speer (1975). It is rather complicated, but fortunately we will not need to delve into its intricacies. All we require are interpolation formulas consistent with gauge and Poincaré invariance, applicable to the evaluation of Feynman integrals. This we accomplish in steps.

We discuss first a convergent integral of the form $(2\pi)^{-D} \int \mathrm{d}^D k \, f(k^2)$, where typically $f(k^2) = (k^2)^r (k^2 - a^2)^{-m}$ and

$$\mathrm{d}^D k = \mathrm{d}k^0 \mathrm{d}k^1 \dots \mathrm{d}k^{D-1}, \quad k^2 = (k^0)^2 - (k^1)^2 - \dots - (k^{D-1})^2.$$

---

[1] Dimensional regularization had been introduced independently, and in fact slightly earlier, by Bollini and Giambiagi (1972).

We will start by considering integer r, m, but it will be obvious at the end that the formulation is valid for arbitrary, even complex, values of these parameters. Because f is analytic in the k^0 plane,[2] we can rotate the integration from $(-\infty, +\infty)$ to $(-i\infty, +i\infty)$, a so-called *Wick rotation*. We can recover an integration over $(-\infty, +\infty)$ by then defining a new variable $k^0 \to k^D = ik^0$. Thus we obtain an ordinary Euclidean integral in D dimensions,

$$i \int_{-\infty}^{+\infty} \frac{dk^1}{2\pi} \cdots \int_{-\infty}^{+\infty} \frac{dk^D}{2\pi} f(-k_E^2), \quad k_E^2 \equiv (k^1)^2 + \cdots + (k^D)^2 \equiv |k_E|^2.$$

We let $d^D k_E = dk^1 \ldots dk^D$, and introduce polar coordinates, $d^D k_E = d|k_E| |k_E|^{D-1} d\Omega$. Using the formula $\int d\Omega = 2\pi^{D/2}/\Gamma(D/2)$, we finally find

$$\int \frac{d^D k}{(2\pi)^D} f = \frac{i}{(2\pi)^{D/2} \Gamma(D/2)} \int_0^\infty d|k_E| |k_E|^{D-1} f(-|k_E|^2).$$

The manipulations we have carried out are only valid for $D = $ positive integer, and $D/2 + \operatorname{Re} r < \operatorname{Re} m$, but we can use the last formula to *define* the integral for arbitrary complex D, r, m.

Consider next the integral of a polynomial in k^μ times $f(k^2)$; we can reduce this to the former situation by symmetric integration writing, for example,

$$\int d^D k\, f(k^2) k^\mu k^\nu = \frac{g^{\mu\nu}}{D} \int d^D k\, f(k^2) k^2.$$

Likewise,

$$\int d^D k\, f(k^2) k^{\mu_1} \ldots k^{\mu_n} = 0, \quad \text{if } n = \text{ odd integer}.$$

Finally, the general case is treated by expanding in a tensor basis in the k^μ. In this way, the integrals of Appendix B, and many more, can be worked out for arbitrary D. For example, by straightforward application of the methods above and use of standard one-dimensional integration formulas we find

$$\int \frac{d^D k}{(2\pi)^D} \frac{(k^2)^r}{(k^2 - a^2)^m} = i \frac{(-1)^{r-m}}{(16\pi^2)^{D/4}} \frac{\Gamma(r + D/2)\Gamma(m - r - D/2)}{\Gamma(D/2)\Gamma(m)(a^2)^{m-r-D/2}}.$$

If the left-hand side was divergent in, say, the physical case $D = 4$, which occurs (for real parameters) when $m - r - D/2 \leq 0$, this is reflected in poles in the right-hand side due to the poles of the function $\Gamma(m - r - D/2)$, as shown by the above formula. An arbitrariness in the method is already apparent; we might have multiplied the right-hand side above by any function $\varphi(D)$, provided that it is analytic in D and $\varphi(4) = 1$. This will be useful later on.

---

[2] If the denominator of f has poles, one first has to regulate them by replacing $k^2 - a^2$ by $k^2 - a^2 + i\eta$, $\eta \to 0$, and also take care that the deformation of the integration contour is made avoiding the poles.

Now we will consider spin. It will be convenient to distinguish between external and internal lines in Feynman graphs. Later on we will show that, after renormalization, Green's functions with their external legs amputated are finite in perturbation theory, in the limit $D \to 4$. Since spin factors in external legs (i.e., factors u, v, $\bar{u}$, $\bar{v}$, $\epsilon^\mu \ldots$; cf. Appendix D) are certainly finite for $D = 4$, we may already take them in physical dimension. As for spin effects in internal lines, we have to take $g^{\mu\nu}$ in dimension D so that, for example, $g^{\mu\nu}g_{\mu\nu} = D$, etc. Likewise, we must consider that we have D gamma matrices, $\gamma^0, \ldots, \gamma^{D-1}$. To be totally consistent, we would have to admit that the γ^μ were $2^{D/2} \times 2^{D/2}$ matrices (which is the dimension of a D-dimensional Clifford algebra). But this is not necessary. We are still consistent with gauge invariance if we take the γ^μ to be 4×4 matrices so that, e.g., $\mathrm{Tr}\, \gamma^\mu \gamma^\nu = 4g^{\mu\nu}$, and this is what will be done here. Spin is, however, treated in another manner in a different, but related, method of regularization called *dimensional reduction*, useful especially for supersymmetry. The interested reader is referred to Siegel (1979) for details.

Thus, the extension of integrals and Dirac algebra to arbitrary D is fairly simple; a set of useful formulas is collected in Appendices A and B. Only the introduction of γ_5 is a bit trickier. If, for example, we write $\gamma_5 = i\gamma^0\gamma^1\gamma^2\gamma^3$, it is clear that this will not exist for $D < 4$. The definition $\gamma_5 = i\gamma^1 \ldots \gamma^{D-1}$ may be shown to be inconsistent in particular with gauge invariance (see Sect. 7.5, especially between Eqs. (7.5.18, 20) for a discussion). We will choose

$$\gamma_5 = \frac{i}{4!} \epsilon_D^{\mu\nu\rho\sigma} \gamma_\mu \gamma_\nu \gamma_\rho \gamma_\sigma,$$

where ϵ_D coincides with the antisymmetric tensor only for $D = 4$. We do not specify it further beyond requiring that it be such that, for arbitrary D,

$$\gamma_5^2 = 1 \text{ and } \mathrm{Tr}\, \gamma_5 \gamma^\mu \gamma^\nu = 0.$$

It can be shown that this is enough to give a meaning to all calculations involving γ_5. A possible specific choice for ϵ_D is that of 't Hooft and Veltman (1972),

$$\epsilon_D^{\mu\nu\rho\sigma} = \epsilon^{\mu\nu\rho\sigma} \text{ for } \mu\nu\rho\sigma = 0 \text{ to } 3;$$

and $\epsilon_D^{\mu\nu\rho\sigma} = 0$ if any of the indices is larger than 3. Here one assumes that $D \geq 4$.

We have thus fully built dimensional regularization; as long as D is not an integer we see that, with it, all integrals appearing in Feynman graphs are finite. The regularization preserves gauge and Poincaré invariance, but it breaks scale invariance. Indeed, a Feynman integral like the one in (2.3.4b) becomes altered:

$$\int \frac{\mathrm{d}^4 k}{(2\pi)^4} \to \int \frac{\mathrm{d}^D k}{(2\pi)^D}.$$

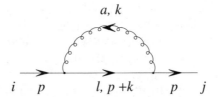

Fig. 3.1.1. One loop correction to the quark propagator.

One could correct this by rescaling all fields and coupling constants accordingly, but it is perhaps more transparent to use instead the prescription

$$\int \frac{\mathrm{d}^4 k}{(2\pi)^4} \to \int \mathrm{d}^D \hat{k} \equiv \int \frac{\mathrm{d}^D k \, \nu_0^{4-D}}{(2\pi)^D}, \quad D = 4 - \epsilon, \qquad (3.1.1a)$$

where

$$\hat{k}^\mu = \nu_0^{4/D-1} k^\mu / (2\pi), \qquad (3.1.1b)$$

thereby explicitly introducing the scale-invariance-breaking arbitrary (but fixed) parameter ν_0 with dimensions of $[\text{Mass}]^1$.

As a first example of the methods of dimensional regularization, let us calculate the propagator of a quark to second order, in momentum space:

$$S_\xi^{ij}(p) = \int \mathrm{d}^4 x \, \mathrm{e}^{\mathrm{i} p \cdot x} \langle \mathrm{T} q^i(x) \bar{q}^j(0) \rangle_0. \qquad (3.1.2)$$

This is given by the graph of Fig. 3.1.1. We have, in an arbitrary gauge, and for dimension $D = 4 - \epsilon$,

$$S_{D\xi}^{ij}(p) = \delta^{ij} \frac{\mathrm{i}}{\not{p} - m + \mathrm{i}0}$$

$$- \frac{1}{\not{p} - m + \mathrm{i}0} g^2 \sum_{l,a} t_{jl}^a t_{li}^a \Sigma_{D\xi}^{(2)}(p) \frac{\mathrm{i}}{\not{p} - m + \mathrm{i}0} \qquad (3.1.3a)$$

$$+ \text{ higher orders,}$$

where

$$\Sigma_{D\xi}^{(2)}(p) = -\mathrm{i} \int \mathrm{d}^D \hat{k} \, \frac{\gamma_\mu (\not{p} + \not{k} + m) \gamma_\nu}{(p+k)^2 - m^2} \frac{-g^{\mu\nu} + \xi k^\mu k^\nu / k^2}{k^2}, \qquad (3.1.3b)$$

and we omit the terms $+\mathrm{i}0$ in the denominators, which are to be understood implicitly.

Writing identically

$$\not{k}(\not{p} + \not{k} + m) = (p+k)^2 - m^2 - (p^2 - m^2) - (\not{p} - m)\not{k},$$

we find

$$\Sigma_{D\xi}^{(2)}(p) = -i \int d^D\hat{k} \left\{ \frac{(D-2)(\not{p}+\not{k}) - Dm - \xi(\not{p}-m)}{k^2\left[(p+k)^2 - m^2\right]} \right.$$

$$\left. - \xi(p^2 - m^2)\frac{\not{k}}{k^4\left[(p+k)^2 - m^2\right]} \right\}.$$

After standard manipulations, this gives (neglecting terms that will vanish as $\epsilon \to 0$)

$$\Sigma_{D\xi}^{(2)}(p) = (\not{p} - m)A_{D\xi}(p^2) + mB_{D\xi}(p^2), \qquad (3.1.4a)$$

where

$$A_{D\xi} = \frac{1}{16\pi^2}\left\{(1-\xi)N_\epsilon - 1 - \int_0^1 dx\,[2(1-x) - \xi]\log\frac{xm^2 - x(1-x)p^2}{\nu_0} \right.$$

$$\left. - \xi(p^2 - m^2)\int_0^1 dx\,\frac{x}{m^2 - xp^2} \right\},$$

$$(3.1.4b)$$

$$B_{D\xi} = \frac{1}{16\pi^2}\left\{-3N_\epsilon + 1 + 2\int_0^1 dx\,(1+x)\log\frac{xm^2 - x(1-x)p^2}{\nu_0} \right.$$

$$(3.1.4c)$$

$$\left. - \xi(p^2 - m^2)\int_0^1 dx\,\frac{x}{m^2 - xp^2} \right\}.$$

Here we have introduced the notation, to be used systematically in what follows,

$$N_\epsilon = \frac{2}{\epsilon} - \gamma_E + \log 4\pi.$$

To one loop, all poles in dimensional regularization appear in this combination. Noting that (see Appendix C) $\sum t_{il}^a t_{lj}^a = C_F\delta_{ij} = \frac{4}{3}\delta_{ij}$, we insert (3.1.4) into (3.1.3) and rearrange it to read

$$S_{D\xi}(p) = i\left\{\not{p} - m + g^2 C_F\Sigma_{D\xi}^{(2)}\right\}^{-1} \qquad (3.1.5a)$$

or

$$S_{D\xi}(p) = i\frac{1 - C_F g^2 A_{D\xi}(p^2)}{\not{p} - m\left[1 - C_F g^2 B_{D\xi}(p^2)\right]} + \text{higher orders.} \qquad (3.1.5b)$$

Actually, we have summed into (3.1.5b) the contribution of the iteration of the one loop correction, Fig. 3.1.2. It is easy to verify that (3.1.5a) is the

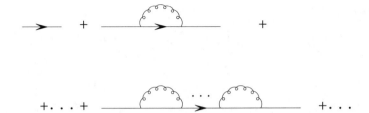

Fig. 3.1.2. Iteration of the one loop correction to the quark propagator.

more general expression for $S_{D\xi}$ if we replace $\Sigma_{D\xi}^{(2)}$ by the corresponding exact expression, $\Sigma_{D\xi}$.

As we see from this calculation, there are two divergences:

$$1 - C_F \frac{g^2}{16\pi^2}(1 - \xi)N_\epsilon \quad \text{(from } A_{D\xi}) \tag{3.1.6}$$

which multiplies the entire $S_{D\xi}$, and

$$1 + 3C_F \frac{g^2}{16\pi^2} N_\epsilon \quad \text{(from } B_{D\xi}) \tag{3.1.7}$$

which multiplies m; but both $A_{D\xi}$, $B_{D\xi}$ are finite provided that we keep $\epsilon \neq 0$.

We end this section with a comment on infrared singularities. In this work we will be mainly concerned with *ultraviolet singularities*, which are connected with the behaviour of integrals as $k \to \infty$ and give divergences proportional to $\Gamma(\epsilon/2)$; but dimensional regularization also regulates *infrared singularities*, caused by divergence of the integrals for $k \to 0$ and producing terms in $\Gamma(-\epsilon/2)$. For details, see Gastmans and Meuldermans (1973).

3.2 Renormalization – Generalities

Let us consider the following process: a photon hits a u quark (for example, in a proton) and u subsequently decays weakly into $d + e + \nu$ (Fig. 3.2.1). To the lowest order in weak and electromagnetic interactions, and to zero order in g, we have the diagram of Fig. 3.2.1a. Gluon radiative corrections then intervene(see the diagrams of Figs. 3.2.1b).[3] In particular, we see that

---

[3] In this diagram we have represented gluon lines by *curly* lines; we reserve *wavy* lines for photons or W, Z lines. This practice will be followed consistently in this text.

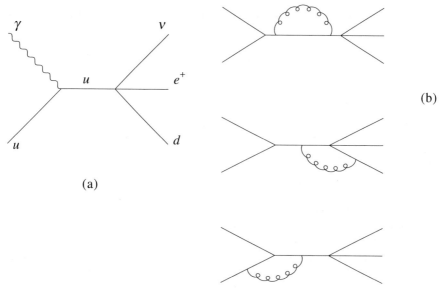

Fig. 3.2.1. The process $\gamma + u \to \nu + e^+ + d$ and some divergent radiative corrections.

$S(p)$ will enter into the amplitude with, in obvious notation, $p = p_\gamma + p_u$. Therefore, it looks as if the result for the amplitude is divergent and no sense can be extracted from the theory, at least in a perturbative expansion.

Of course, this is not so. We have been somewhat lax in our formulation. To show the problem at hand, and its solution, let us consider as a simple example a scalar interaction $\bar{\psi}\psi\phi$ with massless field ϕ. The Lagrangian is

$$\mathcal{L} = \bar{\psi}(i\slashed{\partial} - m)\psi + \tfrac{1}{2}\partial_\mu \phi \partial^\mu \phi + g\bar{\psi}\psi\phi. \tag{3.2.1}$$

As stated previously, the S matrix is given by

$$S = \mathrm{T} \exp \mathrm{i} \int \mathrm{d}^4 x \, \mathcal{L}^0_{\text{int}}(x)$$

$$= 1 + \sum_{n=1}^{\infty} \frac{\mathrm{i}^n}{n!} \int \mathrm{d}^4 x_1 \dots \mathrm{d}^4 x_n \, \mathrm{T}\mathcal{L}^0_{\text{int}}(x_1) \dots \mathcal{L}^0_{\text{int}}(x_n), \tag{3.2.2}$$

where the fields in $\mathcal{L}^0_{\text{int}}$ are to be taken as free, and in normal order; we identified $\mathcal{L}^0_{\text{int}}$ with the trilinear term in (3.2.1) after replacing $\psi \to \psi^0$, $\phi \to \phi^0$:

$$\mathcal{L}^0_{\text{int}} = g : \bar{\psi}^0 \psi^0 : \phi^0. \tag{3.2.3}$$

However, this is incorrect. Clearly, the fields in (3.2.1) are not free; and it is also conceivable that the mass that appears there is not the mass one would have if there were no interactions. This should be apparent from (3.1.5): the mass has been shifted by the amount

$$m\{1 - \tfrac{4}{3}g^2 B_D\},$$

and the normalization multiplied by

$$1 - \tfrac{4}{3}g^2 A_D.$$

On the grounds of invariance, all possible changes are of two types: of the multiplicative type,

$$\psi \to Z_\psi^{1/2}\psi, \quad \phi \to Z_\phi^{1/2}\phi; \quad g \to Z_g g, \quad m \to Z_m m, \tag{3.2.4}$$

or of the type obtained by adding some invariant extra term (or terms) to $\mathcal{L}$. A theory is *renormalizable* if the number of such terms is finite; if infinitely many terms are needed, the theory is *non–renormalizable*. The theory described by (3.2.1) is renormalizable and, besides the replacements (3.2.4), also requires the addition of a term $\lambda\phi^4$ to $\mathcal{L}$. We will neglect this term here; so, taking into account only (3.2.4), we see that the Lagrangian (3.2.1) becomes the so-called "renormalized" Lagrangian,

$$\mathcal{L}^R = Z_\psi \bar{\psi} i \partial\!\!\!/ \, \psi - Z_\psi Z_m m \bar{\psi}\psi + Z_\phi \partial_\mu \phi \partial^\mu \phi + Z_g Z_\psi Z_\phi^{1/2} g \bar{\psi}\psi\phi; \tag{3.2.5}$$

we find that the interaction Lagrangian, defined as $\mathcal{L}_{\text{int}} \equiv \mathcal{L} - \mathcal{L}_{\text{free}}$, is really

$$\begin{aligned}
\mathcal{L}_{\text{int}}^{R0} = {}& : g\bar{\psi}^0\psi^0\phi^0 + (Z_g^{1/2}Z_\psi Z_\phi^{1/2} - 1)g\bar{\psi}^0\psi^0\phi^0 \\
& (Z_\psi - 1)\bar{\psi}^0 i\partial\!\!\!/ \, \psi^0 - (Z_\psi Z_m - 1)m\bar{\psi}^0\psi^0 \\
& + (Z_\phi - 1)\partial_\mu\phi^0\partial^\mu\phi^0 :,
\end{aligned} \tag{3.2.6}$$

where ψ^0, ϕ^0 are *free* fields, satisfying canonical commutation relations. The constants Z are called *renormalization constants*, and the terms containing the factors $(Z - \ldots)$ are the *counterterms*. If we expand them in a power series in g, this series has to begin at unity for, if g were zero, all Z would equal unity. So we write

$$Z_j = 1 + \sum_{n=1}^{\infty} C_j^{(n)} \left(\frac{g^2}{16\pi^2}\right)^n, \tag{3.2.7}$$

where, if we want these counterterms to cancel the divergences of the theory, the $C_j^{(n)}$ are expected to be singular as $\epsilon \to 0$, and actually to be of the form $\sum_{k=0}^n a_k^{(n)} \epsilon^{-k} + O(\epsilon)$.

There is another way in which the necessity of counterterms may be seen (Bogoliubov and Shirkov, 1959). If we look at the expansion (3.2.2), it turns out that, because the fields are singular, the product

$$\mathrm{T}\mathcal{L}_{\text{int}}^0(x_1)\ldots\mathcal{L}_{\text{int}}^0(x_n) \tag{3.2.8a}$$

is undefined for equal arguments, $x_i = x_j$. Therefore, one can add arbitrary terms

$$p(\partial)\delta(x_1 - x_2) \ldots \delta(x_i - x_j) \ldots \delta(x_{n-1} - x_n), \qquad (3.2.8b)$$

with p a polynomial in the derivatives, to each of (3.2.8a). On analysis, the terms (3.2.8b) are seen to correspond to the counterterms.

How arbitrary are the values of the Z? A first condition on them is that $\mathcal{L}^R$ produce finite answers in the limit $\epsilon \to 0$. This, however, does not completely fix all the $C_j^{(n)}$ in (3.2.7), as one can add to them constant (in ϵ) terms. To have a unique theory we have to fix these constants, specifying sufficiently many independent amplitudes as there are renormalization constants, Z.

Let us now return to the QCD Lagrangian. Since QCD is a gauge theory, and we have seen that gauge invariance is essential to keep the theory meaningful, the possible counterterms are strongly restricted: they must respect gauge invariance. A look at the expression for $\mathcal{L}^\xi_{QCD}$, Eq. (2.3.11) shows that the only modifications allowed are the following:[4]

$$q^i(x) \to Z_F^{1/2} q^i(x), \quad \bar{q}^i(x) \to Z_F^{1/2} \bar{q}^i(x),$$
$$\omega_a(x) \to Z_\omega^{1/2} \omega_a(x), \quad \bar{\omega}_a(x) \to Z_\omega^{1/2} \bar{\omega}_a(x),$$
$$B_a^\mu(x) \to Z_B^{1/2} B_a^\mu(x),$$
$$g \to Z_g g, \qquad\qquad\qquad (3.2.9)$$
$$m_q \to Z_{m,q} m_q,$$
$$\lambda \to Z_\lambda \lambda.$$

Gauge invariance forces the Z for all the quarks to be equal to one single Z_F and, likewise, there is one common Z_B for all the gluons. In addition, the potentially different renormalization of the trilinear $\bar{q}qB$, the trilinear BBB, the quadrilinear $BBBB$ and the ghost $\bar{\omega}\omega B$ couplings must be induced by the same Z_g. That this very specific set of Zs is sufficient to render all Green's functions finite is a consequence of the identities (Ward identities for Abelian, Slavnov–Taylor identities for non-Abelian theories) that gauge invariance forces on these Green's functions. As stated earlier, these identities may be generated by the BRST transformations; later on, we will explicitly check a few representative ones.

Let us next introduce a bit of notation. After the replacements of (3.2.9), the renormalized QCD Lagrangian becomes

$$\mathcal{L}^\xi_R = \sum_q \left(i\widetilde{\bar{q}} \widetilde{\slashed{D}} \widetilde{q} - m_q \widetilde{\bar{q}} \widetilde{q} \right) - \tfrac{1}{4} \left(\widetilde{D} \times \widetilde{B} \right)^2 - \frac{\widetilde{\lambda}}{2} (\partial \cdot \widetilde{B})^2 + (\partial_\mu \widetilde{\bar{\omega}}) \widetilde{D}^\mu \widetilde{\omega}, \quad (3.2.10a)$$

---

[4] Note that not all the Z are independent; the Slavnov–Taylor identities, that we will discuss in Sect. 3.3, give $Z_\lambda = Z_B$. Detailed studies of these identities may be found in the treatises of Lee (1976), Taylor (1976) and Fadeyev and Slavnov (1980).

where the tilde means that the corresponding objects embody the appropriate Z factors:

$$\tilde{q} = Z_F^{1/2}q, \quad \tilde{g} = Z_g g, \quad \tilde{m} = Z_m m, \ldots,$$
$$\widetilde{\slashed{D}}\tilde{q} = (\slashed{\partial} - i\tilde{g}t\widetilde{\slashed{B}})\tilde{q}, \ldots, \text{etc.}$$

(3.2.10b)

Thus, $\mathcal{L}_R^\xi$ is formally identical with $\mathcal{L}^\xi$, with the replacement of all objects by renormalized objects. We may also split $\mathcal{L}_R^\xi$, explicitly exhibiting the counterterms:

$$\mathcal{L}_R^\xi = \mathcal{L}_{uD}^\xi + \mathcal{L}_{\text{ct}D}^\xi,$$

(3.2.11a)

where $\mathcal{L}_{uD}^\xi$ is the unrenormalized or "bare" part

$$\mathcal{L}_{uD}^\xi = \sum_q \left(i\bar{q}\slashed{D}q - m_q\bar{q}q \right) - \tfrac{1}{4}(D \times B)^2 - \frac{\lambda}{2}(\partial \cdot B)^2 + (\partial_\mu\bar{\omega})D^\mu\omega, \quad (3.2.11b)$$

and

$$\mathcal{L}_{\text{ct}D}^\xi = \mathcal{L}_R^\xi - \mathcal{L}_{uD}^\xi$$
$$= (Z_F - 1)i\sum_q \bar{q}\slashed{\partial}q + (Z_F Z_B^{1/2}Z_g - 1)g\sum \bar{q}\gamma_\mu t^a q B_a^\mu + \cdots. \quad (3.2.11c)$$

We see that, in perturbation theory, the interaction contains not only the terms $g\sum \bar{q}^0\gamma_\mu t q^0 B^{0\mu}, \ldots$, but also $i(Z_F - 1)\sum \bar{q}^0\slashed{\partial}q^0$, etc., where the fields q^0, B^0, ω^0 are the ones that verify free-field canonical commutation relations, and thus generate the Feynman rules of Appendix D. In the renormalization procedure one expects that, while $\mathcal{L}_{uD}^\xi$, $\mathcal{L}_{\text{ct}D}^\xi$ each require regularization, provided by the value $D \neq 4$, the singularities must compensate in such a way that $\mathcal{L}_R^\xi$ produce finite answers in the limit $D \to 4$. It is far from obvious that there exists a choice of Z values that achieves this, and indeed (at least in perturbation theory) only a limited numbers of field theories are renormalizable. The proof of the renormalizability of non-Abelian field theories, in particular QCD, was given by 't Hooft (1971). Here we will not go so far, but only check explicitly that $\mathcal{L}_R^\xi$ produces finite answers to the lowest orders in perturbation theory.[5]

---

[5] For a very readable account of renormalization, see Lee and Zinn-Justin (1972). A more recent account is that in Fadeyev (1976). For *practical* applications in QCD, the treatise of Pascual and Tarrach (1984) is useful.

3.3 Renormalization of QCD (One Loop)

In our presentation of renormalization theory, based essentially on the discussion of Bogoliubov and Shirkov (1959), finite (*renormalized*) Green's functions are obtained for the VEVs

$$\langle 0|T \dots q(x) \dots B(y) \dots \omega(z) \dots |0\rangle$$

in perturbation theory, calculating with the full interaction Lagrangian, including counterterms, of (3.2.11). The multiplicative character of renormalization for QCD, however, allows us to follow a different path. We may neglect the counterterms and simply rescale the fields and couplings in Green's functions according to (3.2.9). Also, one should be aware that we will be renormalizing *perturbatively*. This means that we have to be consistent, and work to the same order both in the "primitive" interaction and in the counterterms. In this and the following sections, we will see detailed examples of this.

i μ–Renormalization

Consider the renormalized QCD Lagrangian. In order to specify it, we have to give the values of the Z. To do so, we begin by defining the unrenormalized Green's functions,

$$G_{uD}(x_1, \dots, x_N),$$

which are calculated with $\mathcal{L}_{uD}^\xi$. If G corresponds to the VEV of a field product,

$$\langle T\Phi_1(x_1) \dots \Phi_N(x_N)\rangle_0 = G_{uD}(x_1, \dots, x_N), \qquad (3.3.1)$$

where the Φ_k are the q, ω, B or, more generally, local operators built from these, then in perturbation theory,

$$G_{uD}(x_1, \dots, x_N) =$$
$$\sum_{n=0}^\infty \frac{i^n}{n!} \int d^4 z_1 \dots d^4 z_n \, \langle T\Phi_1^0(x_1) \dots \Phi_N^0(x_N) \mathcal{L}_{uD,\mathrm{int}}^\xi(z_1) \dots \mathcal{L}_{uD,\mathrm{int}}^\xi(z_n)\rangle_0.$$
$$(3.3.2)$$

The G_{uD} are, generally speaking, divergent as $D \to 4$. The renormalized Green's functions are defined as

$$G_R(x_1, \dots, x_N) =$$
$$\sum_{n=0}^\infty \frac{i^n}{n!} \int d^4 z_1 \dots d^4 z_n \, \langle T\Phi_1^0(x_1) \dots \Phi_N^0(x_N) \mathcal{L}_{R,\mathrm{int}}^\xi(z_1) \dots \mathcal{L}_{R,\mathrm{int}}^\xi(z_n)\rangle_0.$$
$$(3.3.3)$$

What we then require is that G_R be finite, i.e., that the modifications that counterterms introduce in (3.3.3) cancel the singularities of (3.3.2). In QCD, we have six values of Z; to fix them it will be sufficient to fix *six* independent Green's functions. That the result is independent of the choice we make for these six functions is a consequence of the Ward–Slavnov–Taylor identities among Green's functions, and this is actually a highly nontrivial part of the renormalization program. For the moment, we shall make a specific choice. We work in momentum space and start with the quark propagator

$$S_{R\xi}(p) = \mathrm{i}\left\{\not{p} - m + \Sigma_{R\xi}(p)\right\}^{-1},$$
$$\Sigma_{R\xi}(p) = (\not{p} - m)A_{R\xi}(p^2) + mB_{R\xi}(p^2). \tag{3.3.4a}$$

The μ–renormalization, also called *momentum renormalization*, is defined as follows. Let us choose a spacelike momentum, $\bar{p}$, $\bar{p}^2 = -\mu^2 < 0$; this condition being imposed to avoid confusion between the singularities as $D \to 4$ and the discontinuities of Green's functions as functions of the momenta, which in our case occur for timelike p, $p^2 \geq m^2$. We may specify the values of

$$A_{R\xi}(\bar{p}^2), \quad B_{R\xi}(\bar{p}^2). \tag{3.3.4b}$$

The first will fix Z_F, the second a combination of Z_F, Z_m, Z_λ. Then we turn to the gluon propagator,

$$D_{R\xi}^{\mu\nu} = \left(-q^2 g^{\mu\nu} + q^\mu q^\nu\right)D_{R,\mathrm{tr}}(q^2) + g^{\mu\nu}D_{RL}(q^2), \tag{3.3.5a}$$

which, also choosing $q = \bar{p}$ for simplicity and fixing

$$D_{R,\mathrm{tr}}(\bar{p}^2), \quad D_{RL}(\bar{p}^2), \tag{3.3.5b}$$

allows us to obtain Z_B and a combination of Z_B, Z_λ. The ghost propagator,

$$G_R(p) = \int \mathrm{d}^4 x\, \mathrm{e}^{-\mathrm{i}p\cdot x} \langle T\omega(x)\bar{\omega}(0)\rangle_0, \tag{3.3.6a}$$

when chosen at $p = \bar{p}$,

$$G_R(\bar{p}), \tag{3.3.6b}$$

yields Z_ω.

The missing condition that will allow us to fix Z_g is provided by any vertex: $\bar{q}qB$, BBB, $BBBB$ or $\bar{\omega}\omega B$. Here we will select the first. If we define the "amputated" vertex V by

$$\int \mathrm{d}^4 x\, \mathrm{d}^4 y\, \mathrm{e}^{-\mathrm{i}p_1\cdot x}\mathrm{e}^{\mathrm{i}p_2\cdot y} \langle q_\beta^k(y) B_\mu^a(0)\bar{q}_\alpha^j(x)\rangle_0$$
$$= \sum D_{\mu\nu}^{ab}(p_2 - p_1)S_{\beta\alpha'}^{ki}(p_2)V_{R\xi;\alpha'\beta'}^{il;b,\nu}(p_1, p_2)S_{\beta'\alpha}^{lj}(p_1),$$
$$V_{R\xi;\alpha'\beta'}^{il;b,\nu}(p_1, p_2) = \mathrm{i}t_{il}^b\gamma_{\alpha'\beta'}^\nu + V_{R\xi;\alpha'\beta'}^{(2)il;b,\nu}(p_1, p_2) + \cdots, \tag{3.3.7a}$$

(α, β Dirac indices, and by S we denote to the quark propagators) then we can specify the value of the vertex

$$V_{R\xi}\big|_{p_1^2=p_2^2=(p_1-p_2)^2=-\mu^2}. \tag{3.3.7b}$$

The implementation of the renormalization group program is greatly facilitated by the fact, already noted, that $\mathcal{L}_R^\xi$ may be obtained from $\mathcal{L}_{uD}^\xi$ by simply effecting the replacements (3.2.9). To calculate any renormalized Green's function, we begin by writing it more explicitly, in momentum space and after so-called *amputation* (eliminating the external legs), as

$$\Gamma_R \equiv \Gamma(p_1,\ldots,p_{N-1};m,g,\lambda),$$

where, denoting by Φ to generic fields,

$$\Gamma(p_1,\ldots,p_{N-1};m,g,\lambda)\delta\left(\sum p\right)$$
$$= K_1(p_1)\ldots K_N(p_N)\int \mathrm{d}^4 x_1 \ldots \mathrm{d}^4 x_N \, \mathrm{e}^{\mathrm{i}\sum x_k\cdot p_k}\langle \mathrm{T}\Phi_1(x_1)\ldots\Phi_N(x_N)\rangle_0,$$
$$\tag{3.3.8}$$

and where the K_i are the appropriate inverse propagators, $\mathrm{i}K(p) = S_R^{-1}(p)$ for fermion fields, $\mathrm{i}K(p) = D_R^{-1}(p)$ for gluons, etc. Note that amputation has the virtue of removing the poles associated with the external legs. Moreover, since S_R, D_R are *renormalized* propagators, Γ will contain a factor $Z_\Phi^{-1/2}$ for each field and a factor Z_Φ for each K_Φ, and hence an effective factor $Z_\Phi^{1/2}$ for each field Φ.

Next, we calculate

$$\Gamma_{uD}(p_1,\ldots,p_{N-1};m,g,\lambda)$$

by using $\mathcal{L}_{uD;\mathrm{int}}^\xi$, cf. Eq. (3.3.2). Then, $\Gamma \equiv \Gamma_R$ is obtained from Γ_{uD} as

$$\Gamma(p_1,\ldots,p_{N-1};m,g,\lambda)$$
$$= Z_{\Phi_1}^{1/2}\ldots Z_{\Phi_N}^{1/2}\Gamma_{uD}(p_1,\ldots,p_{N-1};Z_m m, Z_g g, Z_\lambda \lambda). \tag{3.3.9}$$

This equation takes on a more transparent appearance if we define the bare couplings,[6]

$$m_{quD} = Z_{mq}m_q, \quad \lambda_{uD} = Z_\lambda\lambda, \quad g_{uD} = Z_g g, \tag{3.3.10}$$

for then (3.3.9) reads

$$\Gamma(p_1,\ldots,p_{N-1};m,g,\lambda)$$
$$= Z_{\Phi_1}^{1/2}\ldots Z_{\Phi_N}^{1/2}\Gamma_{uD}(p_1,\ldots,p_{N-1};m_{uD},g_{uD},\lambda_{uD}), \tag{3.3.11a}$$

---

[6] It is at times convenient to think of the mass and gauge parameter as coupling constants.

or, written in terms of nonamputated Green's functions, and using the notation $p = (p_1, \ldots, p_{N-1})$ for short,

$$G_R(p, m, g, \lambda) = Z_{\Phi_1}^{-1/2} \ldots Z_{\Phi_N}^{-1/2} G_{uD}(p, m_{uD}, g_{uD}, \lambda_{uD}). \qquad (3.3.11b)$$

To see how this works, consider the (*non–amputated*) quark propagator. According to the general discussion,

$$S_R(p; g, m, \lambda) = Z_F^{-1/2} Z_F^{-1/2} S_{uD}(p; Z_g g, Z_m m, Z_\lambda \lambda).$$

We shall calculate to second order only. In this case, we may replace Z_g, Z_λ by unity, since the corrections will be of higher order in g. Using the expression computed in (3.1.4) and (3.1.5),

$$S_R(p; g, m, \lambda) = i Z_F^{-1} \frac{1 - C_F g^2 A_{D\xi}(p^2)}{\not{p} - Z_m m \left[1 - C_F g^2 B_{D\xi}(p^2)\right]}.$$

As stated, to determine the Z we have to specify S_R at a given $p = \bar{p}$. We will do so by requiring that, at this point, S_R equal the free propagator:

$$S_R(\bar{p}; g, m, \lambda) = \frac{i}{\not{p} - m}. \qquad (3.3.12)$$

Thus we find, with $\bar{p}^2 = -\mu^2$,

$$Z_F \equiv Z_{FD}^\xi(\mu^2, m^2) = 1 - C_F \frac{g^2}{16\pi^2} \left\{ (1 - \xi)N_\epsilon - 1 - \int_0^1 dx \, [2(1 - x) - \xi] \right.$$

$$\left. \times \log \frac{xm^2 + x(1 - x)\mu^2}{\nu_0^2} + \xi(\mu^2 + m^2) \int_0^1 dx \, \frac{x}{m^2 + \mu^2 x} \right\},$$

$$(3.3.13)$$

and

$$Z_m \equiv Z_m(\mu^2, m^2) = 1 - C_F \frac{g^2}{16\pi^2} \left\{ 3N_\epsilon - 1 - 2 \int_0^1 dx \, (1 + x) \right.$$

$$(3.3.14)$$

$$\left. \times \log \frac{xm^2 + x(1 - x)\mu^2}{\nu_0^2} + \xi(\mu^2 + m^2) \int_0^1 dx \, \frac{x}{m^2 + \mu^2 x} \right\}.$$

An important fact to be noted is that, while the divergent part of Z_F depends on the gauge, that of Z_m is gauge independent (although in this renormalization scheme, the finite parts of Z_m are still gauge dependent). The gauge dependence of Z_F implies that there will exist gauge choices where it is *finite*. From (3.3.13) it is clear that, to second order, this will be the case for the Landau gauge, $\xi = 1$.

In QED there is a natural renormalization scheme: we take electrons and photons on their mass shells, i.e., we choose $\bar{p}^2 = m^2$ for S and $\bar{q}^2 = 0$ for $D^{\mu\nu}$. Because confinement is a feature of QCD, no such natural scheme exists in our case; in fact, the residue of $D^{\mu\nu}(q)$ at $q^2 = 0$ is infinite. So we are free to choose renormalization schemes that simplify the calculations as much as possible. This occurs in particular for the so-called minimal subtraction schemes, that we subsequently discuss.

ii The Minimal Subtraction Scheme

It was noted by 't Hooft (1973) that the simplest way to eliminate divergences in Green's functions is to chop off the poles in $1/\epsilon$ that appear in dimensional regularization. This procedure is known as *minimal subtraction*. Subsequently, a number of people realized that these poles appear in the combination

$$N_\epsilon = \frac{2}{\epsilon} - \gamma_{\mathrm{E}} + \log 4\pi. \tag{3.3.15}$$

Therefore, if one simply cancels the $2/\epsilon$ one introduces, somewhat artificially, the transcendentals γ_{E}, $\log 4\pi$. These, it will be recalled, appear because of the specific way in which we continued to the dimension $D = 4 - \epsilon$, which yielded the terms

$$(4\pi)^{\epsilon/2} \Gamma(\epsilon/2) = N_\epsilon + O(\epsilon).$$

It seems natural to eliminate these by introducing the *modified minimal subtraction scheme*, or $\overline{\mathrm{MS}}$, in which the entire N_ϵ is subtracted. In this scheme we find

$$\bar{Z}_F = 1 - C_F \frac{g^2}{16\pi^2}(1 - \xi)N_\epsilon, \tag{3.3.16}$$

$$\bar{Z}_m = 1 - 3C_F \frac{g^2}{16\pi^2}N_\epsilon. \tag{3.3.17}$$

In this book we will mainly work with this $\overline{\mathrm{MS}}$ scheme and shall therefore drop from now on the bar over the Z.

 An interesting property of this scheme is that here Z_m is fully gauge invariant. This is due to the gauge independence of the mass term, $m\bar{q}q$, and has been checked explicitly to four loops. Another interesting property is that Z_m is actually independent of m.

 From (3.3.16) and (3.3.17) we see that if we write

$$Z_F = 1 + \frac{g^2}{16\pi}c_F^{(1)}N_\epsilon + \cdots,$$

$$Z_m = 1 + \frac{g^2}{16\pi}c_m^{(1)}N_\epsilon + \cdots,$$

then we have

$$c_F^{(1)} = -C_F(1 - \xi) \tag{3.3.18}$$

$$c_m^{(1)} = -3C_F. \tag{3.3.19}$$

The coefficients of Z_m of second, third and fourth order have been calculated; they are given in Sect. 10.1.

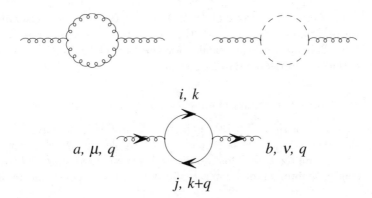

Fig. 3.3.1A. One loop corrections to the gluon propagator.

Next we evaluate, in the $\overline{\text{MS}}$ scheme, the remaining renormalization constants of QCD. We begin with the gluon propagator. Its transverse part may be written as

$$D^{\mu\nu}_{uD;\text{tr};ab} = i\,\frac{-g^{\mu\nu} + q^\mu q^\nu/q^2}{q^2}\,\delta_{ab}$$

$$+ \sum \frac{-g^{\mu\mu'} + q^\mu q^{\mu'}/q^2}{q^2}\,\delta_{aa'}\,\Pi^{a'b'}_{\mu'\nu'}(q^2)i\,\frac{-g^{\nu'\nu} + q^{\nu'}q^\nu/q^2}{q^2}\,\delta_{b'b} + \cdots.$$

$$(3.3.20)$$

The relevant diagrams are shown in Fig. 3.3.1A. The part of Π originating from gluons and ghosts in Fig. 3.3.1A has already been calculated earlier, Eq. (2.3.9).[7] To second order, the renormalization of the gauge parameter, mass or g do not intervene. The part arising from a quark loop, with the routing of momenta given in Fig. 3.3.1A, is, for each flavour f of quark,

$$\Pi^{\mu\nu}_{f\,\text{quark};ab} = -ig^2 \sum_{ij} t^a_{ij} t^b_{ji} \int \frac{d^D k}{(2\pi)^D} \nu_0^{4-D}\,\frac{\text{Tr}(\slashed{k} + m_f)\gamma^\mu(\slashed{k} + \slashed{q} + m_f)\gamma^\nu}{(k^2 - m_f^2)[(k+q)^2 - m_f^2]}.$$

---

[7] Equation (2.3.9) was calculated without taking into account a factor ν_0^{4-D}. When we include it, the only difference in the result is the replacement of $\log(-q^2)$ by $\log(-q^2/\nu_0^2)$.

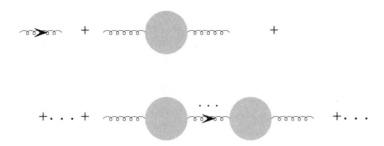

Fig. 3.3.1B. Iterations of the one loop corrections to the gluon propagator.

The calculation may be carried out with the standard techniques. The result is like that for the photon vacuum polarization in QED, apart from the colour factor $\mathrm{Tr}\, t^a t^b$. The result is, with n_f the total number of quark flavours,

$$
\Pi^{\mu\nu}_{\text{all quark};ab} = -2T_F\delta_{ab}\frac{g^2}{16\pi^2}(-g^{\mu\nu}q^2 + q^\mu q^\nu)
$$
$$
\times\left\{\tfrac{2}{3}N_\epsilon n_f - 4\int_0^1 dx\, x(1-x)\sum_{f=1}^{n_f}\log\frac{m_f^2 - x(1-x)q^2}{\nu_0^2}\right\}.
\tag{3.3.21}
$$

We can sum all the graphs of Fig. 3.3.1B, where the hazy disk represents any gluon, ghost or quark loop. If we write

$$
\Pi^{\mu\nu}_{\text{all};ab} = -\delta_{ab}(-g^{\mu\nu}q^2 + q^\mu q^\nu)\Pi_{\text{all}},
\tag{3.3.22a}
$$

where "all" means that we have summed gluons, ghosts and quarks, we obtain the expression for the transverse part of the gluon propagator, correct to one loop,

$$
D^{\mu\nu}_{uD;\text{tr};ab}(q) = i\delta_{ab}\frac{-g^{\mu\nu} + q^\mu q^\nu/q^2}{[1 - \Pi_{\text{all}}(q^2)]q^2}.
\tag{3.3.22b}
$$

This equation is the analogue of (3.1.5).

Let us introduce the notation

$$
f \overset{\text{div}}{=} g,
$$

to mean that the coefficients of N_ϵ of f and g are equal (*equal divergent parts*). The renormalized nonamputated transverse part of D is

$$
D^{\mu\nu}_{R\,\text{tr};ab} = Z_B^{-1}D^{\mu\nu}_{uD;\text{tr};ab};
$$

using Eqs. (2.3.9), (3.3.20) and (3.3.21) we see that

$$
1 - \Pi_{\text{all}} \overset{\text{div}}{=} 1 + \frac{g^2}{16\pi^2}\left\{\frac{10C_A}{6} - \frac{8T_F n_f}{6}\right\}N_\epsilon,
$$

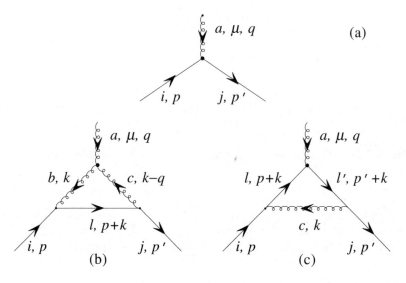

Fig. 3.3.2. The (amputated) quark–gluon vertex. (a) Bare vertex.
(b,c) One loop corrections.

and therefore, in the $\overline{\text{MS}}$ scheme and in the Fermi–Feynman gauge to one
loop,

$$Z_B = 1 + \frac{g^2}{16\pi^2} \left\{ \tfrac{5}{3}C_A - \tfrac{4}{3}T_F n_f \right\} N_\epsilon. \tag{3.3.23}$$

In an arbitrary gauge, Z_B has been calculated by Gross and Wilczek (1973a)
and Politzer (1973). The corresponding $c_{B\xi}^{(1)}$ is

$$c_{B\xi}^{(1)} = \tfrac{1}{2} \left\{ 10 + 3\xi - \frac{4n_f}{3} \right\}. \tag{3.3.24}$$

We shall not calculate Z_λ explicitly; the Slavnov–Taylor identity which we
proved in Sect. 2.4 implies that, to all orders, $Z_B = Z_\lambda$, so in particular

$$c_{\lambda\xi}^{(1)} = c_{B\xi}^{(1)}, \tag{3.3.25}$$

which the reader may verify easily.

To complete this section, let us calculate Z_g. This we do from the $\bar{q}qB$
vertex. With the choice of momenta of Fig. 3.3.2, we write the (amputated)
vertex as

$$V^\mu_{uD,ija} = ig\gamma^\mu t^a_{ji} + i\Gamma^{(2)\mu}_{uD,ija} + \cdots \tag{3.3.26a}$$

(cf. Eq. 3.3.7) with

$$\Gamma_{uD,ija}^{(2)\mu}(p,p') = \left(\Gamma^{(b)} + \Gamma^{(c)}\right)_{uD,ija}^{\mu}. \tag{3.3.26b}$$

The indices in $\Gamma^{(b)}$, $\Gamma^{(c)}$ refer to the contributions of the diagrams in Figs. 3.3.2b,c. Note that the three diagrams where the gluon is radiated and absorbed by the *same* external leg are not included, because the vertex is taken to be *amputated*. The diagram of Fig. 3.3.2a gives the term $ig\gamma^\mu t_{ji}^a$ in Eq. (3.3.26a). As should be obvious from the previous examples, the quark masses play no role in the evaluation of the Z (except, of course, Z_m) so we will simplify the calculation by taking $m = 0$. Also, we will only calculate the divergent part of the Γ. Then, in the Fermi–Feynman gauge,

$$i\Gamma_{uD,ija}^{(b)\mu} \overset{\text{div}}{=} ig^3 C_{ij}^a$$
$$\times \int d^D\hat{k}\, \frac{\gamma^\beta[(2k-q)^\mu g_{\alpha\beta} - (k+q)_\beta g_\alpha^\mu + (2q-k)_\alpha g_\beta^\mu](\slashed{p} + \slashed{k})\gamma^\alpha}{[(p+k)^2+i0][(k-q)^2+i0](k^2+i0)}$$
$$\overset{\text{div}}{=} ig^3 C_{ij}^a \gamma^\mu \lim_{\eta\to 0} \int d^D\hat{k}\, \frac{2(2-D)/D}{(k^2-i\eta)^2} \overset{\text{div}}{=} 3C_{ij}^a g N_\epsilon \gamma^\mu \frac{g^2}{16\pi^2}. \tag{3.3.27a}$$

We have used the customary notation

$$d^D\hat{k} = \frac{d^D k}{(2\pi)^D}\nu_0^{4-D},$$

and the colour factor C_{ij}^a is

$$C_{ij}^a = -\sum t_{jl}^c t_{li}^b f^{bca} = \tfrac{1}{2}[t^c, t^b]_{ji} f^{bca}$$
$$= \frac{i}{2} C_A t_{ji}^a = \tfrac{3}{2} i t_{ji}^a.$$

We have profited from the antisymmetry of the f^{bca} to replace $t^c t^b f^{bca} \to \tfrac{1}{2}[t^c, t^b] f^{bca}$.

Likewise,

$$i\Gamma_{uD,ija}^{(c)\mu} \overset{\text{div}}{=} -i^2 g^3 C_{ij}'^a \int d^D\hat{k}\, \frac{\gamma_\beta(\slashed{p}' + \slashed{k})\gamma^\mu(\slashed{p} + \slashed{k})\gamma_\alpha g^{\alpha\beta}}{[(p+k)^2+i0][(p'+k)^2+i0](k^2+i0)}$$
$$\overset{\text{div}}{=} ig C_{ij}'^a N_\epsilon \gamma^\mu \frac{g^2}{16\pi^2}, \tag{3.3.27b}$$

and the colour factor is now

$$C_{ji}'^a = \sum_c (t^c t^a t^c)_{ji} = \sum_c ([t^c, t^a]t^c)_{ji} + (t^a \sum_c t^c t^c)_{ji} \tag{3.3.27c}$$
$$= \{-\tfrac{1}{2}C_A + C_F\} t_{ji}^a;$$

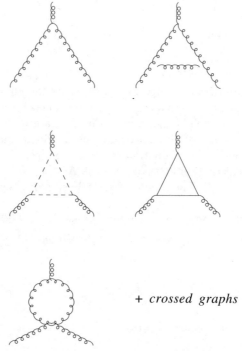

+ *crossed graphs*

Fig. 3.3.3. The gluon–gluon (amputated) vertex: bare vertex and one loop corrections.

we have used repeatedly formulas of Appendix C. Adding $\Gamma^{(a)}$, $\Gamma^{(b)}$,

$$\Gamma^{(2)\mu}_{uD,ija} \overset{\text{div}}{=} igt^a_{ji}\gamma^\mu N_\epsilon\{C_A + C_F\}\frac{g^2}{16\pi^2}. \qquad (3.3.28)$$

Now, the renormalization of the vertex involves Z_g and Z_F, Z_B,

$$V^\mu_{R,ija} = Z_F^{-1}Z_B^{-1/2}Z_g V^\mu_{uD,ija}; \qquad (3.3.29)$$

using the values for the Z_F, Z_B that we found before, and the expression just calculated for the divergent part of $\Gamma^{(2)}$, we find, to second order,

$$Z_g = 1 - \frac{g^2}{16\pi^2}\left\{\tfrac{11}{6}C_A - \tfrac{2}{3}T_F n_f\right\}N_\epsilon. \qquad (3.3.30)$$

Thus,

$$c_g^{(1)} = -\left\{\tfrac{11}{2} - \tfrac{1}{3}n_f\right\}.$$

It is interesting to watch the cancellation of the C_F terms. This is necessary because we could have calculated Z_g for a pure Yang–Mills theory. We see that the cancellation is due to the gauge structure; cf. Eq. (3.3.27c).

The original calculation was actually performed using the gluon–gluon vertex (Gross and Wilczek, 1973a; Politzer, 1973). This involves the diagrams of Fig. 3.3.3. One could also have used the $\bar{\omega}\omega B$ vertex. The result in all cases is the same, which is of course a reflection of gauge invariance.

A result that is worth noting is that, to all orders, Z_g is gauge invariant, in the $\overline{\text{MS}}$ scheme of renormalization (Caswell and Wilczek, 1974).

iii Renormalization in the Background Field Formalism

As an example of the uses of the background gauge method that was developed in Sect. 2.7, we will compute the coupling constant renormalization using it.

In the background gauge formalism the renormalization is performed as in other gauges; but, because the Lagrangian now also involves the background field ϕ, we have another renormalization constant, Z_ϕ: to Eqs. (3.2.9) we have to add

$$\phi_a^\mu(x) \to Z_\phi^{1/2} \phi_a^\mu(x). \tag{3.3.31}$$

The background field method simplifies the calculation of Z_g, at least to one and two loops.[8] The reason is that, as shown in Sect. 2.7, the Lagrangian is invariant under gauge transformations. This means that the ϕ field strength tensor,

$$G_{\phi,a}^{\mu\nu} = \partial^\mu \phi_a^\nu - \partial^\nu \phi_a^\mu + g \sum f_{abc} \phi_b^\mu \phi_c^\nu, \tag{3.3.32}$$

must be the same in terms of renormalized and unrenormalized fields. In the presence of the quadratic piece of (3.3.32), this is only possible if

$$Z_g = Z_\phi^{-1/2}, \tag{3.3.33}$$

because only thus will G be proportional to $\widetilde{G}$:

$$G_{\phi,a}^{\mu\nu} \to Z_\phi^{1/2}(\partial^\mu \phi_{uD,a}^\nu - \partial^\nu \phi_{uD,a}^\mu + Z_g Z_\phi^{1/2} g \sum f_{abc} \phi_{uD,b}^\mu \phi_{uD,c}^\nu) = \widetilde{G}_{\phi,a}^{\mu\nu}.$$

Therefore, to evaluate Z_g with the background field method, we only have to find Z_ϕ, which is accomplished by calculating the graphs of Fig. 3.3.4, a much simpler task than the calculation of Z_B, Z_g which was necessary with the previous methods.

The calculation is straightforward with the Feynman rules for the background gauge, given in Appendix D. It is here interesting to watch two effects: first, both diagrams (Fig. 3.3.4a and Fig. 3.3.4b) give, independently, transverse polarization functions. Secondly, the graph of Fig. 3.3.4a gives a result that does not depend on the gauge parameter, a or ξ. This is a reflection of

---

[8] However, to three and, especially, four loops, the complication caused by the extra couplings of ϕ offsets the gains we make. Nevertheless, the background field method has other applications.

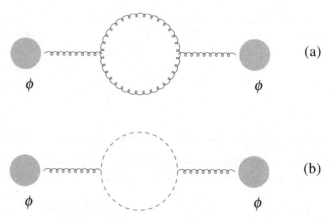

Fig. 3.3.4. Renormalization of the background field, ϕ.

the gauge invariance (with respect to ϕ) of the Lagrangian. The results are, for the divergent pieces,

$$
\begin{aligned}
Z_\phi^{\text{fig. }a} &= \frac{g^2}{16\pi^2}\frac{10C_A}{3}N_\epsilon, \\
Z_\phi^{\text{fig. }b} &= \frac{g^2}{16\pi^2}\frac{C_A}{3}N_\epsilon,
\end{aligned}
\tag{3.3.34a}
$$

and therefore Eq. (3.3.33) gives

$$
Z_g = Z_\phi^{-1/2} = \left[1 + Z_\phi^{\text{fig. }a} + Z_\phi^{\text{fig. }b}\right]^{-1/2} = 1 - \frac{g^2}{16\pi^2}\frac{11C_A}{6}N_\epsilon + O(g^4),
\tag{3.3.34b}
$$

which is, of course, in agreement with our previous result, Eq. (3.3.30), up to fermion contributions which we have not included now.

3.4 The Renormalization Group

Consider, as a first example for what will follow, the renormalized quark propagator. In the μ–scheme and in the Fermi–Feynman gauge,

$$
S_R^{(\mu)}(p; g, m) = \mathrm{i}\,\frac{1 - C_F g^2 A_R^{(\mu)}(p^2)}{\not{p} - m[1 - C_F g^2 B_R^{(\mu)}(p^2)]},
\tag{3.4.1a}
$$

where

$$A_R^{(\mu)}(p^2) = \frac{2}{16\pi^2} \int_0^1 dx \, (1-x) \log \frac{xm^2 + x(1-x)\mu^2}{xm^2 - x(1-x)p^2},$$

$$B_R^{(\mu)}(p^2) = -\frac{2}{16\pi^2} \int_0^1 dx \, (1+x) \log \frac{xm^2 + x(1-x)\mu^2}{xm^2 - x(1-x)p^2}.$$

(3.4.1b)

In the $\overline{\text{MS}}$ scheme,

$$S_R^{(\nu_0)}(p; g, m) = i \frac{1 - C_F g^2 \bar{A}_R^{(\nu_0)}(p^2)}{\not{p} - m[1 - C_F g^2 \bar{B}_R^{(\nu_0)}(p^2)]},$$

(3.4.2a)

and

$$\bar{A}_R^{(\nu_0)}(p^2) = \frac{1}{16\pi^2} \left\{ -1 - 2 \int_0^1 dx \, (1-x) \log \frac{xm^2 - x(1-x)p^2}{\nu_0^2} \right\};$$

$$\bar{B}_R^{(\nu_0)}(p^2) = \frac{1}{16\pi^2} \left\{ 1 + 2 \int_0^1 dx \, (1+x) \log \frac{xm^2 - x(1-x)p^2}{\nu_0^2} \right\}.$$

(3.4.2b)

We see that the renormalization introduces an arbitrary mass parameter in the Green's functions of the theory: it is the renormalization point μ in the μ-scheme or the mass ν_0 in the $\overline{\text{MS}}$ scheme.

Let us begin the discussion with the μ scheme. Suppose we change the renormalization point to μ'. If we merely replace μ by μ' in Eq. (3.4.1b), we would find a value $S_R^{(\mu')}$ different from $S_R^{(\mu)}$. However, we want to have a unique theory; we must therefore compensate for this change. This we do by allowing a dependence of the parameters of the theory on μ. So we rewrite (3.4.1a) as

$$S_R^{(\mu)}(p; g(\mu), m(\mu)) = i \frac{1 - C_F g(\mu)^2 A_R^{(\mu)}(p^2)}{\not{p} - m(\mu)[1 - C_F g(\mu)^2 B_R^{(\mu)}(p^2)]}.$$

(3.4.3)

That such $m(\mu)$, $g(\mu)$ exist is clear from the expression of S_R in terms of the unrenormalized propagator:

$$S_R^{(\mu)}(p; g(\mu), m(\mu)) = Z_F^{-1}(\mu) S_{uD}(p; g_{uD}, m_{uD});$$
$$m_{uD} = Z_m(\mu) m(\mu), \quad g_{uD} = Z_g g(\mu).$$

(3.4.4)

All we thus have to do is to appropriately choose the μ dependence of Z_F, Z_m, Z_g etc. to get a unique theory. Thus we take a new renormalization point μ', $S_R^{(\mu')}$ such that now

$$S_R^{(\mu')}(\bar{p}'; g(\mu'), m(\mu')) = \frac{i}{\not{p}' - m(\mu')},$$
$$p'^2 = -\mu'^2$$

and we fix[9] $m(\mu')$, $Z_F(\mu')$, $Z_m(\mu')$ by requiring equality of the propagators, $S_R^{(\mu)}(p; g(\mu), m(\mu))$ and $S_R^{(\mu')}(p; g(\mu'), m(\mu'))$, for all p. This gives, for example,

$$m(\mu') = m(\mu) \left\{ 1 - \frac{g^2}{6\pi^2} \int_0^1 dx\, (1+x) \log \frac{xm + x(1-x)\mu'^2}{xm + x(1-x)\mu^2} \right\}.$$

In the $\overline{\text{MS}}$ scheme the argument is simpler, but also subtler.[10] We regularized in such a manner that an arbitrary mass scale, ν_0, was introduced. If we want to obtain Green's functions independent of ν_0 we can make them so by cancelling not only the divergence of the term $(4\pi)^{\epsilon/2}\Gamma(\epsilon/2)\nu_0^\epsilon$, but the ν_0 dependence as well. This can only be achieved at the expense of introducing a new mass scale, ν, so that we replace $Z \to Z(\nu) = (\nu_0/\nu)^\epsilon Z$, which thus cancels the quantity $N_\epsilon^{\nu_0} = 2/\epsilon - \gamma_E + \log 4\pi + \log \nu_0$. The renormalized Green's functions will depend on ν, but no longer on ν_0. Now, we want the theory to be independent of the value of ν that we choose: it will be sufficient to allow for a ν dependence of g, m, ξ (besides Z). For amputated Γ,

$$\begin{aligned} &\Gamma_R(p_1, \ldots, p_{N-1}; g(\nu), m(\nu), \xi(\nu); \nu) \\ &= Z_{\Phi_1}^{1/2}(\nu) \ldots Z_{\Phi_N}^{1/2}(\nu) \Gamma_{uD}(p_1, \ldots, p_{N-1}; g_{uD}, m_{uD}, \xi_{uD}); \end{aligned} \tag{3.4.5}$$

and

$$\begin{aligned} g_{uD} &= Z_g(\nu)g(\nu), \quad m_{uD} = Z_m(\nu)m(\nu), \\ \lambda_{uD} &= Z_\lambda(\nu)\lambda(\nu), \quad \xi = 1 - \lambda^{-1}. \end{aligned} \tag{3.4.6}$$

It is not difficult to see what ν dependence we need. We recall that ν_0 entered in the combination

$$d^D\hat{k} = \frac{d^D k}{(2\pi)^D} \nu_0^{4-D},$$

so the only dependence on ν_0 lies in the divergent part:

$$(4\pi)^{\epsilon/2}\Gamma(\epsilon/2)\nu_0^\epsilon.$$

Therefore, the $Z_j(\nu)$ will be of the form

$$\begin{aligned} Z_j(\nu) &= 1 + C_j^{(1)}(\nu)\frac{g^2}{16\pi^2} + \ldots, \\ C_j^{(1)}(\nu) &= c_j^{(1)} \left\{ \frac{2}{\epsilon} - \gamma_E + \log 4\pi + \log \frac{\nu_0^2}{\nu^2} \right\}: \end{aligned} \tag{3.4.7}$$

---

[9] g does not intervene for S to the order at which we are working.

[10] Our version of the $\overline{\text{MS}}$ scheme is slightly different (although equivalent) to the standard one.

the coefficients of the $\log \nu^2$ terms are the same $c_j^{(1)}$ as the coefficients of the divergent term N_ϵ we have calculated, up to a change of sign. It is easy to show that the same is true, to lowest order, for the coefficients of $\log \mu^2$ in the μ-scheme.

The set of transformations $\mu \to \mu'$ (or $\nu \to \nu'$) constitutes the *renormalization group*,[11] first introduced by Stückelberg and Peterman (1953); see also Gell-Mann and Low (1954) and Bogoliubov and Shirkov (1959). The invariance of physical quantities under this group of transformations may be used, as will be done here, to study the asymptotic behaviour of Green's functions. This is best accomplished by using the Callan (1970) and Symanzik (1970) equation, which will be the subject of the next section.

3.5 The Callan–Symanzik Equation

The Callan–Symanzik equation is more simply derived by noting that the Γ_{uD}, g_{uD}, m_{uD}, λ_{uD} are ν-independent (for definiteness we work in the $\overline{\text{MS}}$ scheme). Therefore we have

$$\frac{\nu \mathrm{d}}{\mathrm{d}\nu}\Gamma_{uD}(p_1,\ldots,p_{N-1};g_{uD},m_{uD},\lambda_{uD}) = 0,$$

and using (3.4.5) and (3.5.6) we immediately obtain

$$\left\{\frac{\nu\partial}{\partial\nu} + g\beta\frac{\partial}{\partial g} + (-\lambda)\delta\frac{\partial}{\partial\lambda} + \sum_q m_q\gamma_{m,q}\frac{\partial}{\partial m_q} - \gamma_\Gamma\right\}$$

$$\times \Gamma_R(p_1,\ldots,p_{N-1};g(\nu),m(\nu),\lambda(\nu);\nu) = 0. \tag{3.5.1}$$

We have defined the universal functions β, γ, δ:

$$\frac{\nu\mathrm{d}}{\mathrm{d}\nu}g(\nu) = g(\nu)\beta,$$

$$\frac{\nu\mathrm{d}}{\mathrm{d}\nu}m_q(\nu) = m_q(\nu)\gamma_{m,q}, \tag{3.5.2}$$

$$\frac{\nu\mathrm{d}}{\mathrm{d}\nu}\lambda(\nu) = [-\lambda(\nu)]\delta,$$

and, moreover,

$$Z_\Gamma = Z_{\Phi_1}^{1/2}\ldots Z_{\Phi_N}^{1/2}, \quad Z_\Gamma^{-1}\frac{\nu\mathrm{d}}{\mathrm{d}\nu}Z_\Gamma = \gamma_\Gamma. \tag{3.5.3}$$

---

[11]Actually, a group structure is obtained only within a given renormalization scheme.

The functions α, β and γ can again be calculated using the same trick that was used to derive (3.5.1): Eq. (3.3.10), and the fact that the g_{uD}, m_{uD}, λ_{uD} are independent of ν. We get,

$$\beta = -Z_g^{-1}(\nu) \frac{\nu \mathrm{d}}{\mathrm{d}\nu} Z_g(\nu),$$

$$\gamma_{m,q} = -Z_m^{-1}(\nu) \frac{\nu \mathrm{d}}{\mathrm{d}\nu} Z_m(\nu), \qquad (3.5.4)$$

$$\delta = -Z_\lambda^{-1}(\nu) \frac{\nu \mathrm{d}}{\mathrm{d}\nu} Z_\lambda(\nu).$$

Equation (3.5.1) is not very useful as it stands, because it contains the partial derivative $\partial/\partial\nu$. However, we can transform it into a more useful form by using dimensional analysis. Suppose ρ_Γ is the dimension[12] of Γ_R; then $\nu^{-\rho_\Gamma}\Gamma_R$ is dimensionless, and thus it can only depend on ratios of dimensional parameters. We scale the momenta,

$$\nu^{-\rho_\Gamma}\Gamma_R(\lambda p_1,\ldots,\lambda p_{N-1};g,m,a^{-1};\nu)$$
$$= F(\lambda p_1/\nu,\ldots,\lambda p_{N-1}/\nu;g,m/\nu,a^{-1}),$$

and we replaced the gauge parameter λ by $a = \lambda^{-1}$ to avoid confusion with the scale λ. If we now trade the $\nu\partial/\partial\nu$ for $-\lambda\partial/\partial\lambda$, we obtain the Callan–Symanzik equation:

$$\left\{-\frac{\partial}{\partial\log\lambda} + g\beta\frac{\partial}{\partial g} + a^{-1}\delta\frac{\partial}{\partial(a^{-1})} + \sum_q m_q(\gamma_{m,q}-1)\frac{\partial}{\partial m_q} - \gamma_\Gamma + \rho_\Gamma\right\}$$
$$\times \Gamma_R(\lambda p_1,\ldots,\lambda p_{N-1};g,m,a^{-1};\nu) = 0.$$
$$(3.5.5)$$

To solve this equation, we define effective, or "running", parameters given implicitly by the equations

$$\frac{\mathrm{d}\bar{g}(\lambda)}{\mathrm{d}\log\lambda} = \bar{g}(\lambda)\beta(\bar{g}(\lambda)), \quad \frac{\mathrm{d}\bar{m}(\lambda)}{\mathrm{d}\log\lambda} = \bar{m}(\lambda)\gamma_{m,q}, \quad \frac{\mathrm{d}\bar{a}^{-1}(\lambda)}{\mathrm{d}\log\lambda} = \bar{a}^{-1}(\lambda)\delta,$$
$$(3.5.6a)$$

with the boundary conditions

$$\bar{g}|_{\lambda=1} = g(\nu), \quad \bar{m}|_{\lambda=1} = m(\nu), \quad \bar{a}|_{\lambda=1} = a(\nu). \qquad (3.5.6b)$$

---

[12]The dimension of a field is easily deduced by noting that the action $\mathcal{A} = \int \mathrm{d}^4x\,\mathcal{L}$ must be dimensionless. Hence, $[q] = [M]^{3/2}$, $[\omega] = [M]^1$ and $[B] = [M]^1$. The dimension of Γ is obtained from those of the fields it contains. For example, for the quark propagator (nonamputated) one has $\rho_S = -1$; that is, $3/2 + 3/2$ from the quark fields and -4 from the d^4x.

Then,

$$\Gamma_R(\lambda p_1, \ldots, \lambda p_{N-1}; g(\nu), m(\nu), a(\nu)^{-1}; \nu)$$
$$= \lambda^{\rho_\Gamma} \Gamma_R(p_1, \ldots, p_{N-1}; \bar{g}(\lambda), \bar{m}(\lambda), \bar{a}(\lambda)^{-1}; \nu)$$
$$\times \exp\left\{ -\int_0^{\log \lambda} \mathrm{d}\log \lambda' \, \gamma_\Gamma \left(\bar{g}(\lambda'), \bar{m}(\lambda'), \bar{a}(\lambda')^{-1} \right) \right\}.$$

$$(3.5.7)$$

We see that, when we scale the momenta by λ, Γ_R does not simply scale as λ^{ρ_Γ}: corrections to this have developed, incorporated into the exponential term on the right hand side of Eq. (3.5.7). For this reason, γ_Γ is usually called the *anomalous dimension* of Γ_R. In this respect the renormalization group may be interpreted as the realization of scale invariance in local quantum field theory; this realization is nontrivial due to the infinite character of renormalization, which introduces an extraneous mass scale. The development of this point of view may be found in Callan, Coleman and Jackiw (1970) and, for QCD, Collins, Duncan and Joglekar (1977).

There is an additional point to be made about (3.5.7). In principle it is valid independently of perturbation theory; for example, in lattice formulations (see later, Chap. 9). However, most of the applications involve perturbative expansions.

3.6 Renormalization of Composite Operators

Because we probe hadronic structure mostly with external currents, both weak and electromagnetic, we must discuss not only Green's functions, but matrix elements of composite operators as well. These operators may, from the point of view of renormalization, be classified into two categories: those which are conserved or partially conserved, and those that are not conserved (nor partially conserved).

A conserved operator is the electromagnetic current of any number of quark flavours: $J^\mu_{\mathrm{em}} = \sum_q Q_q V^\mu_{qq}$ with

$$V^\mu_{qq}(x) =: \bar{q}(x)\gamma^\mu q(x) :;$$

it satisfies

$$\partial_\mu V^\mu_{qq}(x) = 0 \qquad (3.6.1a)$$

to all orders in perturbation theory. A partially conserved current is, for example, the axial weak current

$$A^\mu_{qq'}(x) =: \bar{q}(x)\gamma^\mu \gamma_5 q'(x) : .$$

Using the equations of motion (2.1.6), we see that it verifies

$$\partial_\mu A^\mu_{qq'}(x) = \mathrm{i}(m_q + m_{q'})J^5_{qq'}(x), \quad J^5_{qq'}(x) =: \bar{q}(x)\gamma_5 q'(x) :, \qquad (3.6.1b)$$

so it will be asymptotically conserved in the limit of high energies, where masses can be neglected.

In general, the matrix elements of any composite operator are divergent. However, if we take into account the full QCD Lagrangian, i.e., including counterterms, then the conserved and quasi-conserved currents have finite matrix elements.[13] Physically this is obvious; a formal proof will be given later.

Nonconserved operators, on the other hand, generally require renormalization; that is, they have to be defined including a specific Z to make them finite. To see this, we begin with a simple example, the operator defined as $M(x) \equiv: \bar{q}(x)q(x):$, with sum over the implicit colour indices understood so that M is a colour singlet. As discussed in Sects. 3.2 and 3.3, we may either work with $\bar{q}q$ and calculate taking counterterms into account, or use (for amputated Green's functions) $Z_F \bar{q}q$, replace $g \to g_{uD} = Z_g g$, $m \to m_{uD} = Z_m m$ and neglect counterterms. This, however, is insufficient to make M finite: to obtain finite matrix elements we must still multiply by an extra Z_M, called the operator renormalization: we thus define

$$M_R(x) = Z_M M(x). \tag{3.6.2}$$

Let us evaluate Z_M. We use the formulas of Sect. 1.2: letting the subscript or superscript 0 mean free fields, $q_0 = q^0$, $B_0 = B^0$,

$$M_R(x) = Z_M \mathrm{T} : \bar{q}_0(x)q_0(x) : \exp \mathrm{i} \int \mathrm{d}^4 z \, \mathcal{L}_{\mathrm{int}}^0(z).$$

To lowest order in g, this is

$$\begin{aligned}
M_R(x) = {}& Z_M Z_F^{-1} : \bar{q}_0(x)q_0(x) : \\
& - \frac{g^2}{2!} Z_M \sum \int \mathrm{d}^4 z_1 \mathrm{d}^4 z_2 \, \mathrm{T}\Big\{ : \bar{q}_0(x)q_0(x) : : : \bar{q}_0(z_1)t^a \gamma_\mu q_0(z_1) : \\
& \times : \bar{q}_0(z_2)t^b \gamma_\nu q_0(z_2) : B_{0a}^\mu(z_1) B_{0b}^\nu(z_2) \Big\}.
\end{aligned} \tag{3.6.3}$$

Because we expect $Z_M = 1 + O(g^2)$, we can neglect the Z_M in the second term on the right hand side of (3.6.3). We next consider matrix elements between quark states with the same momentum, that we denote by $\langle M \rangle_p$, $\langle M_R \rangle_p$; it is not difficult to see that, in our case, the divergence is the same

---

[13]This is true to lowest order in weak and electromagnetic interactions. If we include higher orders in these, we would have to add weak and electromagnetic counterterms, Z_F^{Weak}, Z_F^{em}, etc.

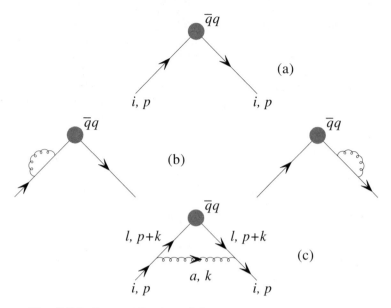

Fig. 3.6.1. Renormalization of the operator $\bar{q}q$.
(a) Bare vertex. (b) Radiative corrections to external legs.
(c) Vertex correction.

for diagonal or off–diagonal matrix elements. Then Eq. (3.6.3) gives, after simple manipulations and without amputation now,

$$\langle M_R \rangle_p = Z_M Z_F^{-1} \langle M_0 \rangle_p$$
$$+ \, \mathrm{i} \langle M_0 \rangle_p \left\{ g^2 C_F \int \mathrm{d}^D \hat{k} \, \frac{-\gamma^\mu (\not{p} + \not{k})(\not{p} + \not{k})\gamma_\mu}{k^2 (p+k)^4} + S_{uD}(p) + S_{uD}(p) \right\},$$

$$M_0 \equiv \; : \bar{q}_0 q_0 : \; .$$

$$(3.6.4)$$

The calculation has been performed in the Fermi–Feynman gauge, and neglecting the quark masses, which does not affect the divergence; the Feynman diagrams that intervene in the calculation are shown in Fig. 3.6.1. Clearly, the divergent part of one of the S_{uD} in (3.6.3) (see Fig. 3.6.1b) is exactly cancelled by that of Z_F; we thus need only the formula

$$-\mathrm{i} C_F g^2 \int \mathrm{d}^D \hat{k} \, \frac{\gamma^\mu \gamma_\mu}{k^2 (p+k)^2} \overset{\mathrm{div}}{=} \frac{4 g^2 C_F}{16\pi^2} \Gamma(\epsilon/2)(4\pi)^{\epsilon/2} \nu_0^\epsilon,$$

which, when added to the remaining S_{uD} gives us the desired result:

$$Z_M(\nu) = 1 - 3 C_F \frac{g^2}{16\pi^2} \left\{ \frac{2}{\epsilon} + \log 4\pi - \gamma_{\mathrm{E}} - \log \nu^2 / \nu_0^2 \right\}. \qquad (3.6.5)$$

We leave it as an exercise to show that Z_M is actually independent of the gauge, the reason being that M is a gauge invariant operator. The quantity γ_M, $Z_M = 1 - (g^2/16\pi^2)N_\epsilon\gamma_M + \ldots$ is called the anomalous dimension of the operator, to one loop (this name is also used for the whole of Z_M).

If we had carried out the calculation for $\bar{q}\gamma^\mu q'$ or $\bar{q}\gamma^\mu\gamma_5 q'$ we would have obtained zero for the anomalous dimension. As stated before, this is a special case of a general result, which we now prove. Let J^μ be a quasi-conserved operator, i.e., for zero masses $\partial_\mu J^\mu(x) = 0$. Consider any T-product with arbitrary fields Φ_i:

$$\mathrm{T}J^\mu(x)\Phi_1(y_1)\ldots\Phi_N(y_N).$$

Then, using $\partial_0\theta(x^0 - y^0) = \delta(x^0 - y^0)$, we find the identity

$$\partial_\mu \mathrm{T}J^\mu(x)\Phi_1(y_1)\ldots\Phi_N(y_N) = \mathrm{T}(\partial_\mu J^\mu(x))\Phi_1(y_1)\ldots\Phi_N(y_N)$$

$$+ \sum_{k=1}^{N} \delta(x^0 - y_k)\mathrm{T}\Phi_1(y_1)\ldots[J^0(x), \Phi_k(y_k)]\ldots\Phi_N(y_N). \tag{3.6.6}$$

Now let

$$\delta(x^0 - y_k^0)[J^0(x), \Phi_k(y_k)] = \Phi_k'(y_k)\delta(x - y).$$

If Z_J and Z_D are the renormalization constants for J^μ and $\partial \cdot J$, and γ_J and γ_D the anomalous dimensions (coefficients of $-(g^2/16\pi^2)N_\epsilon$) we find, from (3.6.6) and by differentiating $\nu d/d\nu$, that

$$\gamma_J\partial_\mu\mathrm{T}J^\mu(x)\Phi_1(y_1)\ldots\Phi_N(y_N)$$

$$= \mathrm{T}\left\{\sum \gamma_m m\frac{\partial}{\partial m}\partial \cdot J(x)\right\}\Phi_1(y_1)\ldots\Phi_N(y_N) \tag{3.6.7}$$

$$+ \gamma_D\mathrm{T}(\partial \cdot J(x))\Phi_1(y_1)\ldots\Phi_N(y_N).$$

This is possible only if $\gamma_J = 0$ and, moreover,

$$\gamma_D\partial \cdot J = -\sum \gamma_m m\frac{\partial}{\partial m}\partial \cdot J. \tag{3.6.8}$$

This relation may be verified for the case $J^\mu = \bar{q}\gamma^\mu q'$ with the help of our previous calculation, because

$$\partial \cdot J = \mathrm{i}(m' - m)\bar{q}q';$$

we may then use the γ_m (whose explicit value will be given in Sect. 3.7). Alternatively, one can take into account (3.3.17) and (3.4.6) to verify that, to second order,

$$m_{uD}(\bar{q}q)_{uD} = mZ_m(\bar{q}q)_{uD} = mZ_M(\bar{q}q)_{uD} = m(\bar{q}q)_R,$$

as indeed Z_m equals the Z_M we have just calculated.

3.7 The Running Coupling Constant
and the Running Mass in QCD: Asymptotic Freedom

Let us now turn to Eqs. (3.5.6) and (3.5.7). To solve (3.5.6), we assume that for some ν the renormalized coupling constant is sufficiently small that we can expand the functions β, γ_m, δ in power series in $g(\nu)$:

$$\beta = -\left\{ \beta_0 \frac{g^2(\nu)}{16\pi^2} + \beta_1 \left(\frac{g^2(\nu)}{16\pi^2} \right)^2 + \beta_2 \left(\frac{g^2(\nu)}{16\pi^2} \right)^3 + \cdots \right\},$$

$$\gamma_m = \gamma_m^{(0)} \frac{g^2(\nu)}{16\pi^2} + \gamma_m^{(1)} \left(\frac{g^2(\nu)}{16\pi^2} \right)^2 + \gamma_m^{(2)} \left(\frac{g^2(\nu)}{16\pi^2} \right)^3 + \cdots, \qquad (3.7.1)$$

$$\delta = \delta^{(0)} \frac{g^2(\nu)}{16\pi^2} + \delta^{(1)} \left(\frac{g^2(\nu)}{16\pi^2} \right)^2 + \delta^{(2)} \left(\frac{g^2(\nu)}{16\pi^2} \right)^3 + \cdots.$$

The value of β_0 can be read off from (3.3.30) and (3.5.4):[14]

$$\beta_0 = \tfrac{1}{3} \left\{ 11 C_A - 4 T_F n_f \right\} = 11 - \tfrac{2}{3} n_f. \qquad (3.7.2a)$$

Using the calculations of Z_g to second order (two loops) by Caswell (1974) and Jones (1974) we also have:[15]

$$\beta_1 = \tfrac{34}{3} C_A^2 - \tfrac{20}{3} C_A T_F n_f - 4 C_F T_F n_f = 102 - \tfrac{38}{3} n_f. \qquad (3.7.2b)$$

We then calculate the corresponding expressions for $\bar{g}$. We introduce the standard notation

$$\alpha_s(\nu^2) \equiv \frac{\bar{g}(\nu)^2}{4\pi},$$

and then Eqs. (3.5.6a) give immediately

$$\frac{d\bar{g}}{d\log\lambda} = -\beta_0 \frac{\bar{g}^3}{16\pi^2},$$

so that, defining $\lambda^2 = Q^2/\nu^2$,

$$\int_{\alpha_s(\nu^2)}^{\alpha_s(Q^2)} \frac{d\alpha_s}{\alpha_s^2} = -\frac{\beta_0}{2\pi} \int_0^{\frac{1}{2}\log Q^2/\nu^2} d\log\lambda'$$

---

[14] Gross and Wilczek (1973), Politzer (1973). Actually, β_0 had been calculated before by 't Hooft (unpublished) and Khriplovich (1969). See also Terentiev and Vanyashin (1965).

[15] β_0, β_1 are scheme independent; the value for β_2 given here is that in the $\overline{\text{MS}}$ scheme. The coefficients β_2, β_3 have also been calculated; the first by Tarasov, Vladimirov and Zharkov (1980) and the second recently by Larin, van Ritbergen and Vermaseren (1997a); see Sect. 10.1 in our text for details.

with the solution

$$\alpha_s(Q^2) = \frac{\alpha_s(\nu^2)}{1 + \alpha_s(\nu^2)\beta_0(\log Q^2/\nu^2)/4\pi}. \tag{3.7.3}$$

It is customary to re–express this in terms of an invariant mass parameter, denoted by Λ, so that (3.7.3) becomes

$$\alpha_s(Q^2) = \frac{4\pi}{\beta_0 \log(Q^2/\Lambda^2)}, \quad \Lambda^2 = \nu^2 e^{-4\pi/\beta_0\alpha_s(\nu^2)}. \tag{3.7.4a}$$

With the explicit value of β_0,

$$\alpha_s(Q^2) = \frac{12\pi}{(33 - 2n_f) \log(Q^2/\Lambda^2)}. \tag{3.7.4b}$$

The renormalization group equation may be solved by iteration to any arbitrary order. For example, to two loops,

$$\alpha_s(Q^2) = \frac{12\pi}{(33 - 2n_f) \log Q^2/\Lambda^2} \left\{ 1 - 3\frac{153 - 19n_f}{(33 - 2n_f)^2} \frac{2 \log\log Q^2/\Lambda^2}{\log Q^2/\Lambda^2} \right\}, \tag{3.7.4c}$$

and to three loops we find,

$$\alpha_s(Q^2) = \frac{4\pi}{\beta_0 L} \left\{ 1 - \frac{\beta_1 \log L}{\beta_0^2 L} + \frac{\beta_1^2 \log^2 L - \beta_1^2 \log L + \beta_2\beta_0 - \beta_1^2}{\beta_0^4 L^2} \right\} \tag{3.7.4d}$$

where

$$L = \log \frac{Q^2}{\Lambda^2}, \quad \beta_2 = \tfrac{2857}{2} - \tfrac{5033}{18}n_f + \tfrac{325}{54}n_f^2.$$

Denoting temporarily by $\alpha_s^{(n)}$ to α_s calculated to n-loop accuracy (so that $\alpha_s^{(1)}$ is given by (3.7.4a) and $\alpha_s^{(3)}$ by (3.7.4d)), one can easily verify that, as $Q^2 \to \infty$,

$$\alpha_s^{(k)}(Q^2)/\alpha_s^{(n)}(Q^2) \to 1, \quad \alpha_s^{(n)}(Q^2) \to 0$$

for all $n, k \geq 1$, provided that $n_f < 16$, a bound comfortably satisfied, as there seem to exist only six flavours of quark. The vanishing of $\alpha_s(Q^2)$ for large momenta, Q^2, is the celebrated property of *asymptotic freedom*, first discussed by Gross and Wilczek (1973a) and Politzer (1973). Recalling (3.5.7) this means that, at large spacelike momenta $\lambda p_i \sim q$, $q^2 = -Q^2$, $Q^2 \to \infty$, the theory will behave as a free field theory, modulo logarithmic corrections. Moreover, and because $\alpha_s(Q^2)$ vanishes for large Q^2, it follows that for sufficiently large momenta we will be able to calculate the corrections to free-field theory in a power series in $\alpha_s(Q^2)$.

The running mass is also easily calculated. To lowest order, we require (3.5.2), (3.5.6) and (3.3.17). We then have

$$\frac{d\bar{m}}{m d \log \lambda} = \gamma_m^{(0)} \frac{\bar{g}^2}{16\pi^2} = \frac{\gamma_m^{(0)}}{2\beta_0 \log \lambda}.$$

Using (3.7.4a) with $\log Q^2/\Lambda^2 = 2 \log \lambda$, and introducing the integration constant $\hat{m}$ (which is the mass analogue of Λ), this gives,

$$\bar{m}(Q^2) = \hat{m} \left(\tfrac{1}{2} \log \frac{Q^2}{\Lambda^2} \right)^{\gamma_m^{(0)}/\beta_0}, \qquad \gamma_m^{(0)} = -3C_F, \tag{3.7.5a}$$

which we also write as

$$\bar{m}(Q^2) = \hat{m} \left(\tfrac{1}{2} \log \frac{Q^2}{\Lambda^2} \right)^{-d_m}, \qquad d_m = \frac{12}{33 - 2n_f}. \tag{3.7.5b}$$

d_m (or $\gamma_m^{(0)}$) are called at times the *anomalous dimension of the mass*. One checks that, as $Q^2 \to \infty$, also $\bar{m}(Q^2)$ vanishes logarithmically. The higher order coefficients $\gamma_m^{(n)}$ are known to four loops; they may be found in Sect. 10.1. To two loops,

$$m(t) = \hat{m} \left(\tfrac{1}{2} \log t/\Lambda^2 \right)^{-d_m} \left[1 - d_1 \frac{\log \log t/\Lambda^2}{\log t/\Lambda^2} + d_2 \frac{1}{\log t/\Lambda^2} \right];$$

$$d_m = \frac{4}{\beta_0}, \quad d_1 = 8 \frac{51 - \tfrac{19}{3} n_f}{\beta_0^3}, \quad d_2 = \frac{8}{\beta_0^3} \left[\left(\tfrac{101}{12} - \tfrac{5}{18} n_f \right) \beta_0 - 51 + \tfrac{19}{3} n_f \right],$$

$$\beta_0 = 11 - \tfrac{2}{3} n_f. \tag{3.7.5c}$$

The running gauge parameter can be similarly calculated; the details may be found in Narison (1989). One finds, to two loops,

$$\bar{\xi}(Q^2) = 1 - \frac{1}{\hat{\lambda}(\tfrac{1}{2} \log Q^2/\Lambda^2)^{d_\xi}} \left\{ 1 + \frac{9}{(39 - 4n_f)\hat{\lambda}(\log Q^2/\Lambda^2)^{d_\xi}} \right\}^{-1},$$

$$d_\xi = \frac{39 - 4n_f}{66 - 4n_f}. \tag{3.7.6}$$

As a first example of the techniques developed, we present an evaluation of the behaviour of the renormalized, nonamputated quark propagator, at large momentum:

$$S_R(p, g(\nu), m(\nu), \xi(\nu); \nu), \quad p^2 = -Q^2, \ Q^2 \gg \Lambda^2.$$

The naive dimension of S_R is $\rho_S = -1$. Hence, Eq. (3.5.7) gives, with $p = \lambda n$, $n^2 = -\Lambda^2$, and noting that $Z = Z_F$ now,

$$S_R(p, g(\nu), m(\nu), \xi(\nu); \nu) = S_R(n, \bar{g}(\lambda), \bar{m}(\lambda), \bar{\xi}(\lambda); \nu) \left(\frac{Q^2}{\Lambda^2}\right)^{-1/2}$$

$$\times \exp\left\{-\frac{1-\xi}{3\pi} \int_0^{\log Q/\Lambda} d\log\lambda' \, \alpha_s(\lambda'^2)\right\}.$$

To zero order in g,

$$S_R(n, \bar{g}(\lambda), \bar{m}(\lambda), \bar{\xi}(\lambda); \nu) \underset{Q^2 \to \infty}{\simeq} \frac{i}{\not{n}},$$

so, using (3.7.4a),

$$S_R(p, g(\nu), m(\nu), \xi(\nu); \nu) \underset{Q^2 \to \infty}{\simeq} \frac{i}{\not{p}} \frac{1}{(\frac{1}{2}\log Q^2/\Lambda^2)^{d_{F\xi}}}, \tag{3.7.7a}$$

and the anomalous dimension of the propagator is

$$d_{F\xi} = \frac{3(1-\xi)C_F}{22C_A - 8T_F n_f}. \tag{3.7.7b}$$

S_R behaves, at large momenta, as a free propagator, except for the logarithmic correction $(\log Q^2/\Lambda^2)^{-d_{F\xi}}$. Note that $d_{F\xi}$ depends on the gauge, as expected, and it vanishes in the Landau gauge $\xi = 1$, where S_R thus has canonical dimension.

3.8 Heavy and Light Quarks: the Decoupling Theorem. Effective n_f, Λ

The singularities in the $\overline{\text{MS}}$ scheme are independent of quark masses; therefore, when calculating the β_n or the $\gamma_m^{(n)}$ one has to take into account all existing quark flavours. For simplicity, let us concentrate on the β function and work in an axial gauge so that the entire Q^2 evolution may be obtained with just the gluon propagator. Furthermore, we will simplify the discussion by considering a model with only two quark flavours: one essentially massless, $\hat{m}_l = 0$, and a heavy one, $\hat{m}_h \gg \Lambda$. In the $\overline{\text{MS}}$ scheme to one loop, we have

$$\alpha_s(Q^2) = \frac{12\pi}{(33 - 2n_f)\log Q^2/\Lambda^2}, \tag{3.8.1}$$

and we have to take $n_f = 2$. However, it stands to reason that, while $n_f = 2$ will be a good choice for $Q \gg \hat{m}_h$, there will be a region $\hat{m}_h \gg Q \gg \Lambda$ where one would expect that (3.8.1) with $n_f = 1$ would be a more reasonable choice. This is made more dramatic if we set $\hat{m}_h$ extremely large, for instance,

1 gram. Clearly, GeV–TeV physics can hardly depend on the existence or nonexistence of that particle.

This is basically the content of a theorem proved by Symanzik (1973) and rediscovered[16] and elaborated by Appelquist and Carrazzone (1975) which states that in the particular case of QCD, we can neglect the existence of the heavy quark when $\hat{m}_h \gg Q$; and this neglect only produces errors of order m_h^2/Q^2. In fact, Eq. (3.8.1) is valid as it stands only if $Q^2 \gg m^2$, where m is any relevant mass, in particular $\hat{m}_h$. In intermediate regions we will have a dependence on Q^2 besides that in $\log Q^2/\Lambda^2$.

Since the problem arises because we neglected masses, we must re-derive (3.8.1), but now with masses taken into account; we simplify even further by taking only one quark flavour. We recall that the running coupling constant was defined as $\alpha_s = \bar{g}^2/4\pi$, with $\bar{g}$ a solution of (3.5.6a),

$$\frac{\mathrm{d}\bar{g}}{\mathrm{d}\log Q/\nu} = \bar{g}\beta(\bar{g}), \quad \bar{g}|_{Q=\nu} = g(\nu), \tag{3.8.2a}$$

with

$$\beta = -Z_g^{-1}\frac{\nu\mathrm{d}}{\mathrm{d}\nu}Z_g. \tag{3.8.2b}$$

We now consider the behaviour of the transverse part of the gluon propagator, which we write as in Eq. (2.4.9). From (3.5.1), (3.5.7), and with $Q/\nu = \lambda$,

$$D_{\mathrm{tr}}(q^2; g(\nu), m(\nu); \nu^2)$$
$$= D_{\mathrm{tr}}(\nu^2; \bar{g}(\lambda), \bar{m}(\lambda); \nu^2)\exp\left\{-\int_0^{\log\lambda}\mathrm{d}\log\lambda'\,\gamma_D\left(\bar{g}(\lambda')\right)\right\}. \tag{3.8.3}$$

In the physical gauge we are using, $\gamma_D = 2\beta_0 g^2/16\pi^2$; see Eq. (3.3.18). Hence,

$$D_{\mathrm{tr}}(q^2; g(\nu), m(\nu); \nu^2) = \frac{2}{\log Q^2/\nu^2}D_{\mathrm{tr}}(\nu^2; \bar{g}(\lambda), \bar{m}(\lambda); \nu^2). \tag{3.8.4}$$

Next we require $D_{\mathrm{tr}}(\nu^2; \bar{g}(\lambda), \bar{m}(\lambda); \nu^2)$ exactly in $\bar{m}$. We have

$$D_{\mathrm{tr}}(\nu^2; \bar{g}(\lambda), \bar{m}(\lambda); \nu^2) =$$
$$K_\nu + \frac{2T_F\alpha_s(Q^2)}{\pi}\int_0^1\mathrm{d}x\,x(1-x)\log\frac{\bar{m}^2 + x(1-x)\nu^2}{\nu^2},$$

---

[16]Actually, the result is essentially contained (for Abelian theories) in the basic paper of Kinoshita (1962). For a discussion using functional methods, see Weinberg (1980).

where K_ν is a constant. To begin with, we choose $\nu \sim \Lambda$: then,

$$
D_{\rm tr}(q^2; g(\nu), m(\nu); \nu^2) = \frac{2}{\log Q^2/\nu^2} \left\{ K_\Lambda \right.
$$
$$
\left. + \frac{2T_F \alpha_s(Q^2)}{\pi} \int_0^1 dx\, x(1-x) \log\left[\frac{\bar{m}^2(Q^2)}{\Lambda^2} + x(1-x) \right] \right\}.
$$
$$
(3.8.5)
$$

We will assume $m \gg \Lambda$. This then becomes

$$
D_{\rm tr}(q^2; g(\nu), m(\nu); \nu^2) \approx \left\{ K_\Lambda + \frac{T_F \alpha_s(Q^2)}{\pi} \log \frac{\bar{m}(Q^2)}{\Lambda^2} \right\} \frac{2}{\log Q^2/\nu^2}. \quad (3.8.6)
$$

It is clear that, in the momentum region where $m^2 \gg Q^2 \gg \Lambda^2$, the correction to K_Λ in (3.8.6) is large: although nominally of higher order, it will dominate in the limit of very large quark mass. The same will of course be the case for higher order corrections so the approach is not very useful here. This was to be expected. The $\overline{\rm MS}$ scheme, or any other mass independent scheme (like that of Weinberg, 1973b) must necessarily destroy convergence when there is a mass much larger than the momentum scale.

A simple solution to this problem is to modify the $\overline{\rm MS}$ scheme taking an *effective* number of flavours; for example,

$$
n_f(Q^2) = \sum_{f=1}^{n_f} \theta(Q^2 - m_f^2), \quad (3.8.7)
$$

i.e., we consider particles with masses much larger than the momentum to be effectively decoupled.[17]

We now show that this procedure is consistent; we will do this for the transverse part of the gluon propagator, which will indicate how to extend it to the general case. That (3.8.7) is right for Q^2 much larger than all quark masses we already know; the corrections are easily checked to be $O(\hat{m}^2/Q^2)$. Then we consider $Q^2 \ll m^2$. Because the contributions of quarks and gluons to $D_{\rm tr}$ are additive, we need only consider the former. To one loop, then,

$$
D_{\rm tr}^{\rm (quark)} = 1 - \frac{g^2}{4\pi^2} \int_0^1 dx\, x(1-x) \log \frac{x(1-x)Q^2 + \hat{m}^2}{\nu^2}, \quad (3.8.8)
$$

---

[17]Other interpolation formulas are possible; cf. Weinberg (1973b) or the papers of Coquereaux (1980, 1981). It should be remarked that a simple choice such as (3.8.7) can only be an approximation and that, particularly when there are several momentum scales involved, there is no substitute for a detailed calculation taking into account all masses which do not decouple.

and to this order we need not consider the renormalization of g or m. Now, for $Q^2 \ll \hat{m}^2$, we obtain

$$D_{\text{tr}}^{(\text{quark})} \simeq 1 - \frac{\alpha_s}{6\pi} \log \frac{\hat{m}^2}{\nu^2} - \frac{\alpha_s}{30\pi} \frac{Q^2}{\hat{m}^2}, \tag{3.8.9}$$

i.e., constant up to terms $O(Q^2/\hat{m}^2)$. Therefore, it coincides (up to these corrections) with the gluon propagator calculated with *zero* flavours, but for a different value of ν^2; namely, $\nu^2 \to \nu^2\{1 + \log \hat{m}^2/\nu^2\}$. Because physical observables are independent of ν, it follows that we can drop this heavy quark, which only contributes corrections $O(Q^2/\hat{m}^2)$.

The case of the gluon propagator is simple; in general, the corrections can include logarithms and so they are of the type

$$O\left(\frac{Q^{2j}}{\hat{m}^{2j}} \log^k(Q^2/\hat{m}^2)\right).$$

The decoupling theorem is particularly transparent in the μ scheme of renormalization. Consider again the quark contribution to the gluon propagator. We work to second order and then, recalling (3.3.21),

$$D_{uD,\text{tr}}^{(\text{quark})}(q^2) = \frac{i}{q^2}$$

$$+ \frac{T_F g^2}{16\pi^2} \left\{ \frac{2N_\epsilon n_f}{3} - 4 \int_0^1 dx\, x(1-x) \sum_{f=1}^{n_f} \log \frac{m_f^2 - x(1-x)q^2}{\nu_0^2} \right\} + \cdots.$$

It will be remembered that in the μ scheme we renormalize by requiring the equality $D_{R,\text{tr}}^{(\text{quark})}(q^2 = -\mu^2) = D_{\text{free,tr}}^{(\text{quark})}(-\mu^2)$. Therefore,

$$D_{R,\text{tr}}^{(\text{quark})}(q^2) = \frac{i}{q^2} + \frac{T_F g^2}{16\pi^2} \left\{ -4 \int_0^1 dx\, x(1-x) \sum_f \log \frac{m_f^2 - x(1-x)q^2}{m_f^2 + x(1-x)\mu^2} \right\}.$$

Take $Q^2 = -q^2$. For $Q^2, \mu^2 \gg m_f^2$,

$$\int_0^1 dx\, x(1-x) \log \frac{m_f^2 + x(1-x)Q^2}{m_f^2 + x(1-x)\mu^2} \simeq \tfrac{1}{6} \log \frac{Q^2}{\mu^2} + O\left(\frac{m_f^2}{\mu^2}, \frac{m_f^2}{Q^2}\right);$$

for $m_f^2 \gg Q^2, \mu^2$,

$$\int_0^1 dx\, x(1-x) \log \frac{m_f^2 + x(1-x)Q^2}{m_f^2 + x(1-x)\mu^2} \simeq O\left(\frac{\mu^2}{m_f^2}, \frac{Q^2}{m_f^2}\right),$$

from which expressions the decoupling theorem is apparent.

Let us now return to the $\overline{\text{MS}}$ scheme, with a momentum-dependent effective n_f. When using this modified version of the $\overline{\text{MS}}$ scheme, some care has to be exercised because the QCD parameters will become *effective* parameters, and they will vary across thresholds. To see what this means, take for example the expression (3.7.4b) for the running coupling constant. When crossing e.g. the $\bar{c}c$ quark threshold we will have two expressions: one for $Q \gg m_c^2$, and another for $Q^2 \ll m_c^2$. In the first, $n_f = 4$; in the second, $n_f = 3$. Because we want to have the same coupling constant, we will have to admit that Λ really depends on n_f. Writing $\Lambda(n_f)$ we thus have (Marciano, 1984)

$$\alpha_s(Q^2) = \frac{12\pi}{27 \log Q^2/\Lambda^2(3)}, \quad Q^2 \ll m_c^2,$$

$$\alpha_s(Q^2) = \frac{12\pi}{25 \log Q^2/\Lambda^2(4)}, \quad Q^2 \gg m_c^2.$$

Matching these two expressions we find the relation between the values of Λ; for example, matching at m_c^2 gives

$$\Lambda(4) = [\Lambda(3)/m_c]^{2/25} \, \Lambda(3). \tag{3.8.10a}$$

The dependence on n_f is weak, but noticeable. A more accurate result is obtained using higher order formulas for α_s. For example, to two loops a simple calculation gives

$$\Lambda(4) = \Lambda(5) \left(\frac{m_b}{\Lambda(5)}\right)^{2/25} \left(\log \frac{m_b^2}{\Lambda^2(5)}\right)^{963/14375},$$

$$\Lambda(4) = \Lambda(3) \left(\frac{\Lambda(3)}{m_c}\right)^{2/25} \left(\log \frac{m_b^2}{\Lambda^2(3)}\right)^{-107/1875}, \tag{3.8.10b}$$

etc.

The running masses also run differently at different momenta. For example, to leading order, Eq. (3.7.5b) should be replaced by

$$\bar{m}(Q^2) = \frac{\hat{m}(3)}{[\frac{1}{2} \log Q^2/\Lambda^2(3)]^{d_m(3)}}, \quad d_m(n_f) = \frac{12}{33 - 2n_f}, \tag{3.8.11a}$$

for $Q^2 \ll m_c^2$, and

$$\bar{m}(Q^2) = \frac{\hat{m}(4)}{[\frac{1}{2} \log Q^2/\Lambda^2(4)]^{d_m(4)}},$$

$$\frac{\hat{m}(4)}{\hat{m}(3)} = \frac{[\log m_c/\Lambda(4)]^{d_m(4)}}{[\log m_c/\Lambda(3)]^{d_m(3)}}, \tag{3.8.11b}$$

when $m_c^2 \ll Q^2 \ll m_b^2$, and so forth.

3.9 The Operator Product Expansion (OPE) at Short Distances. Nonperturbative Effects in Quark and Gluon Propagators

i Short Distance Expansion

The tool to analyze rigorously the product of operators at short or lighlike distances is the *operator product expansion* (OPE).[18] To discuss it, we begin with free fields and the simplest possible case of a time ordered product of scalar fields:

$$T\phi(x)\phi(y).$$

As $x \to y$ this is singular; but this singularity is a c-number. We may separate it and write

$$T\phi(x)\phi(y) = \underline{\phi(x)\phi(y)}+ : \phi(x)\phi(y) := \Delta(x-y)\,1+ : \phi(x)\phi(y) :$$

where the contraction is a c-number, coinciding with the propagator:

$$\Delta(x) = \frac{i}{(2\pi)^2} \int d^4k\, e^{-ik\cdot x} \frac{1}{k^2 + i0} = -\frac{1}{(2\pi^2)} \frac{1}{x^2 - i0}.$$

The operator $: \phi(x)\phi(y) :$ is regular as $x \to y$, and so is of course the unit operator. In general, we can write the product of the local (elementary or composite) operators A, B as the *short distance*, or *Wilson expansion*

$$TA(x)B(y) = \sum_t C_t(x-y)N_t(x,y), \tag{3.9.1}$$

where the C_t are c-numbers, and the $N_t(x,y)$ are bilocal operators, regular as $x \to y$; the use of the letter N for them is a reminder that they will be normal-ordered composite operators. The expansion (3.9.1) is simply a generalization of the free-field case. We write

$$TA(x)B(y) = \sum \frac{i^n}{n!} \int dz_1 \dots dz_n\, TA^0(x)B^0(y)\mathcal{L}^0_{\text{int}}(z_1) \dots \mathcal{L}^0_{\text{int}}(z_n),$$

where the superscript 0 means that the fields have to be taken as free. Systematic application of Wick's theorem then produces (3.9.1). However, it is seldom necessary to write the above expression in complete generality; if we are

---

[18]The operator product expansion was first introduced by Wilson (1969) and further developed (for short distances) by Zimmermann (1970), Wilson and Zimmermann (1972) and others. For the light cone expansions, see Brandt and Preparata (1971), Fritzsch and Gell-Mann (1971). The general use of this tool for deep inelastic scattering was developed by Christ, Hasslacher and Muller (1972); its application to QCD was discussed first by Gross and Wilczek (1973b, 1974) and Georgi and Politzer (1974). We consider here expansions of T-products in view of applications, but the same techniques may be used to expand ordinary products.

interested in the behaviour as $x \to y$ only, there is a simpler way to proceed. One considers a basis formed by all the operators with the same quantum numbers and transformation properties as the product AB. In particular, if A and B are scalars, and gauge invariant, only scalar, gauge-invariant operators have to be considered. In this case, we have the operators (sum over omitted colour indices understood)

$$1, \; : \bar{q}(x)q(y) :, \; \bar{q}\not{D}q(y) :, \; \ldots, \; : (\bar{q}(x)q(y))^2 :, \; \ldots, \; : G(x)G(y) :, \; \ldots, \tag{3.9.2}$$

in fact, an infinite array: but, in the limit $x \to y$, only a few (at times only one for the leading behaviour) are required. This may be seen as follows. Let ρ_N be the naive dimension of the operator N. The lowest dimensional operators in (3.9.2) are 1, with $\rho_1 = 0$, $: \bar{q}q :$ with $\rho_{:\bar{q}q:} = 3$, $: \bar{q}\not{D}q :$ with $\rho_{:\bar{q}\not{D}q:} = 4$ and $: GG :$ with $\rho_{:GG:} = 4$. If we suppose that the dimension of each A, B is 3, simple counting tells us that the Wilson coefficient C_1 has dimension 6, $C_{:\bar{q}q:}$ has dimension 3, and both $C_{:\bar{q}\not{D}q:}$, $C_{:GG:}$ have dimension 4. Therefore, extracting explicitly a mass from $C_{:\bar{q}q:}$,

$$\begin{aligned}
C_1(x-y) &\underset{x \to y}{\approx} (x-y)^{-6}, \quad C_{:\bar{q}q:}(x-y) \underset{x \to y}{\approx} m(x-y)^{-2}, \\
C_{:\bar{q}\not{D}q:} &\underset{x \to y}{\approx} (x-y)^{-2}, \quad C_{:GG:} \underset{x \to y}{\approx} (x-y)^{-2},
\end{aligned} \tag{3.9.3}$$

where x^6 means $(x \cdot x)^3$, x^{-2} means $1/(x \cdot x)$, etc. The coefficients accompanying other, higher dimensional operators will be finite as $x \to y$, so (3.9.3) exhausts the list of singular coefficients.

Clearly, (3.9.3) will only be exactly true in free field theory; however in QCD, and because of asymptotic freedom, high energies and thus short distances behaviour only get *logarithmic* corrections, which do not substantially alter the analysis. If we now take any matrix element of the expansion (3.9.1), with N_t the operators in (3.9.2),

$$\begin{aligned}
\langle \Phi | TA(x)B(0) | \Psi \rangle \underset{x \to 0}{\simeq} & \; C_1(x)\langle \Phi | \Psi \rangle \\
& + C_{:\bar{q}q:}(x)\langle \Phi | : \bar{q}(0)q(0) : | \Psi \rangle \\
& + C_{:\bar{q}\not{D}q:}(x)\langle \Phi | : \bar{q}(0)\not{D}q(0) : | \Psi \rangle \\
& + C_{:GG:}(x)\langle \Phi | : G^2(0) : | \Psi \rangle + \cdots,
\end{aligned} \tag{3.9.4}$$

then, because the normal operators are regular, we find that the $x \to 0$ behaviour of the left hand side in (3.9.4) is given by that of the Wilson coefficients up to the finite constants $\langle \Phi | N_t | \Psi \rangle$. Thus, the leading behaviour as $x \to 0$ of $TA(x)B(0)$ is given by $C_1(x)$, and the subleading one by $C_{:\bar{q}q:}$, $C_{:\bar{q}\not{D}q:}$ and $C_{:GG:}$.

Let us return to (3.9.1). Since the operators $N_t(x,y)$ are regular, we can expand them in $x - y$. With $y = 0$, we then write

$$N_t(x,0) = \sum_n x_{\mu_1} \ldots x_{\mu_n} N_t^{(n)\mu_1,\ldots,\mu_n}(0,0).$$

For example,

$$: \bar{q}(0)q(-x) := \sum_n x_{\mu_1} \dots x_{\mu_n} \frac{(-1)^n}{n!} : \bar{q}(0)\partial^{\mu_1} \dots \partial^{\mu_n} q(0) : . \qquad (3.9.5)$$

In a gauge theory such as QCD, we should replace the derivatives in (3.9.5) by covariant derivatives (see Appendix I). So we finally obtain the expansion

$$T A(x)B(0) \underset{x \to 0}{\simeq} C_1(x) \, 1$$
$$+ C_{:\bar{q}q:}(x) \sum_n x_{\mu_1} \dots x_{\mu_n} \frac{(-1)^n}{n!} : \bar{q}(0)D^{\mu_1} \dots D^{\mu_n} q(0) : + \cdots . \qquad (3.9.6)$$

As $x \to 0$, the derivatives in (3.9.6) are (in general) subleading because of the extra x_μ factors. This is not, however, the case in the *light–cone expansion*. In this expansion we are interested in the behaviour of operator products as $x^2 \to 0$ for otherwise arbitrary x. Because of this, in the light–cone limit, all derivatives on the right hand side of (3.9.6) contribute equally.

The details of the light–cone expansion are deferred to Sects. 4.4 ff. We now pass to an application of the short distance expansion.

ii Nonperturbative Effects in Quark and Gluon Propagators

To all orders in perturbation theory, the vacuum expectation values of the operators $: \bar{q}(0)q(0) :$ and

$$\alpha_s : G^2(0) := \frac{g^2}{4\pi} \sum_a : G_a^{\mu\nu}(0)G_{a\mu\nu}(0) : \qquad (3.9.7)$$

vanish. However, later in this book we will give arguments that indicate that, in the *physical* (nonperturbative) vacuum, that we denote by $|\mathrm{vac}\rangle$, both the *quark condensate* and the *gluon condensate*,

$$\langle \bar{q}q \rangle \equiv \langle \bar{q}q \rangle_\mathrm{vac} \equiv \langle \mathrm{vac}| : \bar{q}(0)q(0) : |\mathrm{vac}\rangle \qquad (3.9.8)$$

and

$$\langle \alpha_s G^2 \rangle \equiv \langle \alpha_s G^2 \rangle_\mathrm{vac} \equiv \langle \mathrm{vac}|\alpha_s : G^2(0) : |\mathrm{vac}\rangle, \qquad (3.9.9)$$

are nonzero.[19] This will induce effects in all Green's functions, in particular in the quark and gluon propagator, that we now study.

We begin with the quark propagator,

$$S_\xi^{ij}(p) = \int \mathrm{d}^D x \, \mathrm{e}^{\mathrm{i}p \cdot x} \langle T q^i(x)\bar{q}^j(0) \rangle_\mathrm{vac}, \qquad (3.9.10)$$

---

[19]Sum over omitted colour indices is understood in (3.9.8) and (3.9.9).

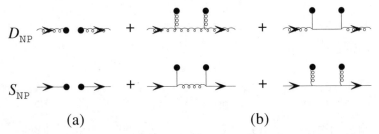

Fig. 3.9.1. First nonperturbative contributions to the quark and gluon propagators. The black blobs represent the vacuum expectation values of the operators associated to the lines that terminate in the blobs.
(a) Disconnected diagrams. (b) Connected, higher order corrections.

which will be evaluated for large p. We write an OPE for it, neglecting terms that give zero when sandwiched with $|\text{vac}\rangle$:

$$T q^i(x) \bar{q}^j(0) = \delta_{ij} \left\{ C_1(x) \mathbf{1} - C_{\bar{q}q}(x) \sum_l : \bar{q}^l(0) q^l(x) : + \cdots \right\}; \qquad (3.9.11)$$

only the perturbative coefficient, C_1, was considered in Sects. 3.1 and 3.3. To zero order in g,

$$\sum_l : \bar{q}^l_\beta(0) q^l_\alpha(x) : \underset{x \to 0}{\approx} \frac{1}{4} \left\{ \delta_{\alpha\beta} - \frac{imx_\mu}{D} \gamma^\mu_{\alpha\beta} \right\} \sum_l : \bar{q}^l(0) q^l(0) :;$$

α, β are Dirac indices, and we still neglect terms that vanish between vacuum states.

If we denote by S_P, S_{NP} the perturbative, nonperturbative contributions of (3.9.11) to (3.9.10) (the last represented graphically in Fig. 3.9.1), we find

$$S = S_P + S_{NP},$$

$$S^{(0)ij}_{NP}(p) \approx -(2\pi)^D \frac{\delta_{ij} \langle \bar{q} q \rangle_{\text{vac}}}{4N_c} \left\{ 1 - \frac{m_q}{D} \gamma^\mu \frac{\partial}{\partial p^\mu} \right\} \delta(p), \qquad N_c = 3. \qquad (3.9.12a)$$

This expression vanishes for $p \neq 0$, and has a singularity at the origin. Both effects, as well as similar ones for the gluon case, Eq. (3.9.15) below, are due to neglect of confinement. A regulated version of (3.9.12a) requires a fuller treatment of this, that will be sketched in Sect. 9.5iv; but, even as they stand, we will see that terms like that in (3.9.12a) are important in connection with a number of physical quantities.

The expression (3.9.12a) can be generalized to all orders in the quark mass, m_q, but still for a free propagator, to read

$$\langle \text{vac}| : \bar{q}_i(x)\bar{q}_j(0) : |\text{vac}\rangle = \frac{-(2\pi)^D \Gamma(D/2)\langle \bar{q}(0)q(0)\rangle \delta_{ij}}{4N_c m_q}$$

$$\times \sum_{n=0}^{\infty} \frac{1}{n!\Gamma(n+D/2)} \int d^D \hat{p} \, e^{-ip\cdot x}(\slashed{p}+m_q)\left(\frac{m_q^2}{4}\partial_p^2\right)^n \delta_D(p). \tag{3.9.12b}$$

The second order correction to S_{NP} is evaluated by writing

$$S_{\text{NP}}^{(2)ij}(p) = \sum \frac{1}{\slashed{p}-m_q}$$

$$\times g^2 \int d^D \hat{k} \, i\gamma_\mu t_{il}^a S^{ll'}(p+k) i\gamma_\nu t_{l'j}^b \delta_{ab} \frac{-g^{\mu\nu}+\xi k^\mu k^\nu/k^2}{k^2} \frac{i}{\slashed{p}-m_q},$$

and replacing, on the right hand side, $S^{ll'}$ by $S_{\text{NP}}^{(0)ll'}$. We thus find

$$S_{\text{NP}} = S_{\text{NP}}^{(0)} + S_{\text{NP}}^{(2)} + \cdots ,$$

$$S_{\xi\text{NP}}^{(2)}(p) = -i\delta_{ij}\frac{g^2 C_F \langle \bar{q}q\rangle}{12p^4}\left\{D - \xi - \frac{2(D-2)}{D}(1-\xi)\frac{m_q \slashed{p}}{p^2}\right\} \tag{3.9.13}$$

$$+ \text{ higher orders in } m_q^2/p^2.$$

Note that this is gauge dependent, so one cannot interpret in a direct way the quantity

$$M_\xi(p) = \frac{-\pi\alpha_s C_F \langle \bar{q}q\rangle}{3p^2}(4-\xi)$$

as a physical mass, although it has some mass–like properties, such as breaking chiral invariance.

A similar calculation may be performed for the gluon propagator (see Fig. 3.9.1). We write

$$D_{\xi ab}^{\mu\nu}(k) = \int d^4 x \, e^{ik\cdot x} \langle TB_a^\mu(x)B_b^\nu(0)\rangle_{\text{vac}},$$

$$TB_a^\mu(x)B_b^\nu(0) = \delta_{ab}\left\{C_1^{\mu\nu}(x)\mathbf{1} + C_{G2}^{\mu\nu}(x) : G^{\alpha\beta}(0)G_{\alpha\beta}(0) : + \cdots\right\}, \tag{3.9.14}$$

obtaining

$$D = D_{\text{P}} + D_{\text{NP}},$$

where

$$D_{\text{NP}ab}^{(0)\mu\nu}(k) = (2\pi)^D \frac{\delta_{ab}\langle G^2\rangle}{4(N_c^2-1)D(D-1)(D+2)}\left\{(D+1)g^{\mu\nu}\Box - 2\partial^\mu\partial^\nu\right\}\delta(k). \tag{3.9.15}$$

It is to be noted that $D_{\mathrm{NP}}^{(0)\mu\nu}$ is transverse, $k_\mu D_{\mathrm{NP}}^{(0)\mu\nu}(k) = 0$. $D_{\mathrm{NP}}^{(0)\mu\nu}$ also contributes to $S_{\mathrm{NP}}^{(2)}$; it adds to (3.9.13) the expression

$$S_{G^2;\mathrm{NP}}^{(2)}(p) = \frac{2C_F}{3(N_c^2 - 1)} \frac{\pi\langle\alpha_s G^2\rangle}{p^4} \frac{i}{\not{p}}. \tag{3.9.16}$$

It is also not difficult to evaluate the (second order) contributions of $\langle\bar{q}q\rangle$, $\langle G^2\rangle$ to D. These would give mass–like terms, unfortunately gauge dependent. In fact, as we will see later in this text, the masses of physical particles are *not* directly related to quantities such as M_ξ or the equivalent for gluons; these contribute only at a nonleading order. The main contributions come from the terms in Eqs. (3.9.12), (3.9.15). A detailed discussion of this in connection with $\langle\bar{q}q\rangle$ may be found in Pascual and de Rafael (1982).

4 Perturbative QCD
I. Deep Inelastic Processes

> *"Is there any point to which you would wish to draw my attention?"*
> *"To the curious incident of the dog in the night-time"*
> *"The dog did nothing in the night-time!"*
> *"That was the curious incident,"* remarked Sherlock Holmes.
>
> ARTHUR CONAN DOYLE, 1892

4.1 e^+e^- Annihilation into Hadrons

The Lagrangian for strong and electromagnetic interactions of quarks may be written as

$$
\mathcal{L}_{\text{QCD+em}} = \sum_q \left\{ i\bar{q}\,\slashed{D}q - m_q\bar{q}q \right\} - \tfrac{1}{4}(D \times B)^2
$$
$$
+ e\sum_q Q_q\bar{q}\gamma_\mu q A^\mu - \tfrac{1}{4}F^{\mu\nu}F_{\mu\nu},
$$

(4.1.1)

where Q_q is the charge of quark flavour q in units of the proton charge, e. Sum over omitted colour indices is understood. We have not explicitly written gauge fixing and ghost terms in (4.1.1). The electromagnetic current operator is

$$
J^\mu = \sum_q Q_q : \bar{q}\gamma^\mu q : .
$$

Let us consider a generic hadronic state with the quantum numbers of e^+e^- that we denote by Γ. The (unpolarized) hadron annihilation cross section of e^+e^- is defined as the sum of the cross sections for all the processes[1] $e^+e^- \to \Gamma$, averaged over the spins of the e^+e^-. To calculate it, we consider the matrix element,

$$
\langle\Gamma|S|e^+(p_1,\sigma_1)e^-(p_2,\sigma_2)\rangle
$$
$$
= \langle\Gamma|\text{T}\exp \text{i}\int \text{d}^4x\,\{\mathcal{L}_{\text{int,QCD}}(x) + \mathcal{L}_{\text{int,em}}(x)\}|e^+(p_1,\sigma_1)e^-(p_2,\sigma_2)\rangle.
$$

---

[1] The name "deep inelastic scattering" is at times reserved only for scattering of photons (or W, Z) probes off hadron targets, in the deep inelastic kinematic region; but we include in its study that of hadronic annihilations of e^+e^-, Z, τ both for historical reasons and because of the similarities in the theoretical analysis.

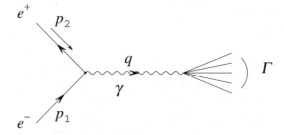

Fig. 4.1.1. Hadronic annihilations of an electron–positron pair.

To lowest order in the electromagnetic interactions, we obtain

$$\langle \Gamma|S|e^+(p_1,\sigma_1)e^-(p_2,\sigma_2)\rangle = -\frac{e^2}{2!}\langle \Gamma|\int d^4x_1\,d^4x_2\,T\mathcal{L}^0_{int,em}(x_1)\mathcal{L}^0_{int,em}(x_2)$$

$$\times \exp i \int d^4x\,\mathcal{L}^0_{int,QCD}(x)|e^+(p_1,\sigma_1)e^-(p_2,\sigma_2)\rangle.$$

Using the Feynman rules for QED we then find, with the kinematics of Fig. 4.1.1,

$$F(e^+e^- \to \Gamma) = \frac{2\pi e^2}{q^2}\,\bar{v}(p_1,\sigma_1)\gamma_\mu u(p_2,\sigma_2)\langle \Gamma|J^\mu(0)|0\rangle;$$

hence, we get the cross section

$$\sigma_h(s) \equiv \sum_\Gamma \sigma(e^+e^- \to \Gamma, s \equiv (p_1+p_2)^2 = q^2)$$

$$= \frac{8\pi^2\alpha^2}{s^3}l_{\mu\nu}\sum_\Gamma (2\pi)^4\delta(p_1+p_2-p_\Gamma)\langle \Gamma|J^\nu(0)|0\rangle^*\,\langle \Gamma|J^\mu(0)|0\rangle,$$

$$(4.1.2)$$

where $l_{\mu\nu}$ is the leptonic tensor

$$l_{\mu\nu} = \tfrac{1}{4}\sum_{\sigma_1\sigma_2} \bar{v}(p_1,\sigma_1)\gamma_\mu u(p_2,\sigma_2)[\bar{v}(p_1,\sigma_1)\gamma_\nu u(p_2,\sigma_2)]^*$$

$$= \tfrac{1}{2}\{q_\mu q_\nu - q^2 g_{\mu\nu} - (p_1-p_2)_\mu(p_1-p_2)_\nu\},$$

and we have neglected the electron mass.

Equation (4.1.2) shows that we have to consider the quantity

$$\Delta^{\mu\nu} = \sum_{\Gamma} (2\pi)^4 \delta(p_1 + p_2 - p_\Gamma) \langle \Gamma | J^\nu(0) | 0 \rangle^* \, \langle \Gamma | J^\mu(0) | 0 \rangle.$$

Using completeness, $\sum_\Gamma |\Gamma\rangle\langle\Gamma| = 1$ and that, because of energy–momentum conservation, the term with a reversed order of Js gives zero, one can rewrite $\Delta^{\mu\nu}$ as

$$\Delta^{\mu\nu}(q) = \int \mathrm{d}^4 x \, \mathrm{e}^{\mathrm{i}q\cdot x} \langle [J^\mu(x), J^\nu(0)] \rangle_0. \tag{4.1.3}$$

It is convenient to define the *hadronic* vacuum polarization tensor, also called current–current *correlator*,

$$\Pi^{\mu\nu}(q) = \mathrm{i} \int \mathrm{d}^4 x \, \mathrm{e}^{\mathrm{i}q\cdot x} \langle \mathrm{T} J^\mu(x) J^\nu(0) \rangle_0; \tag{4.1.4a}$$

then one can see that $\Delta^{\mu\nu} = 2\,\mathrm{Im}\,\Pi^{\mu\nu}$: the e^+e^- annihilation cross section is related to the imaginary part of the photon propagator.

A slight complication is introduced here because of the interplay of strong and electromagnetic interactions. Because $\Pi^{\mu\nu}$ is the coefficient of a second-order term in the electric charge, e, we should, when evaluating $\Pi^{\mu\nu}$, also consider the electromagnetic charge renormalization. The simplest solution to this problem is to neglect it, as it does *not* arise when evaluating the quantity relevant for us here, $\mathrm{Im}\,\Pi^{\mu\nu}$. A second point is that, in the preceding chapter, we have derived the renormalization group equations for *spacelike* momenta, while we now require the quantity $\mathrm{Im}\,\Pi^{\mu\nu}(q^2)$ for timelike q. We can use the analytic properties that $\Pi(t)$ possesses,[2] to evaluate it for spacelike $t < 0$ and obtain $\mathrm{Im}\,\Pi^{\mu\nu}(q^2)$ for positive q^2 by analytic continuation. Alternatively, we can repeat the arguments given in Sects. 3.4, 5 directly for this last quantity. Both methods are of course strictly equivalent and have been used in the existing literature.

The electromagnetic current is conserved; therefore we need not include a Z_J for the renormalization of it. If we explicitly extract the tensor $-g^{\mu\nu}q^2 + q^\mu q^\nu$ from $\Pi^{\mu\nu}$,

$$\Pi^{\mu\nu}(q) = (-g^{\mu\nu}q^2 + q^\mu q^\nu)\Pi(q^2), \tag{4.1.4b}$$

then the general theory gives us the simple result

$$\mathrm{Im}\,\Pi_R\left(q; m(\nu), g(\nu); \nu\right) = \mathrm{Im}\,\Pi_R\left(\nu n; \bar{m}(Q^2), \bar{g}(Q^2); \nu\right),$$
$$Q^2 = -q^2 = s, \quad n^2 = 1. \tag{4.1.5}$$

Therefore, all we have to do is calculate $\mathrm{Im}\,\Pi_R(q; m(\nu), g(\nu); \nu)$ and then replace $q \to \nu$, $m(\nu) \to \bar{m}(Q^2)$, $g(\nu) \to \bar{g}(Q^2)$.

---

[2] It can be proved with great generality that $\Pi(t)$ is analytic in the complex t plane except for a cut along the real axis above the threshold for production of hadrons, $t_0 = 4m_\pi^2$ to ∞, and that the discontinuity there coincides with $\mathrm{Im}\,\Pi$.

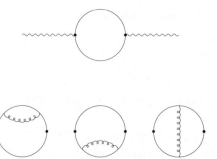

Fig. 4.1.2. Vacuum polarization to order zero in α_s (top figure), and corrections of order α_s.

To zero order the calculation is of course trivial. We have the diagram at the top of Fig. 4.1.2 which gives, to corrections of order $\bar{m}^2/s$,

$$\operatorname{Im} \Pi^{(0)} = \frac{N_c}{12\pi} \sum_{f=1}^{n_f} Q_f^2, \quad N_c = 3, \tag{4.1.6}$$

and n_f is the effective number of flavours, i.e., the flavours excited at the annihilation energy $E = s^{\frac{1}{2}}$. This justifies the old parton model result (Cabibbo, Parisi and Testa, 1970; Feynman, 1972) in which quarks were considered to be free, so that the cross section into hadrons is like that into a $\mu^+\mu^-$ pair, up to the quark charges. Because of this, it is customary to define the ratio

$$R(s) = \frac{\sigma_h(s)}{\sigma_{e^+e^- \to \mu^+\mu^-}}, \tag{4.1.7}$$

where both cross sections are to be calculated to lowest order in the *electromagnetic* interactions. We see that we have obtained

$$R^{(0)}(s) = 3 \sum_{f=1}^{n_f} Q_f^2, \tag{4.1.8}$$

a result that, because of asymptotic freedom, would become exact at infinite energy, $s \to \infty$.

The following correction involves the lower diagrams in Fig. 4.1.2. One may profit from the fact that they are like those in QED with the gluon replaced by a photon, except for the group-theoretic factor

$$\sum_{a,k} t_{ik}^a t_{kj}^a = C_F \delta_{ij}, \quad C_F = 4/3,$$

and that, in QED, they were calculated long ago by Jost and Luttinger (1950). So we find (Appelquist and Georgi, 1973; Zee, 1973)

$$R(s) = 3 \sum_{f=1}^{n_f} Q_f^2 \left\{ 1 + \frac{\alpha_s(s)}{\pi} \right\} + O(\alpha_s^2). \qquad (4.1.9)$$

The following correction is the first where the non-Abelian character of the gluons enters explicitly (and not merely in the running of α_s). It is scheme dependent; in the $\overline{\text{MS}}$ scheme it is[3]

$$R(s) = 3 \sum_{f=1}^{n_f} Q_f^2 \left\{ 1 + \frac{\alpha_s(s)}{\pi} + r_2 \left(\frac{\alpha_s(s)}{\pi} \right)^2 \right\} + O(\alpha_s^3),$$

$$r_2 = \tfrac{365}{24} - 11\zeta(3) + \left[\tfrac{2}{3}\zeta(3) - \tfrac{11}{12} \right] n_f \simeq 2.0 - 0.12 n_f. \qquad (4.1.10)$$

The symbol $\zeta(3)$ stands for Riemann's function, $\zeta(3) \simeq 1.20$, and the two loop expression is to be used for $\alpha_s(s)$. The *four loop* expression for R is also known (Gorishny, Kataev and Larin, 1991; Sugurladze and Samuel, 1991). Also known are the $O(\alpha_s^2)$ expressions, including finite mass corrections: see Chetyrkin, Kuhn and Steinhauser (1997) and Chetyrkin, Harlander, Kuhn and Steinhauser (1997). The fact that in the renormalization group equations $\alpha_s(Q^2)$ is defined at spacelike momenta first enters the calculation at order α_s^3, contributing a term proportional to π^2 (from the analytical continuation) to the coefficient r_3: in the $\overline{\text{MS}}$ scheme

$$R(s) = 3 \sum_{f=1}^{n_f} Q_f^2 \left\{ 1 + \frac{\alpha_s(s)}{\pi} + r_2 \left(\frac{\alpha_s(s)}{\pi} \right)^2 + r_3 \left(\frac{\alpha_s(s)}{\pi} \right)^3 \right\} + O(\alpha_s^4),$$

$$(4.1.11a)$$

where

$$r_3 = -6.637 - 1.200 n_f - 0.005 n_f^2 - 1.240 \left(\sum_1^{n_f} Q_f \right)^2 \left(3 \sum_1^{n_f} Q_f^2 \right)^{-1}$$

$$= 18.243 - 4.216 n_f + 0.086 n_f^2 - 1.240 \left(\sum_1^{n_f} Q_f \right)^2 \left(3 \sum_1^{n_f} Q_f^2 \right)^{-1} + r_3^{\text{an}}.$$

$$(4.1.11b)$$

The three loop expression for $\alpha_s(s)$, Eq. (3.7.4d), is to be used in this formula. The piece coming from the analytical continuation is r_3^{an},

$$r_3^{\text{an}} = - \left(\tfrac{121}{48} - \tfrac{11}{36} n_f + \tfrac{1}{108} n_f^2 \right) \pi^2 \simeq -24.880 + 3.016 n_f - 0.091 n_f^2; \quad (4.1.11c)$$

---

[3] Chetyrkin, Kataev and Tkachov (1979); Dine and Sapiristein (1979).

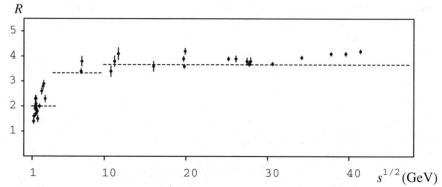

Fig. 4.1.3. Plot of R vs. $s^{1/2}$. We only show a selection of points. The dotted line is the parton model prediction. More data can be found in e.g. the compilations of Wiik and Wolf (1979) and the Particle Data Group (1996).

it partially cancels the remaining terms to give the smaller result for r_3 quoted above. For the record, we mention that the leading terms in the limit in which one takes the number of colours as a variable, and lets $N_c \to \infty$, are, for r_2, r_3,

$$r_2 \underset{N_c \to \infty}{\to} 0.28\,N_c^2, \quad r_3 \underset{N_c \to \infty}{\to} 1.05\,N_c^3.$$

A question to be discussed further is the number of flavours to be taken into account, i.e., the value of n_f in the formulas given. We have already referred to "the number of active flavours": the question is related to the problem of the quark masses. If we have a quark of mass $m_q \neq 0$, then the corrections that this induces, for $s \gg m_q^2$, are of order m_q^2/s and one can thus neglect m_q, as we have done in our calculations. The situation is different, however, for quarks for which $m_q^2 \gg s$, for then there is insufficient energy to create the corresponding $\bar{q}q$ pairs, and these quarks only contribute through virtual loops. Here we will follow the analysis of Sect. 3.8 and conclude that, since these quarks only give corrections at the level $\sim s/m_q^2$, their *contribution* can be forgotten. This is what is meant by *active flavours*: one only considers the contributions of quarks whose mass is much smaller than the energy. As discussed in Sect. 3.8, we also assume the QCD parameter Λ to depend on n_f. Note that the regions $s \approx 4m_q^2$ (*thresholds*) are omitted from the analysis; they will be discussed in Sect. 5.4. In fact, it may be shown that perturbative QCD is not directly applicable there.

A precise comparison of theory and experiment for R requires incorporation of higher order electromagnetic, and of weak interaction effects, important for $s^{\frac{1}{2}} \gtrsim 30$ GeV, incorporation of the dependence on the masses for the heavier quarks, c, b, and careful study of the corresponding thresholds.

Moreover, and as *systematic* experimental errors are rather large, the comparison is not very precise.[4] It is shown in Fig. 4.1.3 here and, for larger values of the energy, in next section (Fig. 4.2.4). If one fits the values of the QCD parameter Λ, a wide range is obtained:

$$\Lambda(n_f = 4, \text{three loops}) = 100 \text{ to } 700 \text{ MeV}. \qquad (4.1.12)$$

Still, the agreement between theory and experiment is nontrivial, holding as it does between $s \sim 2 \text{ GeV}^2$ and the highest energies of LEP 200, $s \sim (180 \text{ GeV})^2$.

4.2 τ and Z Decays Involving Hadrons

i τ Decay

The τ *lepton*, so called because it shares the properties of other, truly light[5] particles such as the electron or muon, has a rather large mass, $m_\tau = 1777.0 \pm 0.3$ MeV and thus can decay not only into other leptons, but into final states involving hadrons as well. The *leptonic* decays are

$$\tau \rightarrow \nu_\tau + W \qquad l = \mu \text{ or } e. \qquad (4.2.1)$$
$$ \ \llcorner\!\rightarrow l^- + \bar{\nu}_l$$

We will write formulas for $\tau = \tau^-$; the formulas for τ^+ are obtained with obvious changes. Besides the decays (4.2.1), we have the *hadronic* decays which may be split between decays that involve only nonstrange particles, or decays involving strange particles. At zero order in strong interactions, these are given, respectively, by the processes

$$\tau \rightarrow \nu_\tau + W \qquad , \qquad \tau \rightarrow \nu_\tau + W \qquad . \qquad (4.2.2)$$
$$ \llcorner\!\rightarrow d + \bar{u} \qquad \llcorner\!\rightarrow s + \bar{u}$$

The Feynman diagrams for Eqs. (4.2.1, 2) are shown in Fig. 4.2.1.

For a precise calculation (besides of course QCD corrections) we need electroweak radiative corrections. These have been evaluated by Marciano and Sirlin (1988). Taking them into account we find, for the leptonic decays,

$$\Gamma(\tau \rightarrow \nu_\tau + l^- + \bar{\nu}_l) = \frac{G_F^2 m_\tau^5}{192\pi^3} f(m_l^2/m_\tau^2) r_{\text{EW}},$$
$$f(x) = 1 - 8x + 8x^2 - x^4 - 12x^2 \log x, \qquad (4.2.3)$$
$$r_{\text{EW}} = \left[1 + \frac{\bar{\alpha}}{2\pi}\left(\tfrac{25}{4} - \pi^2\right)\right]\left\{1 + \tfrac{3}{5}\frac{m_\tau^2}{M_W^2}\right\},$$

---

[4] A detailed discussion of the comparison of theory and experiment, including the analysis of errors, can be found in Barnett, Dine and McLerran (1980), Ali (1981) and, especially, Marshall (1989).

[5] *Lepton* comes from the Greek word for *light*.

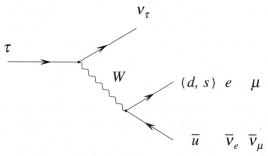

Fig. 4.2.1. Leptonic and hadronic decays of τ at zero order in α_s.

where $\bar{\alpha}$ is the QED running coupling evaluated at m_τ^2, $\bar{\alpha} \simeq 1/133.3$, G_F is the Fermi coupling constant as measured, for example, in μ decay, and the mass of the final lepton is neglected in $r_{\rm EW}$, which is then only evaluated to corrections $O(m_l^2/M_W^2)$. The prediction (4.2.3) is in perfect agreement with the experimental leptonic decay rates.

The hadronic decays, which are the ones that interest us here, involve the piece of the weak Lagrangian,

$$\mathcal{L}_{W,\rm eff} = \frac{g_W}{2\sqrt{2}} W_\mu^+ \bar{u}\gamma^\mu(1 - \gamma_5)d_C + \text{Hermitian conjugate}, \qquad (4.2.4)$$

and d_C is the Cabibbo rotated d quark field,

$$d_C = (\cos\theta_C)d + (\sin\theta_C)s.$$

In a more precise analysis we would take into account the full Kobayashi–Maskawa mixing matrix: this is effected by replacing, in all subsequent formulas, the mixing factors according to $\cos^2\theta_C \to |V_{ud}|^2$, $\sin^2\theta_C \to |V_{us}|^2$.

Inclusive hadronic decays of the τ are very similar to e^+e^- annihilation into hadrons, the main difference being that we now have axial as well as vector current correlators. We define the ratio

$$R_\tau \equiv \frac{\Gamma(\tau \to \nu_\tau + \text{hadrons})}{\Gamma(\tau \to \nu_\tau + e + \bar{\nu}_e)}, \qquad (4.2.5)$$

and the current correlators

$$\begin{aligned}
\Pi_{ij}^{(V)\mu\nu}(k) &= i\int d^4x\, e^{ik\cdot x}\langle 0|TV_{ij}^\mu(x)V_{ij}^\nu(0)^\dagger|0\rangle, \\
\Pi_{ij}^{(A)\mu\nu}(k) &= i\int d^4x\, e^{ik\cdot x}\langle 0|TA_{ij}^\mu(x)A_{ij}^\nu(0)^\dagger|0\rangle;
\end{aligned} \qquad (4.2.6)$$

i, j are flavour indices and the currents are those entering Eq. (4.2.4), viz.,

$$V_{ij}^{\mu} = \bar{q}_i \gamma^{\mu} q_j, \quad A_{ij}^{\mu} = \bar{q}_i \gamma^{\mu} \gamma_5 q_j.$$

We may split the Π into a *transverse* and a *longitudinal* part, writing

$$\Pi_{ij}^{(V,A)\mu\nu}(k) = (-g^{\mu\nu}k^2 + k^{\mu}k^{\nu})\Pi_{ij}^{(V,A;1)}(k^2) + k^{\mu}k^{\nu}\Pi_{ij}^{(V,A;0)}(k^2) \quad (4.2.7)$$

where the indices $J = 0, 1$ in $\Pi^{(V,A;J)}$ refer to the total spin carried by the correlator.

The hadronic decay rates are evaluated by an extension of the calculation performed for the e^+e^- annihilation into hadrons. We find, for decays into (respectively) nonstrange, strange final states, and integrating over the energy of the neutrino ν_τ,

$$R_\tau^{\text{n.st.}} = \frac{12\pi\cos^2\theta_C}{m_\tau^2} \int_{t_0}^{m_\tau^2} dt \left(1 - \frac{t}{m_\tau^2}\right)$$
$$\times \left\{\left(1 + \frac{2t}{m_\tau^2}\right) \operatorname{Im} \Pi_{\text{n.st.}}^{(1)}(t) + \operatorname{Im} \Pi_{\text{n.st.}}^{(0)}(t)\right\}, \quad (4.2.8a)$$

$$R_\tau^{\text{st.}} = \frac{12\pi\sin^2\theta_C}{m_\tau^2} \int_{t_0}^{m_\tau^2} dt \left(1 - \frac{t}{m_\tau^2}\right)$$
$$\times \left\{\left(1 + \frac{2t}{m_\tau^2}\right) \operatorname{Im} \Pi_{\text{st.}}^{(1)}(t) + \operatorname{Im} \Pi_{\text{st.}}^{(0)}(t)\right\}. \quad (4.2.8b)$$

Here $t = p_h^2$ (Fig. 4.2.2) is the invariant mass of the hadronic final state with production threshold at $t_0 = m_\pi^2$, $t_0 = m_K^2$ for the n.st, st. cases respectively. Moreover,

$$\Pi_{\text{n.st.}}^{(J)} = \Pi_{ud}^{(V;J)} + \Pi_{ud}^{(A;J)},$$
$$\Pi_{\text{st.}}^{(J)} = \Pi_{us}^{(V;J)} + \Pi_{us}^{(A;J)}. \quad (4.2.8c)$$

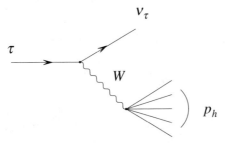

Fig. 4.2.2. Hadronic decay of τ. The similarity with Fig. 4.1.1 is obvious.

The main differences with e^+e^- annihilation are as follows. First, the appearance of the correlators $\Pi^{(V;0)}$ and $\Pi^{(A;J)}$, $J = 0, 1$; secondly, we now have an *integral* over the energies of the hadronic states.

It is convenient to consider separately the strange and nonstrange decays. In the first, and because the s quark intervenes and one has $m_s \sim 200$ MeV, the approximation of neglecting the quark masses is not as good as for the nonstrange case where $m_{u,d}$ are really minute in comparison with the scale of the problem, m_τ. In fact, one can use the *experimental* data for strange τ decays to get estimates of the s quark mass; see Sect. 10.4 for the resulting figures. Nevertheless, we will here neglect the mass of the s quark. The reader interested in the corrections due to inclusion of m_s can find them in the recent (and excellent) review of Pich (1997) and, for the determinations of m_s, in the review of Chen (1998).

If we neglect quark masses, then the analysis simplifies a lot. The reason is that the divergences of the currents, $\partial_\mu V_{ij}^\mu$, $\partial_\mu A_{ij}^\mu$, are proportional to $m_i \pm m_j$; see Eq. (2.8.5). Therefore we can neglect the correlators[6] $\Pi^{(V,A;0)}$: indeed, contracting (4.2.7) with k_μ, k_ν it immediately follows that these correlators are proportional to $(m_i \pm m_j)^2$. Another simplification obtained when neglecting masses is that now

$$\Pi_{ij}^{(V,J)} = \Pi_{ij}^{(A,J)}.$$

The reason is that, when the mass of quark i is zero, the QCD Lagrangian is invariant under the chiral transformation $q_i \rightarrow \gamma_5 q_i$. If we apply these transformations to the u quark (but not to the d, s) then under it $V_{ud} \rightarrow A_{ud}$, $V_{us} \rightarrow A_{us}$ and the announced result follows. It is to be noted that this equality $\Pi_{ij}^{(V,J)} = \Pi_{ij}^{(A,J)}$ only requires that we neglect the mass of the u quark.

In view of these simplifications, we see that we have reduced the calculation of R_τ to that of the integral

$$F(m_\tau) = \int_{t_0}^{m_\tau^2} dt \left(1 - \frac{t}{m_\tau}\right)^2 \left(1 + \frac{2t}{m_\tau}\right) \text{Im}\, \Pi(t), \qquad (4.2.9)$$

with the nonstrange and strange decays ratios given by

$$R_\tau(\text{n.st.}) = \frac{12\pi \cos^2 \theta}{m_\tau^2} F, \quad R_\tau(\text{st.}) = \frac{12\pi \sin^2 \theta}{m_\tau^2} F,$$

---

[6] Except for the A case when $k^2 \simeq 0$, where we have a pole due to the pion (or kaon) intermediate state, which produces a divergence proportional to $1/(M_\pi^2 \sim m_u + m_d)$, say for the nonstrange decays; see Sect. 7.3. This problem can be avoided by subtracting the channels $\tau \rightarrow \nu_\tau \pi$, $\nu_\tau K$ from $\tau \rightarrow \nu_\tau + $ hadrons.

and Π is defined as

$$(-g^{\mu\nu}k^2 + k^\mu k^\nu)\Pi(k^2) = \mathrm{i}\int \mathrm{d}^4x\, \mathrm{e}^{\mathrm{i}k\cdot x}\langle 0|TV^\mu(x)V^\nu(0)|0\rangle. \qquad (4.2.10)$$

V^μ is here a generic current for massless quarks: the ensuing Π is independent, in the limit of massless quarks, of which quarks we choose for V^μ.

In this approximation Π is identical with the quantity of the same name introduced for e^+e^- annihilations in Eq. (4.1.4b), up to the factor $\sum Q_f^2$ due to the charge of the quarks that does not appear here. Therefore, to zero order in α_s we may write

$$\mathrm{Im}\,\Pi(t) = N_c/12\pi, \quad N_c = 3.$$

The main problem with calculating hadronic τ decays lies in the fact that, while QCD predicts $\mathrm{Im}\,\Pi(t)$ for large t, the integral (4.2.9) involves all values of t between threshold and m_τ^2; and the integrand is in fact peaked at *small* values of t. However, not all is lost. We can consider that the quantity $F(m_\tau^2)$ in (4.2.9) does only depend on m_τ^2; therefore we may expand it in powers of $\alpha_s(m_\tau^2)$, and check, *a posteriori*, that the series has reasonable convergence properties. This expansion can be arranged in fancy ways (see Pich, 1997, for example). Some of these are useful in that they simplify the evaluation of nonperturbative corrections, or for resummations; but they can be easily proved to be mathematically equivalent to the following simple procedure. First, one computes $\mathrm{Im}\,\Pi(t)$ in an expansion in powers of $\alpha_s(t)$. The expansion, up to the quark charges, is identical to that for the e^+e^- annihilation (see the previous section), so we have

$$\mathrm{Im}\,\Pi(t) = \frac{1}{4\pi}\left\{1 + \frac{\alpha_s(t)}{\pi} + r_2\left(\frac{\alpha_s(t)}{\pi}\right)^2 + r_3\left(\frac{\alpha_s(t)}{\pi}\right)^3 + \cdots\right\}. \qquad (4.2.11)$$

Then one writes $\alpha_s(t)$ in terms of $\alpha_s(m_\tau^2)$ using the renormalization group equations so that, to two loops, which is what we need for (4.2.11),

$$\begin{aligned}
\alpha_s(t) = \alpha_s(m_\tau^2)\Bigg\{&1 - \frac{\beta_0 \log t/m_\tau^2}{4\pi}\alpha_s(m_\tau^2) \\
&+ \frac{\beta_0[\log^2 t/m_\tau^2 - (\beta_1/\beta_0^2)\log t/m_\tau^2]}{16\pi^2}\alpha_s^2(m_\tau^2)\Bigg\}.
\end{aligned} \qquad (4.2.12)$$

One then substitutes (4.2.12) into (4.2.11) and the result into (4.2.9). Integrating taking care that, at the perturbative level, the threshold should be taken as $t_0 = (m_i + m_j)^2$ with m the *quark* masses, the desired result is obtained. For $n_f = 3$,

$$F(m_\tau^2) = F^{(0)}\left\{1 + \frac{\alpha_s(m_\tau^2)}{\pi} + 5.2023\left(\frac{\alpha_s(m_\tau^2)}{\pi}\right)^2 + 26.366\left(\frac{\alpha_s(m_\tau^2)}{\pi}\right)^3 + \cdots\right\}, \qquad (4.2.13)$$

and $F^{(0)}$ is the zero order result,

$$F^{(0)} = \frac{1}{4\pi} \int_{t_0}^{m_\tau^2} dt \left(1 - \frac{t}{m_\tau}\right)^2 \left(1 + \frac{2t}{m_\tau}\right).$$

The series (4.2.13) is not a prodigy of fast convergence; but it is not too bad either. With the current values of $\alpha_s(m_\tau^2) \sim 0.3$, $\alpha_s(m_\tau^2)/\pi \sim 0.1$, the various terms inside the bracket in (4.2.13) are

$$1, \ 0.1, \ 0.05, \ 0.026.$$

It is quite obvious that the series has to diverge, and there is already a hint of this in the last term here; but it can probably be used to the order given in (4.2.13). In fact, a number of estimates has been made of higher order terms, nonperturbative corrections, etc. (Pich, 1997, and work quoted there) and it would appear that the expected error is something like ~ 0.02.

An alternate method uses *moments* of the differential decay rate. One defines (Le Diberder and Pich, 1992)

$$R_\tau^{nl} = \frac{1}{\Gamma_{\text{leptons}}} \int_{t_0}^{m_\tau^2} dt \left(1 - \frac{t}{m_\tau}\right)^n \left(\frac{t}{m_\tau^2}\right)^l \frac{d\Gamma_{\text{hadrons}}}{dt},$$

which have the advantage that, for large l, they weight the large t region. One then uses these moments for comparison with experiment, as has been done by e.g. the ALEPH collaboration (Buskulic et al., 1997).

Having obtained the theoretical expression for the decay, one can take $\alpha_s(m_\tau^2)$ from other sources and *predict* the decay rate; or, better still, one can use the *experimental* decay rate to find a very precise determination of $\alpha_s(m_\tau^2)$. If we do this, we find

$$\alpha_s(m_\tau^2) = 0.330 \pm 0.046 \quad \text{(moments)}, \tag{4.2.14a}$$

$$\alpha_s(m_\tau^2) = 0.350 \pm 0.020 \quad \text{(from } R_\tau\text{)}. \tag{4.2.14a}$$

ii Z Decay

The piece of the weak interaction Lagrangian responsible for Z decay into hadrons is

$$\mathcal{L}_Z = \frac{e}{\sin 2\theta_W} Z_\mu \sum_q \bar{q}\gamma^\mu(v_q + a_q\gamma_5)q,$$

$$v_u = \tfrac{1}{2} - \tfrac{4}{3}\sin^2\theta_W, \; a_u = \tfrac{1}{2}, \quad v_d = -\tfrac{1}{2} + \tfrac{2}{3}\sin^2\theta_W, \; a_d = -\tfrac{1}{2},$$

and identical expressions for v, a for the c, s and t, b weak doublets. At tree level a simple calculation using the diagram of Fig. 4.2.3 gives

$$\Gamma^{(0)}(Z \to \bar{q}q) = \frac{G_F M_Z^3}{6\pi\sqrt{2}} (v_q^2 + a_q^2). \qquad (4.2.15)$$

A precise calculation involves vector and axial currents, and is similar in (almost[7]) all respects to the calculation for τ decay with the simplification that now the vector currents are diagonal, V_{qq}^μ, so the correlator $\Pi^{(V;0)}$ is zero. Neglecting the masses of the quarks as compared to M_Z, an approximation that even for the b quark only produces an error of $\bar{m}_b^2(M_Z^2)/M_Z^2 \sim 10^{-4}$, the axial correlators may be identified with the vector ones and we find that the ratio of full QCD to zero order is for Z decays as for e^+e^-. So,

$$\frac{\Gamma(Z \to \bar{q}q)}{\Gamma^{(0)}(Z \to \bar{q}q)} = 1 + \frac{\alpha_s(M_Z^2)}{\pi} + r_2\left(\frac{\alpha_s(M_Z^2)}{\pi}\right)^2 + r_3\left(\frac{\alpha_s(M_Z^2)}{\pi}\right)^3 + \cdots.$$

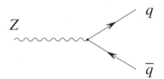

Fig. 4.2.3. Z decay into a $\bar{q}q$ pair.

---

[7] The only important difference is the appearance now of a diagram $Z \to \bar{t}t \to GG \to \bar{q}q$, which produces a contribution of order $m_t^2\alpha_s^2$ with m_t the top quark mass. This effect involves a diagram similar to that of the axial anomaly that we will study in Sect. 7.5. For a detailed review of Z decay calculations see, for example, Chetyrkin, Kuhn and Kwiatowski (1996).

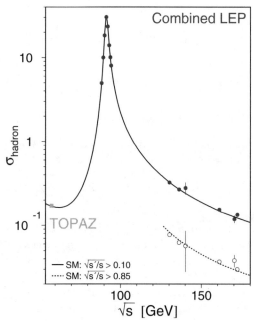

Fig. 4.2.4. The ratio $\sigma(e^+e^- \to$ hadrons$)/\sigma(e^+e^- \to \mu^+\mu^-)$ in the vicinity of the Z as a function of the c.m. energy of the e^+e^- pair, $s^{1/2}$. $s'^{1/2}$ is the visible hadron energy, which would equal $s^{1/2}$ if there was no radiation from the initial state. To identify the events, cuts are made requiring that $\sqrt{s'/s}$ be larger than a certain pre-assigned number. According to the value of this number we find the solid or hollow dots in the figure. *Data:* combined LEP data. *Continuous and dotted lines:* theoretical calculation with the standard model. (Communicated by E. Fernández, 1997.)

The more precise values for the Z decay come from the experiments made at the accelerator LEP, where one produces Z in e^+e^- annihilations, subsequently observing its decays according to the scheme $e^+e^- \to Z \to$ hadrons. The cross section, normalized to $e^+e^- \to \mu^+\mu^-$, is depicted in Fig. 4.2.4, where the comparison with the theoretical predictions is also shown.

One can look at the total hadronic decay width, as we have done here, or to the decays into jets, which will be discussed in Sect. 5.4. As for τ decays, we can take $\alpha_s(M_Z^2)$ from other sources, and obtain theoretical predictions for Z decays, or *deduce* $\alpha_s(M_Z^2)$ from the Z width. If we do the last, we find

$$\alpha_s(M_Z^2) = 0.123 \pm 0.006. \qquad (4.2.16)$$

The variation between (4.2.14) and (4.2.16) dramatically exhibits the running of the strong coupling constant. The compatibility between the two determinations is highly nontrivial. Indeed, if we evolve the weighted average of (4.2.14) from $\mu^2 = m_\tau^2$ to $\mu^2 = M_Z^2$ with the renormalization group, we get

$$\alpha_s(M_Z^2) = 0.119 \pm 0.0025, \qquad (4.2.17).$$

in impressive agreement with (4.2.16).[8]

4.3 Kinematics of Deep Inelastic Scattering.
The Parton Model

i Kinematics. Structure Functions

We consider the process $l + h \to l' + \text{all}$, where l and l' are leptons, h is a hadron target, and "all" means that we sum over all possible final hadron states, Γ (see Fig. 4.3.1). We will assume that the scattering takes place at high energy, and that also the virtuality of the exchanged particle is large (*deep inelastic scattering*, or DIS for short). If $l = l' = e$ or μ we are probing h with electromagnetic interactions to which one has to add, if the energy is large enough, neutral weak currents. In this case, and working (as we will throughout this section) to lowest order in weak or electromagnetic interactions, the vector particle represented with the wavy line in Fig. 4.3.1 is a virtual photon γ^* or Z^*. Neglecting this last contribution, we have that the relevant interaction will be the electromagnetic one, and the relevant current the e.m. current:

$$\mathcal{L}_{\text{int,em}} = e J_{\text{em}}^\mu A_\mu, \quad J_{\text{em}}^\mu = \sum_q Q_q \bar{q} \gamma^\mu q,$$

and Q_q is the charge of flavour q in units of the proton charge, e.

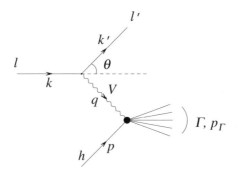

Fig. 4.3.1. Deep inelastic scattering process, and its kinematics.

---

[8] Note, however, that the errors given in this and the preceding subsection for α_s are only *indicative*; see Chap. 10 for a full discussion about determinations of α_s.

If $l = \nu_{\mu,e}$, $l' = \mu, e$ we have charged weak interactions; then the exchanged particle in Fig. 4.3.1 would be a virtual W. The interaction and current are now

$$\mathcal{L}_{\text{int},W} = \frac{1}{2\sqrt{2}} g_W W_\mu J_W^\mu \quad J_W^\mu = \bar{u}\gamma^\mu(1-\gamma_5)d' + \bar{c}\gamma^\mu(1-\gamma_5)s' + \bar{t}\gamma^\mu(1-\gamma_5)b',$$

where

$$g_W^2/M_W^2 = 4\sqrt{2}G_F, \quad G_F \simeq 1.027\, m_{\text{proton}}^{-2},$$

and d', s', b' are the Cabibbo–Kobayashi–Maskawa rotated quarks. Finally, for $l = l' = \nu$, or for the neutral weak interactions of e, μ, we take

$$\mathcal{L}_{\text{int},NC} = \frac{e}{2\cos\theta_W \sin\theta_W} Z_\mu J_Z^\mu,$$

with

$$J_{Z;ud}^\mu = \left(\frac{1}{2} - \frac{4\sin\theta_W}{3}\right)\bar{u}\gamma^\mu u + \left(-\frac{1}{2} + \frac{2\sin\theta_W}{3}\right)\bar{d}\gamma^\mu d + \tfrac{1}{2}\bar{u}\gamma^\mu u - \tfrac{1}{2}\bar{d}\gamma^\mu d,$$

for the ud part of the current, identical expressions replacing $u \to c \to t$, $d \to s \to b$ for the other quarks, and corresponding ones for the leptonic part of the current. The weak mixing angle is $\sin^2\theta_W \simeq 0.22$.

With the kinematics of Fig. 4.3.1, we introduce the Bjorken variables,

$$Q^2 = -q^2, \quad \nu = p\cdot q, \quad x = Q^2/\nu;$$

note that, in terms of these,

$$s = p_\Gamma^2 = -Q^2 + m_h^2 + 2\nu = 2\nu\{1 + m_h^2/2\nu - x\},$$

where s coincides with the square of the energy of the final hadron system. The deep inelastic, or *Bjorken, limit* will be defined as

$$Q^2, \nu \gg \Lambda^2, \quad x = Q^2/2\nu = \text{fixed}.$$

Using the standard rules, the scattering matrix may be written for e.g., the electromagnetic electron scattering case, as

$$\mathcal{T}(e + h \to e + \Gamma) = \frac{2\alpha}{q^2}\bar{u}(k',\sigma')\gamma^\mu u(k,\sigma)(2\pi)^2\delta(p + q - p_\Gamma)\langle\Gamma|J_\mu(0)|p,\tau\rangle,$$

$$(4.3.1)$$

and we have suppressed the tag "em" in the electromagnetic current, J. Here σ, σ' are the spins of the incoming/outgoing electrons, and τ that of the target, h. Note that we normalize the states covariantly (cf. Appendix G):

$$\langle p',\tau'|p,\tau\rangle = 2p^0\delta_{\tau\tau'}\delta(\mathbf{p} - \mathbf{p}').$$

The unpolarized cross section for $e + h \to e + \text{all}$ will thus involve the tensors

$$L^{\mu\nu} = \tfrac{1}{2} \sum_{\sigma\sigma'} [\bar{u}(k', \sigma')\gamma^\mu u(k, \sigma)]^* \bar{u}(k', \sigma')\gamma^\nu u(k, \sigma)$$
$$= 2(k^\mu k'^\nu + k^\nu k'^\mu - k \cdot k' g^{\mu\nu}),$$

where we neglect the lepton masses (as we will do systematically) and

$$W^{\mu\nu}(p, q) = \tfrac{1}{2}\tfrac{1}{2} \sum_\tau \sum_\Gamma (2\pi)^6 \delta(p + q - p_\Gamma)\langle p, \tau | J^\mu(0)^\dagger | \Gamma \rangle \langle \Gamma | J^\nu(0) | p, \tau \rangle.$$
(4.3.2a)

Of course, $J^\dagger = J$, but we have written the general expression that also holds for weak currents. The factors of $1/2$ both in $L^{\mu\nu}$ and $W^{\mu\nu}$ are introduced to average the spin of the initial electron and hadron, assumed to be of spin one-half, and over the "helicity" of the virtual photon: as we will see, the process may be related to (off-shell) Compton scattering.

Equation (4.3.2a) may be recast in the form

$$W^{\mu\nu}(p, q) = \tfrac{1}{2}(2\pi)^2 \int d^4z \, e^{iq \cdot z} \langle p | [J^\mu(z)^\dagger, J^\nu(0)] | p \rangle,$$
(4.3.2b)

where the average over the spin of the target, τ, is understood. The equivalence of (4.3.2a) and (4.3.2b) may be verified by inserting a sum over a complete set of states, $\sum_\Gamma |\Gamma\rangle\langle\Gamma|$ in (4.3.2b), replacing

$$J^\mu(z) \to U(z)J^\mu(0)U^{-1}(z),$$

with $U(z)$ a translation by the vector z so that $U(z)|p\rangle = e^{ip \cdot z}|p\rangle$, and noting that the second term in the commutator does not contribute because of energy–momentum conservation. In this form, the relation with forward Compton scattering $\gamma^* + h \to \gamma^* + h$ is obvious (see Fig. (4.3.2)); the tensor $W^{\mu\nu}$ is the same that appears in the expression for the imaginary part of this scattering amplitude, $\mathcal{T}(\gamma^* + h \to \gamma^* + h)$, as a simple calculation shows.

Let us consider the general case of weak or electromagnetic currents. The general expression for $W^{\mu\nu}(p, q)$ in terms of invariants is

$$W^{\mu\nu}(p, q) \overset{\text{eff}}{=} \left(-g^{\mu\nu} + \frac{q^\mu q^\nu}{q^2}\right) W_1 + \frac{1}{m_h^2}\left(p^\mu - \frac{\nu q^\mu}{q^2}\right)\left(p^\nu - \frac{\nu q^\nu}{q^2}\right) W_2$$
$$+ i\epsilon^{\mu\nu\alpha\beta}\frac{p_\alpha q_\beta}{2m_h^2} W_3.$$
(4.3.3)

The notation $\overset{\text{eff}}{=}$ means that there are other terms in the expansion, but they will give zero when contracted with the leptonic tensor $L_{\mu\nu}$. We can express

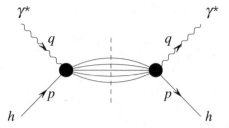

Fig. 4.3.2. Deep inelastic (electromagnetic) scattering as virtual Compton scattering.

the corresponding cross sections in terms of these W_i. In the lab. system of reference (the target h at rest) and with $\theta = \angle(\mathbf{k}, \mathbf{k}')$, and $d\Omega = d\cos\theta d\phi$,

$$\frac{d\sigma^e}{d\Omega dk_0'} = \frac{\alpha^2}{4m_h k_0^2 \sin^4(\theta/2)} \left\{ W_2^e \cos^2\frac{\theta}{2} + 2W_1^e \sin^2\frac{\theta}{2} \right\}. \qquad (4.3.4a)$$

We write all formulas for electron scattering; for μ scattering they are identical, if m_μ is neglected. For neutrino scattering,

$$\frac{d\sigma^{\nu/\bar{\nu}}}{d\Omega dk_0'} = \frac{G_F^2 k_0'^2}{2\pi^2 m_h} \left\{ W_2^{\nu/\bar{\nu}} \cos^2\frac{\theta}{2} + 2W_1^{\nu/\bar{\nu}} \sin^2\frac{\theta}{2} \mp \frac{k_0 + k_0'}{2m_h} W_3^{\nu/\bar{\nu}} \right\};$$
$$(4.3.4b)$$

in the last, the signs $\pm$ hold for ν, $\bar{\nu}$ respectively.

The W_i are invariant and thus depend only on Q^2 and ν or, equivalently, Q^2 and x. It is convenient to define the *structure functions*,[9]

$$F_1^a(x, Q^2) = W_1^a, \quad F_2^a(x, Q^2) = \frac{\nu}{m_h^2} W_2^a, \quad F_3^a(x, Q^2) = \frac{\nu}{m_h^2} W_3^a, \quad (4.3.5)$$

where the superscript a refers to the process, $(e/\mu)h$, νh, $\bar{\nu} h$. The *longitudinal structure function*,

$$F_L^a(x, Q^2) \equiv F_2^a(x, Q^2) - 2x F_1^a(x, Q^2), \qquad (4.3.6)$$

---

[9] We return here to the standard definition of structure functions; this is at variance with the older editions of this book, where we used

$$f_1^{\text{old}} = 2x F_1^{\text{now}}, \quad f_2^{\text{old}} = F_2^{\text{now}}, \quad f_3^{\text{old}} = x F_3^{\text{now}}.$$

Although more rational, the old conventions were different from what has become ingrained use.

is also used in lieu of F_1^a. Expressed in terms of the structure functions, Eqs. (4.3.2), (4.3.3) become

$$W_a^{\mu\nu}(p,q) = \tfrac{1}{2}(2\pi)^2 \int d^4z\, e^{iq\cdot z} \langle p|[J_a^\mu(z)^\dagger, J_a^\nu(0)]|p\rangle$$

$$\stackrel{\text{eff}}{=} -g^{\mu\nu} F_1^a + \frac{p^\mu p^\nu}{\nu} F_2^a + i\epsilon^{\mu\nu\alpha\beta}\frac{p_\alpha q_\beta}{2\nu} F_3^a, \tag{4.3.7}$$

and we have neglected terms proportional to q^μ, q^ν which will give zero when contracted with the leptonic tensor. Note that, throughout this, and some of the coming section, we use z for the space–time variable to avoid confusion with Bjorken's variable x.

As for e^+e^- annihilations, we will find it convenient to consider a T-product of currents:

$$T_a^{\mu\nu}(p,q) = i(2\pi)^3 \int d^4z\, e^{iq\cdot z} \langle p|TJ_a^\mu(z)^\dagger J_a^\nu(0)|p\rangle; \tag{4.3.8a}$$

if we now write the decomposition analogous to (4.3.7),

$$T_a^{\mu\nu}(p,q) = -g^{\mu\nu}T_1^a(x,Q^2) + \frac{p^\mu p^\nu}{\nu}T_2^a(x,Q^2) + i\epsilon^{\mu\nu\alpha\beta}\frac{p_\alpha q_\beta}{2\nu}T_3^a(x,Q^2), \tag{4.3.8b}$$

then it follows that

$$F_i^a = \frac{1}{2\pi} \operatorname{Im} T_i^a, \quad i = 1,2,3. \tag{4.3.8c}$$

ii The Parton Model

let us now consider the Bjorken limit in the so-called *infinite momentum frame*, neglecting the mass of the target, that we will for definiteness suppose to be a proton:

$$p = (p_0, 0, 0, p_0), \quad q = (\nu/2p_0, \sqrt{Q^2}, 0, -\nu/2p_0), \quad p_0 \approx \nu^{1/2} \to \infty. \tag{4.3.9}$$

Rewriting $q \cdot z$ as

$$q \cdot z = \tfrac{1}{2}(q_0 + q_3)(z_0 - z_3) + \tfrac{1}{2}(q_0 - q_3)(z_0 + z_3) - q_1 z_1,$$

we see that $z \cdot z \simeq 0$ corresponds, in the Bjorken limit, to

$$z_0 \pm z_3 \simeq 1/\nu^{1/2}, \quad z_1 \simeq 1/\nu^{1/2},$$

i.e., $z^2 \to 0$. Actually, we could still have z_2 as large as we wished. But in this case, one has $z \cdot z < 0$ where, by locality, the commutator in Eq. (4.3.2b) or Eq. (4.3.7) vanishes: we get no contribution to the structure functions except if $z_2^2 \sim z_0^2$, i.e., still $z^2 \sim 0$. Because of a well-known property of Fourier transforms, it follows that the fixed x, large ν behaviour of the transforms in

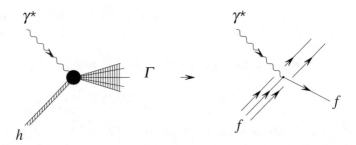

Fig. 4.3.3. The parton model: resolution of the scattering $\gamma^* + h$ into scattering off the individual partons.

(4.3.2b), or (4.3.7), are given by the values of the argument for z such that $z^2 = O(1/Q^2) \to 0$; that is to say, by the light-cone behaviour of

$$[J_a^\mu(z)^\dagger, J_a^\nu(0)] \quad \text{or} \quad \mathrm{T} J_a^\mu(z)^\dagger J_a^\nu(0). \qquad (4.3.10)$$

Because of asymptotic freedom, we expect that, up to logarithmic corrections, we will be able to calculate these commutators, or T-ordered products, neglecting strong interactions, and treating the hadron target as a bunch of free quarks. This suggests the *parton model* (Feynman, 1969). Let us specify it. As stated, we consider that the hadron consists of a collection of "parts", quarks and (as we will see) also gluons, that do not interact among themselves during the process of scattering by the external probe, γ^*; see Fig. 4.3.3. Then we take it that they have each a certain fraction of the hadron momentum. To be precise, if we denote this fraction by x, we will have a certain probability of finding each individual parton, f, with momentum xp. Let us denote this probability, also called *density*, by $q_f(x)$. If we could solve the equations that govern the formation of the hadron as a bound state of the partons, we could calculate $q_f(x)$, which is easily seen to be related to the wave function of the parton in the hadron. As it is, we have to take the densities from experiment.

Within this model, it is not difficult to calculate the cross section for e.g. $e + h \to e + \text{all}$. Because h is a bunch of free partons, this cross section is the sum of the elementary cross sections $e + f \to e + f$, weighted with q_f, and summed to all f. To the order at which we are working (neglect of strong interactions except in that they bind the partons in h), we need only consider the quarks among the f, since the gluons do not have electric charge. However, we have to consider not only quarks, but *antiquarks* as well. Indeed, one would expect that even a hadron such as the proton would contain light quark–antiquark pairs. Thus we find a formula like (4.3.4a), viz.

$$\frac{d\sigma^e}{d\Omega dk_0'} = \frac{\alpha^2}{4m_h k_0^2 \sin^4(\theta/2)} \left(\frac{m_h^2 x}{\nu} \cos^2 \frac{\theta}{2} + \sin^2 \frac{\theta}{2} \right) \sum_f Q_f^2 q_f(x).$$

Therefore, within the model, we have calculated the structure functions in terms of the densities:

$$F_2^{\text{parton}}(x, Q^2) = x \sum_f Q_f^2 q_f(x), \qquad (4.3.11a)$$

and, moreover, we find the relation[10]

$$F_2^{\text{parton}}(x, Q^2) = 2x F_1^{\text{parton}}(x, Q^2). \qquad (4.3.11b)$$

Also we verify that x actually coincides with the Bjorken variable, thus providing a physical interpretation for it: x may be considered either as the ratio $Q^2/2\nu$, or as the fraction of momentum carried by the parton struck by the photon. Later, we will rewrite (4.3.11) in a more detailed form, specifying some of the properties of the parton densities q_f.

A remarkable feature of (4.3.11) is *scaling*. Scaling was proposed by Bjorken (1969) before the parton model – which in fact was devised to explain it. Scaling means that as $Q^2 \to \infty$ the structure functions F_i should become independent of Q^2:

$$F_i(x, Q^2) \underset{\substack{Q^2 \to \infty \\ x \text{ fixed}}}{\to} F^{(0)}(x). \qquad (4.3.12)$$

We shall see that a rigorous treatment justifies this in the sense that one gets scaling up to logarithmic variations of the type $(\log Q^2/\Lambda^2)^d$. What is more, the corrections can be calculated, and the results of the calculations are in full agreement with experimental verifications.

---

[10]This relation is the co-called Callan–Gross relation; we will study it in detail later on.

4.4 Light Cone Expansion of Products of Currents

With the parton model we have evaluated the structure functions F_i neglecting the strong interactions, except in as much as they bind quarks. This calculation can be made more precise using the operator product expansion, which will also allow us later to take interactions into account. We will then consider T-ordered products of currents

$$TJ(x)J'(0) \tag{4.4.1}$$

building them first from *free* fields, and later taking into account interactions; from the beginning we will express (4.4.1) as combinations, with known coefficients,

$$TJ(x)J'(0) = \sum_i C_i(x)N_i(0), \tag{4.4.2}$$

of a set of local operators $N_i(0)$. All bound state complexities will be buried in the matrix elements

$$\langle p|N_i(0)|p\rangle. \tag{4.4.3}$$

We will present a detailed derivation: the techniques are important not only for deep inelastic scattering, but also for the analysis of a large class of other processes where the dynamics is governed by the light cone behaviour of field products.

Before entering into the specific calculations, a number of points are worth discussing. First of all, when forgetting interactions one may write Taylor expansions like

$$: \bar{q}(0)q(x) := \sum_n \frac{1}{n!} x_{\mu_1} \dots x_{\mu_n} : \bar{q}(0)\partial^{\mu_1} \dots \partial^{\mu_n} q(0) :; \tag{4.4.4}$$

when interactions are taken into account one should replace derivatives by covariant derivatives,

$$\partial^\mu \to D^\mu, \tag{4.4.5}$$

as discussed in Sect. 3.9 (see also Appendix I). We will do this as a matter of course. Secondly, we remark that, in the free field case, a product of operators containing only quark fields will only produce operators N with quark fields; but, when interactions are taken into account, *gluon* operators such as

$$: G^{\mu\nu}(0)G_{\nu\lambda}(x) : \tag{4.4.6}$$

will also appear. We will see this in the next section.

We will be interested in T-products of vector and axial vector currents like those considered in Sect. 2.8. They will correspond to observable quantities, and thus will be represented by Hermitian operators. Take n_f to be the number of flavours that are excited in the range of energies in which we are interested. We will consider the corresponding group of transformations $SU_F(n_f)$. Denote its generators by T^a; thus, for $n_f = 2$, $T^a = \sigma^a$ (σ^a the

Pauli matrices) and if $n_f = 3$, the T^a are $T^a = \frac{1}{2}\lambda^a$, with λ^a the Gell-Mann matrices. In general, there are $n_f^2 - 1$ matrices T^a. We will let flavour indices $a, b, c \ldots$ run from 1 to $n_f^2 - 1$ and we will introduce the unit matrix $T^0 \equiv 1$. We may unify them by letting Greek indices $\alpha, \beta \ldots$ run from 0 to $n_f^2 - 1$. Then we define the currents,

$$
\begin{aligned}
V_\alpha^\mu(x) &= \sum_{ff'} : \bar{q}_f(x) T_{ff'}^\alpha \gamma^\mu q_{f'}(x) :, \\
A_\alpha^\mu(x) &= \sum_{ff'} : \bar{q}_f(x) T_{ff'}^\alpha \gamma^\mu \gamma_5 q_{f'}(x) : .
\end{aligned}
\tag{4.4.7}
$$

Sums over omitted colour indices are understood. Other currents can be built from these. For example, for $n_f = 3$, the electromagnetic current is

$$
J_{\text{em}}^\mu = V_3^\mu + \frac{1}{\sqrt{3}} V_8^\mu.
$$

We first discuss the flavour algebra. Considering the product of two Ts, we can write it as

$$
T^\alpha T^\beta = c_{\alpha\beta}^S T^0 + \sum_{c=1}^{n_f^2-1} c_{\alpha\beta c}^{NS} T^c = c_{\alpha\beta}^S + \sum_{c=1}^{n_f^2-1} c_{\alpha\beta c}^{NS} T^c.
\tag{4.4.8a}
$$

The superindices in the coefficients c^S, c^{NS} refer to the transformation properties under the flavour group: the unit matrix is a *singlet*, while the T^c transform as *nonsinglet*.

We will make explicit calculations for the case in which $\alpha = a \neq 0$, $\beta = b \neq 0$; if one of the α, β equals zero, the calculations are trivial. First of all, we have $c_{ab}^S = (1/2n_f)\delta_{ab}$, as is easily checked by taking traces. Writing

$$
T^a T^b = \frac{1}{2} \left(\{T^a, T^b\} + [T^a, T^b] \right)
$$

we can then split

$$
c_{abc}^{NS} = \frac{1}{2} \left(d_{abc}^{NS} + i f_{abc}^{NS} \right),
$$

where d_{abc}^{NS} is symmetric, and f_{abc}^{NS} antisymmetric, in the first two indices. For $n_f = 3$, we have

$$
d_{abc}^{NS} = d_{abc}, \qquad f_{abc}^{NS} = f_{abc},
$$

and the d, f are now as in Appendix C.

As an example, consider the electromagnetic current in the case $n_f = 4$. Here, it is more convenient to write it as

$$
J_{\text{em}}^\mu = \sum_{ff'} : \bar{q}_f \gamma^\mu Q_{ff'}^e q_{f'} :
\tag{4.4.8b}
$$

where $Q_e = Q^e$ is the matrix, in flavour space, of the charges of the quarks, u, d, s, c:

$$Q_e = \begin{pmatrix} 2/3 & 0 & 0 & 0 \\ 0 & -1/3 & 0 & 0 \\ 0 & 0 & -1/3 & 0 \\ 0 & 0 & 0 & 2/3 \end{pmatrix}. \tag{4.4.8c}$$

Then

$$Q_e^2 = \tfrac{5}{18} + T^{NS}, \tag{4.4.8d}$$

and

$$T^{NS} = \begin{pmatrix} 1/6 & 0 & 0 & 0 \\ 0 & -1/6 & 0 & 0 \\ 0 & 0 & -1/6 & 0 \\ 0 & 0 & 0 & 1/6 \end{pmatrix}$$

is a combination of the matrices T^3, T^8 and T^{15}.

For the complete calculation we begin, as stated, with the free field case and consider the T-product of two vector currents. With i, k, j, l flavour indices and α, β, δ, ρ Dirac ones, we can use Wick's theorem to get

$$
\begin{aligned}
TV_a^\mu(x)V_b^\nu(y) = & \sum T : \bar{q}_{i\alpha}(x)T_{ik}^a\gamma_{\alpha\beta}^\mu q_{k\beta}(x) :: \bar{q}_{j\delta}(y)T_{jl}^b\gamma_{\delta\rho}^\nu q_{l\rho}(y) : \\
& \underset{z^2\to 0}{\simeq} \frac{N_c\delta_{ab}(g^{\mu\nu}z^2 - 2z^\mu z^\nu)}{2n_f\pi^4(z^2 - i0)^4}\, 1 \\
& + \tfrac{1}{2}\sum(d_{abc}^{NS} + if_{abc}^{NS})\gamma_{\alpha\beta}^\mu S_{\beta\delta}(x-y)\gamma_{\delta\rho}^\nu : \bar{q}_\alpha(x)T^c q_\rho(y) : \\
& + \tfrac{1}{2}\sum(d_{abc}^{NS} - if_{abc}^{NS})\gamma_{\alpha\beta}^\nu S_{\beta\delta}(y-x)\gamma_{\delta\rho}^\mu : \bar{q}_\alpha(y)T^c q_\rho(x) : \\
& + \cdots,
\end{aligned}
\tag{4.4.9}
$$

where we have written the unit operator, 1, explicitly; $z = x - y$ and the dots stand for operators with four quark fields, $: \bar{q}q\bar{q}q :$. As explained in Sect. 3.9, these will give subleading contributions on the light cone, and for the moment we are only interested in leading effects, which is why we omit them. To obtain (4.4.9) we have repeatedly used the relation

$$Tq_{i\beta}(x)\bar{q}_{j\alpha}(y) = - : \bar{q}_{j\alpha}(y)q_{i\beta}(x) : +\delta_{ij}S_{\beta\alpha}(x-y),$$

and properties of the γ matrices; see Appendix A.

Next, we replace the propagator by its light cone behaviour,

$$S(z) \underset{z^2\to 0}{\simeq} \frac{2i\slashed{z}}{(2\pi)^2(z^2 - i0)^2},$$

which is easily obtained from the explicit expression

$$S(z) = \frac{i}{(2\pi)^4}\int d^4p\, e^{-ip\cdot z}\frac{\slashed{p} + m}{p^2 - m^2 + i0}.$$

After some manipulations of γ matrices, Eq. (4.4.9) can be simplified to

$$
TV_a^\mu(x)V_b^\nu(y) \underset{z^2 \to 0}{\simeq} i \sum (d_{abc}^{NS} + i f_{abc}^{NS})
$$

$$
\times \left\{ S^{\mu\alpha\nu\beta} \frac{z_\alpha}{(2\pi)^2(z^2 - i0)^2} : \bar{q}(x)T^c\gamma_\beta q(y) : \right.
$$

$$
\left. + i\epsilon^{\mu\alpha\nu\beta} \frac{z_\alpha}{(2\pi)^2(z^2 - i0)^2} : \bar{q}(x)T^c\gamma_\beta\gamma_5 q(y) : \right\}
$$

$$
+ (x \leftrightarrow y,\, a \leftrightarrow b,\, \mu \leftrightarrow \nu) + \text{(constant term)}.
$$

(4.4.10)

By "constant term" we denote the term

$$
\text{(constant term)} = \frac{N_c\delta_{ab}(g^{\mu\nu}z^2 - 2z^\mu z^\nu)}{2n_f\pi^4(z^2 - i0)^4}\,1.
$$

This term is actually dominating (as is clear on dimensional grounds) for the T-product $TV_a^\mu(x)V_b^\nu(y)$ itself, and indeed in other cases that we will see in the future; but for *deep inelastic scattering*, because the cross section involves the *discontinuity* of $TV_a^\mu V_b^\nu$, it simply does not contribute (being real) and can thus be neglected. Taking now $y = 0$, and expanding the regular operators $: \bar{q}\ldots q :$ in powers of z, we find the light-cone expansion:

$$
TV_a^\mu(z)V_b^\nu(0) \underset{z^2 \to 0}{\simeq} -\frac{i}{2} \sum_{n=\text{odd}} d_{abc}^{NS} \frac{S^{\mu\alpha\nu\beta}z_\alpha}{\pi^2(z^2 - i0)^2} \frac{z_{\mu_1}\ldots z_{\mu_n}}{n!}
$$

$$
\times : \bar{q}(0)T^c\gamma_\beta D^{\mu_1}\ldots D^{\mu_n} q(0) :
$$

$$
+ \frac{i}{2} \sum_{n=\text{odd}} f_{abc}^{NS} \frac{\epsilon^{\mu\alpha\nu\beta}z_\alpha}{\pi^2(z^2 - i0)^2} \frac{z_{\mu_1}\ldots z_{\mu_n}}{n!}
$$

(4.4.11)

$$
\times : \bar{q}(0)T^c\gamma_\beta\gamma_5 D^{\mu_1}\ldots D^{\mu_n} q(0) :
$$

$$
+ \text{constant term} + \text{gradient terms}
$$

$$
+ \text{terms odd under } (x \leftrightarrow y,\, a \leftrightarrow b,\, \mu \leftrightarrow \nu).
$$

We have not written explicitly those terms that are odd under the exchange $(x \leftrightarrow y,\, a \leftrightarrow b,\, \mu \leftrightarrow \nu)$. Also, we have brought all derivatives to act to the right by adding, if necessary, a gradient:

$$
(\bar{q}\overleftarrow{\partial})q = \partial(\bar{q}q) - \bar{q}(\overrightarrow{\partial}q).
$$

All of these terms give zero for the structure functions because we take diagonal matrix elements, $\langle p|TV_a^\mu V_b^\nu|p\rangle$. However, in other processes they have to be retained; in Sect. 5.7 we will study a situation in which the gradient terms have to be taken into account because nondiagonal matrix elements are involved.

It will prove convenient to rearrange (4.4.11) in a way that will lend itself to easiest comparison with the expression in terms of structure functions. This we will do in the case where we consider the T-product of two electromagnetic currents. In this situation, the term proportional to f^{NS} in (4.4.11) vanishes. If we do not show explicitly the terms that will give zero for deep inelastic scattering, we rewrite (4.4.10) simply as

$$
T J_{\text{em}}^{\mu}(z) J_{\text{em}}^{\nu}(0) \underset{z^2 \to 0}{\simeq} i \sum_{n=\text{odd}} \frac{S^{\mu\alpha\nu\beta} z_{\alpha}}{\pi^2 (z^2 - i0)^2} \frac{z_{\mu_1} \dots z_{\mu_n}}{n!}
$$
$$
\times : \bar{q}(0) Q_e^2 \gamma_{\beta} D^{\mu_1} \dots D^{\mu_n} q(0) : .
$$
(4.4.12)

Separating a term proportional to $g^{\mu\nu}$, whose matrix element will thus be identified with F_1, from another that will yield F_2, and changing slightly the notation, we find

$$
T J_{\text{em}}^{\mu}(z) J_{\text{em}}^{\nu}(0) \underset{z^2 \to 0}{\simeq} i \Bigg\{ g^{\mu\nu} \frac{1}{\pi^2 (z^2 - i0)^2} \sum_{n=\text{even}} z_{\mu_1} \dots z_{\mu_n} \frac{1}{(n-1)!}
$$
$$
\times : \bar{q}(0) Q_e^2 \gamma^{\mu_1} D^{\mu_2} \dots D^{\mu_n} q(0) :
$$
$$
+ \frac{-1}{2\pi^2 (z^2 - i0)^2} \sum_{n=\text{even}} z_{\mu_1} \dots z_{\mu_n} \frac{1}{n!}
$$
(4.4.13)
$$
\times [: \bar{q}(0) Q_e^2 \gamma^{\mu} D^{\nu} D^{\mu_1} \dots D^{\mu_n} q(0) : + (\mu \leftrightarrow \nu)] \Bigg\}.
$$

For the second term here we have written

$$
z_{\alpha}/(z^2 - i0)^2 = -\tfrac{1}{2} \partial_{\alpha} (z^2 - i0)^{-1}
$$

transferring then the derivative to act on the z_{μ_j}. Now only remains the flavour algebra. We separate the squared mass matrix Q_e^2 into a part proportional to the unit matrix, and hence singlet under flavour transformations, and a traceless matrix, hence nonsinglet. We write $Q_e^2 = c^S \mathbb{1} + T^{NS}$. For $n_f = 3$, a simple calculation gives $c^S = 2/9$, $T^{NS} = (1/3)T^3 + (1/3^{\frac{3}{2}})T^8$; for $n_f = 4$, see Eq. (4.4.8). Then,

$$
T J_{\text{em}}^{\mu}(z) J_{\text{em}}^{\nu}(0) \underset{z^2 \to 0}{\simeq} - g^{\mu\nu} \frac{i}{\pi^2 (z^2 - i0)^2} \sum_{n=\text{even}} z_{\mu_1} \dots z_{\mu_n} \frac{i^{n-1}}{n-1}
$$
$$
\times \left\{ N_{NS}^{(e)\mu_1 \dots \mu_n}(0) + c^S N_S^{(e)\mu_1 \dots \mu_n}(0) \right\}
$$
$$
+ \frac{i}{2\pi^2 (z^2 - i0)^2} \sum_{n=\text{even}} z_{\mu_1} \dots z_{\mu_n} i^{n-1}
$$
$$
\times \left\{ N_{NS}^{(e)\mu_1 \dots \mu_n}(0) + c^S N_S^{(e)\mu_1 \dots \mu_n}(0) + (\mu \leftrightarrow \nu) \right\},
$$
(4.4.14a)

and we have defined the operators

$$N_{NS}^{(e)\mu_1\cdots\mu_n}(x) = \frac{i^{n-1}}{(n-2)!} \sum_{ff'} : \bar{q}_f(x)\gamma^{\mu_1} D^{\mu_2} \ldots D^{\mu_n} T_{ff'}^{NS} q_{f'}(x) :,$$

$$N_S^{(e)\mu_1\cdots\mu_n}(x) = \frac{i^{n-1}}{(n-2)!} \sum_f : \bar{q}_f(x)\gamma^{\mu_1} D^{\mu_2} \ldots D^{\mu_n} q_f(x) : .$$

(4.4.14b)

To end this section, we will rederive the property of scaling from the light cone expansion. Consider, for example, Eq. (4.4.12). Taking matrix elements we have

$$T_{\text{em}}^{\mu\nu}(p,q) \overset{\text{Bj}}{=} (2\pi)^3 \Bigg\{ -\frac{g^{\mu\nu}}{\pi^2} \int d^4z \, e^{iq\cdot z} \sum_{n=\text{even}} \frac{iz_{\mu_1}\ldots iz_{\mu_n}}{(z^2 - i0)^2(n-1)} A_n^{\mu_1\cdots\mu_n}(p)$$
$$-\frac{1}{2\pi^2} \int d^4z \, e^{iq\cdot z} \sum_{n=\text{even}} \frac{iz_{\mu_1}\ldots iz_{\mu_n}}{z^2 - i0} [A_n^{\mu\nu\mu_1\cdots\mu_n}(p) + (\mu \leftrightarrow \nu)] \Bigg\},$$

(4.4.15a)

where the tag "Bj" means that equality holds asymptotically in the Bjorken limit, and we have defined the matrix elements

$$A_n^{\mu_1\cdots\mu_n}(p) = \frac{i^n}{(n-2)!} \langle p| : \bar{q}(0) Q_e^2 \gamma^{\mu_1} D^{\mu_2} \ldots D^{\mu_n} q(0) : |p\rangle.$$
(4.4.15b)

We may write the A in terms of invariants. Of these, there will be only one proportional to $p^{\mu_1}\ldots p^{\mu_n}$; all the others will contain the metric tensor $g^{\mu_i\mu_j}$ at least once. We will call these "trace terms', so we have

$$A_n^{\mu_1\cdots\mu_n}(p) = -ip^{\mu_1}\ldots p^{\mu_n} a_n + \text{trace terms}.$$
(4.4.15c)

The reason why we do not specify trace terms is that they will give contributions proportional to $p^2 = m_h^2$ (*target mass corrections*) that, on dimensional grounds, must be of the order m_h^2/Q^2, m_h^2/ν, and hence negligible in the Bjorken limit. Note that the a_n are pure numbers, as they only depend on p^2, which is constant. Then,

$$T_{\text{em}}^{\mu\nu}(p,q) \overset{\text{Bj}}{=} i(2\pi)^3 \Bigg\{ \frac{g^{\mu\nu}}{\pi^2} \int d^4z \, e^{iq\cdot z} \frac{1}{(z^2 - i0)^2} \sum_{n=\text{even}} (iz\cdot p)^n \frac{a_n}{n-1}$$
$$+ \frac{p^\mu p^\nu}{\pi^2} \int d^4z \, e^{iq\cdot z} \frac{1}{z^2 - i0} \sum_{n=\text{even}} (iz\cdot p)^n a_{n+2} \Bigg\}.$$

Comparing with (4.3.8b), we find

$$T_1^{\text{em}}(x, Q^2) = 8i\pi \int d^4z \, e^{iq\cdot z} \frac{1}{(z^2 - i0)^2} \sum_{n=\text{even}} (iz\cdot p)^n \frac{a_n}{n-1},$$

$$T_2^{\text{em}}(x, Q^2) = 8i\nu\pi \int d^4z \, e^{iq\cdot z} \frac{1}{z^2 - i0} \sum_{n=\text{even}} (iz\cdot p)^n a_{n+2}.$$

(4.4.16)

The final formula that we will require is

$$\frac{\partial}{\partial q_{\mu_1}} \cdots \frac{\partial}{\partial q_{\mu_n}} = 2^n q^{\mu_1} \cdots q^{\mu_n} \left(\frac{\partial}{\partial q^2}\right)^n + \text{trace terms}, \qquad (4.4.17)$$

for then, replacing the iz_μ by $\partial/\partial q^\mu$ and using for these (4.4.17), it follows that Eq. (4.4.16) can be written as

$$T_1^{\text{em}}(x, Q^2) \overset{\text{Bj}}{=} 8i\pi \sum_{\text{even}} \frac{2^n a_n}{n-1} q_{\mu_1} \cdots q_{\mu_n} p^{\mu_1} \cdots p^{\mu_n} \left(\frac{\partial}{\partial q^2}\right)^n \int d^4 z \, \frac{e^{iq \cdot z}}{(z^2 - i0)^2}$$

$$\overset{\text{Bj}}{=} - (2\pi)^3 \sum_{n=\text{even}} (2\nu)^n \frac{a_n}{n-1} \left(\frac{\partial}{\partial q^2}\right)^n \log q^2$$

$$= (2\pi)^3 \sum_{n=\text{even}} \frac{(n-2)! a_n}{x^n} = \text{independent of } Q^2.$$

$$(4.4.18a)$$

Likewise, we find

$$T_2^{\text{em}}(x, Q^2) \overset{\text{Bj}}{=} 2x T_1^{\text{em}}(x, Q^2). \qquad (4.4.18b)$$

Taking the imaginary part, we therefore obtain scaling and, moreover, $F_2 = 2xF_1$. This last relation, which implies that to the order at which we are working the longitudinal structure function vanishes, is known as the Callan–Gross (1969) relation; see also Bjorken and Paschos (1969).

Another derivation of scaling that makes apparent that $F_2(x)/x$ is the probability that the quark has fraction x of the total momentum may be found in Gross (1976).

4.5 The OPE for Deep Inelastic Scattering in QCD: Moments

Throughout the discussions of the previous section, the underlying field theory was unspecified except when it was free-field theory. Now we shall add substance to that earlier discussion.

We again consider a two-current T-product:

$$T J_P^\mu(x)^\dagger J_P^\nu(y), \qquad (4.5.1)$$

where P labels any current or combination of currents like those in (4.4.7). In our calculations we will still neglect terms that are suppressed by powers of M^2/Q^2, where M is any mass like, for example, the target mass or the quark masses. The OPE may be written by specifying a basis consisting of operators that give leading contributions, in powers of M^2/Q^2, for free fields: in QCD this will only be modified by logarithmic corrections. If one classifies operators according to their *twist* τ, where $\tau = \rho - j$ with ρ the (free field) dimension and j the spin of the operator, it is not difficult to see by mere dimensional analysis that the leading operators are those of twist 2. Operators of twist $\tau = 2 + 2n$ are suppressed by powers $(M^2/Q^2)^n$ with respect to the former.

Now, the only operators of twist 2 that can be formed, and which can be connected to (4.5.1), are

$$N_{NS,a\pm}^{\mu_1\ldots\mu_n} = \frac{\mathrm{i}^{n-1}}{(n-2)!} \, \mathcal{S} : \bar{q}(0)T^a\gamma^{\mu_1}(1\pm\gamma_5)D^{\mu_1}\ldots D^{\mu_n}q(0) :,$$

$$N_{S\pm}^{\mu_1\ldots\mu_n} = \frac{1}{2}\frac{\mathrm{i}^{n-1}}{(n-2)!} \, \mathcal{S} : \bar{q}(0)\gamma^{\mu_1}(1\pm\gamma_5)D^{\mu_1}\ldots D^{\mu_n}q(0) :, \qquad (4.5.2)$$

$$N_G^{\mu_1\ldots\mu_n} = \frac{\mathrm{i}^{n-2}}{(n-2)!} \, \mathcal{S}\,\mathrm{Tr} : G^{\mu_1\alpha}(0)D^{\mu_2}\ldots D^{\mu_{n-1}}G_\alpha{}^{\mu_n}(0) : \, .$$

The labels S/G denote fermion/vector boson (gluon) singlet operators; $\mathcal{S}$ stands for symmetrization, i.e., sum over permutations of the Minkowski indices divided by the number of such permutations. The trace refers to the colour indices and, finally,

$$D_\mu G_{\alpha\beta}^a = \sum_c \left\{ \partial_\mu \delta^{ac} + g \sum f^{abc} B_\mu^b \right\} G_{\alpha\beta}^c.$$

Among the operators in (4.5.2), we have already encountered the first two types, cf. Eq. (4.4.14) with $N_{NS}^e = \frac{1}{2}(N_{NS+} + N_{NS-})$. As for the third type, it is obvious that the only way in which currents made up of quarks can have nonzero projection on purely gluon operators is to take interactions into account, which is the reason why the N_G only appear now.

If we work in a gauge that requires ghosts, there are other operators beyond (4.5.2) that have to be considered; they are made up of ghosts. We can consider working in a ghostless gauge, say a lightlike gauge. For leading order calculations, it can be shown that one can also work in a gauge with ghosts, but these may be *neglected*. This is because the mixing matrix of the ghosts and other operators is triangular (cf. Kluberg-Stern and Zuber, 1975; Dixon and Taylor, 1974); so we forget about ghosts. With this, we can write the OPE for the product of currents (4.5.1) as follows:

$$\mathrm{T}J_P^\mu(z)^\dagger J_P^\nu(0) = -\sum_{j,n} \bar{C}_{1Pj}^n(z^2)g^{\mu\nu}\,\mathrm{i}^{n-1}z_{\mu_1}\ldots z_{\mu_n}N_j^{\mu_1\ldots\mu_n}(0)$$

$$-\sum_{j,n} \bar{C}_{2Pj}^n(z^2)\mathrm{i}^{n-1}z_{\mu_1}\ldots z_{\mu_n}N_j^{\mu\nu\mu_1\ldots\mu_n}(0)$$

$$+\sum_{j,n} \bar{C}_{3Pj}^n(z^2)\epsilon^{\mu\nu\alpha\beta}\,\mathrm{i}^{n-2}z_\beta z_{\mu_1}\ldots z_{\mu_n}N_j^{\alpha\mu_1\ldots\mu_n}(0),$$

$$(4.5.3)$$

where the sum over j runs over all the operators in (4.5.2) that have the same quantum numbers as the T-product of currents. In this context, it is worth noting that the flavour symmetries are preserved by the QCD interaction and therefore the flavour algebra can be carried over as in the free field case.

In the particularly important case of two electromagnetic currents, (4.5.3) becomes

$$
\begin{aligned}
\mathrm{T} J^{\mu}_{\mathrm{em}}(z) J^{\nu}_{\mathrm{em}}(0) = g^{\mu\nu} \Bigg\{ & \sum_{n=\text{even}} \bar{C}^{n}_{1NS}(z^2) N^{(e)\mu_1\cdots\mu_n}_{NS}(0) \\
& + c^{S}\bar{C}^{n}_{1S}(z^2) N^{(e)\mu_1\cdots\mu_n}_{S}(0) \Bigg\} \mathrm{i}^{n-1} z_{\mu_1}\cdots z_{\mu_n} \\
& + \sum_{n=\text{even}} \Bigg\{ \bar{C}^{n}_{2NS}(z^2) N^{(e)\mu\nu\mu_1\cdots\mu_n}_{NS}(0) \\
& + c^{S}\bar{C}^{n}_{2S}(z^2) N^{(e)\mu\nu\mu_1\cdots\mu_n}_{S}(0) \Bigg\} \mathrm{i}^{n-1} z_{\mu_1}\cdots z_{\mu_n} \\
& + \Bigg\{ g^{\mu\nu} \sum_{n=\text{even}} c^{S}\bar{C}^{n}_{1G}(z^2) N^{\mu_1\cdots\mu_n}_{G}(0) \\
& + \sum_{n=\text{even}} c^{S}\bar{C}^{n}_{2G}(z^2) N^{\mu\nu\mu_1\cdots\mu_n}_{G}(0) \Bigg\} \mathrm{i}^{n-2} z_{\mu_1}\cdots z_{\mu_n}
\end{aligned}
\tag{4.5.4}
$$

and the $N^{(e)}_{S,NS}$ are as in Eq. (4.4.14b). Thus, we have symmetrized the N in the Minkowski indices; this is permissible if, as occurs in our case, only diagonal matrix elements are required and terms $O(m_h^2/Q^2)$ are neglected; cf. Eqs. (4.4.15b,c).

In fact, both (4.5.3) and (4.5.4) have been written somewhat sketchily. When taking into account interactions, renormalization will occur. This causes, among others, two important effects. First, because the operators N_S, N_G have the same quantum numbers (those of a flavour singlet), they will mix under renormalization: only the N_{NS} operators are renormalized by themselves. Secondly, renormalization introduces a dependence of the $\bar{C}$, N on a dimensional parameter that we will temporarily denote by μ to avoid confusion with the variable $\nu = p \cdot q$.

The currents J of the form

$$
J^{\mu}(x) = a V^{\mu}(x) + b A^{\mu}(x),
\tag{4.5.5}
$$

do not require specific renormalization because the operators V, A are conserved or quasi–conserved (see Sect. 3.6). However, except in special instances, the operators N require renormalization, and so do the Wilson coefficients $\bar{C}$.

For the nonsinglet operators, which do not mix, renormalization reads[11]

$$N_{NS,a\pm R}^{\mu_1\cdots\mu_n} = Z_{n-2}^{a\pm}(\mu)N_{NS,a\pm}^{\mu_1\cdots\mu_n}. \tag{4.5.6a}$$

Actually, the Z are independent of a and of whether we have $\pm$.

For the singlet operators, on the other hand, we have matrix renormalization

$$\mathbf{N}_R^{\mu_1\cdots\mu_n} = \mathbf{Z}_{n-2}(\mu)\mathbf{N}^{\mu_1\cdots\mu_n}, \tag{4.5.6b}$$

and we have defined the matrices

$$\mathbf{N} = \begin{pmatrix} N_S \\ N_G \end{pmatrix}, \tag{4.5.6c}$$

and

$$\mathbf{Z} = \begin{pmatrix} Z_{SS} & Z_{SG} \\ Z_{GS} & Z_{GG} \end{pmatrix}. \tag{4.5.6d}$$

With this, we define the anomalous dimension and anomalous dimension matrix for the operators N,

$$
\begin{aligned}
\gamma_{NS}(n,g) &= -Z_n(\mu)^{-1}\frac{\mu\partial}{\partial\mu}Z_n(\mu), \\
\boldsymbol{\gamma}(n,g) &= -\mathbf{Z}_n(\mu)^{-1}\frac{\mu\partial}{\partial\mu}\mathbf{Z}_n(\mu),
\end{aligned}
\tag{4.5.7}
$$

and their expansions:

$$
\begin{aligned}
\gamma_{NS}(n,g) &= \sum_{k=0}^{\infty}\gamma_{NS}^{(k)}\left(\frac{g^2}{16\pi^2}\right)^{k+1}, \\
\boldsymbol{\gamma}(n,g) &= \sum_{k=0}^{\infty}\boldsymbol{\gamma}^{(k)}\left(\frac{g^2}{16\pi^2}\right)^{k+1}.
\end{aligned}
\tag{4.5.8}
$$

We will pursue this matter of renormalization later; for the moment we return to the formal machinery, that follows closely the free field derivation of the previous section. Consider momentum space and write the part of the OPE that contributes to the nonsinglet piece of the structure function F_2, i.e., to the part of F_2 that contains the nonsinglet operators. This is the quantity for which we will carry out detailed calculations; later, we will present results for other structure functions as well as for the singlet piece.

Selecting the appropriate portion of (4.5.3), we have

$$
\begin{aligned}
&\mathrm{i}\int \mathrm{d}^4z\, \mathrm{e}^{\mathrm{i}q\cdot z}\mathrm{T}J^\mu(z)J^\nu(0)\Big|_{p^\mu p^\nu}^{NS} \\
&= \sum_n \int \mathrm{d}^4z\, \mathrm{e}^{\mathrm{i}q\cdot z}\bar{C}_{2NS}^n(z^2)\mathrm{i}^n z_{\mu_1}\ldots z_{\mu_n}N_{NS}^{\mu\nu\mu_1\cdots\mu_n}(0);
\end{aligned}
\tag{4.5.9}
$$

---

[11]Note that, as in Sect. 3.6, the quark or gluon fields entering the N are assumed to be *renormalized*.

so, if we take the matrix element relevant to deep inelastic scattering, as in Eq. (4.3.8a), we find

$$\frac{p^\mu p^\nu}{\nu} T_{2NS} = (2\pi)^3 \sum_n \int d^4z \, e^{iq\cdot z} \bar{C}_{2NS}^n(z^2) i^n z_{\mu_1} \ldots z_{\mu_n} \langle p|N_{NS}^{\mu\nu\mu_1\ldots\mu_n}(0)|p\rangle.$$

(4.5.10)

We can write, up to trace terms,

$$i\langle p|N_{NS}^{\mu\nu\mu_1\ldots\mu_n}(0)|p\rangle = p^\mu p^\nu p^{\mu_1} \ldots p^{\mu_n} \bar{A}_{NS}^n$$

(4.5.11)

and replace

$$i^n z_{\mu_1} \ldots z_{\mu_n} \to \frac{\partial}{\partial q_{\mu_1}} \ldots \frac{\partial}{\partial q_{\mu_n}} = 2^n q_{\mu_1} \ldots q_{\mu_n} \left(\frac{\partial}{\partial q^2}\right)^n + \text{trace terms},$$

(4.5.12)

so that (4.5.10) becomes

$$T_{2NS}(x, Q^2; g, \mu)$$

$$= (2\pi)^3 \nu \sum_{n=\text{even}} 2^n \bar{A}_{NS}^n \left(\frac{\partial}{\partial q^2}\right)^n \int d^4z \, e^{iq\cdot z} \frac{1}{i} \bar{C}_{2NS}^n(z^2)(q\cdot p)^n$$

(4.5.13)

$$= \tfrac{1}{2}(2\pi)^3 \sum_{n=\text{even}} (2\nu)^{n+1} \bar{A}_{NS}^n \left(\frac{\partial}{\partial q^2}\right)^n \int d^4z \, e^{iq\cdot z} \frac{1}{i} \bar{C}_{2NS}^n(z^2).$$

Because of the calculations of Sect. 4.4 we know that, in the free field case, $\bar{C}_{2NS}^n(z^2)$ behaves on the light cone as

$$i\bar{C}_{2NS}^n(z^2)\big|_{g=0} \underset{z^2\to 0}{=} \frac{1}{\pi^2(z^2 - i0)},$$

(4.5.14)

so we will define new coefficients taking this into account; in momentum space, we therefore set

$$C_{2NS}^n(Q^2/\mu^2, g^2/4\pi) \equiv 4(Q^2)^{n+1} \left(\frac{\partial}{\partial q^2}\right)^n \int d^4z \, e^{iq\cdot z} \frac{1}{i} \bar{C}_{2NS}^n(z^2). \quad (4.5.15)$$

We thus arrive at the expression, the analogue of (4.4.18),

$$T_{2NS}(x, Q^2; g, \mu) = 2 \sum \frac{1}{x^{n+1}} A_{NS}^n C_{2NS}^n(Q^2/\mu^2, g^2/4\pi), \quad A^n \equiv (2\pi)^3 \bar{A}^n.$$

(4.5.16)

As we will see later, asymptotic freedom allows us to calculate the Wilson coefficients C in (4.5.16); but in general the A are unknown constants. To be able to extract physical information, we have to single out the individual terms in Eq. (4.5.16). This is done as follows. From the known analyticity properties of the T, it follows that we can write a dispersion relation[12] for T_2 at fixed Q^2, in the variable ν:

$$T_{2NS}(x, Q^2; g, \mu) = \frac{1}{\pi}\left\{\int_{Q^2/2}^{\infty} \frac{d\nu'}{\nu' - \nu} \operatorname{Im} T_{2NS}\left(\frac{Q^2}{2\nu'}, Q^2; g, \mu\right)\right.$$
$$\left. - \int_{-\infty}^{-Q^2/2} \frac{d\nu'}{\nu' - \nu} \operatorname{Im} T_{2NS}\left(\frac{Q^2}{2\nu'}, Q^2; g, \mu\right)\right\}.$$

$$(4.5.17)$$

One can only relate this to the physical structure functions if T has a definite signature, i.e., is even or odd under the exchange $q \to -q$; this is the case for T_2 in electroproduction as $T_2(x, \ldots) = T_2(-x, \ldots)$. In the general case we would have to consider symmetric or antisymmetric combinations of T-products, and structure functions. Then we change variables in (4.5.17), to get

$$T_{2NS}(x, Q^2; g, \mu) = \frac{1}{\pi}\int_0^1 \frac{dx'}{x'(1 - x'^2/x^2)} \operatorname{Im} T_{2NS}(x', Q^2; g, \mu).$$

It only remains to expand in powers of x'/x to obtain (Cornwall and Norton, 1969)

$$T_{2NS}(x, Q^2; g, \mu) = 2\sum_n \frac{1}{x^n}\mu_{2NS}(n + 1, Q^2; g, \mu), \qquad (4.5.18)$$

where the *moments* μ_{2NS} are defined by

$$\mu_{2NS}(n, Q^2; g, \mu) = \int_0^1 dx\, x^{n-2} F_{2NS}(x, Q^2; g, \mu). \qquad (4.5.19)$$

One also says that the $\mu(n)$ is the *Mellin transform* of $F(x)$. Comparing with (4.5.16) we immediately obtain the expression for the moments:

$$\mu_{2NS}(n, Q^2; g, \mu) = A_{NS}^n C_{2NS}^n(Q^2/\mu^2, g^2/4\pi). \qquad (4.5.20)$$

It should be kept in mind that we have derived Eqs. (4.5.19) and (4.5.20) under the assumption of evenness for T: otherwise, we cannot replace the integral $\int_{-1}^0 dx'$ by $\int_0^1 dx'$. Therefore, Eqs. (4.5.19, 20) are only valid for $n =$ even if T is even, as is the case for T_2 in electron scattering, or for $n =$ odd if T was odd, as occurs for T_3 in neutrino scattering by a singlet target. The corresponding equations for other n have to be obtained by analytic (Regge–Carlson) continuation. This is rather trivial for the leading order calculations

---

[12]In principle, the dispersion relation should be written with subtractions, but it may be seen that these alter nothing of what follows, provided that the integral in Eq. (4.5.19) below is convergent. For information on dispersion relations, see the treatise of Eden, Landshoff, Olive and Polkinghorne (1966).

(see Sect. 4.6) but less straightforward for higher order ones. Another point is that, as already noted, we have to restrict the above equations to values of n such that the integral in (4.5.19) converges. From Regge theory, and sum rule considerations (topics which will be discussed later on), we expect that this will occur for $\mathrm{Re}\, n \geq 1$ for the nonsinglet structure functions, and for $\mathrm{Re}\, n \geq 2$ for singlet ones.

4.6 Renormalization Group Analysis:
the QCD Equations for the Moments

We will now write a renormalization group equation for the moments. Since these are integrals over the structure functions, they are physical observables and hence are independent of the renormalization point. As a result of this and of Eqs. (4.5.6), (4.5.11) and (4.5.20), it follows that the renormalization constant of the Wilson coefficients C must be precisely the inverse of that of the operators N. Thus we obtain renormalization group equations. For the nonsinglet case, we have the Callan–Symanzik type equation,

$$\left\{ \mu\frac{\partial}{\partial\mu} + \beta(g)g\frac{\partial}{\partial g} - \gamma_{NS}(g,n) \right\} C^n_{2NS}(Q^2/\mu^2, g^2/4\pi) = 0, \qquad (4.6.1)$$

with solution

$$C^n_{2NS}(Q^2/\mu^2, g^2/4\pi) = e^{-\int_0^t dt'\gamma_{NS}(g(Q'^2),n)} C^n_{2NS}\left(1, \alpha_s(Q^2)\right),$$
$$t = \tfrac{1}{2}\log Q^2/\mu^2, \quad t' = \tfrac{1}{2}\log Q'^2/\mu^2. \qquad (4.6.2)$$

For the singlet case there are complications due to the coupled character of the equations. It is necessary to introduce an extra structure function, denoted by $F_G(x, Q^2)$ or $xG(x, Q^2)$ whose physical interpretation is that it describes the (momentum) density of the *gluons* one finds in the target:[13]

$$\mathbf{F} = \begin{pmatrix} F_S \\ F_G \end{pmatrix}, \quad \mathbf{C}^n = \begin{pmatrix} C^n_S \\ C^n_G \end{pmatrix},$$
$$\boldsymbol{\mu}_2(n, Q^2) = \int_0^1 dx\, x^{n-2} \mathbf{F}_2(x, Q^2). \qquad (4.6.3)$$

The analogue of (4.6.2) is now (Gross and Wilczek, 1974; Gross, 1976)

$$\mathbf{C}^n(Q^2/\mu^2, g^2/4\pi) = \mathrm{Te}^{-\int_0^t dt'\boldsymbol{\gamma}(g(Q'^2),n)} \mathbf{C}^n_{2NS}\left(1, \alpha_s(Q^2)\right). \qquad (4.6.4)$$

The operator T is formally identical to time ordering, except that now it orders in the variable $t = \tfrac{1}{2}\log Q^2/\mu^2$.

---

[13] The structure function G is not uniquely defined beyond the leading order. We will see this in some detail when we discuss higher order calculations in Sect. 4.7iii.

Because of asymptotic freedom we can use perturbation theory if Q^2 is large enough that $\alpha_s(Q^2)$ is small, and calculate the Wilson coefficients from Eqs. (4.6.2) and (4.6.4). However, since the A_n are still unknown, we will only be able to predict the *evolution* of the moments with Q^2. To see this, consider (4.6.2) to lowest order. We obtain

$$C_{2NS}^n(Q^2/\mu^2, g^2/4\pi) = C_{2NS}^n(1, 0) \left(\frac{\log Q^2/\Lambda^2}{\log \mu^2/\Lambda^2} \right)^{d_{NS}(n)}, \qquad (4.6.5)$$

where the anomalous dimension[14] d_{NS} is

$$d_{NS}(n) = -\gamma_{NS}^{(0)}(n)/2\beta_0. \qquad (4.6.6a)$$

$C_{2NS}^n(1, 0)$ is merely the free field value of the Wilson coefficient, which we calculated in Sect. 4.4. One can then eliminate the A^n by normalizing to a reference Q_0^2 sufficiently large that $\alpha_s(Q_0^2)$ will also be small. We then obtain the QCD evolution equations for the moments to leading order: dropping unnecessary labels,

$$\mu_{NS}(n, Q^2) = \left[\frac{\alpha_s(Q_0^2)}{\alpha_s(Q^2)} \right]^{d_{NS}(n)} \mu_{NS}(n, Q_0^2). \qquad (4.6.6b)$$

For the singlet,

$$\boldsymbol{\mu}(n, Q^2) = \left[\frac{\alpha_s(Q_0^2)}{\alpha_s(Q^2)} \right]^{\mathbf{D}(n)} \boldsymbol{\mu}(n, Q_0^2),$$

$$\mathbf{D}(n) = -\boldsymbol{\gamma}^{(0)}(n)/2\beta_0. \qquad (4.6.7)$$

It only remains for us to calculate the anomalous dimensions $\gamma_{NS}^{(0)}$ and $\boldsymbol{\gamma}^{(0)}$. To do this, we first have to deduce the Feynman rules for the operators N. This is straightforward (see Sect. 2.6); they are collected in Appendix E. Then we have to evaluate the renormalization constants for the N. The singlet case may be found in Gross and Wilczek (1974) and Georgi and Politzer (1974); here we will concentrate on the $N_{NS}^{\mu_1\cdots\mu_n}$, which involves the diagrams of Fig. 4.6.1. In the Feynman gauge, the diagram of Fig. 4.6.1a gives

$$V_{Aij} = i^5 g^2 \int d^D\hat{k} \, \frac{\gamma^\mu \slashed{k} \slashed{\Delta}(\Delta\cdot k)^{n-1}\slashed{k}\gamma^\nu(-g_{\mu\nu})}{k^4(k-p)^2} \sum_{a,l} t_{jl}^a t_{li}^a$$

$$= ig^2 \delta_{ij} C_F$$

$$\times \int_0^1 dx\,(1-x) \int d^D\hat{l} \, \frac{-2\gamma^\mu(\slashed{l}+x\slashed{p})\slashed{\Delta}(\slashed{l}+x\slashed{p})\gamma_\mu[\Delta\cdot(l+xp)]^{n-1}}{(l^2+x(1-x)p^2)^3}.$$

---

[14]The name "anomalous dimension" is used for both the γ and the $d \equiv \gamma/2\beta_0$.

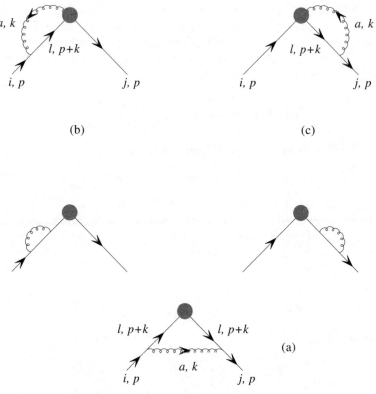

Fig. 4.6.1. Diagrams involved in the calculation of Z_n^{NS}.

To calculate Z we only require the divergent part of the coefficient of

$$(\Delta \cdot p)^{n-1} \slashed{\Delta}.$$

Let us use the notation $a \overset{\text{eff}}{=} b$ to mean that a and b have equal such divergent parts. Then, identifying the coefficient of $(\Delta \cdot p)^{n-1} \slashed{\Delta}$ in V,

$$V_{Aij} \overset{\text{eff}}{=} ig^2 C_F \delta_{ij} \int_0^1 dx \, (1-x) \int \frac{d^D \hat{l}}{(l^2 + x(1-x)p^2)^3}$$
$$\times \left\{ -\frac{2l^2}{D} \gamma^\alpha \gamma^\beta \slashed{\Delta} \gamma_\beta \gamma_\alpha x^{n-1} \right\} (\Delta \cdot p)^{n-1} \slashed{\Delta} \qquad (4.6.8)$$
$$= \frac{g^2}{16\pi^2} N_\epsilon C_F \delta_{ij} \frac{2}{n(n+1)} (\Delta \cdot p)^{n-1} \slashed{\Delta}.$$

The diagram of Fig. 4.6.1b gives

$$V_{Bij} = -i^3 g^2 C_F \delta_{ij} \int d^D \hat{k} \frac{\Delta^\mu \slashed{\Delta} \sum_{l=0}^{n-2} (\Delta \cdot p)^l [\Delta \cdot (p+k)]^{n-l-2} (\slashed{p} + \slashed{k}) \gamma_\mu}{k^2 (k+p)^2}.$$

Here we also have to extract the coefficient of the $(\Delta \cdot p)^{n-1} \slashed{\Delta}$ term; so,

$$V_{Bij} \stackrel{\text{eff}}{=} 2i g^2 C_F \delta_{ij} \slashed{\Delta} \int_0^1 dx \int d^D \hat{r} \frac{\sum_{l=0}^{n-2} (\Delta \cdot p)^l [\Delta \cdot r + x \Delta \cdot p]^{n-l-1}}{(r^2 + x(1-x)p^2)^2}$$

$$\stackrel{\text{eff}}{=} -2 \frac{g^2 N_\epsilon}{16\pi^2} C_F \delta_{ij} (\Delta \cdot p)^{n-1} \slashed{\Delta} \int_0^1 dx \sum_{l=1}^{n-1} x^l$$

$$= \frac{g^2}{16\pi^2} N_\epsilon C_F \delta_{ij} \left(-2 \sum_{j=2}^n \frac{1}{j} \right) (\Delta \cdot p)^{n-1} \slashed{\Delta}.$$

$$(4.6.9)$$

The diagram in Fig. 4.6.1c gives the same result as that we have just calculated. The two diagrams without tags in Fig. 4.6.1 merely give, for non-amputated matrix elements, contributions equivalent to the wave function renormalization, Z_F. To obtain γ_{NS} we still have to add the counterterm contribution, to obtain $Z_n Z_F^{-1} N$ finite. Therefore, using the value of Z_F calculated in Sect. 3.3, we find,

$$Z_n^{NS} = 1 + \frac{g^2 N_\epsilon}{16\pi^2} C_F \left\{ 4S_1(n) - 3 - \frac{2}{n(n+1)} \right\}, \qquad (4.6.10)$$

$$S_1(n) \equiv \sum_{j=1}^n \frac{1}{j}, \qquad (4.6.11)$$

and thus

$$\gamma_{NS}^{(0)}(n) = 2C_F \left\{ 4S_1(n) - 3 - \frac{2}{n(n+1)} \right\} \qquad (4.6.12)$$

and

$$d_{NS}(n) = \frac{16}{33 - 2n_f} \left\{ \frac{1}{2n(n+1)} + \frac{3}{4} - S_1(n) \right\}. \qquad (4.6.13)$$

Likewise, for the singlet case we obtain the matrices

$$\mathbf{D}(n) = -\frac{\boldsymbol{\gamma}^{(0)}(n)}{2\beta_0}, \qquad \boldsymbol{\gamma}^{(0)} = -\frac{32}{3} \times$$

$$\times \begin{pmatrix} \frac{1}{2n(n+1)} + \frac{3}{4} - S_1(n) & \frac{3}{8} n_f \frac{n^2 + n + 2}{n(n+1)(n+2)} \\ \frac{n^2 + n + 2}{2n(n^2 - 1)} & \frac{33 - 2n_f}{16} + \frac{9}{4} \left[\frac{1}{n(n-1)} + \frac{1}{(n+1)(n+2)} - S_1(n) \right] \end{pmatrix}.$$

$$(4.6.14)$$

The function $S_1(n)$ can be continued analytically to complex n. Due to Carlson's theorem (see, e.g., Titchmarsh, 1939) there is only one such continuation with the property that Eqs. (4.5.19), (4.6.3), (4.6.6) and (4.6.7) remain valid for complex n; it is

$$S_\nu(n) = \sum_{k=1}^{\infty} \left(\frac{1}{k^\nu} - \frac{1}{(k+n)^\nu} \right), \qquad (4.6.15\text{a})$$

and we have profited to give a generalized definition which will cover other functions that will appear later. For the case of interest to us now, one has the relation to the digamma function,

$$S_1(n) = \psi(n+1) + \gamma_{\rm E}, \quad \psi(z) \equiv \frac{{\rm d}\log \Gamma(z)}{{\rm d}z}, \qquad (4.6.15\text{b})$$

where $\gamma_{\rm E} \simeq 0.5772$ is Euler's constant.

To this order, there is no problem with even/odd structure functions, nor with the corresponding validity of the original equations for only even/odd values of n, because the continuations of the $\gamma^{(0)}(n)$ starting from even or odd values of n coincide.

4.7 QCD Equations for the Moments to Second and Higher Orders

i Nonsinglet

In the previous section we derived the QCD equations for the evolution of the moments to leading order. Now we will turn to second and higher order corrections. Because, as we will see immediately, in deep inelastic scattering one and two loop corrections (for example) get mixed, it is customary to speak of *leading order* (LO) calculations, precisely the ones carried over in the preceding section; next-to-leading order (NLO), or evaluations pushed to one extra order in α_s; next-to-next-to-leading order (NNLO), etc. From LO onwards the calculations are very lengthy and complicated, and the results long and uninspiring. We will here present explicitly some of the simplest results, and the simplest calculation, just to give a flavour of what is involved; detailed references will be quoted so that the interested reader may find explicit results and evaluations.

From Eqs. (4.6.2) and (4.6.4) we see that to calculate next-to-leading contributions we have to consider two separate effects besides, of course, using the two loop expression for $\alpha_s(Q^2)$, Sect. 3.7, and taking into account the finite parts of the LO diagrams that we calculated in the previous section. First, we have the effect of the anomalous dimensions to two loops,

$\gamma_{NS}^{(1)}(n)$ and $\boldsymbol{\gamma}^{(1)}(n)$. Then, we must calculate the next term (one loop) in the expansion of the Wilson coefficients; for the nonsinglet,

$$C_{NS}^n(1, \alpha_s(Q^2)) = C_{NS}^n(1,0) \left\{ 1 + C_{NS}^{(1)n}(1,0) \frac{\alpha_s(Q^2)}{4\pi} + \cdots \right\}. \qquad (4.7.1)$$

The calculation of the nonsinglet anomalous dimensions was carried out first by Floratos, Ross and Sachrajda (1977); the results were simplified, and some errors corrected, by González–Arroyo, López and Ynduráin (1979). These results were checked by Curci, Furmanski and Petronzio (1980). The NNLO anomalous dimension $\gamma_{NS}^{(2)}(n)$ is not known for all n, although some specific values have been evaluated recently by Larin, Nogueira, van Ritbergen and Vermaseren (1997). Letting the indices $\pm$ refer to even/odd structure functions and with ζ being Riemann's funtion,

$$\zeta(n) = \sum_{j=1}^{\infty} 1/j^n,$$

and the definitions

$$S_l^+(x/2) \equiv S_l(x/2), \quad S_l^-(x/2) \equiv S_l(x/2 - 1/2),$$

$$\widetilde{S}^{\pm}(x) \equiv \tfrac{5}{8}\zeta(3) \mp \sum_{k=1}^{\infty} \frac{(-1)^k}{(k+x)^2} S_1(k+x),$$

we have

$$\gamma_{NS}^{(1)\pm}(n) = \tfrac{32}{9} S_1(n) \left[67 + 8\frac{2n+1}{n^2(n+1)^2} \right] - 64 S_1(n) S_2(n)$$

$$- \tfrac{32}{9} \left[S_2(n) - S_2^{\pm}(n/2) \right] \left\{ 2S_1(n) - \frac{1}{n(n+1)} \right\}$$

$$- \tfrac{128}{9} \widetilde{S}^{\pm}(n) + \tfrac{32}{3} S_2(n) \left\{ \frac{3}{n(n+1)} - 7 \right\} + \tfrac{16}{9} S_3^{\pm}(n/2)$$

$$- 28 - 16\frac{151n^4 + 260n^3 + 96n^2 + 3n + 10}{9n^3(n+1)^3}$$

$$\pm \tfrac{32}{9} \frac{2n^2 + 2n + 1}{n^3(n+1)^3} + \frac{32n_f}{27} \left\{ 6S_2(n) - 10S_1(n) + \tfrac{3}{4} + \frac{11n^2 + 5n - 3}{n^2(n+1)^2} \right\}.$$
$$(4.7.2a)$$

Let us now turn to the Wilson coefficients. Since these are constants, they can be calculated by taking matrix elements of $TJ^\mu J^\nu$ between any states. We are at liberty to take whichever make the calculation simplest, and, of course, we choose quark and gluon states. A point to be kept in mind is that, unlike for the anomalous dimension, the Wilson coefficients also depend on the process under consideration. The coefficients to one loop

(NLO calculation) have been evaluated by a number of people.[15] Here, we give the values for the electroproduction on proton targets:

$$C_{NS}^{(1)}(n) =$$

$$C_F \left\{ 2S_1(n)^2 + 3S_1(n) - 2S_2(n) - \frac{2S_1(n)}{n(n+1)} + \frac{3}{n} + \frac{4}{n+1} + \frac{2}{n^2} - 9 \right\}.$$
(4.7.2b)

For other processes, see the compilation of Buras (1980). The *two loop* Wilson coefficients would enter the NNLO calculation; they have been evaluated by van Neerven and Zijlstra (1991a,b,c), (1992a,c), to which we refer for the explicit expressions. They are checked for particular values of n in the evaluation of Larin, Nogueira, van Ritbergen and Vermaseren (1997).

Having calculated the anomalous dimension (to two loops) and the coefficient (to one loop), the NLO evolution equation for the moments is immediately written:

$$\mu_{NS}(n, Q^2) = \frac{1 + C_{NS}^{(1)}(n)\alpha_s(Q^2)/4\pi}{1 + C_{NS}^{(1)}(n)\alpha_s(Q_0^2)/4\pi} \left\{ \frac{1 + (\beta_1/\beta_0)\alpha_s(Q^2)/4\pi}{1 + (\beta_1/\beta_0)\alpha_s(Q_0^2)/4\pi} \right\}^{p(n)}$$

$$\times \left[\frac{\alpha_s(Q_0^2)}{\alpha_s(Q^2)} \right]^{d_{NS}(n)} \mu_{NS}(n, Q_0^2),$$

$$p(n) = \tfrac{1}{2} \left[\frac{\gamma^{(1)}(n)}{\beta_1} - \frac{\gamma^{(0)}(n)}{\beta_0} \right].$$

(4.7.3)

ii Longitudinal Structure Function

To LO the two structure functions F_2 and $2xF_1$ are equal and hence to this order the longitudinal structure function $F_L = F_2 - 2xF_1$ vanishes. This was shown in Sect. 4.4 for free fields; but since leading order QCD corrections only multiply $C_L^n(1,0)$ by a factor of $(\log Q^1/\Lambda^2)^\delta$, $\delta = d_{NS}$ or $\mathbf{D}$, it follows that all moments of F_L vanish to this order, as claimed. This means that for the longitudinal case (4.7.1) should actually read

$$C_L^n(1, \alpha_s) = C_L^n(1, 0) \left\{ \frac{\alpha_s}{4\pi} + \cdots \right\}.$$

---

[15]Kingsley (1973); Walsh and Zerwas (1973); Zee, Wilczek and Treiman (1974); Witten (1976); De Rújula, Georgi and Politzer, (1977a); Calvo (1977); Hinchliffe and Llewellyn Smith (1977); Altarelli, Ellis and Martinelli (1978); Abad and Humpert (1978); Kubar-André and Paige (1979); Floratos, Ross and Sachrajda (1979),etc. The values reported by Bardeen, Buras, Duke and Muta (1978) or Buras (1980, 1981) have all been checked by at least two independent calculations.

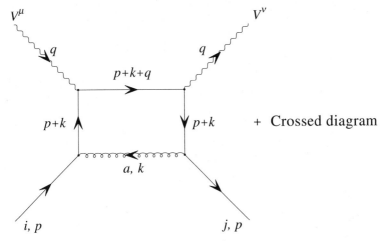

Fig. 4.7.1. Diagrams contributing to F_L^{NS}. The "Crossed diagram" is a diagram with the photon lines exchanged.

To NLO, however, we get a nonzero F_L and hence violations of the Callan–Gross relation. To evaluate this it is convenient to extract a factor that depends on the process, and a process-independent part, writing

$$C_{PL}^{(1)n}(1,0) = \delta_P B_L^{(1)n}. \qquad (4.7.4)$$

The factors are, denoting a proton or a neutron by N, and an "isoscalar" nucleon (average of p, n) by I,

$$\delta_{PNS} = \begin{cases} \frac{1}{6}, & \text{for } F_2^{eN} \\ 1, & \text{for } F_2^{(\nu,\bar{\nu})I} \end{cases} ; \quad \delta_{PF} = \begin{cases} \frac{5}{18}, & \text{for } F_{2F}^{eN} \\ 1, & \text{for } F_2^{(\nu,\bar{\nu})I} \end{cases} , \quad n_f = 4. \quad (4.7.5)$$

Let us then consider the nonsinglet piece of F_L, F_L^{NS}. The only diagram that enters the calculation is that of Fig. 4.7.1; all other diagrams either contribute equally to the terms of which F_L^{NS} is the difference, or are singlet. In fact, for the singlet component of the longitudinal structure function we have two contributions: that of the quark singlet which, to the present order of accuracy, coincides with the nonsinglet one, and the *gluon* contribution, evaluated with the help of the diagram of Fig. 4.7.2.

To NLO, and since F_L begins at order α_s, we do not have to worry about the contribution of the renormalization of the operators N which will, in the present case, give effects of order α_s^2. The calculation is further simplified by

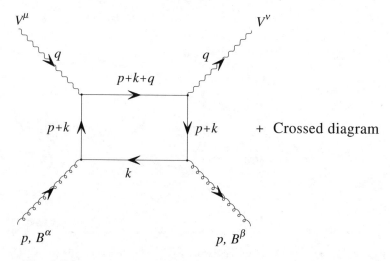

Fig. 4.7.2. Diagram contributing to the gluon component of F_L.

noting that, if we keep terms proportional to $q^\mu q^\nu$ in $T^{\mu\nu}$, then F_L is the only invariant amplitude which is multiplied by them: for, say, vector currents

$$T^{\mu\nu} = \left(g^{\mu\nu} - \frac{q^\mu q^\nu}{q^2}\right) T_L + \left(g^{\mu\nu} - p^\mu p^\nu \frac{q^2}{\nu} + \frac{p^\mu q^\nu + p^\nu q^\mu}{\nu}\right) T_2, \quad (4.7.6)$$

$F_L = (1/2\pi) \operatorname{Im} T_L$. In general, we have to carry out the calculation for $p^2 < 0$ to regulate infrared divergences; but, again, this is unnecessary for F_L to the order at which we are working, as it remains finite in the limit $p^2 \to 0$.

The amplitude for the diagram of Fig. 4.7.1 is then

$$\frac{i}{2}(2\pi)^3 \sum_\sigma \int d^4z\, e^{iq\cdot z} \langle p,\sigma; j| T J^\mu(z) J^\nu(0) | p,\sigma; i\rangle$$

$$\equiv T'^{\mu\nu}_{ij} + \text{crossed term},$$

$$T'^{\mu\nu}_{ij} = -\frac{i}{4} C_F \delta_{ij} g^2$$

$$\times \sum_\sigma \bar{u}(p,\sigma) \int d^D\hat{k}\, \frac{\gamma_\alpha(\slashed{p} + \slashed{k})\gamma^\mu(\slashed{p} + \slashed{k} + \slashed{q})\gamma^\nu(\slashed{p} + \slashed{k})\gamma^\alpha}{(p+k)^4(p+k+q)^2 k^2}\, u(p,\sigma).$$

Using

$$\sum_\sigma \bar{u}(p,\sigma)\mathcal{M}u(p,\sigma) = \operatorname{Tr}\slashed{p}\mathcal{M}, \quad p^2 = 0,$$

extracting the term proportional to $q^\mu q^\nu$, and introducing Feynman parameters, we find

$$T'^{NS}_L = \frac{g^2}{16\pi^2} C_F \frac{8}{x} \int_0^1 d\alpha\, \alpha \int_0^1 d\beta\, \frac{(1-u_2)u_1}{[1 - u_2 - (1-(u_1+u_2)/x]^2},$$

where $u_1 = \alpha\beta$ and $u_2 = 1 - \alpha$. Expanding in powers of $1/x$ and integrating,

$$T'^{NS}_L = \frac{g^2}{16\pi^2} 4C_F \sum_{n=1}^{\infty} \frac{1}{n+1} \left(\frac{1}{x}\right)^n.$$

The crossed diagram doubles even and cancels odd powers of $1/x$, so

$$T^{NS}_L = \frac{2g^2}{16\pi^2} C_F \sum_{n=\text{even}}^{\infty} \frac{4}{n+1} \left(\frac{1}{x}\right)^n; \tag{4.7.7}$$

writing the analogue of Eq. (4.5.18), we therefore find

$$B^{(1)n,NS}_L = \frac{4}{n+1} C_F, \quad n = \text{even},$$

and comparing with the corresponding expression for the NS piece of F_2,

$$\mu^L_{NS}(n, Q^2) = \delta^{NS}_2 \frac{\alpha_s(Q^2)}{4\pi} \frac{4C_F}{n+1} \mu_{2NS}(n, Q^2). \tag{4.7.8}$$

It is not difficult to invert the Mellin transforms in the definitions of μ_2, μ^L and in (4.7.8) to find directly a relation between *structure functions*. For electroproduction on proton targets, we have

$$F^L_{NS}(x, Q^2) = \int_x^1 dy\, C^L_{NS}(y, Q^2) F_{2NS}\left(\frac{x}{y}, Q^2\right) \tag{4.7.9a}$$

and

$$C^L_{NS}(y, Q^2) = [4C_F x] \frac{\alpha_s(Q^2)}{4\pi} + c^{(1)L}_{NS}(x) \left(\frac{\alpha_s(Q^2)}{4\pi}\right)^2 + \cdots; \tag{4.7.9b}$$

the first term corresponds to the moments relation calculated in (4.7.8), as may be easily checked. The function $c^{(1)L}_{NS}(x)$ was evaluated by Sánchez-Guillén et al. (1991).

For the singlet, the calculation is similar to the one just performed, but using now also the diagram of Fig. 4.7.2 for the leading gluonic piece. One has, for electroproduction on protons,

$$F^L_S(x, Q^2) = \int_x^1 dy\, \left\{ C^L_S(y, Q^2) F_{2S}\left(\frac{x}{y}, Q^2\right) + C^L_G(y, Q^2) F_G\left(\frac{x}{y}, Q^2\right) \right\}, \tag{4.7.10a}$$

where the kernels C^L are now

$$C_S^L(x, Q^2) = C_{NS}^L(x, Q^2) + C_{PS}^L(x, Q^2),$$

$$C_{NS}^L(x, Q^2) = [4C_F x]\frac{\alpha_s(Q^2)}{4\pi} + c_{NS}^{(1)L}(x)\left(\frac{\alpha_s(Q^2)}{4\pi}\right)^2 + \cdots,$$

$$C_{PS}^L(x, Q^2) = c_{PS}^{(1)L}(x)\left(\frac{\alpha_s(Q^2)}{4\pi}\right)^2 + \cdots,$$

$$C_G^L(x, Q^2) = [16n_f T_F x(1 - x)]\frac{\alpha_s(Q^2)}{4\pi} + c_G^{(1)L}(x)\left(\frac{\alpha_s(Q^2)}{4\pi}\right)^2 + \cdots.$$

$$(4.7.10b)$$

The kernel $c_{PS}^{(1)L}(x)$ was also calculated by Sánchez-Guillén et al. (1991). The kernel $c_G^{(1)L}(x)$ was evaluated by van Neerven and Zijlstra (1991b,c), who also checked the calculations of Sánchez-Guillén et al. (and corrected an error in the evaluation by these authors of $c_G^{(1)L}(x)$). The results are confirmed by a calculation of the first moments by Larin and Vermaseren (1993). All three kernels may be found collected in Adel, Barreiro and Ynduráin (1997).

The full longitudinal function is the sum of the nonsinglet and the quark singlet component:

$$F_L = F_2 - 2xF_1 = F_S^L + F_{NS}^L.$$

iii Singlet

The singlet calculations are much more difficult than the nonsinglet, or longitudinal structure function, ones. For the anomalous dimension, Floratos, Ross and Sachrajda (1979) and González-Arroyo and López (1980) made the first two-loop calculation; these contained errors in the γ_{GG} term. A correct calculation of this was provided by Furmanski and Petronzio (1980) in the so-called Altarelli–Parisi formalism (see López and Ynduráin, 1981, for the *moments* equations). The collected $\gamma_{ij}^{(1)}(n)$ may be found in Adel, Barreiro and Ynduráin (1997). The three-loop calculation has not been completed, although there exists an evaluation for the first few moments by Larin, Nogueira, van Ritbergen and Vermaseren (1997), who also checked the previous evaluations to two loops.

The coefficients were first calculated to one loop by Bardeen, Buras, Duke and Muta (1978), and Bardeen and Buras (1979); see also Buras (1980). To two loops they were evaluated by van Neerven and Zijlstra (1991a,c; 1992a,c), in the Altarelli–Parisi formalism; these calculations have also been checked by the moments evaluation of Larin, Nogueira, van Ritbergen and Vermaseren (1997).

The *equations* themselves are now not trivial to obtain, because of their matrix character. To derive them, we let

$$a = \alpha_s(Q^2)/4\pi, \quad a' = \alpha_s(Q'^2)/4\pi, \quad t = \tfrac{1}{2}\log\frac{Q^2}{\nu^2}.$$

We also define, temporarily suppressing the variable n to lighten the notation,

$$\mathbf{D}(a) = \frac{\boldsymbol{\gamma}(a)}{2\beta(a)},$$

and the series expansions

$$\begin{aligned}
\mathbf{C}(a) &= 1 + \mathbf{C}^{(1)}\,a + \mathbf{C}^{(2)}\,a^2 + \dots, \\
\boldsymbol{\gamma}(a) &= \boldsymbol{\gamma}^{(0)}\,a + \boldsymbol{\gamma}^{(1)}\,a^2 + \boldsymbol{\gamma}^{(2)}\,a^3 + \cdots, \\
-\beta(a) &= \beta_0 a^2 + \beta_1 a^3 + \beta_2 a^4 + \dots, \\
\mathbf{D}(a) &= \frac{1}{a}\mathbf{D}^{(0)} + \mathbf{D}^{(1)} + \mathbf{D}^{(2)}a + \dots;
\end{aligned} \tag{4.7.11a}$$

$$\mathbf{D}^{(0)} = \frac{-1}{2\beta_0}\boldsymbol{\gamma}^{(0)}, \quad \mathbf{D}^{(1)} = \frac{-1}{2\beta_0}\left(\boldsymbol{\gamma}^{(1)} - \frac{\beta_1}{\beta_0}\boldsymbol{\gamma}^{(0)}\right),$$

$$\mathbf{D}^{(2)} = \frac{-1}{2\beta_0}\left[\boldsymbol{\gamma}^{(2)} - \frac{\beta_1}{\beta_0}\boldsymbol{\gamma}^{(1)} + \left(\frac{\beta_1^2}{\beta_0} - \frac{\beta_2}{\beta_0}\right)\boldsymbol{\gamma}^{(0)}\right], \cdots. \tag{4.7.11b}$$

The matrix $\mathbf{C}$ is built from the Wilson coefficients. As already remarked, the gluon structure function is not unique, beyond the LO. This is because we may alter the mixing by adding pieces proportional to ϵ, $\epsilon^2,\dots$, $\epsilon = 4 - D$, shifting pieces from coefficients to matrix elements. A way to make the $\mathbf{C}$ unique is to require that it commute with the anomalous dimension matrix:

$$[\mathbf{C}(a), \boldsymbol{\gamma}(a)] = 0.$$

Expanding, this implies

$$\begin{aligned}
&[\boldsymbol{\gamma}^{(0)}, \mathbf{C}^{(1)}] = 0, \\
&\boldsymbol{\gamma}^{(1)}\mathbf{C}^{(1)} + \boldsymbol{\gamma}^{(0)}\mathbf{C}^{(2)} = \mathbf{C}^{(1)}\boldsymbol{\gamma}^{(1)} + \mathbf{C}^{(2)}\boldsymbol{\gamma}^{(0)}.
\end{aligned} \tag{4.7.12a}$$

The solution to these equations is

$$C_{21}^{(1)} = \frac{\gamma_{21}^{(0)}}{\gamma_{12}^{(0)}}\,C_{12}^{(1)},$$

$$C_{22}^{(1)} = C_{11}^{(1)} + \frac{\gamma_{22}^{(0)} - \gamma_{11}^{(0)}}{\gamma_{12}^{(0)}}\,C_{12}^{(1)};$$

$$C_{21}^{(2)} = \frac{C_{12}^{(1)}\gamma_{21}^{(1)} + C_{12}^{(2)}\gamma_{21}^{(0)} - C_{21}^{(1)}\gamma_{12}^{(1)}}{\gamma_{12}^{(0)}},$$

$$C_{22}^{(2)} = C_{11}^{(2)} + \frac{\gamma_{22}^{(0)} - \gamma_{11}^{(0)}}{\gamma_{12}^{(0)}}\,C_{12}^{(2)} + \frac{\gamma_{22}^{(1)} - \gamma_{11}^{(1)}}{\gamma_{12}^{(0)}}\,C_{12}^{(1)} + \frac{C_{11}^{(1)} - C_{22}^{(1)}}{\gamma_{12}^{(0)}}\,\gamma_{12}^{(1)}. \tag{4.7.12b}$$

The $\boldsymbol{\mu}$ satisfy the differential equation

$$\frac{\partial}{\partial a}\boldsymbol{\mu}(a) = \left\{ \frac{\partial \mathbf{C}(a)}{\partial a}\mathbf{C}^{-1}(a) - \mathbf{C}(a)\mathbf{D}(a)\mathbf{C}^{-1}(a) \right\} \boldsymbol{\mu}(a). \qquad (4.7.13)$$

We now seek $\mathbf{M}(a)$ such that

$$\frac{\partial}{\partial a}\left\{ a^{\mathbf{D}^{(0)}} \mathbf{M}(a)\mathbf{C}^{-1}(a)\boldsymbol{\mu}(a) \right\} = 0; \qquad (4.7.14)$$

this gives the condition

$$\frac{1}{a}\mathbf{D}^{(0)}\mathbf{M}(a) + \frac{\partial \mathbf{M}(a)}{\partial a} - \mathbf{M}(a)\mathbf{D}(a) = 0, \qquad (4.7.15a)$$

which may be solved iteratively. Write

$$\mathbf{M}(a) = 1 + \mathbf{M}^{(1)}a + \mathbf{M}^{(2)}a^2 + \cdots. \qquad (4.7.15b)$$

Then, to NLO and NNLO we obtain the equations

$$\begin{aligned} \mathbf{M}^{(1)} + [\mathbf{D}^{(0)}, \mathbf{M}^{(1)}] &= \mathbf{D}^{(1)}, \\ 2\mathbf{M}^{(2)} + [\mathbf{D}^{(0)}, \mathbf{M}^{(2)}] &= \mathbf{D}^{(2)} + \mathbf{M}^{(1)}\mathbf{D}^{(1)}. \end{aligned} \qquad (4.7.15c)$$

To solve them we define the matrix $\mathbf{S}$ that diagonalizes $\mathbf{D}^{(0)}$:

$$\mathbf{S}^{-1}\mathbf{D}^{(0)}\mathbf{S} = \hat{\mathbf{D}}^{(0)} = \begin{pmatrix} d_+ & 0 \\ 0 & d_- \end{pmatrix}, \quad d_+ > d_-. \qquad (4.7.16a)$$

We take it to be

$$\mathbf{S} = \begin{pmatrix} 1 & \dfrac{D_{12}^{(0)}}{d_- - d_+} \\ \dfrac{d_+ - D_{11}^{(0)}}{D_{12}^{(0)}} & \dfrac{d_- - D_{11}^{(0)}}{d_- - d_+} \end{pmatrix}, \quad S_{11} = \det \mathbf{S} = 1, \qquad (4.7.16b)$$

and we also define

$$\begin{aligned} \mathbf{S}^{-1}\mathbf{D}^{(N)}\mathbf{S} &\equiv \bar{\mathbf{D}}^{(N)}, \quad \mathbf{S}^{-1}\mathbf{M}^{(N)}\mathbf{S} \equiv \bar{\mathbf{M}}^{(N)}, \quad \mathbf{S}^{-1}\boldsymbol{\gamma}^{(N)}\mathbf{S} \equiv \bar{\boldsymbol{\gamma}}^{(N)} \\ (\bar{\mathbf{D}}^{(0)} &= \hat{\mathbf{D}}^{(0)}). \end{aligned} \qquad (4.7.17)$$

From Eq. (4.7.15c),

$$\bar{\mathbf{M}}^{(1)} = \begin{pmatrix} \bar{D}_{11}^{(1)} & \dfrac{1}{1 + d_+ - d_-}\bar{D}_{12}^{(1)} \\ \dfrac{1}{1 + d_- - d_+}\bar{D}_{21}^{(1)} & \bar{D}_{22}^{(1)} \end{pmatrix}, \qquad (4.7.18a)$$

$$\bar{\mathbf{M}}^{(2)} = \begin{pmatrix} \frac{1}{2}\left[\bar{D}_{11}^{(2)} + (\bar{M}^{(1)}\bar{D}^{(1)})_{11}\right] & \dfrac{\bar{D}_{12}^{(2)} + (\bar{M}^{(1)}\bar{D}^{(1)})_{12}}{2 + d_+ - d_-} \\ \dfrac{\bar{D}_{21}^{(2)} + (\bar{M}^{(1)}\bar{D}^{(1)})_{21}}{2 + d_- - d_+} & \frac{1}{2}\left[\bar{D}_{22}^{(2)} + (\bar{M}^{(1)}\bar{D}^{(1)})_{22}\right] \end{pmatrix}. \quad (4.7.18b)$$

Because of Eq. (4.7.14) it follows that one can write

$$a^{\mathbf{D}^{(0)}}\mathbf{M}(a)\mathbf{C}^{-1}(a)\boldsymbol{\mu}(a) \equiv \widetilde{\mathbf{b}} = \text{independent of } a;$$

hence, inserting the moment index n explicitly,

$$\boldsymbol{\mu}(n,a) = \mathbf{C}(n,a)\mathbf{S}(n)\bar{\mathbf{M}}(n,a)^{-1}a^{-\hat{\mathbf{D}}^{(0)}(n)}\mathbf{b}(n),$$
$$\mathbf{b} = \mathbf{S}\widetilde{\mathbf{b}} = \text{independent of } a, \tag{4.7.19a}$$

from which the evolution of the moments is obtained directly: considering (4.7.19a) for Q^2 and Q_0^2, and eliminating from the equations the unknown $\mathbf{b}$, we find the evolution equation

$$\boldsymbol{\mu}(n,a) = \mathbf{C}(n,a)\mathbf{S}(n)\bar{\mathbf{M}}(n,a)^{-1}(a_0/a)^{\hat{\mathbf{D}}^{(0)}(n)}\bar{\mathbf{M}}(n,a_0)\mathbf{S}^{-1}\mathbf{C}(n,a_0)^{-1}\boldsymbol{\mu}(n,a_0), \tag{4.7.19b}$$

where $a = \alpha_s(Q^2)/4\pi$, $a_0 = \alpha_s(Q_0^2)/4\pi$ and α_s is to be calculated to as many loops as the anomalous dimensions.

4.8 The Altarelli–Parisi, or DGLAP, Method

The OPE method for analysis of deep inelastic scattering is fairly rigorous and not too difficult to use; but it does not, perhaps, appeal to physical intuition. In particular, its connection with the parton model is not particularly transparent. This is one of the reasons for the success of the Altarelli–Parisi method,[16] in which close contact is maintained with the parton model at each step.

Before discussing the partonic interpretation, let us further elaborate the equations that we have. For the sake of definiteness we will consider in detail the nonsinglet part of F_2; in fact, we will concentrate on the contribution of a given quark flavour f to F_2. This contribution is proportional to the quark density q_f which, in the free parton model, is independent of Q^2. When interactions are taken into account, q_f will acquire a momentum dependence. If we let μ be a fixed reference momentum, and define $t = \frac{1}{2}\log Q^2/\mu^2$; then we generalize (4.3.11) to

---

[16] Altarelli and Parisi 1977; see also Dokshitzer, Dyakonov and Troyan (1980). The method is at times also called the DGLAP method, because Dokshitzer (1977), Lipatov (1975) and Gribov and Lipatov (1972) had derived similar, or even equivalent equations independently.

$$F_2(x, Q^2) = \sum_f \delta_f x q_f(x, t); \qquad (4.8.1)$$

the δ_f are known constants, depending on the particular process we are considering.

What the QCD equations give us is the *evolution* of the moments with t; thus, we recast (4.6.6) in differential form, which for the q_f reads

$$\frac{d\tilde{q}_f(n, t)}{dt} = -\frac{\gamma_{NS}^{(0)}(n)\alpha_s(t)}{4\pi}\tilde{q}_f(n, t); \qquad (4.8.2)$$

we have written $\alpha_s(t)$ for $\alpha_s(Q^2)$, $t = \frac{1}{2}\log Q^2/\mu^2$, and defined the moments of the densities

$$\tilde{q}_f(n, t) \equiv \int_0^1 dx\, x^{n-1} q_f(x, t). \qquad (4.8.3)$$

Equations (4.8.3) and (4.6.6) are fully equivalent, one being the integrated form of the other. Then, we invert the Mellin transform in (4.8.3). If we define the so-called *splitting function* $P_{NS}^{(0)}(z)$ by

$$\int_0^1 dz\, z^{n-1} P_{NS}^{(0)}(z) = -\tfrac{1}{4}\gamma_{NS}^{(0)}(n), \qquad (4.8.4a)$$

then the convolution theorem for Mellin transforms tells us that

$$\frac{\partial q_f(x, t)}{\partial t} = \frac{\alpha_s(t)}{\pi} \int_x^1 \frac{dy}{y}\, q_f(y, t) P_{NS}^{(0)}(x/y). \qquad (4.8.4b)$$

This is the Altarelli–Parisi equation for the singlet densities, to LO. Its equivalence to (4.6.6) may be easily verified by projecting it into moments, and integrating. Equation (4.8.4b) can also be written in infinitesimal form as

$$q_f(x, t) + dq_f(x, t) = \int_0^1 dy \int_0^1 dz\, \delta(zy - x) q_f(y, t)$$
$$\times \left\{ \delta(z - 1) + \frac{\alpha_s(t)}{\pi} P_{NS}^{(0)}(z)dt \right\}. \qquad (4.8.5)$$

We see that $P_{NS}^{(0)}(z)$ can be interpreted as governing the rate of change of the parton distribution probability with t. We will elaborate on this presently.

Consider the scattering of an off-shell probe, say a photon, off a parton. In the parton model (Fig. 4.8.1a), quarks are assumed to be free, with a certain probability of having a fraction of the proton momentum, $q_f(x)$. We now allow for a dependence on t, which is due to the fact that the quark may radiate gluons. The various processes in which this radiation occurs, either of real or virtual gluons, are depicted in Fig. 4.8.1b,c. If we work in an axial gauge, the calculation is simplified enormously. Indeed, in this gauge only the diagram of Fig. 4.8.1b will give a term proportional to t: hence, when calculating

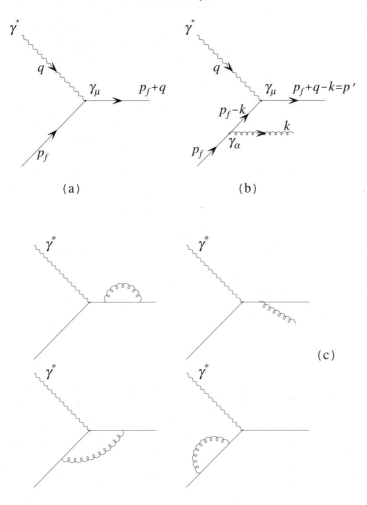

Fig. 4.8.1. Diagrams involved in scattering of a photon off a quark.
(a) Free quark. (b) Radiation of a gluon, producing a term depending on t
in the axial gauge. (c) Other radiation diagrams.

the variation with t it is the only one that will contribute. Moreover, QCD
corrections to the coupling are, in this gauge, taken into account to LO by
simply replacing $g^2/4\pi \to \alpha_s(t)$. This last is easily understood if we recall
our calculation of Eq. (2.3.18) and compare it to (3.3.29) and (3.3.30): the
entire Z_g comes from the gluon propagator in this gauge.

To zero order in g we have the diagram of Fig. 4.8.1a, as stated. Let us take quarks as massless, and work in the reference frame where

$$q = (0, 0, 0, -Q), \quad p = \frac{Q}{2x}(1, 0, 0, 1).$$

The structure function F_2 is proportional to a cross section, and will thus be the sum of the point-like cross sections for each quark weighted with the q_f, as we saw in Sect. 4.3. With the obvious changes of notation, and with w proportional to the cross section,

$$\frac{1}{x}F_{2NS}(x, t) = \sum_f \delta_{NS,f} \int_0^1 \frac{dy}{y}\, q_f(y, t) w_{\text{pointlike}}(p_f, q). \qquad (4.8.6)$$

The origin of each piece is clear. We have defined y by $p_f = yp$.

Now, because quarks are massless, we must have $(p_f + q)^2 = 0$ and therefore,

$$w_{\text{pointlike}}(p_f, q) = \delta(y/x - 1):$$

substituting, we recover (4.8.1), as could have been expected. We will then rewrite (4.8.6) as

$$q_f(x, t) = \int_0^1 \frac{dy}{y}\, \delta(y/x - 1) q_f(y, t). \qquad (4.8.7)$$

This is only valid to zero order in g (free parton model). We require the corrections to it due to the gluon interactions, shown in Fig. 4.8.1b,c. We may split these into two sets: vertex corrections and radiation of real gluons. We will discuss the first later on. For the diagrams with radiation of real gluons, we need only consider that of Fig. 4.8.1b (as discussed) if we work in a lightlike gauge and are only interested in t-dependent terms. Its corresponding amplitude is

$$\mathcal{A}^\mu = (2\pi)^{-2} \bar{u}(p_f - k + q, \sigma') \gamma^\mu \frac{i}{\not{p}_f - \not{k}}\, i\gamma^\alpha g t_{ij}^a u(p_f, \sigma) \epsilon_\alpha^*(k, \lambda),$$

and this is is normalized so that, if γ^* were a real photon, we would have the scattering amplitude

$$F(\gamma_q^* \to G + q') = \epsilon_\mu \mathcal{A}^\mu.$$

The probability for the process is therefore proportional to

$$w^{\mu\nu} = \tfrac{1}{2} \int \frac{d^3\mathbf{k}}{2k^0} \frac{d^3\mathbf{p'}}{2p'^0}\, \delta(p_f + q - k - p') \sum_{\text{spins}} \mathcal{A}^{\mu*}\mathcal{A}^\nu$$

$$= \tfrac{1}{2} \sum_{\sigma\sigma'\lambda} \sum_{a,j} \int d^4k\, \theta(k^0)\delta(k^2)\theta(p_f^0 - k^0 + q^0)\delta\left((p_f - k + q)^2\right) \mathcal{A}^{\mu*}\mathcal{A}^\nu.$$

Note that the gluon is real and we then have to take

$$\sum_\lambda \epsilon_\alpha(k,\lambda)\epsilon_\beta^*(k,\lambda) = -g_{\alpha\beta} + \frac{k_\alpha u_\beta + k_\beta u_\alpha}{k \cdot u};$$

recall that we have chosen a lightlike gauge,

$$k \cdot \epsilon = u \cdot \epsilon = 0, \quad u^2 = 0.$$

Defining $\delta_+(v^2) \equiv \delta(v^2)\theta(v^0)$, we obtain

$$w^{\mu\nu} = \frac{g^2 C_F}{2(2\pi)^2}\Phi^{\mu\nu}, \tag{4.8.8a}$$

$$\begin{aligned}
\Phi^{\mu\nu} &= \int d^4k\, \delta_+(k^2)\delta_+\left((p_f - k + q)^2\right)\left(-g_{\alpha\beta} + \frac{k_\alpha u_\beta + k_\beta u_\alpha}{k \cdot u}\right) \\
&\times \frac{\mathrm{Tr}(\not p_f - \not k)\gamma^\mu(\not p_f - \not k + \not q)\gamma^\nu(\not p_f - \not k)\gamma^\beta \not p_f \gamma^\alpha}{(p_f - k)^4}.
\end{aligned} \tag{4.8.8b}$$

Expression (4.8.8) is divergent for massless quarks and gluons, so it has to be regulated. One could use dimensional regularization for this, but it is simpler (and more physical: the initial quark is in a bound state, hence off–shell) to take $p_f^2 = -\mu^2$. Because the region of integration in (4.8.8b) is compact, the divergence as $\mu^2 \to 0$ is the only way we may obtain a logarithm which, as we will see, is of the form $\log Q^2/\mu^2$. Since it is only the logarithmic term that is of interest to us, we can greatly simplify the calculation.

First of all, throughout (4.8.8b), except in the denominator, we may take $p_f^2 = 0$: the corrections will be of order μ^2/Q^2 (eventually with an extra logarithm). Thus,

$$\begin{aligned}
&\left(-g_{\alpha\beta} + \frac{k_\alpha u_\beta + k_\beta u_\alpha}{k \cdot u}\right)\mathrm{Tr}(\not p_f - \not k)\gamma^\mu(\not p_f - \not k + \not q)\gamma^\nu(\not p_f - \not k)\gamma^\beta \not p_f \gamma^\alpha \\
&= -2(p_f - k)^2\Bigg\{ \mathrm{Tr}\,\gamma^\mu(\not p_f - \not k + \not q)\gamma^\nu \not k \\
&+ \mathrm{Tr}\,\gamma^\mu(\not p_f - \not k + \not q)\gamma^\nu\left[(p \cdot u)(\not p_f - \not k) + (p_f - k)\cdot u\, \not p_f + 2k\cdot p_f \not u\right]\frac{1}{u \cdot k}\Bigg\}.
\end{aligned}$$

Since $p_f^2 = k^2 = 0$, $2k \cdot p_f = -(p_f - k)^2$; hence, the last term of the above equation is proportional to $(p_f - k)^4$ and it does not contribute to the logarithm. We then obtain

$$\begin{aligned}
\Phi^{\mu\nu} &\stackrel{\log}{=} -2\int \frac{d^3\mathbf{k}}{2k^0}\delta_+\left((p_f - k + q)^2\right)\frac{1}{(p_f - k)^2}\mathrm{Tr}\Bigg\{\gamma^\mu(\not p_f - \not k + \not q)\gamma^\nu \not k \\
&+ \gamma^\mu(\not p_f - \not k + \not q)\gamma^\nu\left[(\not p_f - \not k)\frac{p_f \cdot u}{k \cdot u} + \not p_f \frac{(p_f - k)\cdot u}{k \cdot u}\right]\Bigg\},
\end{aligned} \tag{4.8.9}$$

where the expression $\overset{\log}{=}$ means "equal logarithmic terms". Next, we write the denominator in (4.8.9) as

$$(p_f - k)^2 = -\mu^2 - 2k^0 p_f^0 + 2k^3 p_f^3 \cos\theta.$$

It only vanishes (for $\mu \to 0$) when $\cos\theta = 1$; that is, when k and p_f are *collinear*: this, incidentally, identifies the gluons that give corrections to scaling. Thus, for the logarithmic term, we may take $\cos\theta = 1$ everywhere except in the denominator. In particular, the delta function in (4.8.9) becomes

$$\delta_+\left((p_f - k + q)^2\right) \to \delta(2\nu - Q^2 - 2Qk^0) \to \frac{1}{2\nu}\delta(\rho - x), \qquad (4.8.10a)$$

and we have defined

$$1 - Qk^0/\nu \equiv \rho. \qquad (4.8.10b)$$

Moreover, for $\cos\theta = 1$,

$$k_{\cos\theta=1} = (1 - \rho)p_f,$$

and we can then easily complete the calculation:

$$\Phi^{\mu\nu} \overset{\log}{=} -2\pi \int_{-1}^{+1} d\cos\theta \int_0^\infty \frac{dk^0\, k^0}{2\nu} \delta(\rho - x)$$

$$\times \frac{1 + \rho^2}{1 - \rho} \frac{\mathrm{Tr}\,\gamma^\mu(\rho p\!\!\!/_f + q\!\!\!/)\gamma^\nu p\!\!\!/_f}{2k^0 p_f^0 \cos\theta - (\mu^2 + 2k^0 p_f^0)}$$

$$\overset{\log}{=} \frac{\pi}{2\nu}\left(\log\frac{Q^2}{\mu^2}\right)\int d\rho\, \frac{1 + \rho^2}{1 - \rho}\delta(\rho - x)\,\mathrm{Tr}\,\gamma^\mu(\rho p\!\!\!/_f + q\!\!\!/)\gamma^\nu p\!\!\!/_f.$$

Therefore, for F_2 and with self-explanatory notation,

$$w_2 = 4C_F \frac{g^2}{16\pi^2}\int d\rho\, \frac{1 + \rho^2}{1 - \rho}\rho\delta(x - \rho)\log\frac{Q^2}{\mu^2}. \qquad (4.8.11)$$

This equation does not give the full answer: it is undefined at $\rho = 1$. This corresponds to a zero energy gluon, which is a typical infrared singularity. In fact, it may be seen that this singularity is exactly cancelled by the vertex and propagator corrections that we have not yet taken into account. Since in these no real gluon is emitted, their contribution to w_2 has to be like that in (4.8.11), but with $\lambda\delta(\rho - 1)$ instead of the term $(1 + \rho^2)/(1 - \rho)$. With these terms included, we thus write

$$w_2 = 4C_F \frac{g^2}{16\pi^2}\left(\log\frac{Q^2}{\mu^2}\right)\int d\rho\, \left\{\frac{1 + \rho^2}{1 - \rho} + \lambda\delta(1 - \rho)\right\}\rho\delta(x - \rho). \qquad (4.8.12)$$

Taking into account the correct value of λ (see below), we find the desired correction to (4.8.7); it is

$$q_f(x,t) = \int_0^1 dy \int_0^1 dz\, \delta(zy - 1) q_f(y,t) \left\{ \delta(z-1) + \frac{t\alpha_s}{\pi} P_{NS}^{(0)}(z) \right\},$$

$$P_{NS}^{(0)}(z) = C_F \left\{ \frac{1+z^2}{(1-z)_+} + \tfrac{3}{2}\delta(1-z) \right\},$$

$$(4.8.13\text{a})$$

where we define, for any function φ,

$$\int_0^1 dz\, \frac{1}{(1-z)_+}\, \varphi(z) \equiv \int dz\, \frac{\varphi(z) - \varphi(1)}{1-z}. \qquad (4.8.13\text{b})$$

If we identify this $P_{NS}^{(0)}$ with the splitting function introduced previously, we may check that Eq. (4.8.4) is indeed satisfied. It is because of this equivalence that we did not bother to calculate the coefficient λ of $\delta(1-\rho)$: the condition[17] $\gamma_{NS}^{(0)}(n=1) = 0$ fixes λ directly. The comparison of (4.8.13) with (4.8.5) may be carried out at once. It is sufficient to take α_s to be defined at μ^2 and take $t \to dt$ to be infinitesimal.

It is still possible to use a different procedure to rederive the evolution equations, which is perhaps more interesting than the former method. We consider that an arbitrary number of gluons can be emitted; thus, we may sum all the diagrams where gluons are radiated. Of course, this is an impossible task; but it simplifies enormously if we only consider *leading logarithms*. In this case, it may be shown (see, e.g., Gribov and Lipatov, 1972) that only the ladder graphs contribute (Fig. 4.8.2). It then turns out that we can calculate the diagrams, and even sum them. In this way, we recover the results of the standard analysis, with two bonuses. First, we see that the LO in the running coupling constant is equivalent to summing all the leading logarithms in $g^2/16\pi^2$:

$$\left(\frac{g^2}{16\pi^2} \right)^n \log^n \frac{Q^2}{\mu^2}.$$

Secondly, it gives a hint as to how to treat processes where the operator product method is not applicable. We will not delve further into this matter, but refer to the lectures of Sachrajda (1979) and work quoted there.

Let us return to (4.8.4b). Choosing the scale of Q^2 to be Λ^2, so that we can take $t = \tfrac{1}{2} \log Q^2/\Lambda^2$, we replace $\alpha_s(t)$ by the running coupling constant $\alpha_s(Q^2)$. Moreover, $\partial/\partial t = 2Q^2 \partial/\partial Q^2$. We can thus write the Altarelli–Parisi equation as

$$\frac{Q^2 \partial}{\partial Q^2} q_f(x, Q^2) = \frac{\alpha_s(Q^2)}{2\pi} \int_x^1 \frac{dy}{y} P_{NS}^{(0)}\left(\frac{x}{y} \right) q_f(y, Q^2). \qquad (4.8.14)$$

---

[17]For the singlet, the corresponding condition is $\det \boldsymbol{\gamma}^{(0)}(n=2) = 0$; see Sect. 4.9i.

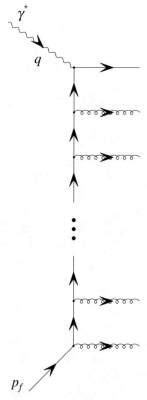

Fig. 4.8.2. Ladder graph for emission of gluons in deep inelastic scattering.

For the *singlet* case, the corresponding equations will involve the density of gluons, which we denote by $G(x, Q^2)$, so that $F_G = xG$. These equations may be obtained from the moments equations, Eqs. (4.6.6), or with partonic methods similar to the ones we have just employed (Altarelli and Parisi, 1977).

However we choose to derive them, the equations are

$$\frac{Q^2 \partial}{\partial Q^2} \begin{pmatrix} q^S(x, Q^2) \\ G(x, Q^2) \end{pmatrix} = \frac{\alpha_s(Q^2)}{2\pi} \int_x^1 \frac{dy}{y} \, \mathbf{P}^{(0)} \left(\frac{x}{y} \right) \begin{pmatrix} q^S(y, Q^2) \\ G(y, Q^2) \end{pmatrix}; \quad (4.8.15a)$$

the kernel is now

$$\mathbf{P}^{(0)} = \begin{pmatrix} P_{qq}^{(0)} & P_{qG}^{(0)} \\ P_{Gq}^{(0)} & P_{GG}^{(0)} \end{pmatrix}, \quad (4.8.15b)$$

with

$$P_{qq}^{(0)}(x) = C_F \left[\frac{1+x^2}{(1-x)_+} + \tfrac{3}{2}\delta(1-x) \right] = P_{NS}^{(0)}(x),$$

$$P_{qG}^{(0)}(x) = \frac{x^2 + (1-x)^2}{2} n_f,$$

$$P_{Gq}^{(0)}(x) = C_F \left[\frac{1+(1-x)^2}{x} \right],$$

$$P_{GG}^{(0)}(x) = 2C_A \left[\frac{x}{(1-x)_+} + \frac{1-x}{x} + x(1-x) \right] + \frac{11C_A - 2n_f}{6}\delta(1-x).$$

$$(4.8.15c)$$

The second order (NLO) kernels have been calculated by Curci, Furmanski and Petronzio (1980) and Furmanski and Petronzio (1980). The NNLO kernels are not known.

The Altarelli–Parisi method allows a physically transparent decomposition of structure functions for various processes in terms of a few "quark densities", $q(x, Q^2)$, for quarks with flavour q. For easy reference, we collect here the expressions for a few important processes. We let I be an isoscalar target, and p a proton target. Then, writing f for q_f,

$$F_{2S}^{ep} = \begin{cases} \tfrac{2}{9}x(u + \bar{u} + d + \bar{d} + s + \bar{s}), & n_f = 3, \\[2mm] \tfrac{5}{18}x(u + \bar{u} + d + \bar{d} + s + \bar{s} + c + \bar{c}), & n_f = 4, \end{cases}$$

$$(4.8.16a)$$

$$F_{2NS}^{ep} = \begin{cases} \tfrac{1}{3}x(\tfrac{2}{3}u - \tfrac{1}{3}d - \tfrac{1}{3}s + \tfrac{2}{3}\bar{u} - \tfrac{1}{3}\bar{d} - \tfrac{1}{3}\bar{s}), & n_f = 3, \\[2mm] \tfrac{1}{6}x(u - d - s + c + \bar{u} - \bar{d} - \bar{s} + \bar{c}), & n_f = 4; \end{cases}$$

$$F_{2S}^{eI} = F_{2S}^{ep}; \quad F_{2NS}^{eI} = \begin{cases} \tfrac{1}{18}x(u + \bar{u} + d + \bar{d} - 2s - 2\bar{s}), & n_f = 3, \\[2mm] \tfrac{1}{6}x(c - s + \bar{c} - \bar{s}), & n_f = 4; \end{cases}$$

$$(4.8.16b)$$

$$F_{2NS}^{\nu I} = 0, \quad F_2^{\nu I} = F_{2S}^{\nu I} = \begin{cases} \tfrac{9}{2}F_{2S}^{ep}, & n_f = 3, \\[2mm] \tfrac{18}{5}F_{2S}^{ep}, & n_f = 4; \end{cases}$$

$$(4.8.16c)$$

$$F_S^{\nu I} = 0, \quad F_3^{\nu I} = F_{NS}^{\nu I} = \begin{cases} x(u - \bar{u} + d - \bar{d} + s - \bar{s}), & n_f = 3, \\[2mm] x(u - \bar{u} + d - \bar{d} + s - \bar{s} + c - \bar{c}), & n_f = 4. \end{cases}$$

$$(4.8.16d)$$

Some of these we have presented before. Furthermore, one can define the "valence" quarks q_v as the excess of quarks over antiquarks in a hadron (so a proton has $\int_0^1 dx\, u_v = 1$, $\int_0^2 dx\, d_v = 1$) and the "sea" as the rest, etc. Detailed treatments may be found, for example, in the excellent reviews of Buras (1980) and Altarelli (1982).

4.9 General Consequences of QCD
for Structure Functions

i Sum Rules

We have stated repeatedly that the matrix elements A^n cannot be calculated in perturbation theory; but there are cases where the corresponding composite operators are related to symmetry generators. In this situation they correspond to observable quantities and thus their matrix elements are measurable, at least in principle. As discussed in Sect. 3.6, such operators do not require renormalization, and the corresponding anomalous dimensions vanish; therefore, for $Q^2 \to \infty$, these A^n can be calculated with the free quark–parton model.[18]

Such operators are those with $n = 1$ for the nonsinglet, and a combination of those with $n = 2$ for the singlet; no others may exist because it is only for these that $\gamma_{NS}(n)$ and $\det \boldsymbol{\gamma}(n)$ vanish. This means that the integrals

$$\int_0^1 \mathrm{d}x\, x^{-1} F_{2NS}(x, Q^2), \tag{4.9.1a}$$

and a combination of

$$\int_0^1 \mathrm{d}x\, F_{2i}(x, Q^2), \quad i = S,\, G, \tag{4.9.1b}$$

can, at least in principle, be calculated in absolute value. In practice, this is useful in favourable cases where the integrals in (4.9.1) can be related to observables on which information is available; this gives rise to sum rules, many of which had been discovered already with the parton model, and which become exact theorems in QCD. Here we will discuss a few typical cases.

We begin with nonsinglet structure functions. For $F_{2,3;NS}$ the operators that appear for $n = 1$ are combinations of the

$$N^\mu_{NS,a\pm} = \mathrm{i} : \bar{q} T^a \gamma^\mu (1 \pm \gamma_5) q :,$$

which indeed generate chiral symmetry transformations (Sect. 2.8). As expected, $\gamma^{(0)}_{NS}(1) = \gamma^{(1)-}_{NS}(1) = 0$. For electroproduction with three flavours u, d and s (the decomposition is different for four flavours), we have, writing the equation somewhat loosely,

$$\mathrm{i} \mathrm{T} J^\mu_{\mathrm{em}}(z) J^\nu_{\mathrm{em}}(0) \Big|^{NS}_{\substack{p^\mu p^\nu \\ n=1}} \underset{z^2 \to 0}{=} \tfrac{1}{3} \bar{C}^1_{2NS}(z^2) J_{\mathrm{em}}(0)$$

---

[18]In general, we have to go to $Q^2 \to \infty$ because of the residual dependence on the interaction due to the Wilson coefficients, or for a more subtle matter of analytical continuation: for odd n, one has the continued $\gamma^{(k)+}_{NS}(n)$ different from $\gamma^{(k)}_{NS}(n) = \gamma^{(k)-}_{NS}(n)$ for high orders, $k \geq 1$.

(cf. Eq. (4.5.4)) so that, being now more precise, we find

$$\frac{1}{i} \bar{A}^1_{2NS} p^\mu = \langle p | J^\mu_{\text{em}}(0) | p \rangle = 2(2\pi)^{-3} p^\mu Q_h,$$

where Q_h is the charge of the target, in units of e. Therefore, and taking into account NLO corrections,

$$\int_0^1 dx \, x^{-1} F^{eh}_{2NS} = \tfrac{1}{3} Q_h \left\{ 1 + \frac{13 + 8\zeta(3) - 2\pi^2}{33 - 2n_f} \frac{\alpha_s(Q^2)}{3\pi} \right\}. \tag{4.9.2}$$

For neutrino scattering we have the *Adler sum rule*, which is exactly valid for all Q^2:

$$\int_0^1 dx \, x^{-1} \left(F^{\bar{\nu}p}_2 - F^{\nu p}_2 \right) = 2. \tag{4.9.3}$$

The relevant operator here is the isospin one. Eq. (4.9.3) has no corrections because it may be related to an equal-time commutator (Sect. 2.8 and Adler, 1966). For electroproduction, because the function is even, the correction involves $\gamma^{(1)+}_{NS}(1) \neq 0$; see López and Ynduráin (1981).

For the structure function F_3 we have the *Gross–Llewellyn Smith* (1969) *sum rule*:

$$\int_0^1 dx \, \left\{ F^{\bar{\nu}p}_3(x, Q^2) + F^{\nu p}_3(x, Q^2) \right\}$$
$$= 3 \left\{ 1 - \frac{\alpha_s(Q^2)}{\pi} - 3.58 \left(\frac{\alpha_s(Q^2)}{\pi} \right)^2 - 19.0 \left(\frac{\alpha_s(Q^2)}{\pi} \right)^3 \right\}. \tag{4.9.4}$$

The calculation of the higher order corrections is due to Chyla and Kataev (1992) and Larin and Vermaseren (1991). Other nonsinglet sum rules may be found collected in the review of Buras (1980).

We now turn to the singlet. In this case, the conserved operator corresponds to $n = 2$. This is reflected in that $\det \boldsymbol{\gamma}^{(0)} = \det \boldsymbol{\gamma}^{(1)} = \ldots = 0$. (Because singlet structure functions are always even it is unnecessary to distinguish $\boldsymbol{\gamma}^\pm$; only $\boldsymbol{\gamma}^+ \equiv \boldsymbol{\gamma}$ enters). Indeed,

$$\boldsymbol{\gamma}^{(0)}(2) = \tfrac{1}{9} \begin{pmatrix} 64 & -12n_f \\ -64 & 12n_f \end{pmatrix},$$
$$\boldsymbol{\gamma}^{(1)}(2) = \tfrac{1}{243} \begin{pmatrix} 64[367 - 39n_f] & -3666n_f \\ -64[367 - 39n_f] & 3666n_f \end{pmatrix} \tag{4.9.5}$$

and to three loops the same structure may be checked with the help of the calculation of Larin, Nogueira, van Ritbergen and Vermaseren (1997). The normalization of the structure function $F_G(x, Q^2) = xG(x, Q^2)$ is in principle arbitrary; we have chosen it so that the eigenvector of $\boldsymbol{\gamma}$ corresponding to the zero value will be precisely the sum of F_G and F_S. Now the conserved operator is the energy–momentum tensor: from Eq. (2.8.2),

$$\Theta^{\mu\nu} = i \sum_f \bar{q}_f \gamma^\mu D^\nu q_f + g_{\alpha\beta} G^{\mu\alpha} G^{\beta\nu} - g^{\mu\nu} \mathcal{L}.$$

The term $g^{\mu\nu}\mathcal{L}$ contributes only to order M^2/Q^2 and may thus be neglected. We thus find the *momentum sum rule*,

$$\int_0^1 dx \left\{ F_{2S}(x, Q^2) + F_{2G}(x, Q^2) \right\} = \delta \left\{ 1 + c_2 \frac{\alpha_s(Q^2)}{\pi} + \cdots \right\}, \quad (4.9.6)$$

with δ, c_2 depending on the process. For electroproduction,

$$\delta^{ep} = \langle Q_f^2 \rangle = 5/18 \ (n_f = 4) \quad \text{and} \quad c_2^{ep} = -5/9.$$

For νI, νp,

$$\delta^{\nu I} = 1, \quad \delta^{\nu p} = 2/3.$$

In fact, for $Q^2 \to \infty$, one can calculate the individual integrals of each of the F_{2i}, $i = S, G$. This is so because, for $n = 2$,

$$d_+(2) = 0, \quad d_-(2) = \frac{32 + 6n_f}{99 - 6n_f} < 0;$$

hence, to leading order in α_s we can write

$$\boldsymbol{\mu}(2, Q^2) \underset{Q^2 \to \infty}{=} \mathbf{S}\mathbf{b}(2), \quad \mathbf{b}(2) = b \begin{pmatrix} 1 \\ 0 \end{pmatrix},$$

with $\mathbf{S}$ given in (4.7.16b) and b a number independent of Q^2. Therefore, determining b from (4.9.6),

$$\int_0^1 dx \, F_{2S}(x, Q^2) \underset{Q^2 \to \infty}{=} \delta \frac{3n_f}{16 + 3n_f},$$

$$\int_0^1 dx \, F_G(x, Q^2) \underset{Q^2 \to \infty}{=} \delta \frac{16}{16 + 3n_f}. \quad (4.9.7)$$

Unfortunately, the corrections are of the form

$$K \left[\alpha_s(Q^2) \right]^{-d_-(2)}, \quad d_-(2) \sim 0.6,$$

with K a quantity not given by perturbative QCD. Eqs. (4.9.7) are among those that provide the best evidence for the existence of gluons. If they did not exist, one would expect all momentum to be carried by the quarks. Hence, for e.g. neutrino–isoscalar scattering where $\delta = 1$, one would expect

$$\int_0^1 dx \, F_2(x, Q^2) \approx 1,$$

which for, say, $n_f = 4$ is *twice* the experimental value. In fact, one has (De Groot et al., 1979; Berge et al., 1991)

$$\int_0^1 dx\, F_2^{\text{exp}}(x, Q^2) \approx 0.44 \pm 0.003, \quad Q^2 = 30 \text{ to } 200 \text{ GeV}^2,$$

which compares very nicely with the theoretical figure taking gluons into account, (4.9.7) which gives[19]

$$\int_0^1 dx\, F_2^{\text{exp}}(x, Q^2) \to \tfrac{12}{28} = 0.43.$$

The leading order analysis of these relations was performed by Gross and Wilczek (1974), although the momentum sum rule at the partonic level had already been discussed by Llewellyn Smith (1972).

ii Behaviour of Structure Functions as $x \to 1$

The QCD evolution equations take on a particularly simple form at the endpoints, $x = 0, 1$, where they have also implications for questions other than deep inelastic scattering. Here we start by considering the limit as $x \to 1$. Considering first the nonsinglet component of structure functions, we will assume that

$$F_{NS}(x, Q^2) \underset{x \to 1}{\simeq} A(Q^2)(1 - x)^{\nu(\alpha_s)} \tag{4.9.8}$$

with eventual logarithms (see below). Actually, (4.9.8) can be made plausible in QCD from the so-called counting rules, although we will not give the full arguments here (see Brodsky and Lepage, 1980, and references therein). On general grounds, we expect that the $x \to 1$ behaviour of the structure functions would be related to the large n behaviour of the moments. It is easy to verify that

$$d_{NS}(n) \underset{n \to \infty}{\simeq} -\frac{16}{33 - 2n_f} \left(\log n - \tfrac{3}{4} + \gamma_{\mathrm{E}} + O(1/n) \right). \tag{4.9.9}$$

Using (4.9.8) and integrating, we then obtain

$$\mu_{NS}(n, Q^2) \underset{n \to \infty}{\simeq} A(Q^2) \frac{\Gamma(n - 1)\Gamma(1 + \nu(\alpha_s))}{\Gamma(n + \nu(\alpha_s))},$$

while, from (4.9.9), (4.6.6),

$$\frac{\mu_{NS}(n, Q^2)}{\mu_{NS}(n, Q_0^2)} \underset{n \to \infty}{\simeq} \exp \left\{ \frac{16}{33 - 2n_f} \left(\log n - \tfrac{3}{4} + \gamma_{\mathrm{E}} \right) \left[\log \frac{\alpha_s(Q^2)}{\alpha_s(Q_0^2)} \right] \right\}.$$

---

[19]Note that neutrinos or e, μs only probe *quarks*, so the experimentally measured function is, precisely, F_{2S}. To get $xG = F_G$ we would require probes that acted on gluons.

Equating, we find the explicit form of $A(Q^2)$, $\nu(\alpha_s)$; to LO (Gross, 1974),

$$F_{NS} \underset{x \to 1}{\simeq} A_{0NS} \left[\alpha_0(Q^2)\right]^{-d_0} \frac{(1-x)^{\nu_{NS}(\alpha_s)}}{\Gamma(1+\nu_{NS}(\alpha_s))},$$

$$\nu_{NS}(\alpha_s) = \nu_{0NS} - \frac{16}{33-2n_f}\log\alpha_s(Q^2), \quad d_0 = \frac{16}{33-2n_f}\left(\tfrac{3}{4} - \gamma_E\right).$$

(4.9.10)

The constants A_{0NS}, ν_{0NS} are not given by perturbative QCD, although one expects, from counting rules arguments (Sect. 5.7), that $\nu_{0NS} \approx 2$ to 3.

For the singlet, the calculations are somewhat more complicated because of the matrix character of the equations. One finds that for the gluons (4.9.8) must be modified, but for the quarks the behaviour is like that for the nonsinglet (Martin, 1979; López and Ynduráin, 1981) and one then obtains

$$F_S \underset{x \to 1}{\simeq} A_{0S} \left[\alpha_s(Q^2)\right]^{-d_0} \frac{(1-x)^{\nu_S(\alpha_s)}}{\Gamma(1+\nu_S(\alpha_s))},$$

$$F_G = xG \underset{x \to 1}{\simeq} \tfrac{2}{5}A_{0S} \left[\alpha_s(Q^2)\right]^{-d_0} \frac{(1-x)^{\nu_S(\alpha_s)+1}}{\Gamma(2+\nu_S(\alpha_s))|\log(1-x)|}.$$

(4.9.11)

Here d_0 is as before; ν_S is given by a formula similar to that for ν_{NS} above,

$$\nu_S(\alpha_s) = \nu_{0S} - \frac{16}{33-2n_f}\log\alpha_s(Q^2),$$

(4.9.12)

and the *same* constants A_{0S}, ν_{0S} appear for quarks and gluons.

Second order corrections modify these behaviours. We refer to López and Ynduráin (1981) for the explicit NLO results for gluon, quark singlet and nonsinglet, and discuss in greater detail the higher order corrections to the nonsinglet, that coincide with those for the quark singlet apart an from an eventual difference in the constants. This is so because, as can be read directly from the expression for the anomalous dimension matrix, the influence on the quark component of the mixing between quark and gluon structure functions vanishes as $x \to 1$. Using the explicit expression for $\gamma_{NS}^{(1)}(n)$, Eq. (4.7.2), we find (González-Arroyo, López and Ynduráin, 1979)

$$F_{NS} \underset{x \to 1}{\simeq} A_{0NS} \left[\alpha_s(Q^2)\right]^{-d_0} \frac{e^{a(\alpha_s)\alpha_s(Q^2)}}{\Gamma(1+\nu_{1NS}(\alpha_s))}$$

$$\times (1-x)^{\nu_{1NS}(\alpha_s)+2[\log(1-x)]\alpha_s/3\pi}.$$

(4.9.13)

Here,

$$\nu_{1NS}(\alpha_s) = \nu_{NS}(\alpha_s) - \psi(\nu_{NS}(\alpha_s)+1)\frac{4\alpha_s(Q^2)}{3\pi} - a_1\alpha_s(Q^2),$$

$$a(\alpha_s) = a_0 + a_1\psi(\nu_{NS}(\alpha_s)+1)$$

$$+ \frac{2}{3\pi}\left\{[\psi(\nu_{NS}(\alpha_s)+1)]^2 - \psi'(\nu_{NS}(\alpha_s)+1)\right\}.$$

The constants a_0, a_1 can be calculated in terms of $\gamma_{NS}^{(1)}(n)$, $n \to \infty$, and they are $a_0 \approx 1.18$, $a_1 \approx 0.06$. ν_{NS} is as in Eq. (4.9.10), and ψ, ψ' are the digamma function and its first derivative.

It is interesting to note that, because of the term

$$(1 - x)^{2[\log(1-x)]\alpha_s/3\pi} \tag{4.9.14}$$

in (4.9.13), we obtain corrections as large as we wish if x is near enough to unity. Thus perturbation theory fails for $x \to 1$, something that ought to be expected on physical grounds: when $x = 1$ we encounter bound states (the elastic contribution to $\gamma^* + N \to$ all, viz., $\gamma^* + N \to N$ in electroproduction on nucleons). Actually, there are other reasons why the perturbative QCD analysis fails if x is too close to 1, that we will consider in Sect. 4.11.

From (4.9.14) we see that (4.9.13) is valid in an intermediate region,

$$1 - x \ll 1 \quad \text{but} \quad \frac{2\alpha_s}{3\pi}|\log(1 - x)| \ll 1. \tag{4.9.15}$$

It is interesting that the leading behaviour in $|\log(1-x)|$ in the region (4.9.15) can be obtained to all orders; the result is essentially equivalent to replacing, in the evolution equations, $\alpha_s(Q^2)$ by $\alpha_s((1 - x)Q^2)$. This was first conjectured by Amati et al. (1980); see also Ciafalloni and Curci (1981). For the general proof and details see Sterman (1987) and, for a simpler version, Catani and Trentadue (1989).

iii The Limit $x \to 0$, Nonsinglet

The kinematic region corresponding to the limit as $x \to 0$ is that of *fixed* Q^2 and hadronic energy $\nu \to \infty$. As noted by Abarbanel, Goldberger and Treiman (1969), this is the Regge limit,[20] since the structure functions are proportional to cross sections. For example, for electroproduction on proton targets, we can write F_2 in terms of the cross section for a virtual γ^* with invariant mass $-Q^2$ scattering on a proton:

$$\sigma_{\gamma^*(-Q^2)p}(s) = \frac{4\pi^2\alpha}{Q^2}F_2(x, Q^2), \quad \text{with } s = Q^2/x.$$

This limit has been studied extensively in hadron physics and, in the particular case of nonsinglet scattering, has been found to be given by so-called exchange of Regge trajectories, either the ρ-particle trajectory or a trajectory degenerate with it in such a way that one has, for *mass shell* scattering amplitudes, T,

$$T_{NS} \underset{\nu \to \infty}{\simeq} f s^{\alpha_\rho(0)}, \tag{4.9.16}$$

---

[20]For Reggeology, see for example Barger and Cline (1969).

where $\alpha_\rho(0)$ is the *intercept* of the ρ trajectory; from fits to experimental data, $\alpha_\rho(0) \approx 0.5$. This quantity is supposed to be universal and all the dependence on Q^2 is to be found in the constant f. Thus, changing variables to the deep inelastic ones, we assume that, for a given Q_0^2 large enough that perturbation theory be valid, but sufficiently small that one can believe Regge-type arguments (say, $Q_0^2 = $ a few GeV^2),

$$F_{NS}(x, Q_0^2) \underset{x \to 0}{\simeq} B_{NS}(Q_0^2) x^{\lambda_{NS}(Q_0^2)}, \qquad (4.9.17)$$

where

$$\lambda_{NS}(Q_0^2) = 1 - \alpha_\rho(0) \approx 0.5.$$

We allow $\lambda_{NS}(Q^2)$ to be dependent on Q^2; but, as we will see, the QCD evolution equations confirm the Regge theory property that it has to be a constant, independent of Q^2.

From the expression for the moments,

$$\mu_{NS}(n, Q^2) = \int_0^1 dx\, x^{n-2} F_{NS}(x, Q^2),$$

it follows that the behaviour of $F_{NS}(x, Q^2)$ for $x \to 0$ is related to the rightmost singularity of $\mu_{NS}(n, Q^2)$ in the variable n (considered as a continuous variable). As is easily verified, the behaviour (4.9.17) corresponds to a pole of $\mu_{NS}(n, Q_0^2)$ for $n = n_0 \equiv 1 - \lambda_{NS}$. Thus, for n near this value,

$$\mu_{NS}(n, Q_0^2) \underset{n \simeq 1 - \lambda_{NS}(Q_0^2)}{\simeq} \frac{B_{NS}(Q_0^2)}{n - (1 - \lambda_{NS})} \qquad (4.9.18a)$$

For $Q^2 = Q_0^2$, our assumption (4.9.17) implies that this singularity occurs at $n_0 \approx 0.5$. On the other hand, from (4.5.20) we have μ_{NS} a product of two terms. Of these, A_{NS}^n is independent of Q^2, and all the dependence on this variable is in the Wilson coefficient $C_{NS}^n(Q^2/\mu^2, \alpha_s(Q^2))$. Now, the rightmost singularity of the product will be that of whichever of the two terms that is furthest to the right. To LO, (4.6.5, 6) tells us that

$$C_{NS}^n(Q^2/\mu^2, \alpha_s(Q^2)) = \text{const.} \times \left[\alpha_s(Q^2)\right]^{-d_{NS}(n)},$$

$$d_{NS}(n) = -\gamma_{NS}^{(0)}(n)/2\beta_0, \quad \gamma_{NS}^{(0)}(n) = 2C_F \left\{ 4S_1(n) - 3 - \frac{2}{n(n+1)} \right\},$$

from which explicit expressions it follows that the rightmost singularity of $C_{NS}^n(Q^2/\mu^2, \alpha_s(Q^2))$ occurs at $n = 0$. To NLO, the expressions found (cf. Sect. 4.7i) share this property and to NNLO, although we do not have a full calculation (the values of $\gamma_{NS}^{(2)}$ are not known), the coefficients also have the first singularity at $n = 0$ so we will assume this to be valid to all orders. However, from our assumption (4.9.17) it follows that, when $Q^2 = Q_0^2$, the dominating singularity is located at $n \approx 0.5$: it must therefore be due to a singularity of the matrix element,

$$A_{NS}^n \sim \frac{1}{n - (1 - \lambda_{NS})}. \tag{4.9.18b}$$

But this is independent of Q^2: hence, and as we had anticipated (and as Regge theory demanded) we find that λ_{NS} is independent of Q^2, so we can write, for all Q^2 now,

$$F_{NS}(x, Q^2) \underset{x \to 0}{\simeq} B_{NS}(Q^2) x^{\lambda_{NS}}.$$

It only remains to evaluate the Q^2 dependence of $B_{NS}(Q^2)$. This is trivial: from the expression for the singularity near $n = 1 - \lambda_{NS}$, Eq. (4.9.18b), it follows that (for arbitrary Q^2 now)

$$\mu_{NS}(n, Q^2) \underset{n \simeq 1 - \lambda_{NS}}{\simeq} \frac{B_{NS}(Q^2)}{n - (1 - \lambda_{NS})};$$

while, from (4.6.6b),

$$\mu_{NS}(n, Q^2) \underset{n \simeq 1 - \lambda_{NS}}{\simeq} \left[\alpha_s(Q^2)\right]^{-d_{NS}(n=1-\lambda_{NS})}$$

$$\times \left[\alpha_s(Q_0^2)\right]^{+d_{NS}(n=1-\lambda_{NS})} \mu_{NS}(n, Q_0^2).$$

Comparing, we find the very general and explicit result

$$F_{NS}(x, Q^2) \underset{x \to 0}{\simeq} B_{0NS} \left[\alpha_s(Q^2)\right]^{-d_{NS}(1-\lambda_{NS})} x^{\lambda_{NS}}; \tag{4.9.19}$$

B_{0NS} is a constant, not given by perturbative QCD, and $d_{NS}(1 - \lambda_{NS})$ can be evaluated with the explicit expression (4.6.13).

The above analysis is due to Martin (1979) and López and Ynduráin (1981), where the NLO extension may be found. It amounts to replacing Eq. (4.9.12) by

$$F_{NS}(x, Q^2) \underset{x \to 0}{\simeq} B_{0NS} \left\{1 + c_{NS}^{(1)} \frac{\alpha_s(Q^2)}{4\pi}\right\} \left[\alpha_s(Q^2)\right]^{-d_{NS}(1-\lambda_{NS})} x^{\lambda_{NS}}, \tag{4.9.20}$$

and the explicit expression for $c_{NS}^{(1)}$ in terms of the $\gamma_{NS}^{(1)\pm}(n)$, $C_{NS}^{(1)}(n)$ of Eqs. (4.7.2) may be found in the last quoted paper.

iv The Limit $x \to 0$, Singlet

The $x \to 0$ limit of singlet structure functions presents a number of difficulties. First, we now have coupled equations, as the gluon density also intervenes. This is of a technical nature, and is not too complicated; more important is the fact that there is no universally accepted behaviour to be used as input at $Q_0^2 \sim$ a few GeV^2. The Regge singularity that dominates structure functions with the exchange quantum numbers of the vacuum, i.e., singlet, is known as the Pomeron, and there is no consensus as to its nature. Cross sections for on mass–shell, *physical* hadrons behave as constants (modulo logarithmic corrections, that we will neglect here). So one would feel tempted to assume

$$F_S(x, Q_0^2) \underset{x \to 0}{\simeq} C_S, \quad F_G(x, Q_0^2) \underset{x \to 0}{\simeq} C_G. \tag{4.9.21}$$

We will refer to this as the *soft* Pomeron. However, it has been known for some time that the dominance of a soft Pomeron for *off-shell* processes leads to inconsistencies (Migdal, Polyakov and Ter-Martirosian, 1974; Abarbanel and Bronzan, 1974; Lipatov, 1976; for a review, see Moshe, 1978), which led some physicists to postulate a *hard Pomeron*,

$$F_S(x, Q_0^2) \underset{x \to 0}{\simeq} B_S(Q_0^2) x^{-\lambda}, F_G(x, Q_0^2) \underset{x \to 0}{\simeq} B_G x^{-\lambda}. \tag{4.9.22}$$

In principle, one could take the λ as being different for quarks and gluons, and Q^2 dependent; but it may be easily proved, following methods similar to those employed for the nonsinglet, that the exponents of x should be the same for quarks and gluons, and Q^2-independent.

The behaviour that follows from (4.9.21) was first considered by De Rújula et al. (1974); the detailed formulas, to LO, were given by Martin (1979). A partial NLO evaluation is due to by Ball and Forte (1995) and a complete one (including the longitudinal structure function) may be found in Adel, Barreiro and Ynduráin (1997).

The key point in the analysis is that the relevant equation, analogue to (4.5.20), now reads

$$\boldsymbol{\mu}(n, Q^2) = \mathbf{C}^n \mathbf{A}^n,$$

$$\mathbf{A}^n = \langle p | \mathbf{O}_n | p \rangle, \quad \mathbf{O}_n = \begin{pmatrix} \bar{q}_f \gamma \overbrace{\partial \ldots \partial}^{n-1} q_f \\ G \underbrace{\partial \ldots \partial}_{n-2} G \end{pmatrix},$$

and the one corresponding to (4.6.6b) is now (4.6.6c), which we repeat here:

$$\boldsymbol{\mu}(n, Q^2) = \left[\alpha_s(Q_0^2) / \alpha_s(Q^2) \right]^{\mathbf{D}(n)} \boldsymbol{\mu}(n, Q_0^2), \quad \mathbf{D}(n) = -\boldsymbol{\gamma}^{(0)}(n) / 2\beta_0.$$

We have to compare the locations of the singularities in n of the anomalous dimensions and coefficients, $\boldsymbol{\gamma}^{(N)}(n)$, $\mathbf{C}^{(N)}(n)$, on the one hand, and of the matrix elements, $\mathbf{A}^n$, on the other.

The singularities of the $\boldsymbol{\gamma}^{(N)}(n)$, $\mathbf{C}^{(N)}(n)$ lie either to the *left*, or at $n = 1$ for the LO, NLO and the known piece of NNLO. If we now assume a soft Pomeron, the singularity of $\mathbf{A}^n$ lies at the same place, $n = 1$. Since the singularity of $\mathbf{D}^{(0)}(n)$ appears in the exponent, it dominates the other and a rather simple calculation produces the behaviour, to LO,

$$
F_S(x, Q^2) \underset{x \to 0}{\simeq} \frac{c_0}{|\log x|} \left[\frac{9 |\log x| \log[\alpha_s(Q_0^2)/\alpha_s(Q^2)]}{4\pi^2(33 - 2n_f)} \right]^{\frac{1}{4}}
$$
$$
\times \exp\left\{ \sqrt{D_0 \log x|} \left[\log \frac{\alpha_s(Q_0^2)}{\alpha_s(Q^2)} \right] - D_1 \log \frac{\alpha_s(Q_0^2)}{\alpha_s(Q^2)} \right\},
$$
(4.9.23a)

$$
F_G(x, Q^2) \underset{x \to 0}{\simeq} \frac{9c_0}{n_f} \left[\frac{33 - 2n_f}{576\pi^2 |\log x| \log[\alpha_s(Q_0^2)/\alpha_s(Q^2)]} \right]^{\frac{1}{4}}
$$
$$
\times \exp\left\{ \sqrt{D_0 \log x|} \left[\log \frac{\alpha_s(Q_0^2)}{\alpha_s(Q^2)} \right] - D_1 \log \frac{\alpha_s(Q_0^2)}{\alpha_s(Q^2)} \right\},
$$
(4.9.23b)

where

$$
D_0 = \frac{144}{33 - 2n_f}, \quad D_1 = \frac{33 + 2n_f/9}{33 - 2n_f}
$$

are related to $\mathbf{D}^{(0)}(n)$ for $n \simeq 1$, and the constants c_i are different from (although related to) the C_i in (4.9.21). In fact, the *exponent* in (4.9.23) is universal in the sense that it only follows from the singularity of $\mathbf{D}^{(0)}(n)$ and would also be obtained if, for example, the input $F_i(x, Q_0^2)$ vanished as $x \to 0$; see Martin (1979) for the details.

If, on the other hand, we assume a hard Pomeron, we get a very different behaviour:

$$
F_S(x, Q^2) \underset{x \to 0}{\simeq} B_{0S} \left[\alpha_s(Q^2) \right]^{-d_+(1+\lambda)} x^{-\lambda},
$$
$$
F_G(x, Q^2) \underset{x \to 0}{\simeq} \frac{d_+(1 + \lambda) - D_{11}^{(0)}(1 + \lambda)}{D_{12}^{(0)}(1 + \lambda)} B_{0S} \left[\alpha_s(Q^2) \right]^{-d_+(1+\lambda)} x^{-\lambda},
$$
(4.9.24)

where $d_+(1 + \lambda)$ is the largest eigenvalue of the matrix $\mathbf{D}^{(0)}(n)$ at $n = 1 + \lambda$.

The proof is fairly simple. From (4.7.19), and with a slight change of notation,

$$
\boldsymbol{\mu}(n, \alpha_s) = \mathbf{C}(n, \alpha_s)\mathbf{S}(n)\bar{\mathbf{M}}(n, \alpha_s)\alpha_s^{-\hat{\mathbf{D}}^{(0)}(n)}\mathbf{b}(n).
$$

For the behaviour as $x \to 0$, we are interested in the value of $\boldsymbol{\mu}(n, \alpha_s)$ near the singularity, $n_0 = 1 + \lambda$. To LO,

$$
\boldsymbol{\mu}(n, \alpha_s) \underset{n \to 1+\lambda}{\simeq} \mathbf{S}(1 + \lambda)\alpha_s^{-\hat{\mathbf{D}}^{(0)}(1+\lambda)}\mathbf{b}(n), \tag{4.9.25}
$$

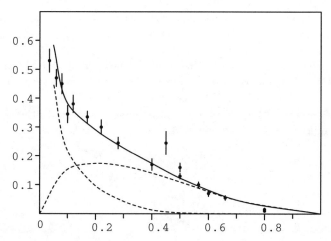

Fig. 4.9.1. Comparison of theory and experiment for F_2, decomposed in $F_{NS} + F_S$, as in the text. The NS and S components are represented by the dashed curves. The full curve is the whole F_2. SLAC data (Bodek et al., 1979) for $Q^2 = 22.4 \text{ GeV}^2$.

and we must have

$$\mathbf{b}(n) \underset{n \to 1+\lambda}{\simeq} \frac{1}{n - (1 + \lambda)} \mathbf{B}_0.$$

Let $d_\pm$ be the eigenvalues of $\mathbf{D}^{(0)}(1 + \lambda)$. For the values of λ that are phenomenologically relevant ($\lambda = 0.3$ to 0.5) one has $d_+ > d_- + 1$; so that, to LO and NLO,[21] we can neglect α^{-d_-} as compared to α^{-d_+}. Hence, (4.9.25) becomes

$$\boldsymbol{\mu}(n, \alpha_s) \underset{n \to 1+\lambda}{\simeq} \mathbf{S}(1 + \lambda) \begin{pmatrix} \alpha^{-d_+} \\ 0 \end{pmatrix} \frac{B_{01}}{n - (1 + \lambda)}.$$

The desired result follows by comparing with the result of taking the moments of (4.9.24); the ratio F_S/F_G equals precisely $S_{21}(1 + \lambda)$; cf. (4.7.16b).

To NLO, (4.9.24) is replaced by multiplying the right hand side by $\{1 + c_i \alpha_s / 4\pi\}$, $i = S, G$ known in terms of $\mathbf{C}^{(1)}(1 + \lambda)$, $\boldsymbol{\gamma}^{(0,1)}(1 + \lambda)$. Further details, the NLO expressions and the implications for the longitudinal structure function, both for the soft and hard Pomerons, may be found in Adel, Barreiro and Ynduráin (1997).

We finish this subsection by showing an explicit example of structure function, viz., F_2 in electron–proton scattering. From the discussions earlier in this section, we expect that the nonsinglet component will vanish when

---

[21]To NNLO, the contribution of d_- should also be taken into account.

$x \to 0$ as $x^{0.5}$, and also when $x \to 1$, as $(1-x)^{2.5}$ now. As for the singlet piece, we would have it *diverging* as $x \to 0$, and vanishing faster than the nonsinglet for $x \to 1$. Moreover, the integral of the singlet, and the integral of the nonsinglet divided by x, are expected to be constant. The pattern is thus like that show in Fig. 4.9.1. The figure has been obtained by writing

$$F_2(x, Q^2) = C_{NS} x^{0.52}(1-x)^{2.3} + C_S x^{-0.37}(1-x)^{6.7},$$

and then fitting the C_i to the data. In later sections we will see more detailed examples of comparisons of theory and experiment.

v The BFKL Pomeron

When studying the small x limit of (singlet) structure functions, we have followed the strategy of assuming a given behaviour at a fixed Q^2 and then evolving with the renormalization group to find the behaviour for arbitrary Q^2. A different approach is that of the BFKL group[22] (Kuraev, Lipatov and Fadin, 1976; Balitskii and Lipatov, 1978), who sum leading terms in the variable $\log x$. They find the behaviour

$$F_2(x, Q^2) \underset{x \to 0}{\simeq} x^{-\omega_0 \alpha_s}, \quad \omega_0 = \frac{4 C_A \log 2}{\pi}. \qquad (4.9.26)$$

It is difficult to decide what (4.9.26) means. If we assume (as would seem natural) that one has $\alpha_s \to \alpha_s(Q^2)$, the resulting behaviour is incompatible with what follows from the OPE and the renormalization group, as we have just seen in the previous subsection that it implies that the exponent of x should be constant. It has been speculated that perhaps there are two regimes, one with x small, but not too small, where (4.9.24) would be valid, and another one, at ultra–small values of x, where one would have (4.9.26). This may be so, but as we will see, the experimental data run quite contrary to (4.9.26) with $\alpha_s(Q^2)$; and this down to the very small x data of HERA that reach to $x \sim 10^{-5}$.

Another possibility, supported by the fact that the region of integration to obtain (4.9.26) includes soft momenta, is that one has $\alpha_s \to \alpha_s(k_0^2)$, with k_0^2 some fixed cut–off momentum of order Λ^2. If this is so, the BFKL analysis would constitute a confirmation of the assumption used in deriving the hard Pomeron behaviour (4.9.22), with $Q_0^2 = k_0^2$.

---

[22]See also Ciafalloni (1988) and Catani, Fiorini and Marchesini (1990), who use a somewhat more comprehensible, but equivalent, approach.

4.10 Target Mass Corrections

Consider a moment μ_{NS} of a nonsinglet structure function. In principle, it depends not only on n and α_s, but also on a set of masses: the masses of the target (that we assume to be a nucleon) m_N, quark masses m_q and, eventually, nonperturbative masses. Let us neglect the latter for now. As will be argued in Sect. 7.4, the u, d and s have small masses, the largest being $m_s \sim 200$ MeV. With the values of Λ that we have, it follows that perturbative QCD will hardly make sense unless $Q^2 \geq 2$ GeV^2; compared with this, even the s quark mass is negligible. Heavy quarks c, b are a different matter, but we will leave them for the moment. It remains the mass of the target that gives corrections $O(m_N^2/Q^2)$, which is quite sizable. In this section, we will show how to take these corrections into account.

The effect of target mass corrections was first studied by Nachtmann (1973); it leads to so-called ξ-scaling. Here we will follow the method of Georgi and Politzer (1976). Recall the expansions (4.5.3) and (4.5.11); in general, these expansions should contain, besides the terms shown there, other terms which are of the two following types. There are terms that correspond to the operators (Wick normal ordering implicit)

$$g^{\mu\nu}\bar{q}\,\slashed{D}D^{\mu_1}\ldots q \quad \text{and} \quad g^{\mu\nu}\bar{q}D^2 D^{\mu_1}\ldots q.$$

Using the equations of motion, $\slashed{D}q = -im_q q$; hence, these terms will give contributions proportional to quark masses, which we are now neglecting.[23] However, terms of the second type,

$$\langle p|N_{NS}^{\mu\nu\mu_1\cdots\mu_n}(0)|p\rangle = (p^2)^m g^{\mu_i\mu_j}\ldots g^{\mu_l\mu_s}p^{\mu_{k_1}}\ldots p^{\mu_{k_r}}\bar{A}'^n_{NS},$$

give, as will be seen shortly, corrections in m_N^2/Q^2. We neglected these corrections earlier, but we will focus on them now. Consider $N^{\mu_1\cdots\mu_n}$, n even; later we will replace $n \to n+2$ and identify $\mu_{n+1} \to \mu$, $\mu_{n+2} \to \nu$. Because N is symmetrized, its matrix elements can be written quite generally as

$$i\langle p|N_{NS}^{\mu_1\cdots\mu_n}(0)|p\rangle = \sum_{j=0}^{n/2}(-1)^j\frac{(n-j)!}{2^j n!}\left\{\sum_{\text{permutations}}g^{\mu_{i_1}\mu_{i'_1}}\ldots g^{\mu_{i_j}\mu_{i'_j}}\right\}(p^2)^j$$

$$\times\left\{\sum_{\text{permutations}}p^{\mu_{k_1}}\ldots p^{\mu_{k_{n-2j}}}\right\}\bar{A}_{NS,j}^{(\text{TMC})n-2},$$

$$N_{NS}^{\mu_1\cdots\mu_n}(0) = \mathcal{S}\bar{q}\gamma^{\mu_1}D^{\mu_2}\ldots D^{\mu_n}q\Big|_{NS}.$$

$$(4.10.1)$$

---

[23]We will not discuss the incorporation of heavy quark (c, b) mass effects, that follows a pattern very similar to that of target mass corrections; see for example Nachtmann (1973), Georgi and Politzer (1976) and Barbieri, Ellis, Gaillard and Ross (1976). For calculations including radiative corrections, see Laenen, Riemersma, Smith and van Neerven (1993).

Now, $g_{\mu_i \mu_j} \langle p | N_{NS}^{\mu_1 \cdots \mu_n}(0) | p \rangle = 0$; we have sufficiently many relations that we can solve for all the A_j^n in terms of A_0^n. Then,

$$T_{2NS}^{(\text{TMC})} = \frac{1}{2} \sum_n x^{-n-1} \sum_{j=0}^{\infty} \left(\frac{p^2}{Q^2} \right)^j \frac{(n+j+2)!(n+2j)!}{j!n!(n+2j+2)!} A_{NS}^{(0)n+2j} C_{NS}^{n+2j},$$

$$A_{NS}^{(0)n} \equiv A_{NS,j=0}^{(\text{TMC})n}.$$

$$(4.10.2)$$

Therefore, we obtain the result

$$\mu_{NS}^{(\text{TMC})}(n, Q^2) = \sum_{j=0}^{\infty} \left(\frac{m_N^2}{Q^2} \right)^j \frac{(n+j)! C_{NS}^{n+2j}}{j!(n-2)!(n+2j)(n+2j-1)} A_{NS}^{(0)n+2j},$$

$$\mu_{NS}^{(\text{TMC})}(n, Q^2) = \int_0^1 dx\, x^{n-2} F_{2NS}^{(\text{TMC})}(x, Q^2).$$

$$(4.10.3)$$

It is convenient to define the function F_2 to be the limit $m_N \to 0$ of $F_2^{(\text{TMC})}$, and set

$$\mu_{NS}(n, Q^2) = \int_0^1 dx\, x^{n-2} F_{2NS}(x, Q^2). \qquad (4.10.4)$$

It is to these μ, F_2 that the equations derived in the previous sections apply. To obtain the moments with the TMC taken into account, we use (4.10.3),

$$\mu_{NS}^{(\text{TMC})}(n, Q^2) = \sum_{j=0}^{\infty} \left(\frac{m_N^2}{Q^2} \right)^j \frac{(n+j)!}{j!(n-2)!(n+2j)(n+2j-1)} \mu_{NS}(n+2j, Q^2),$$

$$(4.10.5)$$

but we do not have to go through the moments. After some simple manipulations we find that (4.10.5) is equivalent to (ξ-scaling)

$$F_{2NS}^{(\text{TMC})}(x, Q^2) = \frac{x^2/\xi^2}{(1 + 4x^2 m_N^2/Q^2)^{3/2}} F_{2NS}(\xi, Q^2)$$

$$+ \frac{6m_N^2}{Q^2} \frac{x^3}{(1 + 4x^2 m_N^2/Q^2)^2} \int_\xi^1 \frac{d\xi'}{\xi'^2} F_{2NS}(\xi', Q^2)$$

$$+ \frac{12m_N^4}{Q^4} \frac{x^4}{(1 + 4x^2 m_N^2/Q^2)^{5/2}} \int_\xi^1 d\xi' \int_{\xi'}^1 \frac{d\xi''}{\xi''^2} F_{2NS}(\xi'', Q^2),$$

$$(4.10.6a)$$

where ξ is Nachtmann's variable,

$$\xi = \frac{2x}{1 + (1 + 4x^2 m_N^2/Q^2)^{1/2}}. \qquad (4.10.6b)$$

A few features of these formulas are worth noting. First, for small x, and since TMCs behave like $x^2 m_N^2/Q^2$, we can neglect them completely. TMCs are relevant for large – but not too large – values of x. Indeed, if applied at

$x \to 1$, inconsistencies develop. There are two reasons for this. Higher twist corrections are also largest for $x \to 1$. Although one expects higher twist corrections to be proportional to $3M^2/Q^2$ with $M \sim \Lambda$ and hence smaller than TMCs by half an order of magnitude, cancellations may (and probably do) occur.[24] Secondly, and as we saw in Sect. 4.9ii, perturbation theory fails for $x \to 1$. Because of this, it is perhaps more consistent to expand (4.10.6) in powers of m_N^2/Q^2 and retain only the leading term. The expression for TMCs then simplifies to

$$F_{2NS}^{(\text{TMC})}(x,Q^2) = F_{2NS}(x,Q^2)$$
$$+ \frac{x^2 m_N^2}{Q^2} \left\{ 6x \int_x^1 dy \, \frac{F_{2NS}(y,Q^2)}{y^2} - \frac{x\partial}{\partial x} F_{2NS}(x,Q^2) - 4F_{2NS}(x,Q^2) \right\}$$

$$(4.10.7)$$

and one stops applying perturbative QCD when the second order corrections,

$$\sim \left\{ \frac{x^3 \nu_{NS}(\alpha_s) m_N^2}{(1-x)Q^2} \right\}, \quad \nu_{NS}(\alpha_s) \sim 3 \text{ to } 5$$

are large, taking these corrections as a measure of the theoretical error of the calculation.

4.11 Nonperturbative Effects in e^+e^- Annihilations and Higher Twists in Deep Inelastic Scattering

We treat these effects in the same section because they are, from our point of view here, clearly related. We begin with the first. As discussed in Sect. 4.1, we have to consider the quantity $\Pi^{\mu\nu}$ given by Eqs. (4.1.4). So we look at the product

$$TJ^\mu(x)J^\nu(0)$$

from the OPE point of view. We write a short distance expansion for it; in momentum space and with $Q^2 = -q^2$,

$$i \int d^4x \, e^{iq \cdot x} TJ^\mu(x)J^\nu(0) = (-g^{\mu\nu}q^2 + q^\mu q^\nu) \Big\{ C_0 \left(Q^2/\nu^2, g(\nu) \right) 1$$
$$+ \sum_f C_f \left(Q^2/\nu^2, g(\nu) \right) m_f : \bar{q}_f(0)q_f(0) :$$
$$+ C_G \left(Q^2/\nu^2, g(\nu) \right) \alpha_s : \sum_a G_a^{\mu\nu}(0)G_{a\mu\nu}(0) : + \cdots \Big\}.$$

$$(4.11.1)$$

----

[24]For a discussion of this, see De Rújula, Georgi and Politzer (1977a,b).

In Sect. 4.1 we only considered the first term, $C_0 1$. This was done for two reasons. First, on purely dimensional grounds,

$$C_f \simeq \frac{\text{(const.)}}{Q^4}, \quad C_G \simeq \frac{\text{(const.)}}{Q^4}, \tag{4.11.2}$$

so they are negligibly small at large momenta. Secondly, to all orders in perturbation theory,

$$\langle : \bar{q}q : \rangle_0 = 0, \quad \langle : G^2 : \rangle_0 = 0. \tag{4.11.3}$$

However, it will be argued later that the physical vacuum is not that of perturbation theory, but must incorporate nonperturbative effects. Using "vac" to denote the physical vacuum, it is very likely that, as already stated in Sect. 3.9,

$$\langle : \bar{q}q : \rangle_{\text{vac}} \neq 0, \quad \langle : G^2 : \rangle_{\text{vac}} \neq 0.$$

Let us return to (4.11.1). At $Q^2 \to \infty$, any power α_s^n decreases *less* rapidly, and hence overwhelms any of the terms in $(M^2/Q^2)^r$. But it is clear that there may exist intermediate regions where, for example, the nonperturbative terms in (4.11.1) are important when compared to the second, or third order corrections to C_0, which is the purely perturbative term. Thus, for practical applications,[25] it is interesting to look at the entire Eq. (4.11.1). We already know C_0

$$C_0(Q^2/\nu^2; g(\nu), \nu) = N_c \sum_f Q_f^2 \frac{-1}{12\pi^2} \left\{ \log \frac{-q^2}{\nu^2} + \frac{3C_F}{\beta_0} \log \log \frac{-q^2}{\nu^2} + \cdots \right\}$$
$$+ O(m_f^2/Q^2). \tag{4.11.4}$$

It should be noted that the calculation of Sect. 4.1 neglected perturbative contributions due to the quark masses; these are the $O(m_f^2/Q^2)$ in (4.11.4). It may seem unjustified to take into account the terms in (4.11.1) while neglecting the $O(m_f^2/Q^2)$. These terms are indeed very important for heavy quarks, and their incorporation does not cause great problems (except that they complicate the calculations). For light quarks, the fact that their masses can be neglected is purely a matter of numerology: it so happens that, with the occasional exception of the s quark, their contribution is, for the relevant values of $Q^2 \sim 2$ GeV^2, much smaller than that of the nonperturbative terms in (4.11.1).

The coefficients C_f, C_G can be calculated using the nonperturbative pieces of the propagators, and a detailed sample calculation similar to this

---

[25]Some of the applications may be found in the extensive and pioneering work of Shifman, Vainshtein and Zakharov (1979a, b).

will be presented in Sect. 5.6. One finds (Shifman, Vainshtein and Zakharov, 1979a, b)

$$C_f = \tfrac{2}{3} Q_f^2 \frac{1}{Q^2}, \quad C_G = N_c \sum_f Q_f^2 \frac{1}{36\pi Q^2}. \tag{4.11.5}$$

The anomalous dimensions of the combinations $m : \bar{q}q :$ and $\alpha_s : G^2 :$ vanish to lowest order, which is why the coefficients C_f and C_G do not depend on ν. This is proved for the first by combining our calculations of Z_m (Sect. 3.7) and of Z_M (Sect. 3.6). For $\alpha_s : G^2 :$, see Kluberg-Stern and Zuber (1975) and Tarrach (1982). With all this, we find

$$\Pi^{\mu\nu} = 3 \sum_f Q_f^2 \left(-q^2 g^{\mu\nu} + q^\mu q^\nu\right) \left\{ \frac{-1}{12\pi^2} \left[\log \frac{Q^2}{\nu^2} + \frac{3C_F}{\beta_0} \log\log \frac{Q^2}{\nu^2} \right.\right.$$
$$\left.+ \cdots + O\left(\frac{m_f^2}{Q^2}\right) \right]$$
$$+ \tfrac{2}{3} \frac{m_f \langle : \bar{q}_f(0) q_f(0) : \rangle_{\mathrm{vac}}}{Q^4} + \frac{1}{36\pi} \frac{\langle \alpha_s : G^2 : \rangle_{\mathrm{vac}}}{Q^2} + O\left(\frac{M^6}{Q^6}\right) \bigg\}. \tag{4.11.6}$$

Let us now turn to deep inelastic scattering. In the operator product expansion of Sect. 4.5, we considered only leading twist operators. As for e^+e^- annihilation, there are likely regions where higher twists compete with, say, NLO perturbative contributions. Some operators of higher dimension are related to kinematical effects, TMCs or quark masses; yet others are genuinely new dynamical effects, related to the "primordial" transverse momentum of partons inside a nucleon, or to the fact that this has a finite radius.

Higher twist operators are much more complicated to handle than the leading twist ones; for example, the potential mixing of ghosts with the gluons operators in (4.5.2), which can be proved not to occur (or be trivial) for leading twist, does occur for higher twists. Moreover, the higher twist operators induce new unknown matrix elements analogous to the A^n of (4.5.11), but now there are many more of them because of mixing. Finally, the existence of renormalon singularities, to be discussed in Sect. 10.2ii, makes even the definition of higher twist effects dubious. This is the reason why treatment of higher twist effects is in its infancy, and likely to remain so for quite some time. All we have are partial theoretical calculations and heuristic arguments (De Rújula, Georgi and Politzer, 1977a,b). The latter indicate that the contribution of higher twist operators is probably of the approximate form

$$F^{(\mathrm{HT})}(x, Q^2) \approx \frac{k_1^2}{Q^2} \frac{x}{1-x} F^{(2)}(x, Q^2) + \frac{k_2^2}{Q^2} F^{(2)}(x, Q^2), \tag{4.11.7}$$

where $F^{(2)}$ is the twist-two only structure functions. The constants k_1, k_2 are phenomenological parameters, expected to be $|k_i| \sim p_t^2$, R_N^2, with p_t the transverse momentum of partons in the nucleon, and R_N the radius of the nucleon.[26] We will not delve further into this matter.

4.12 Comparison of DIS Calculations with Experiment

i Parametrizations

Since the theoretical predictions are simpler for the moments of structure functions, it would seem that we should compare QCD predictions with moments. This, however, is not very convenient for the following reasons. First, to obtain the moments experimentally we would require measurements, at fixed Q^2, over the whole range $0 \leq x \leq 1$: these are only available for a very restricted set of values of Q^2. Secondly, even when good data exist, we have a problem with high moments. In fact, these involve integrals of the structure functions multiplied by x^{n-2}, which for large n is strongly peaked at $x \approx 1$. Since it is here that structure functions are smallest, experimental errors become amplified and, even in the more favourable cases, get out of hand for $n \geq 6$: we loose an enormous amount of experimental information. For these reasons, other methods have been devised.

One possibility is to write reasonable parametrizations of the structure functions, which embody QCD results and which can be fitted to experiment. Although not very rigorous or exact (the QCD equations for an *infinite* set of moments cannot be exactly reproduced with a *finite* number of parameters), this method presents the advantages of simplicity and that of producing an explicit representation of the structure functions which can then be used for other processes such as Drell–Yan scattering or high p_t hadron scattering.

The first parametrizations were introduced by Feynman and Field (1977); they are of the form

$$F_a(x, Q^2) = C_a x^{\lambda_a}(1 - x)^{\nu_a}, \qquad (4.12.1a)$$

$a = 1, 2, 3; S, NS$, or, with two Regge poles,

$$F_a(x, Q^2) = (C_a x^{\lambda_a} + C'_a x^{\mu_a})(1 - x)^{\nu_a}. \qquad (4.12.1b)$$

Buras and Gaemers (1978) noted that, if we allow the λ, ν to depend on α_s,

$$\lambda = \lambda_0 + \lambda_1 \log \alpha_s, \quad \nu = \nu_0 + \nu_1 \log \alpha_s,$$

then we can fix the C by using sum rules (see Sect. 4.9i) as known functions of the λ_0, λ_0, ν_0, ν_1, α_s. One then requires simultaneous fits to the QCD equations for the moments *and* to the experimental values of the F_a, which fixes the parameters λ_0, λ_0, ν_0, ν_1.

---

[26]The same order of magnitude for the ks, $|k|^{1/2} \sim 0.1$ to 0.3 GeV, was obtained in a calculation in the bag model by Jaffe and Soldate (1981).

A further step is taken by remarking that one can use the results of Sect. 4.9 to calculate some of the parameters from QCD. In particular, λ_1 (it actually vanishes) and $\nu_1 = -16/(33 - 2n_f)$, cf. Eqs. (4.9.10, 12). Thus, to LO, one writes

$$F_{2NS}(x, Q^2) = \left\{ B_{0NS} \left[\alpha_s(Q^2) \right]^{-d_N S(1-\lambda_{NS})} \left(x^{\lambda_{NS}} - x^{\mu_{NS}(\alpha_s)} \right) \right.$$

$$\left. + A_{0NS} \left[\alpha_s(Q^2) \right]^{-d_0} \frac{\Gamma(1 + \nu_{0NS})}{1 + \nu_{NS}(\alpha_s)} x^{\mu_{NS}(\alpha_s)} \right\} (1 - x)^{\nu_{NS}(\alpha_s)},$$

$$\tag{4.12.2a}$$

$$F_{2S}(x, Q^2) = \left\{ B_{0S} \left[\alpha_s(Q^2) \right]^{-d_+(1+\lambda_S)} \left(x^{-\lambda_S} - x^{-\mu_S(\alpha_s)} \right) \right.$$

$$\left. + A_{0S} \left[\alpha_s(Q^2) \right]^{-d_0} \frac{\Gamma(1 + \nu_{0S})}{1 + \nu_S(\alpha_s)} x^{-\mu_S(\alpha_s)} \right\} (1 - x)^{\nu_S(\alpha_s)}.$$

$$\tag{4.12.2b}$$

Here λ_{NS} can be calculated in terms of the ρ trajectory intercept, and we find $\lambda_{NS} = 0.5$. The $\mu(\alpha_s)$ can be calculated in terms of the other parameters using the sum rules of Sect. 4.9i, and one then fits the seven remaining parameters: $A_{0NS}, A_{0S}, B_{0NS}, B_{0S}, \nu_{0NS}, \nu_{0S}$ and finally λ_S. Similar equations follow for the longitudinal structure function, and the gluon one, *without* introducing new parameters: see López and Ynduráin (1981), where the NLO corrections are also included. The equations (4.12.2) are not exact, of course, but they deviate only by some 1% from the exact results in a reasonable range of x, Q^2 values.

It is possible to complicate the parametrizations, with greater or lesser fortune; parametrizations with as many as 24 parameters may be found on the market. Generally speaking, this is self-defeating, and indeed the author is unaware of any parametrization that (no matter how sophisticated) is valid much beyond the region where it is fitted to experiment. For this reason, precise results require exact reconstruction methods, to which we now turn.

ii Exact Reconstruction

The QCD evolution equations *predict* structure functions $F(x, Q^2)$ in terms of an input at a fixed Q_0^2. We may effect this by inverting the Mellin transform. We use well-known methods to invert Laplace transforms (to which Mellin transforms reduce by a change of variables) to write, for a nonsinglet structure function,

$$F_{NS}(x, Q^2) = \int_x^1 dy \, b(x, y; Q^2, Q_0^2) F_{NS}(y, Q_0^2), \tag{4.12.3a}$$

where the kernel b can be calculated in terms of anomalous dimensions and coefficients. To LO, this has been given by Gross (1974):

$$b(x, y; Q^2, Q_0^2) = \sum_{j=0}^{\infty} G_j(r) b_0(x, y; r + j), \quad r = \frac{16}{2\beta_0} \log \frac{\alpha_s(Q_0^2)}{\alpha_s(Q^2)}; \quad (4.12.3b)$$

$$G_0 = 1, \quad G_1(r) = -\frac{r}{2}, \quad G_2(r) = r \frac{3r + 14}{24}, \dots,$$
$$b_0(x, y; r + j) = \frac{x}{y} \frac{1}{\Gamma(r + j)} \left(\log \frac{y}{x} \right) e^{(3/4 - \gamma_E)r}. \quad (4.12.3c)$$

To NLO, see González-Arroyo, López and Ynduráin (1979).

Alternatively, one can use the Altarelli–Parisi equations directly: for the nonsinglet (cf. Eqs. (4.8.14)),

$$\frac{Q^2 \partial}{\partial Q^2} q_f(x, Q^2) = \frac{\alpha_s(Q^2)}{2\pi} \int_x^1 \frac{dy}{y} P_{NS}^{(0)} \left(\frac{x}{y} \right) q_f(y, Q^2).$$

The kernels have been calculated to NLO by Curci, Furmanski and Petronzio (1980) and Furmanski and Petronzio (1980). Without doubt, the more explicit character of these equations has influenced the fact that, from the end of the nineties, they have been the ones preferred when making exact QCD calculations.[27]

iii Structure Functions at Small x

In recent years, HERA[28] has produced a remarkable set of results on electro-production deep inelastic scattering, from very small to very large Q^2, and reaching to x as little as 10^{-5} or less. The interest of these measurements, from our point of view here, is that they allow us to test QCD in a region where, as shown in Sect. 4.9iv, there is no theoretical unanimity on the input, and there is even doubt on the applicability of the OPE analysis (Sect. 4.9v).

We will consider the combined H1 and Zeus data, and differentiate between two regions: in the first we have $Q^2 \geq 8.5$ GeV^2; the second comprises very small values of x and we allow Q^2 to vary from 8.5 GeV^2 down to $Q^2 \sim 0.11$ GeV^2.

---

[27]The first comparisons of QCD predictions for violations of scaling with experiment were carried out by Hinchliffe and Llewellyn Smith (1977) and by De Rújula, Georgi and Politzer (1977a,b). To NLO, the first evaluation is that by González-Arroyo, López and Ynduráin (1979). Use of the Altarelli–Parisi method was started by Abbott, Atwood and Barnett (1980). More recently, the calculations have been usually performed by the experimental groups themselves: Aubert et al. (1982) for muon DIS, Berge et al. (1991) for neutrino scattering, and so on.

[28]An excellent review of the HERA data, including comparison with the various theoretical models, is that of Cooper-Sarkar, Devenish and De Roeck (1998).

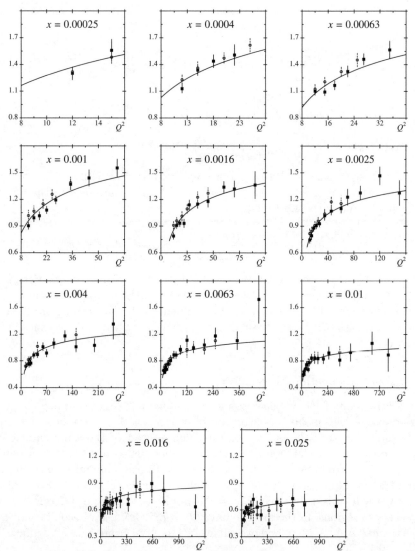

Fig. 4.12.1. Comparison of theory and experiment for low x, large Q^2. Combined H1 and Zeus HERA data.

Starting with the large Q^2 region, we consider first the predictions of the BFKL Pomeron, *with the assumption* that the scale of α_s is Q^2, so we write

$$F_2(x,Q^2) \underset{x\to 0}{\simeq} (\text{const.}) \times x^{-\omega_0 \alpha_s(Q^2)}, \quad \omega_0 = \frac{4C_A \log 2}{\pi}. \qquad (4.12.4)$$

This is in clear disagreement with the data, as shown in Fig. 4.12.1: Eq. (4.12.4) implies that $F_2(x,Q^2)$ should *decrease* as Q^2 increases, contrary to the trend of the experimental data. It has been speculated that NLO corrections[29] could modify (4.12.4) to

$$F_2(x, Q^2) \underset{x \to 0}{\simeq} (\text{const.}) \times x^{-\omega_0 \alpha_s(Q^2) + k_1 \alpha_s(Q^2)^2}$$

and that this could produce the experimentally observed *increase*, at least for subasymptotic Q^2. This is also *not* borne out by the data. In fact, in this case, we would expect that perhaps F_2 would increase for the smaller Q^2; but the trend should be inverted as Q^2 grows, and one could see F_2 decreasing for the larger Q^2. But this is again not seen in experiment. Finally, the trend becomes worse for smaller x. One must conclude that, at presently attainable energies and values of x, the analysis in terms of $\log x$ summations is, to say the least, incomplete.

We then accept the OPE analysis, and consider fitting with either the assumption of a soft or a hard Pomeron, Eqs. (4.9.23) or (4.9.24). If we use the first it is possible to produce a reasonable fit to the data for large Q^2, but (as we will discuss later) the assumption is incompatible with what we find experimentally at small Q^2 and very small x. With the hard Pomeron,

$$F_S(x, Q^2) \underset{x \to 0}{\simeq} B_{0S} \left[\alpha_s(Q^2)\right]^{-d_+(1+\lambda)} x^{-\lambda},$$

we get a very good fit, including the NLO corrections, if $x < 10^{-2}$. For larger x, subleading corrections are important. These may have the following possible origins. First, we may have a subleading Regge trajectory. This can be a P' (Barger and Cline, 1969) or even a soft Pomeron, so that at $Q^2 = Q_0^2$ one had $F_2 \sim (\text{const.}) x^{-\lambda} + \text{const.}$ But even if one had exactly $F_2 = Cx^{-\lambda}$ at a fixed Q_0^2, QCD evolution would give a subleading piece similar to that produced by a soft Pomeron: this is because, as mentioned in connection with the soft Pomeron in Sect. 4.9iv, the exponent of (4.9.23) is universal for any singularity other than $x^{-\lambda}$. If we thus add for background the simplest choice, i.e., a soft Pomeron, and thus write (to LO)

$$F_S(x, Q^2) \underset{x \to 0}{\simeq} B_{0S} \left[\alpha_s(Q^2)\right]^{-d_+(1+\lambda)} x^{-\lambda}$$

$$+ \frac{c_0}{|\log x|} \left[\frac{9|\log x| \log[\alpha_s(Q_0^2)/\alpha_s(Q^2)]}{4\pi^2(33 - 2n_f)}\right]^{\frac{1}{4}} \qquad (4.12.5)$$

$$\times \exp\left\{\sqrt{D_0 \log x|} \left[\log \frac{\alpha_s(Q_0^2)}{\alpha_s(Q^2)}\right] - D_1 \log \frac{\alpha_s(Q_0^2)}{\alpha_s(Q^2)}\right\},$$

then we find the fit shown in Fig. 4.12.1, i.e., we get an excellent reproduction of experimental data. Thus we see that the OPE plus the hard Pomeron

---

[29] Recently evaluated by Fadin and Lipatov (1998) and Camici and Ciafaloni (1998). There is however no consensus that the calculation is complete.

hypothesis suffices to fit experiment down to $x = 10^{-4}$ and up to $Q^2 = 900$ GeV^2. The values of λ found are $\lambda \simeq 0.44 \pm 0.04$, in interesting agreement with the results of the old analyses (López and Ynd.uráin, 1981, and references quoted there) that gave 0.37 ± 0.07.

To finish this very long chapter, we will discuss somewhat the small x, small $Q^2 \leq 8.5$ GeV^2 behaviour of structure functions, following Adel, Barreiro and Yndúrain (1997). This is of interest because it tests the input assumptions, because it permits us to establish contact with the (soft) hadron physics and because, somewhat surprisingly, it produces evidence for a conjectured property of the strong coupling, viz., *saturation*.

Indeed, the quality of the results obtained by assuming that at values of $Q_0^2 \sim 3$ GeV^2 one has a hard singularity, $x^{-\lambda}$, plus a soft (constant) Pomeron term, evolved with QCD to large values $Q^2 \geq 8.5$ GeV^2, leads us naturally to question whether it is possible to extend the analysis to the *low* Q^2 region as well: bearing in mind that, unless we were able to perform a full, non-perturbative calculation, we must content ourselves with *phenomenological* estimates. Here we use approximate, QCD-inspired formulas and assumptions and enquire whether we can still fit the data. We will find that this is indeed the case; in particular, we will see that the extension of the fit of the data to $Q^2 \to 0$ implies self-consistency conditions both for the singlet and the nonsinglet which will allow us to *calculate* the constants λ, λ_{NS}, getting values that are in impressive agreement with the high Q^2 determinations.

The expression for the virtual photon scattering cross section in terms of the structure function F_2 is

$$\sigma_{\gamma(Q^2=0)p}(s) = \frac{4\pi^2\alpha}{Q^2} F_2(x, Q^2), \quad \text{with } s = Q^2/x. \tag{4.12.6}$$

We would like to describe this down to $Q^2 \to 0$. In the low energy region we take the soft-Pomeron dominated expression to be given by an ordinary Pomeron, i.e., behaving as a constant for $x \to 0$ (or, equivalently, $s \to \infty$): the expression for F_2 that will, when evolved to large Q^2, yield (4.12.5) is

$$\begin{aligned}
F_2 = \langle e_q^2 \rangle \Big\{ &B_S[\alpha_s(Q^2)]^{-d_+(1+\lambda)} x^{-\lambda} \\
&+ C + B_{NS}[\alpha_s(Q^2)]^{-d_{NS}(1-\lambda_{NS})} x^{\lambda_{NS}} \Big\}.
\end{aligned} \tag{4.12.7}$$

Because we are interested in a semi-phenomenological description, only LO formulas will be used.

On comparing (4.12.6) and (4.12.7) we see that we have problems if we want to extend the latter to very small Q^2. First of all,

$$\alpha_s(Q^2) = \frac{4\pi}{\beta_0 \log Q^2/\Lambda^2} \tag{4.12.8}$$

diverges when $Q^2 \sim \Lambda^2$. Secondly, Eq. (4.12.6) contains the factor Q^2 in the denominator, so the cross section blows up as $Q^2 \to 0$ unless F_2 were to develop a zero there.

It turns out that there is a simple way to solve both difficulties at the same time. It has been conjectured that the expression (4.12.8) for α_s should be modified for values of Q^2 near Λ^2 in such a way that it *saturates*, producing in particular a finite value for $Q^2 \sim \Lambda^2$. To be precise, one alters (4.12.8) according to

$$\alpha_s(Q^2) \to \frac{4\pi}{\beta_0 \log(Q^2 + M^2)/\Lambda^2},$$

where M is a typical hadronic mass, $M \sim m_\rho \sim \Lambda(n_f = 2)\ldots$. It has been argued that saturation incorporates important nonperturbative effects. Here we will simply set $M = \Lambda = \Lambda_{\text{eff}}$, to avoid a proliferation of parameters. Furthermore, this choice is favoured by quarkonium potential arguments; see Richardson (1979).[30] For the soft Pomeron term we merely replace the constant according to $C \to CQ^2/(Q^2 + \Lambda_{\text{eff}}^2)$. The expression we will use for low Q^2 is thus

$$
\begin{aligned}
F_2 = \langle e_q^2 \rangle \Big\{ & B_S[\widetilde{\alpha}_s(Q^2)]^{-d_+(1+\lambda)} Q^{-2\lambda} s^\lambda \\
& + C \frac{Q^2}{Q^2 + \Lambda_{\text{eff}}^2} + B_{NS}[\widetilde{\alpha}_s(Q^2)]^{-d_{NS}(1-\lambda_{NS})} Q^{2\lambda_{NS}} s^{-\lambda_{NS}} \Big\},
\end{aligned}
\tag{4.12.9a}
$$

where

$$\widetilde{\alpha}_s(Q^2) = \frac{4\pi}{\beta_0 \log(Q^2 + \Lambda_{\text{eff}}^2)/\Lambda_{\text{eff}}^2} \tag{4.12.9b}$$

and we have changed variables, $(Q^2, x) \to (Q^2, s = Q^2/x)$.

We have still not solved our problems: given Eq. (4.12.6) it is clear that a *finite* cross section for $Q^2 \to 0$ will only be obtained if the powers of Q^2 match exactly. This is accomplished by construction for the soft Pomeron term, but for the hard singlet and the nonsinglet piece it will only occur if we have consistency conditions satisfied. With the expression given in (4.12.9b) for $\widetilde{\alpha}_s$, it diverges as const./Q^2 when $Q^2 \to 0$: so we only get a matching of zeros and divergences for $\sigma_{\gamma(Q^2=0)p}(s)$ if $\lambda = \lambda_0$, $\lambda_{NS} = \lambda_{NS0}$ such that

$$d_+(1 + \lambda_0) = 1 + \lambda_0, \quad d_{NS}(1 - \lambda_{NS0}) = 1 - \lambda_{NS0}. \tag{4.12.10}$$

The solution to these equations depends very little on the number of flavours; for $n_f = 2$, probably the best choice at the values of Q^2 that we will be working with, one finds $\lambda_0 = 0.470$, $\lambda_{NS0} = 0.522$. The second is in uncanny agreement with the value obtained with either a Regge analysis in hadron scattering processes, or by fitting structure functions in DIS. The first is larger than the value obtained in the fits to DIS with *only* a hard Pomeron, which gave $\lambda = 0.32$ to 0.38; but falls within the range of values

---

[30]Further discussion of saturation may be found in Sect. 10.2ii.

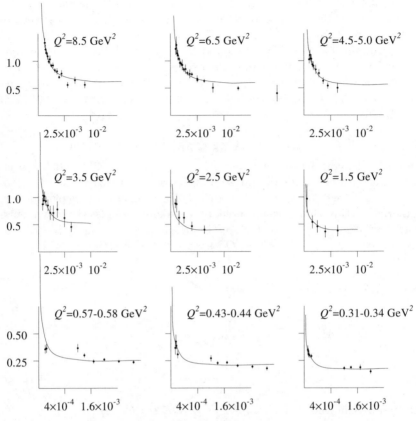

Fig. 4.12.2. Comparison of theory, Eq. (4.12.9), and experiment for low x, small Q^2. HERA data (H1 and Zeus).

obtained with hard plus soft Pomeron, $\lambda = 0.44 \pm 0.04$. The description of experimental data, shown in the comparison of Fig. 4.12.2, is almost too good for what one could expect from such a simple model. It is even more interesting that, as the fits of the HERA groups have shown, this model is the one that produces the best agreement with experiment; other hypotheses yield theoretical curves that deviate substantially from the recent data at very small values of x; see, for example, the quoted review by Cooper-Sarkar, Devenish and De Roeck (1998).

5 Perturbative QCD
II. OZI Forbidden Decays;
Drell–Yan Processes; Jets;
SVZ Sum Rules;
Exclusive Processes and so on.

5.1 OZI Forbidden Decays

The Zweig, or OZI, rule[1] states that decays of heavy resonances that involve disconnected quark graphs (i.e., graphs that can only be connected via gluon lines) are suppressed. The rule works well for resonances such as the ϕ or $f(1270)$, and very well for the J/ψ or Υ: in fact, the heavier the quarks and the resonance, the better the rule works. In QCD this may be understood easily. Consider, for example, the decay of the J/ψ, made up of a $\bar{c}c$ pair of quarks. Because the lightest particles with charm (the D) are too heavy for the J/ψ to decay into them, the process $J/\psi \to$ hadrons has to go via gluons. Due to the quantum numbers of the J/ψ, $J^P = 1^-$, we require at least three gluons (Fig. 5.1.1). Therefore, the width will be

$$\Gamma(J/\psi \to \text{hadrons}) \approx M_{J/\psi}\alpha_s^3,$$

and thus very small. A similar argument holds for the Υ decays. The explanation of this smallness was in fact one of the first successes of QCD (Appelquist and Politzer, 1975; De Rújula and Glashow, 1975).

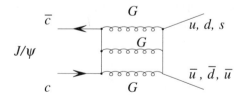

Fig. 5.1.1. Hadron decays of J/ψ.

---

[1] Zweig (1964); Okubo (1963); Izuki, Okada and Shito (1966).

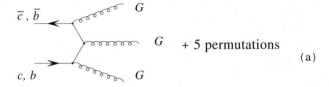

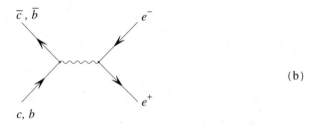

Fig. 5.1.2. The decays of $V = J/\psi$, Υ. (a) Into three gluons. (b) In e^+e^-.

We will now elaborate on this. Let V denote a heavy vector particle, $V = J/\psi$, Υ, Υ', ...; we will, for definiteness, calculate for the J/ψ, and indicate the general results at the end. Because the only mass scale – if we are interested in an inclusive decay – is the mass of the J/ψ, we assume that the running coupling in this formula is to be evaluated at $Q^2 = M_{J/\psi}^2$. The total hadronic width is calculated as follows. Consider the operator $\bar{c}(x)\gamma^\mu c(x)$. It has the quantum numbers of the J/ψ and can thus be used as a field operator for this composite object. The total decay width is then proportional to

$$\sum_\Gamma \langle 0|\bar{c}(x)\gamma^\mu c(x)|\Gamma\rangle \times \langle\Gamma|\bar{c}(x)\gamma^\mu c(x)|0\rangle.$$

In perturbation theory, we may replace the sum over the hadrons $|\Gamma\rangle$ by a sum over gluon and quark states. To lowest order, the first contribution is that of three gluons, so to this order we may replace the decay $J/\psi \to$ hadrons by $J/\psi \to 3G$; see Fig. 5.1.2a. This is identical to the decay of positronium, up to a colour factor. So we have, writing the result in general now,

$$\Gamma_h \equiv \Gamma(V \to \text{hadrons}) = \frac{64(\pi^2 - 9)C_D}{9} \frac{|{}^3S_1(0)|^2}{M_V^2} \left[\alpha_s(M_V^2)\right]^3, \qquad (5.1.1a)$$

and the colour factor is

$$C_D = \frac{1}{16N_c} \sum_{abc} d_{abc}^2 = \tfrac{5}{18}. \qquad (5.1.1b)$$

$^3S_1(0)$ is the wave function for the $\bar{c}c$ inside J/ψ at the origin (we use standard atomic spectroscopic notation), proportional to $\langle J/\psi|\bar{c}(x)\gamma^\mu c(x)|0\rangle$. One can obtain this wave function $^3S_1(0)$ from calculations like the ones we will review in Chap. 6; but a prediction independent of $^3S_1(0)$ may be obtained if we normalize to the leptonic width, $V \to e^+e^-$ (Fig. 5.1.2b) for then, again to LO,

$$\Gamma_l \equiv \Gamma(V \to e^+e^-) = \frac{16\pi Q_q^2 \alpha^2 |^3S_1(0)|^2}{M_V^2},$$

with Q_q the electric charge of the quarks that make up V (in units of e), and α the QED coupling, so that the branching ratio is

$$B_{h/e^+e^-}^V \equiv \frac{\Gamma(V \to \text{hadrons})}{\Gamma(V \to e^+e^-)} = \frac{10(\pi^2 - 9)\alpha_s^3(M_V^2)}{81\pi\alpha^2 Q_q^2}. \qquad (5.1.2)$$

The NLO corrections are very important. They stem from two sources: corrections to the leptonic width, Γ_l (Barbieri et al., 1975 to one loop; Beneke, Signer and Smirnov, 1998 to two loops[2]), which give the expression

$$\Gamma_l^{(\text{NLO})} = \Gamma_l^{(\text{LO})} \left\{ 1 - 4C_F \frac{\alpha_s}{\pi} \right\},$$

and corrections to the hadronic width (Mackenzie and Lepage, 1981),

$$\Gamma_h^{(\text{NLO})} = \Gamma_h^{(\text{LO})} \left\{ 1 + (3.8 \pm 0.5)\frac{\alpha_s}{\pi} \right\},$$

with the $\Gamma^{(\text{LO})}$ given by the previous formulas; the error in $\Gamma_h^{(\text{NLO})}$ is due to the fact that the calculations are made numerically. Taking the ratio, we have

$$B_{h/e^+e^-}^V = \frac{10(\pi^2 - 9)\alpha_s^3(M_V^2)}{81\pi\alpha^2 Q_q^2} \left\{ 1 + (-9.1 \pm 0.5)\frac{\alpha_s(M_V^2)}{\pi} \right\}.$$

To compare with experiment we have to take into account the errors in this formula. Besides the one due to the numerical nature of the computation of the hadronic width, we have finite mass corrections, including phase-space and velocity corrections, and the error due to the value chosen for the renormalization of α_s, $Q^2 = M_V^2$: it may be argued that a more appropriate scale would be m_q^2, or $(M_V/3)^2$, the latter because the decay produces

---

[2] Because only the one loop correction to the hadronic width is know, only the NLO expression for the leptonic decay width is used below.

three gluons.[3] This last uncertainty will only be lifted when the full NNLO corrections are calculated – not an easy task. The finite mass corrections can be estimated by adding a phenomenological piece μ_0^2/M_V^2 to the formula for $B_{h/e^+e^-}^V$ above. In this way, one finds that it is possible to fit the experimental figures (Kobel et al., 1992) for the Υ, Υ', Υ'' simultaneously with a value for the QCD parameter of $\Lambda(n_f = 4, \text{two loops}) = 230 \pm 80$ MeV.

This value is compatible with what one finds from a comparison of the decays $\Upsilon \to 3G$ and $\Upsilon \to GG\gamma$, which can be similarly calculated and which gives $\Lambda \simeq 190$, and the result from J/ψ, ψ' decays that yields a somewhat smaller number, $\Lambda \sim 70^{+100}_{-20}$ MeV. The agreement between these determinations of Λ is a nontrivial test of QCD, as is the fact that the results found are compatible (within errors) with the deep inelastic values obtained in Chap. 4.

It is even possible to extend the analysis to the decays of the ϕ particle, a bound state $\bar{s}s$. In this case, one has to subtract decays into strange particles ($\phi \to \bar{K}K$) from the hadronic decays of the ϕ. The formulas are as in the previous cases; one finds $\Lambda \sim 150$ MeV, a very reasonable number even if the reliability of the result is marred by the large size of the NLO corrections.

The decays of pseudoscalar resonances, such as the η_c or the (as yet undiscovered) η_b, can be treated in a manner similar to that of the vector ones, with a few variations. The relevant diagrams for hadronic decays, which proceed via two gluons, are those of Fig. 5.1.3a. One then normalizes to the two-photon decay, $\eta_c \to \gamma\gamma$ (Fig. 5.1.3b). To leading order,

$$\Gamma^{(\text{LO})}(\eta_q \to 2\gamma) = \frac{48\pi Q_q^4\alpha^2}{M^2(\eta_q)}|^1S_1(0)|^2,$$

$$\Gamma^{(\text{LO})}(\eta_q \to 2G) = \frac{32\pi\alpha_s^2}{3M^2(\eta_q)}|^1S_1(0)|^2. \tag{5.1.3}$$

The next to leading order corrections are fairly large. One has

$$\Gamma^{(\text{NLO})}(\eta_q \to 2\gamma) = \left[1 - \left(5 - \frac{\pi^2}{4}\right)\frac{C_F\alpha_s}{\pi}\right]\Gamma^{(\text{LO})}(\eta_q \to 2\gamma),$$

$$\Gamma^{(\text{NLO})}(\eta_q \to 2G) = \left[1 + \left(\beta_0 \log\frac{\mu}{M_\eta} + \frac{53}{2} - \frac{31\pi^2}{24} - \frac{8}{9}n_f\right)\frac{\alpha_s}{\pi}\right] \tag{5.1.4}$$

$$\times \Gamma^{(\text{LO})}(\eta_q \to 2G).$$

---

[3] It has also been claimed that one should take into account that the decays take place for *timelike* momentum, while the asymptotic freedom formulas are derived for *spacelike* momentum, so one should include the correction due to analytical continuation. This is unclear in that such corrections are of relative order α_s^2, and hence like the uncalculated NNLO corrections. The interested reader may find the results of the analysis in Krasnikov and Pivovarov (1982) and Pennington, Roberts and Ross (1984). The outcome is an increase of the effective value for Λ for J/ψ decays, which improves the agreement of it with the results from Υ decays.

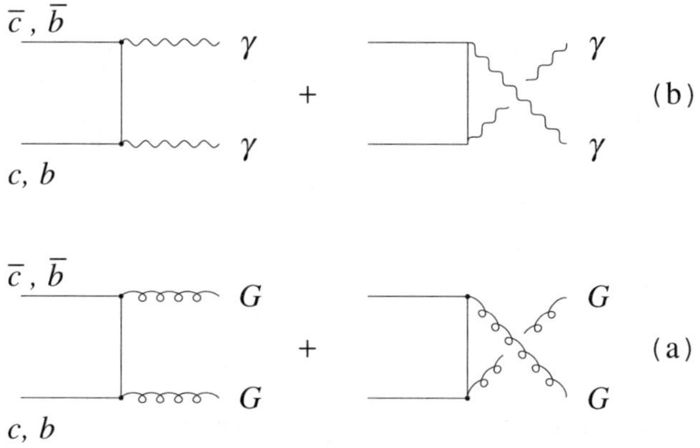

Fig. 5.1.3. The decays of η_c. (a) Into two gluons. (b) In $\gamma\gamma$.

The second expression is valid in the $\overline{\text{MS}}$ scheme, and taking $\alpha_s = \alpha_s(\mu^2)$; it has been obtained by Barbieri, Curci, d'Emilio and Remiddi (1979). The correction to $\Gamma(\eta_q \to 2G)$ is so large that one cannot trust the corresponding prediction very much. If we go ahead anyway, we find the value

$$\Gamma^{(\text{NLO})}(\eta_q \to 2G)/\Gamma^{(\text{NLO})}(\eta_q \to 2\gamma) = 2.6 \times 10^{-4},$$

and we have chosen $\mu = M_\eta$. The experimental figure is $6 \pm 4 \times 10^{-4}$.

For heavy enough quarks, one can obtain rigorously results not only on ratios, but also on exclusive decays (Duncan and Muller, 1980a).

5.2 Drell–Yan Processes

i Partonic Formulation

Consider a collision of two hadrons; then take one quark from one of them, and an antiquark from the other. In the Drell–Yan (1971) mechanism they annihilate into a vector boson (photon, W or Z particles) with large invariant mass squared, Q^2. The vector boson subsequently decays, or materializes, into a lepton pair. In the case of a photon (Fig. 5.2.1) the lepton pair can be e^+e^-, $\mu^+\mu^-$ or $\tau^+\tau^-$; for a Z we have these same pairs, and neutrino–antineutrino pairs as well. For the W mediated processes, we have $e\nu$, $\mu\nu$ or $\tau\nu$ as final lepton pairs. For definiteness, we will consider here the photonic case; the

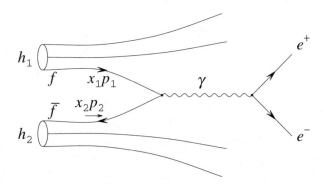

Fig. 5.2.1. Drell–Yan scattering with a photon intermediate state.

extension of the formulation to the W, Z mediated processes only requires minor, and obvious, changes.

Let us denote by $q_f(x)$ the parton distribution functions, defined in Sects. 4.3, 4.7; the indices h, f indicate that the function refers to the distribution of flavour f in hadron h. x is the fraction of the hadron momentum carried by f so that, if p_1, p_2 are the momenta of the hadrons, then f carries momentum xp_1 (assuming f to be in h_1). The total momentum squared of the subprocess, $\bar{f} + f \to e^+e^-$, which coincides with the invariant mass squared of the photon, is thus

$$Q^2 = (x_1p_1 + x_2p_2)^2,$$

and the total energy squared of the hadron–hadron collision is $s = (p_1 + p_2)^2$ so that, neglecting the hadron mass, we have $Q^2 \simeq x_1x_2s$.

In the Drell–Yan process we do not care about the hadronic debris (all partons except $\bar{f}$, f in Fig. 5.2.1); we are only interested in the production of the lepton pair, say e^+e^-. To calculate this, we first evaluate the cross section for the subprocess $\bar{f} + f \to e^+e^-$ as if the quarks were free: thus

$$\hat{\sigma}^{(0)}(\bar{f} + f \to e^+e^-) = \frac{4\pi\alpha^2 Q_f^2}{3N_cQ^2}. \tag{5.2.1}$$

We have neglected the masses of e^+, e^- and of $\bar{f}$, f. The number of colours N_c appears in the denominator because only $\bar{f}$, f of the same colour contribute; Q_f is the charge of f in units of e. To get the cross section for the full process, we multiply $\hat{\sigma}^{(0)}$ by the densities $\bar{q}_{fh_1}(x_1)$, $q_{fh_2}(x_2)$, sum to all flavours and integrate all x_1, x_2 subject to the condition $(x_1p_1 + x_2p_2)^2 = Q^2$. Thus we find the Drell–Yan cross section,

$$\frac{d\sigma^{(0)}}{dQ^2} = \frac{4\pi\alpha^2}{3N_cQ^2s} \sum_f Q_f^2 \int_0^1 \frac{dx_1}{x_1} \int_0^1 \frac{dx_2}{x_2} \delta(1 - \tau/x_1x_2)$$

$$\times \left\{ q_{fh_1}(x_1)\bar{q}_{fh_2}(x_2) + q_{fh_2}(x_2)\bar{q}_{fh_1}(x_1) \right\}, \quad \tau \equiv Q^2/s.$$

(5.2.2)

Note that we have taken into account that you can have f come from h_1 and $\bar{f}$ from h_2 or the converse. The variable τ, customarily used in analyses of Drell–Yan scattering, gives the fraction of the energy (squared) that goes into the lepton pair, e^+e^- in our case. Before entering into a discussion of the QCD corrections, we want to make a few comments. We can have a hadron collision such as $p\bar{p}$ and then both f, $\bar{f}$ can be valence partons; or, in pp collisions, necessarily $\bar{f}$ must come from the sea. The sum in (5.2.2) runs over all the f, valence and sea alike, that contribute. Of the sea we consider that only quarks whose mass is $m_f^2 \ll Q^2$ contribute, and then the neglect of m_f is justified. Which contribution is more important, valence or sea (when both exist), depends on the value of τ. If τ is small, then one of the x_i must be small, thus favouring the sea. If τ allows x_i near the valence maximum ($\sim 1/3$ to $1/4$ at typical energies), then the valence contribution will dominate. In general, we should consider both.

ii Radiative QCD Corrections

The higher order corrections to the lowest order Drell–Yan cross section, Eq. (5.2.2), are of various types. First, we have to admit that the parton densities depend on Q^2 and we thus write $q_f(x, Q^2)$. These densities are to be taken as input and can be obtained from e.g. fits to deep inelastic scattering. Secondly, we have virtual gluon corrections; namely, the radiative corrections to the $\bar{f}f\gamma$ vertex. This interferes with the tree level diagram (Fig. 5.2.2A) and gives thus corrections of order $\alpha_s(Q^2)$.

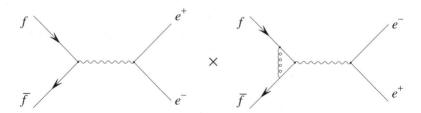

Fig. 5.2.2A. Interference between the tree level Drell–Yan process and the one including the one loop radiative correction to the $\bar{f}f\gamma$ vertex.

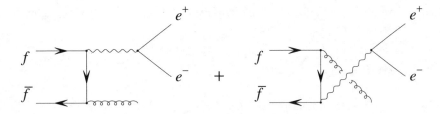

Fig. 5.2.2B. $\bar{f} + f \to e^+ e^- +$ gluon. The two diagrams add coherently.

Then we have another type of correction. Because the hadronic debris is not observed, we have to consider the possibility that in the scattering $\bar{f} f$ an extra gluon is radiated:

$$\bar{f} f \to e^+ e^- + G.$$

This is given by the diagrams of Fig. 5.2.2B, which add coherently one to the other; they also cancel an infrared singularity of the one loop diagram in Fig. 5.2.2A; see below.

Finally, it so happens that the initial hadrons do contain gluons, in addition to quarks and antiquarks. We can thus have gluon initiated processes:

$$G + \bar{f} \to e^+ e^- + \bar{f},$$
$$G + f \to e^+ e^- + f.$$

These processes are shown in Fig. 5.2.2C. The interactions with the spectator partons need not be considered; they are the analogue of the higher twist in deep inelastic scattering, and their contribution is likewise suppressed by powers of $1/Q^2$.

The calculations of the radiative corrections to Drell–Yan scattering are not easy, because of the interplay of infrared singularities and mass singularities.[4] They can be somewhat simplified, but only to NLO, because

---

[4] The NLO corrections were evaluated by Altarelli, Ellis and Martinelli (1978, 1979) and Kubar-André and Paige (1979). See also Humpert and van Neerven (1981) and Harada and Muta (1980). To NNLO, the evaluations are due to van Neerven and Zijlstra (1992b).

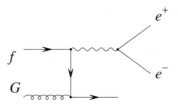

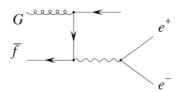

Fig. 5.2.2C. Gluon and f, $\bar{f}$ initiated Drell–Yan scattering. The two diagrams add in cross section (incoherently).

the non-Abelian character of QCD is trivial at this level; no gluon loops appear, so one can regulate infrared singularities by giving the gluon a small mass that is subsequently allowed to go to zero. One finds

$$
\frac{d\sigma^{(\mathrm{NLO})}}{dQ^2} = \frac{4\pi\alpha^2}{3N_cQ^2s}\sum_f Q_f^2 \int_0^1 \frac{dx_1}{x_1} \int_0^1 \frac{dx_2}{x_2} \left\{ \left[\delta(1-z) + \frac{\alpha_s}{\pi}\theta(1-z)\varphi_q(z) \right] \right.
$$
$$
\times \left[q_{fh_1}(x_1)\bar{q}_{fh_2}(x_2) + q_{fh_2}(x_2)\bar{q}_{fh_1}(x_1) \right]
$$
$$
\left. + \frac{\alpha_s}{\pi}\theta(1-z)\varphi_G(z)\left[q_{fh_1}(x_1) + \bar{q}_{fh_2}(x_1) \right]G_{h_2}(x_2,Q^2) + (1 \leftrightarrow 2) \right\}
$$

$$(5.2.3a)$$

and the variable z is $z \equiv Q^2/x_1x_2s$. The functions $\varphi_{q,G}$ are

$$
\varphi_q(z) = \frac{C_F}{2}\left\{ \frac{3}{(1-z)_+} - 6 - 4z + 2(1+z^2)\frac{\log(1-z)}{(1-z)_+} + \left(1 + \frac{4\pi^2}{3} \right)\delta(1-z) \right\}
$$

$$(5.2.3b)$$

and

$$
\varphi_G(z) = \tfrac{1}{2}\left[\left(z^2 + (1-z)^2\right)\log(1-z) + \frac{9z^2}{2} - 5z + \tfrac{3}{2} \right].
$$

$$(5.2.3c)$$

$1/(1-z)_+$ is defined as in (4.8.13b).

From a practical point of view, we distinguish two possibilities. In the first situation, that occurs for example in pp collisions, the antiquark has to be taken from the sea in one of the protons. Then the contribution of both $\bar{q}$ and G terms in (5.2.3) are comparable. In cases such as $\bar{p}p$ collisions the process may be mediated by valence quarks. Then the dominant term in (5.2.3) is the $q\bar{q}$ one.

The corrections due to the processes shown in Figs. 5.2.2B,C are such that (except when the parton radiated is very soft) the energy of the leptons is not equal to that of the originating partons: this is why in the corresponding term in Eq. (5.2.3) the $\delta(1 - z)$ is replaced by a more smooth function. The corrections to the term proportional to $\delta(1-z)$ stem from vertex corrections (see Fig. 5.2.2A) and from radiation of zero momentum gluons, which cancel the infrared singularity of the vertex.

Let us separate off explicitly the piece proportional to $\delta(1 - z)$ in Eq. (5.2.3a), writing

$$\delta(1 - z) + \frac{\alpha_s(Q^2)}{\pi}\varphi_q(z) = K_V(Q^2)\delta(1 - z) + \frac{\alpha_s(Q^2)}{\pi}\varphi_{\text{reg}}(z), \qquad (5.2.4)$$

where $\varphi_{\text{reg}}(z)$ is the regular part of φ_q, so that

$$K_V(Q^2) = 1 + \left(1 + \frac{4\pi^2}{3}\right)\frac{C_F\alpha_s(Q^2)}{2\pi}. \qquad (5.2.5)$$

All the radiative corrections to the partonic formula in (5.2.3) are small, except the term K_V above which, in particular, contains the large coefficient $(2C_F\pi/3)\alpha_s$. This is so large that one may doubt the reliability of the QCD calculation below ISR energies (with $s^{1/2} = 60$ GeV). Actually, the situation is not as bad as it looks at first sight. First, at higher energies (such as the energies for production of W, Z particles and higher) the correction is reasonably small, of 30% or less. Secondly, part of the correction can be combined with higher orders to obtain partial resummations which are exact, plus a remainder with a much smaller second order correction. We will discuss this in the next subsection.

iii The K Factor

Let us take a closer look at the NLO corrections to Drell–Yan scattering, especially at those given in Eq. (5.2.5) and in particular at the piece associated with corrections to the vertex. It has been argued by Parisi (1980) and by Curci and Greco (1980) that at least part of the large correction, proportional to $\pi^2 C_F\alpha_s/\pi$, may be combined with higher orders to give an exponential which represents the exact sum of a class of diagrams, associated with exchange of soft gluons: in this way the corrections, though large, would be under control. Parisi, and Curci and Greco, use the conjectured infrared

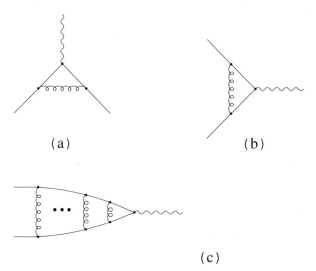

(a)   (b)

(c)

Fig. 5.2.3. Vertex corrections. (a) Spacelike vertex. (b) Timelike vertex. (c) Ladder summation.

properties of the electromagnetic form factor of the quark (see, e.g., Korthals Altes and de Rafael, 1977) to suggest that the piece

$$1 + \pi^2 \frac{C_F \alpha_s}{2\pi} \tag{5.2.6a}$$

of K_V is merely the beginning of the expansion of the exact exponential

$$\exp \frac{\pi C_F \alpha_s(Q^2)}{2} \equiv K(Q^2). \tag{5.2.6b}$$

This is the celebrated K factor (at times the name K factor is used for the full correction; see Eq. (5.2.6c) below). If we accept this, it follows that we should write the whole K_V as

$$1 + \frac{C_F \alpha_s(Q^2)}{2\pi}\left(1 + \frac{4\pi^2}{3}\right) \to e^{\pi C_F \alpha_s / 2}\left\{1 + \left(1 + \frac{\pi^2}{3}\right)\frac{C_F \alpha_s}{2\pi}\right\}; \tag{5.2.6c}$$

the exponential is taken to be exact, and the remainder of the correction, viz., the piece

$$\left(1 + \frac{\pi^2}{3}\right)\frac{C_F \alpha_s}{2\pi}, \tag{5.2.6d}$$

is now comfortably small.

The problem with resummation is that it is not unique, nor is there any guarantee that it will really improve convergence also for the higher orders.

Leaving aside for the moment this last question, we will give *three* derivations of summation whose differences will show the ambiguities very clearly. For the first two, we use renormalization group improvements of the vertex; and for the third, a splitting of the interaction.

We start with the two first methods. Consider the bare electromagnetic vertex of a quark (that we take to be massless) $i\gamma^\mu$. Let q be the momentum of the photon, and we start working in the spacelike region. To one loop we get corrections to the bare vertex given by the diagram in Fig. 5.2.3a, that we write as

$$i\gamma^\mu F_1,$$

and we take into account only the Dirac form factor, F_1, as it is the one that is relevant at high energy. Let p_i be the momenta of the quarks; for the time being we put them off their mass shell, so we have $p^2 \equiv p_i^2 \neq 0$. Moreover, we regulate the infrared divergence by giving a small mass λ to the gluon. Then, renormalizing in the $\overline{\text{MS}}$ scheme at ν^2, and for $|q^2| \gg |p^2|$, λ^2, we have

$$F_1 \simeq 1 - \frac{C_F g^2}{16\pi^2} \int_0^1 dx\, x \int_0^1 dy \left\{ \log \frac{-q^2}{\nu^2 - [(1-u_1)/u_2 - (1-u_2)/u_1]p^2} \right.$$

$$\left. + \frac{x(1-x+xy)}{(1-x)(1-y) + [1-x+(1-y)(1-x+xy)p^2/q^2 - y\lambda^2/q^2]} \right\},$$

$$u_1 = xy, \quad u_2 = 1 - x.$$

The second term in the integrand gives a double logarithm of q^2, and therefore dominates at large q^2. We will henceforth work with double logarithmic precision, so that we need consider only this term. From the above expression for F_1 we find two values for this quantity: if we let $p^2 = 0$, but keep $\lambda \neq 0$, we will be extracting the infrared behaviour, for on shell quarks. If we set $\lambda = 0$ but keep $p^2 \neq 0$ we obtain the (one loop) high momentum behaviour of the form factor, for off-shell quarks. To be precise, one finds

$$F_1 \underset{\substack{\lambda \neq 0, p^2 = 0 \\ q^2 \to \infty}}{\simeq} 1 - \frac{C_F \alpha_s}{4\pi} \log^2 \frac{-q^2}{\lambda^2} \qquad (5.2.7a)$$

and

$$F_1 \underset{\substack{\lambda = 0, p^2 \neq 0 \\ q^2 \to \infty}}{\simeq} 1 - 2\frac{C_F \alpha_s}{4\pi} \log^2 \frac{-q^2}{p^2}: \qquad (5.2.7b)$$

there is a difference of a factor of *two* between the two situations. We may use these results together with the renormalization group, something that may be proved to be a valid procedure in the Abelian case;[5] we would then find the behaviour

$$F_1 \underset{\substack{\lambda \neq 0, p^2 = 0 \\ q^2 \to \infty}}{\simeq} \exp\left(-\frac{C_F \alpha_s}{4\pi} \log^2 \frac{-q^2}{\lambda^2}\right) \qquad (5.2.8a)$$

and

$$F_1 \underset{\substack{\lambda = 0, p^2 \neq 0 \\ q^2 \to \infty}}{\simeq} \exp\left(-2\frac{C_F \alpha_s}{4\pi} \log^2 \frac{-q^2}{p^2}\right). \qquad (5.2.8b)$$

Both expressions depend on the regulator mass, that it be λ^2 or p^2. But, for Drell–Yan scattering, we are interested in *timelike* q^2 (Fig. 5.2.3b), and in fact in the ratio between the timelike form factor (where Drell–Yan scattering occurs) and the form factor at spacelike q^2, which is where the quark and gluon densities are defined in deep inelastic scattering. Thus we obtain the ratios

$$\frac{|F_1^{q^2 > 0}|^2}{|F_1^{q^2 < 0}|^2} \simeq \exp\frac{\pi C_F \alpha_s}{2}, \quad \lambda \neq 0, \ p^2 = 0, \qquad (5.2.9a)$$

and

$$\frac{|F_1^{q^2 > 0}|^2}{|F_1^{q^2 < 0}|^2} \simeq \exp \pi C_F \alpha_s, \quad \lambda = 0, \ p^2 \neq 0. \qquad (5.2.9b)$$

Note that the result is independent of the regulator masses. The first expression is the K factor of Parisi, and of Curci and Greco, Eq. (5.2.6b). As these last authors have shown, the renormalization group analysis presented here is equivalent to summing an infinite ladder of soft gluons (Fig. 5.2.3c) directly in the timelike region.

It is not clear, on physical grounds, which is the more reasonable procedure:[6] the interpretation as sums of ladders of soft gluons seems to point to (5.2.9a) but, on the other hand, we may argue that quarks in Drell-Yan scattering are manifestly off-shell, as they are bound in the colliding hadrons. If we accept this point of view we would use (5.2.9b); extracting it, there would remain for K_V a residual NLO correction of

$$\left(1/2 - \pi^2/3\right) \frac{C_F \alpha_s}{\pi},$$

which is also small.

---

[5] Bogoliubov and Shirkov (1959), pp. 536 ff; Sudakov (1956).

[6] It should perhaps be remarked that we have here an ambiguity only in the choice of the *summation* procedure; to any finite order in perturbation theory, both methods (see Eqs. (5.2.9)) will give the *same* result. This may be checked very easily to NLO: if we take $p^2 = 0$ then in the limit $\lambda \to 0$ we pick the contribution of the *radiation* of a soft gluon, which provides the missing piece to reconcile both ways of calculating.

We will leave this discussion here and introduce the *third* method of summation, giving also a third, and perhaps more physical, approach to the problem. The derivation presents the advantage that it can be proved to be valid rigorously for slowly moving heavy quarks. (This situation is only of academic interest for Drell–Yan scattering; but the same argument and conclusions will hold for the important converse processes, say $e^+e^- \to \bar{b}b$ near the $\bar{b}b$ threshold.)

What one does is to split the corrections to the partonic process into a radiation and a Coulombic part; a separation which is particularly clear in a Coulomb gauge, where the Coulombic part is simply the instantaneous interaction. This can be taken into account by replacing the plane waves of the free quarks in the initial state by relativistic Coulombic wave functions obtained by solving the Dirac equation for a quark moving in the colour field generated by the other. At short distances, the instantaneous interaction may be represented by a Coulomb-type potential,

$$-\frac{C_F \alpha_s}{r};$$

for more details, see Chap. 6. We then take into account the Coulombic corrections by multiplying the cross section $\sigma^{(0)}$ in (5.2.2) by the *Fermi factor*

$$F = \frac{|\Psi_C^{(+)}(r)|^2}{|\Psi^{(0)}(r)|^2}, \quad r \sim 0,$$

where $\Psi^{(0)}$ is the free wave function, and $\Psi_C^{(+)}$ is the wave function of the incoming quarks in the Coulombic potential. This factor is familiar in the theory of electromagnetic corrections to β decay or capture in nuclei (Fermi, 1934). With the known expression for the $\Psi_C^{(+)}$ (Blatt and Weisskopf, 1952; Rose, 1961; Ynduráin, 1996), we obtain, for quarks of equal mass moving with a speed of v in the centre of mass ($v = c = 1$ for massless quarks),

$$F = \left[1 + O(\alpha_s^2)\right] \frac{\pi C_F \alpha_s / 2v}{e^{\pi C_F \alpha_s / 2v} - e^{-\pi C_F \alpha_s / 2v}} e^{\frac{1}{2}\pi C_F \alpha_s / v}. \tag{5.2.10}$$

This is very much like the K factor of Parisi et al., which is not surprising if we notice that solving the Dirac equation in a potential is equivalent to summing an infinite ladder like that in Fig. 5.2.3c; but not exactly equal, thus emphasizing again the dependence of the results on the chosen method of summation.

In this respect, one may also ask why there is no K or F type factor in e^+e^- annihilations. The answer is that there *is* one. It gets cancelled, at the lowest orders, in *inclusive* cross sections for fast moving quarks; but it is certainly there. In particular, it gives the dominant contribution for production of heavy quarks at low speed (Sect. 5.4). However, and except in this case, it is not very useful to perform partial summations. In general, i.e., except for

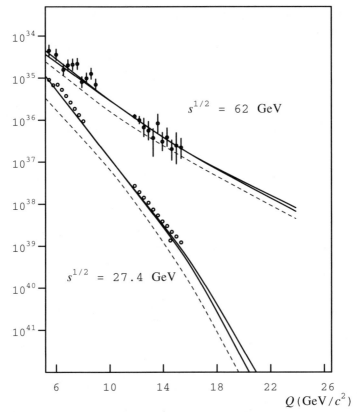

Fig. 5.2.4. Comparison of theory and experiment for Drell–Yan scattering, at energies $s^{1/2}$ of 27.4 GeV and 62 GeV. From the 1987 analysis of G. Altarelli, R. K. Ellis and G. Martinelli. *Dotted lines*: only leading order. *Continuous lines*: NLO theoretical prediction. The separation of the two nearby continuous lines reflects the uncertainty in the theoretical calculation. Data from Angelis et al. (1979), Kourkoumelis et al. (1980) and Ito et al. (1981).

slowly moving quarks, exponentiation into a K or F factor *deteriorates* convergence for $e^+e^- \to$ hadrons. There is also no guarantee that convergence will be improved, even for Drell–Yan scattering, beyond the lowest orders. For this reason, we present results with and without exponentiation below.

Let us return to the K, F, factors. (5.2.10) may be proved to be the correct expression, to relative corrections of order α_s/v, for slowly moving quarks. For fast quarks ($v = 1$) there is, a priori, no reason to prefer one summation to the other among the three presented here. Fortunately, however, the numerical difference among them is very slight. For example, for the Parisi–Curci–Greco and Fermi factors we have the relation

$$F \simeq \left(1 - \frac{(\pi C_F \alpha_s)^2}{24}\right) K,$$

and, numerically, the difference between exponentiation (the numbers inside brackets below) or no exponentiation is also not too large: with $\Lambda = 180$ GeV,

$$K_V \sim 1.60 \ (1.80), \quad \sqrt{Q^2} = 10 \text{ GeV},$$

$$K_V \sim 1.45 \ (1.56), \quad \sqrt{Q^2} = 40 \text{ GeV},$$

$$K_V \sim 1.39 \ (1.47), \quad \sqrt{Q^2} = 90 \text{ GeV}.$$

The agreement between theory and experiment is excellent, down to energies where the exponentiation is important, as shown in Fig. 5.2.4, where the quantity plotted is $d\sigma/d\sqrt{Q^2}dy$ at $y = 0$, in $\text{cm}^2 \text{ GeV}^{-1}$. This is particularly gratifying, because we have here parameter-free predictions, since the parton densities are taken from deep inelastic scattering.

The variable Q^2 is not the only one that can be singled out to analyze Drell–Yan scattering. Two other commonly used variables are the *rapidity*, y, and Feynman's x_F variable. They are defined as

$$x_F = x_1 - x_2, \quad \tanh y = \frac{x_1 - x_2}{x_1 + x_2}.$$

Recalling that $\tau = Q^2/s = x_1 x_2$, one also has $y = \frac{1}{2}\log(x_1/x_2)$. The x_F and y are related to the longitudinal momentum of the e^+e^- pair, in the c.m. of the colliding hadrons. The writing the explicit formulas for the differential cross sections $d\sigma/d\tau dy$, both at partonic and NLO level, is left to the reader as a simple exercise.

One can also measure the *transverse* momentum of the e^+e^- pair. This should be due to radiation of a parton as in Figs. 5.2.2B,C, and will be considered in the sections devoted to jets, particularly Sect. 5.5.

5.3 Jets – Generalities

Consider the lowest order (actually, zero order in α_s) annihilation $e^+e^- \rightarrow$ hadrons. We will assume for definiteness that the energy is substantially less than M_Z, so that we only have to consider the photon-mediated process. If quarks could be produced as free particles, then it would make sense to calculate the cross section for production of an individual pair of quarks $\bar{q}q$,

$$e^+e^- \rightarrow \bar{q}q. \tag{5.3.1}$$

Neglecting the mass of the quarks, letting $s = (p_1 + p_2)^2$, $\theta = \angle(\mathbf{k}_1, \mathbf{p}_1)$ with kinematics as in Fig. 5.3.1, we would get

$$\frac{\mathrm{d}\sigma^{(0)}}{\mathrm{d}\Omega} = \frac{\alpha^2 Q_q^2}{4s}(1 + \cos\theta). \tag{5.3.2}$$

Integrating on angles,

$$\sigma^{(0)}(e^+e^- \rightarrow \bar{q}q) = \frac{4\pi\alpha^2 Q_q}{3s}, \tag{5.3.3}$$

which agrees with the result of the rigorous QCD analysis of Sect. 4.1. Thus, and although (5.3.2) makes no sense as it stands, (5.3.3) indicates that one should be able to connect it with meaningful calculations.

A process such as (5.3.1) does not really exist even in QED. The reason is that there is always the possibility that sufficiently soft gluons (photons, in the case of QED) are radiated: as shown by the analyses of Kinoshita (1962) and Lee and Nauenberg (1964), one has to consider cross sections into bunches of final states, each quark being surrounded by gluons. Mathematically, this is connected to the appearance of infrared divergences when calculating $O(\alpha_s)$ corrections to the lowest order $e^+e^- \rightarrow \bar{q}q$ process.

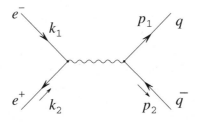

Fig. 5.3.1. Diagram for $e^+e^- \rightarrow \bar{q}q$.

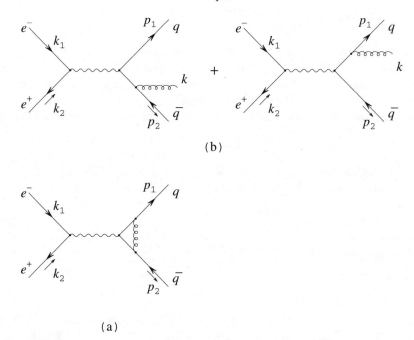

Fig. 5.3.2. Radiative corrections to $e^+ e^- \to \bar{q} q$.
(a) Vertex correction, that gets multiplied by the uncorrected diagram.
(b) Radiation of a gluon.

These corrections are of two types: gluon radiative correction to the $\gamma \bar{q} q$ vertex (Fig. 5.3.2a), which presents an infrared singularity; and emission of an extra gluon, $e^+ e^- \to \bar{q} q + G$, shown in Fig. 5.3.2b, whose divergence for a soft gluon cancels the infrared singularity of the vertex correction, so that the total cross section is finite to order α_s:

$$\sigma^{(1)} = \sigma^{(0)} \left\{ 1 + \frac{3 C_F \alpha_s}{4\pi} \right\}. \tag{5.3.4}$$

This suggests two strategies to make sense of a formula such as (5.3.2). A first possibility is to consider not $d\sigma / d\Omega$ itself, but the expectation value of *infrared finite* observables: an example of which is unity, and then the expectation value is the total cross section.

A second possibility is to mimic the resolution of the infrared catastrophe in QED. Thus, we realize that the processes $e^+ e^- \to \bar{q} q$ and $e^+ e^- \to \bar{q} q + G$ are indistinguishable if either the energy of the gluon, k_0, is below a certain detection threshold, or if its three-momentum $\mathbf{k}$ and one of the momenta of the quarks, $\mathbf{p}_1$, $\mathbf{p}_2$, form an angle smaller than the resolution power of the detector: because, in QCD, quarks and gluons condense into hadrons before reaching the detectors, it is, generally speaking, impossible to know whether the detected hadrons came from a quark or a gluon, or from both. Moreover, we identify (*experimentally*) $\bar{q}q$ and $\bar{q}qG$ when one of the *quark* energies p_{i0} is below the detection threshold: we detect qG (say) that we cannot tell from $q\bar{q}$.

Because of all of this, it follows that what one really measures, and what one thus expects to be finite, is the sum of the cross sections $e^+e^- \to \bar{q}q$ and $e^+e^- \to \bar{q}q + G$ with, in the second case,

$$p_{i0},\ k_0 < \epsilon s^{1/2},$$
$$|\angle(\mathbf{p}_1, \mathbf{k})|,\ |\angle(\mathbf{p}_2, \mathbf{k})|,\ |\angle(\mathbf{p}_1, \mathbf{p}_2)| < \delta, \tag{5.3.5}$$

where the quantities ϵ, δ characterize the detection efficiency. Similar conditions will hold for $e^+e^- \to \bar{q}q + nq + n\bar{q} + n'G$. This is the Sterman–Weinberg (1977) analysis.

The amplitude corresponding to a diagram like those in Fig. 5.3.2b contains the propagator for the virtual parton, let us say the quark, of the form

$$\frac{i}{\not{p}_1 + \not{k} - m_q} \simeq i\frac{\not{p}_1 + \not{k}}{2p_1 \cdot k}, \tag{5.3.6}$$

and we have neglected m_q compared to the energies involved. As stated above, the denominator vanishes for soft partons, p_{10} or $k_0 \simeq 0$; or for collinear momenta, $\mathbf{p}_1 \parallel \mathbf{k}$. Conditions such as (5.3.5) precisely guarantee that this does not happen; under them,

$$p_1 \cdot k > \tfrac{1}{2}s\epsilon^2\delta^2. \tag{5.3.7}$$

Because cross sections involve integrals over all final momenta, the condition (5.3.7) means that we get singularities of the type $(\log \epsilon \log \delta)\alpha_s$. This is negligibly small when α_s becomes small; therefore, the partonic structure of the cross sections will become more and more apparent as the energy increases, because then α_s is tiny and we can afford small ϵ, δ. The details in a few important cases will be found in the coming sections.

The full picture, however, is more complicated. As we have stated before, only *hadrons* reach the detectors. At short distances and small times after the materialization of, say, the photon,[7] we can describe the process in terms of partons, quarks and gluons, as in Figs. 5.3.1, 2. As these partons move apart, the increasing strength of the interaction makes it energetically favoured the creation of quark–antiquark pairs and of gluons profusely from the vacuum. This cloud dresses the original partons, then coalescing

---

[7] We consider photon mediated production for definiteness. There is no difficulty in taking Z mediation into account, or in calculating also W mediated processes, following, with obvious variations, the same arguments as we develop for photons.

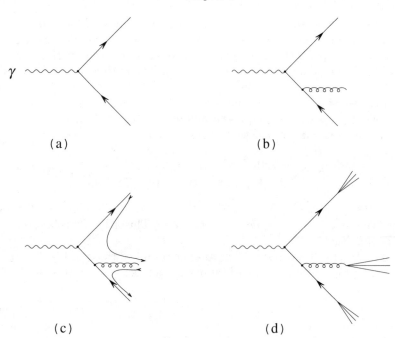

Fig. 5.3.3. Conversion of a photon into three jets. The sequence of times is *abcd*. One expects $|t_a - t_b| \ll \Lambda^{-1}$, but $t_{c,d} - t_{a,b} \sim \Lambda^{-1}$.

into hadrons (*hadronization*). This whole process, depicted schematically in Fig. 5.3.3, leads to the conversion of the original partons into *jets* of hadrons.

Only the first part of the sequence, Fig. 5.3.3a,b, can be treated with perturbative QCD, if the momenta are large enough. What one does to describe the whole process is to split it, somewhat arbitrarily, into two phases: the generation of a certain number of quarks and gluons, described perturbatively; and final hadronization, for which more or less plausible nonperturbative models are used. For example, let **p** be the jet axis, for a given jet, $\mathbf{p} = \sum \mathbf{p}_h$, where the sum runs over all the hadrons in the jet. Feynman and Field (1977) define *fragmentation functions*, $D_{q/h}(x)$ which give the distribution of the fraction x of momentum of hadrons in the jet generated around q (and the same for gluon generated jets). A review of fragmentation, from a phenomenological point of view, is that of Söding (1983). In the present text we will give theoretical arguments in favour of the so-called Lund model of hadronization, first proposed by Andersson, Gustafson and Peterson (1977, 1979), which we will make plausible from the strong coupling limit in lattice QCD in Sect. 9.5.

This is not all. Among the quarks created you can have $\bar{s}s$, $\bar{c}c$ or $\bar{b}b$ pairs. These, particularly the last two, will produce particles that decay before

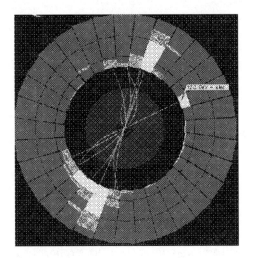

Fig. 5.3.4.
Jets in *ep* deep inelastic
scattering, as observed by
the ZEUS group at HERA.
The straight line is the
electron, with an energy of
12.5 GeV; besides it, two
jets are clearly seen. The
hadronic debris proceeding
in the forward direction is
not detected.
(Communicated by
J. Puga, 1998).

reaching the detectors. So, each original parton becomes a shower of particles
due to hadronization and decay: to the extent that the average number of
particles in a jet is ~ 30 at PETRA or LEP energies, $s^{1/2}$ from 40 to 200
GeV. An example of actual jets, produced in *ep* collisions, may be seen in
Fig. 5.3.4. The theoretical generation of events has to be made with numerical
Monte Carlo programs. We refer to the literature quoted, and to Marchesini
and Webber (1984), Webber (1984), Barreiro (1986) etc. for details. In the
last reference, an excellent review of jets in electron–positron annihilations
can also be found.

5.4 Jets in e^+e^- Annihilations

i Two Jet Events

We first show how to calculate a *physical* two jet cross section in e^+e^- annihilations. As before, and for the sake of definiteness, we will assume the
process to be mediated by a photon. The cross section for $e^+e^- \to \bar{q}q$ is, at
zero order in α_s, given by (5.3.2). However, and as explained in the previous
section, we have to correct this because one counts as two jets processes with
three partons[8] if either two are travelling almost in the same direction, or one
of the three has an energy below the detector's threshold. Thus we should
compute the cross section into "fat" jets, as in Fig. 5.4.1. This we will do
later. To show clearly the mechanism at work, we begin with a somewhat
different method. The total cross section to order α_s is given by the inclusive
result (5.3.4). This includes two *and* three jets, cf. Fig. 5.3.2. Thus the two

---

[8] Or more. Here we will only consider the $O(\alpha_s)$ corrections, so only processes with
three partons (quark–antiquark and a gluon) have to be taken into account.

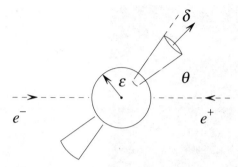

Fig. 5.4.1. Two "fat" jets, with possible extra soft partons (inside sphere). (Compare with Fig. 5.3.4).

jet cross section will be obtained by subtracting, from the piece proportional to α_s in (5.3.4), the cross section into events which are *not* two jets. With obvious notation,

$$\sigma(2j) = \sigma(2j + 3j) - \sigma(\cancel{2}j),$$

$$\sigma(2j + 3j) = \sigma^{(0)} \left\{ 1 + \frac{3C_F \alpha_s}{4\pi} \right\}. \tag{5.4.1}$$

The cross section

$$e^+ e^- \to q(p_1)\bar{q}(p_2)G(k),$$

where we put the momenta in brackets, can be easily calculated with the diagrams of Fig. 5.3.2. Letting $s = (p_1 + p_2 + k)^2$, $x_i = 2p_i^0/s^{1/2}$, we have

$$\frac{1}{\sigma^{(0)}} \frac{d\sigma}{dx_1 dx_2} = \tfrac{1}{2} C_F \frac{\alpha_s}{\pi} \frac{x_1^2 + x_2^2}{(1 - x_1)(1 - x_2)}, \quad x_1 + x_2 \geq 1; \quad 0 \leq x_i \leq 1. \tag{5.4.2}$$

This equation exhibits very clearly the singularities at $x_i = 1$, corresponding to **k** proportional to $\mathbf{p}_i$, including **k** = 0 as a particular case.

Now, this process will *not* be classed as a *two* jet event if the angle θ between the quark momenta is *smaller* than a given $\pi - \eta_0$ (Fig. 5.4.2) with η_0 related to the resolution of the detector. For, if $|\theta| < \pi - \eta_0$, the detectors will disentangle the three jets. Therefore, the not-two-jet cross section will be

$$\sigma(\cancel{2}j) = \int^f \int dx_1 \, dx_2 \, \frac{d\sigma}{dx_1 dx_2},$$

where the upper limit of the integrals is deduced from Fig. 5.4.2 to be given by the curve

$$f : x_1 + x_2 = 1 + \frac{x_1 x_2}{2}(1 + \cos \eta_0).$$

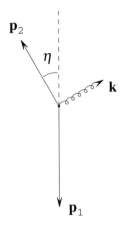

Fig. 5.4.2. The geometry of a $q\bar{q}G$ state.

The integral is easily calculated if we are only interested in terms that do not vanish in the limit $\eta_0 \to 0$, i.e., as we have better and better experimental precision. We find

$$\frac{1}{\sigma^{(0)}}\sigma(2\,j) \underset{\eta_0 \to 0}{\simeq} \tfrac{1}{2}C_F\frac{\alpha_s}{\pi}\left\{\log^2\frac{4}{\eta_0^2} - 3\log\frac{4}{\eta_0^2} + \frac{\pi^2}{3} + \tfrac{7}{2}\right\},$$

so the cross section into two observable jets will be found by subtracting this as in (5.4.1):

$$\sigma_{\rm obs}^{\eta_0}(2j) = \sigma^{(0)}\left\{1 - \frac{C_F\alpha_s}{\pi}\left(\tfrac{1}{2}\log^2\frac{4}{\eta_0^2} - \tfrac{3}{2}\log\frac{4}{\eta_0^2} + \frac{\pi^2}{6} + 1\right)\right\}. \qquad (5.4.3)$$

The angular distribution of the jets is as in (5.3.2) inasmuch as we can neglect η_0 against θ. As expected, $\sigma_{\rm obs}^{\eta_0}$ depends on the resolution η_0.

This result may be compared with what one gets with the Sterman–Weinberg method. We let δ be the half-angle defining the jet, and let $\epsilon s^{1/2}$ be the energy threshold for detection (Fig. 5.4.1). Then a simple calculation, essentially like the one before, gives

$$\frac{d\sigma^{\delta\epsilon}}{d\Omega} = \frac{d\sigma^{(0)}}{d\Omega}\left\{1 - \frac{C_F\alpha_s}{\pi}\left(4\log\delta\log 2\epsilon + 3\log\delta + \frac{\pi^2}{3} - \tfrac{5}{2}\right)\right\}. \qquad (5.4.4)$$

The details of the derivation, including the full result (i.e., without the approximation δ, $\epsilon \to 0$) may be found in Weeks (1979). It is instructive to note that (5.4.3) and (5.4.4) differ: the cross section depends on the definition of what is a two jet event.

It is also interesting to discuss the analogue of K or F factors here. As for Drell–Yan processes (Sect. 5.2) we could sum a ladder of soft gluons, and rewrite (5.4.4) extracting an exponential. For fast moving quarks, little is gained by doing so. The situation is different for slowly moving heavy quarks. Here one can use nonrelativistic quantum mechanics to write the cross section of, for example, $e^+e^- \to \bar{b}b$, as

$$\frac{\mathrm{d}\sigma(e^+e^- \to \bar{b}b)}{\mathrm{d}\Omega} \underset{v\to 0}{\simeq} F\frac{\mathrm{d}\sigma^{(0)}}{\mathrm{d}\Omega},$$

where the factor F is given by a formula similar to (5.2.8), so that one has $F = \pi C_F \alpha_s/[v(1 - \exp(-\pi C_F \alpha_s/v))]$ and v is the velocity, $v = |\mathbf{p}_i|/p_{i0}$. We can take the limit $v \to 0$ to obtain the threshold cross section

$$\sigma(e^+e^- \to \bar{b}b) \underset{v\to 0}{\to} \frac{2\pi^2 N_c C_F}{4m_b^2}Q_b^2\alpha^2\alpha_s. \qquad (5.4.5)$$

The fact that we still have enhancement is not intuitively obvious; it can be traced to the influence of the $\bar{b}b$ bound states. The NLO corrections to (5.4.5) are known (Adel and Ynduráin, 1995); they amount to multiplying the right hand side of (5.4.5) by

$$\left\{1 + \left[\frac{\beta_0}{2}\left(\log\frac{\mu}{mC_F\alpha_s} - 1\right) + a_1\right]\frac{\alpha_s}{\pi}\right\},$$

with $a_1 = (31C_A - 20T_F n_f)/36$.

The equations like (5.4.3, 4) do not give the whole story. Because of *fragmentation* (the breaking of single partons in jets consisting of several hadrons, and eventual decay of some of these) there is some chance that an event $q\bar{q}$ be counted as an event with three, or more jets, if for example some of the transverse momenta of the final hadrons are so large that $p_t/p_{i0} > \sin\delta$. This will give corrections to (5.4.3, 4) of order

$$\frac{\langle N\rangle\langle p_t^2\rangle}{s},$$

with $\langle N\rangle$ the average number of particles. These corrections are very important in some situations, but not for two jet events, so we will leave them for the time being.

Let us return to (5.4.3, 4). We will be able to take these formulas seriously when the $O(\alpha_s)$ corrections there are substantially smaller than the leading term. Moreover, we want to differentiate the two jets, i.e., we want to have $\eta_0 \sim \delta \ll \pi/2$. Suppose, for example, that we require $\eta_0 \sim \pi/8$, and that the correction $O(\alpha_s)$ should be at most $1/4$ of the leading term. Neglecting all but the dominant correction in (5.4.3) we obtain the condition

$$\frac{C_F\alpha_s}{2\pi}\log^2\frac{4}{\eta_0^2} \leq 1/4, \quad \text{i.e.,} \quad \alpha_s \leq 0.11.$$

This implies LEP energies, $s^{1/2}$ from 90 to 180 GeV. Even with a more exact evaluation (the term $\log^2 4/\eta_0^2$ is partially compensated at finite η_0) a value

$\alpha_s \lesssim 0.2$ is required. This explains why the jet structure of the cross section for $e^+e^- \to$ hadrons is only clearly seen at PETRA energies and above, for $s^{1/2} \geq 20$ GeV.

ii Three Jet Events

The three parton cross section, which we have already evaluated in (5.4.2), is proportional to α_s. This is why its study is particularly interesting: it affords a *direct* determination of the quark–gluon coupling.

For a three jet event we define, with the kinematics of Fig. 5.4.2,

$$x_1 = 2p_1^0/s^{1/2}, \quad x_2 = 2p_2^0/s^{1/2}, \quad x_3 = 2k^0/s^{1/2} = 2 - x_1 - x_2. \quad (5.4.6)$$

To analyze the three jet events we will use three representative variables. The first one Y we define,[9] for *physical* particles, as

$$Y \equiv \frac{4}{3} \frac{\sum_{i<j} |\mathbf{p}_i \times \mathbf{p}_j|^2}{\{\sum_{ij} p_i^0 p_j^0\}^2}. \quad (5.4.7a)$$

For partons, this definition gives immediately

$$Y = (x_1 + x_2 - 1)(1 - x_1)(1 - x_2). \quad (5.4.7b)$$

Clearly, Y vanishes for a two jet event, and also for three jets when any of the $x_i = 1$, thus cancelling the infrared and collinear singularities of the cross section. The maximum of Y is $1/27$, and its average is immediately obtained integrating with (5.4.2); we get

$$\langle Y \rangle_{\bar{q}qG} = \frac{1}{\sigma^{(0)}} \int d\sigma \, Y = \frac{2\alpha_s}{15\pi}. \quad (5.4.8)$$

The differential cross section with respect to Y is also easily calculated. One finds

$$\frac{1}{\sigma_0} \frac{d\sigma}{dY} = \frac{C_F \alpha_s}{2\pi} I(Y), \quad (5.4.9a)$$

with

$$I = \int_0^1 dx_1 \int_0^1 dx_2 \frac{x_1^2 + x_2^2}{(1 - x_1)(1 - x_2)} \delta((x_1 + x_2 - 1)(1 - x_1)(1 - x_2) - Y); \quad (5.4.9b)$$

the condition $x_1 + x_2 - 1 > 0$ is automatically fulfilled thanks to the delta function.

---

[9] This variable is similar to the variable V introduced in the 1993 edition of this text; actually, it is identical for partons, but slightly more convenient for physical particles and especially for more than three jets. I am grateful to R. Akhoury and J. Vermaseren for pointing this out.

Measuring Y via (5.4.7a) immediately gives the value of α_s. A complete evaluation, however, would require the calculation of the radiative corrections, which involves in particular four jet events, and of hadronization. For the last, a simple calculation gives

$$\langle Y \rangle_{2\text{jet}} \simeq \frac{4}{3\pi^2} \frac{\langle N \rangle \langle p_t^2 \rangle}{s}.$$

At $s^{1/2} = 35$ GeV and with the experimental values of $\langle N \rangle$, $\langle p_t^2 \rangle$ this gives

$$\frac{\langle Y \rangle_{2\text{jet}}}{\langle Y \rangle_{\bar{q}qG}} \simeq 0.30,$$

which emphasizes the far from negligible effects of hadronization.

A popular variable is *thrust* (Farhi, 1977). For physical particles, it is defined as

$$T = 2 \max \frac{\widetilde{\sum} |\mathbf{p}_{i\parallel}|^2}{s^{1/2}}. \tag{5.4.10a}$$

The sum in the numerator runs over all the particles in a hemisphere; the $\mathbf{p}_{i\parallel}$ are the components of the momenta of the particles along the jet axis contained in the hemisphere. The plane defining the hemisphere is chosen perpendicular to the jet axis; and the latter is found by requiring T to be maximum. That is to say, one chooses a direction characterized by the polar angles (θ, ϕ) as arbitrary jet axis, and evaluates $T(\theta, \phi)$. Then one varies θ and ϕ until a maximum is found: these are the polar angles of the jet axis, for the most energetic jet. In terms of partonic variables, one obviously has

$$T = \max\{x_1, x_2, x_3\}. \tag{5.4.10b}$$

Integrating the cross section (5.4.2) at fixed T one finds the differential cross section

$$\frac{1-T}{\sigma^{(0)}} \frac{d\sigma}{dT} = \frac{C_F \alpha_s}{2\pi} \left\{ 9T^2 - 24T + 12 + \frac{6T^2 - 6T + 4}{T} \log \frac{2T-1}{1-T} \right\}. \tag{5.4.11}$$

For a two jet event, $T = 1$: this is the reason why we multiplied by $1 - T$ in (5.4.11). T varies between 1 and $1/3$, and its average value is (De Rújula, Ellis, Floratos and Gaillard, 1978)

$$\langle 1 - T \rangle_{q\bar{q}G} = \frac{C_F \alpha_s}{2\pi} \left\{ -\frac{3}{4} \log 3 - \frac{1}{18} + 4 \int_{2/3}^{1} \frac{dT}{T} \log \frac{2T-1}{1-T} \right\} \simeq 1.05 \frac{\alpha_s}{\pi}. \tag{5.4.12}$$

Finally, the *energy–energy correlation*, or EEC,[10] is defined as follows. Choose an angle χ different from 0 and π. Then the EEC is

$$\frac{1}{\sigma^{(0)}}\frac{d\Sigma}{d\cos\chi} = \frac{1}{Ns}\frac{2}{\Delta\chi\sin\chi}\sum_{A=1}^{N}\sum_{\text{pairs in }\Delta\chi} E_{Aa}E_{Ab}. \tag{5.4.13}$$

The symbols here are as follows. A labels the events. In each event, E_{Aa} and E_{Ab} are the energies of two particles separated by an angle $\chi \pm \frac{1}{2}\Delta\chi$.

To calculate the EEC we notice that, for small resolution $\Delta\chi$, the condition

$$\chi - \tfrac{1}{2}\Delta\chi \leq \theta_{ab} \leq \chi + \tfrac{1}{2}\Delta\chi,$$

with θ_{ab} the angle between the momenta of particles a, b, can be imposed including a factor

$$\delta(\chi - \theta_{ab})\Delta\chi = \delta(\cos\chi - \cos\theta_{ab})\Delta\chi\sin\chi.$$

Moreover, and with a, b, c varying from 1 to 3,

$$\cos\theta_{ab} = (x_c^2 - x_a^2 - x_b^2)/2x_a x_b, \quad c \neq a, b; \quad E_a E_b = sx_a x_b/4,$$

and $x_3 = 2 - x_1 - x_2$. Therefore, and substituting (5.4.2) for the jet cross section,

$$\frac{1}{\sigma^{(0)}}\frac{d\Sigma}{d\cos\chi} = \frac{C_F\alpha_s}{4\pi}$$

$$\times \int dx_1 \int dx_2 \frac{x_1^2 + x_2^2}{(1-x_1)(1-x_2)}\sum_{a<b} x_a x_b \delta(\cos\theta_{ab} - \cos\chi)$$

$$= \frac{1}{\zeta}\int_0^1 dx\left\{\frac{x^3+x}{(1-\zeta x)^2} - 2\zeta\frac{x^2-x^3}{(1-\zeta x)^3} + \zeta^2\frac{x^3(1-x)^2}{(1-\zeta x)^4}\right\}$$

$$+ \frac{1}{1-\zeta}\int_0^1 dx\frac{x^3-x^4}{(1-\zeta x)^2},$$

$$\zeta = \frac{1-\cos\chi}{2}.$$

The remaining integration is elementary. We find the expression, valid to leading order and for three jet events,

$$\frac{1}{\sigma^{(0)}}\frac{d\Sigma}{d\cos\chi} = \frac{C_F\alpha_s}{8\pi}\frac{3-2\zeta}{\zeta^5(1-\zeta)}\left[2(3-6\zeta+2\zeta^2)\log(1-\zeta) + 3\zeta(2-3\zeta)\right].$$

$$\tag{5.4.14}$$

---

[10]Basham, Brown, Ellis and Lowe (1978). Second order corrections calculated by Ali and Barreiro (1982) and Ellis, Richards and Stirling (1982).

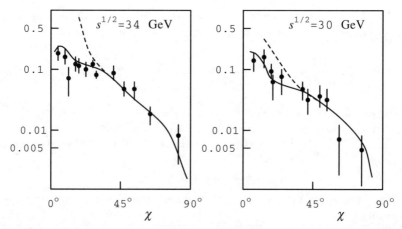

Fig. 5.4.3. Comparison of energy-energy anticorrelation with experiment.
Broken lines: perturbative QCD, $O(\alpha_s^2) + O(\alpha_s^3)$.
Solid lines: including fragmentation. CELLO and Pluto data.
(Communicated by F. Barreiro, 1991).

At times, one uses the *anticorrelation*,

$$\frac{1}{\sigma^{(0)}} \left\{ \frac{\mathrm{d}\Sigma(\pi - \chi)}{\mathrm{d}\cos\chi} - \frac{\mathrm{d}\Sigma(\chi)}{\mathrm{d}\cos\chi} \right\},$$

for which radiative corrections are particularly small. A comparison of the
QCD prediction with experiment at PETRA is shown in Fig. 5.4.3. The
importance of hadronization at small χ is apparent there. The corresponding
value for the QCD parameter is

$$\Lambda = \begin{cases} 280^{+110}_{-90} \text{ MeV} & \text{(CELLO data)}, \\ 170^{+150}_{-100} \text{ MeV} & \text{(PLUTO data)}. \end{cases}$$

The difference between the two determinations is a measure of the *systematic*
differences in the data. The importance of hadronization decreases at the
very high energies of LEP where, in particular using the large number of
events at the Z particle peak, one finds what are probably the more precise
determinations of the QCD coupling; see Sect. 10.3ii.

iii Multijet Events

Four jet events present the novel feature that the tree level amplitude involves diagrams with gluon self-couplings. Calculations of the amplitudes and cross sections are very involved; they may be found in Ali et al. (1981), Ellis, Ross and Terrano (1981), Gaemers, Oldham and Vermaseren (1981), Danckaert et al. (1982), etc. We will only say that the experimental analysis is correspondingly complicated, but the evidence for the necessity of the triple gluon coupling appears clearly.

iv Gluon Jets in Quarkonium Decays

We discuss this topic here because the study of quarkonium decays is mostly performed with e^+e^- accelerators.

Let us consider a vector resonance of heavy quarks, say the J/ψ or the Υ. The last is sufficiently heavy that perturbative QCD may be applied to it; not only to the calculation of the total decay rate (as in Sect. 5.1) but also to the characteristics of the three jets into which the three gluons of its dominating decay evolve. After a straightforward, but long, calculation, essentially identical to that for positronium decay (Akhiezer and Berestetskii, 1963) we can write the differential decay rate as

$$\frac{1}{\Gamma_{3G}^{(0)}}\frac{\mathrm{d}\Gamma_{3G}}{\mathrm{d}x_1\mathrm{d}x_2} = \frac{1}{\pi^2-9}\left\{\left(\frac{1-x_1}{x_2x_3}\right)^2 + \left(\frac{1-x_2}{x_1x_3}\right)^2 + \left(\frac{1-x_3}{x_1x_2}\right)^2\right\},$$

$$(5.4.15)$$

where $x_i = 2k_i^0/M_V$, the k_i are the gluon momenta, and M_V is the mass of the vector resonance. $\Gamma_{3G}^{(0)}$ is the lowest order decay rate, Eq. (5.1.1):

$$\Gamma_{3G}^{(0)} = \frac{160\alpha_s^3|^3S_1(0)|^2}{81M_V^2}.$$

One can analyze this much as we did for $q\bar{q}G$ final states. For example, letting T be the thrust, and integrating with (5.4.15), we find

$$\frac{1}{\Gamma_{3G}^{(0)}}\frac{\mathrm{d}\Gamma_{3G}}{\mathrm{d}T} = \frac{3}{\pi^2-9}\left\{\frac{4(1-T)}{T^2(2-T)^3}(5T^2-12T+8)\log\frac{2(1-T)}{T}\right.$$

$$\left. + \frac{2(3T-2)(1-T)^2}{T^3(2-T)^2}\right\},$$

and the average thrust is

$$\langle T\rangle_{3G} = \frac{3}{\pi^2-9}\left[6\log(2/3) - \frac{3}{2} + \frac{4\pi^2}{3} + 20\int_0^1\mathrm{d}x\frac{\log x}{2+x}\right] \simeq 0.889.$$

We refer to De Rújula, Ellis, Floratos and Gaillard (1978) for more details and references.

5.5 Jets in Hadron Physics

We first briefly describe two situations in which, besides hadrons, leptons also intervene, and concentrate later on jets produced in pure hadronic collisions.

The first process is deep inelastic scattering. When we studied it in Chap. 4, we were interested in the *inclusive* cross section. It so happens, however, that we can say more. Specifically, consider the parton struck by the virtual photon (or W, or Z). One would expect that it would tend to separate itself from the rest (Fig. 5.5.1) and then a jet of hadrons would coalesce around this struck parton. If its transverse momentum component, $\mathbf{p}_{Jt}$ (transverse with respect to the momentum of the debris formed by the partons that have not been struck, $\mathbf{p}_{\Gamma'}$),

$$\mathbf{p}_{Jt} = \frac{1}{|\mathbf{p}_{\Gamma'}|}|\mathbf{p}_J \times \mathbf{p}_{\Gamma'}|,$$

is large, $|\mathbf{p}|_{Jt} \gg \Lambda$, we expect that the process will be calculable in perturbative QCD. (Of course, we have to allow for nondetection of soft partons radiated in addition to the jet, and of partons emitted in a certain cone around the direction of the jet.) This is indeed the case; the details of the LO calculation may be found in Méndez (1978). With the advent of the HERA electron–proton collider, operating at huge energies, the jets produced in deep inelastic collision have been studied in great detail, including processes in which two or even more jets are produced.

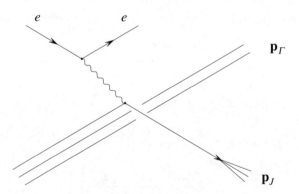

Fig. 5.5.1. Deep inelastic production of a jet.

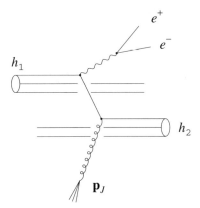

Fig. 5.5.2. Drell–Yan production of e^+e^- and a gluon jet.

The second process is Drell–Yan production of lepton pairs. When evaluating QCD radiative corrections, we referred to the eventual radiation of a parton, as in the diagrams of Fig. 5.2.2B,C. If the transverse momentum of the parton (also here, transverse means with respect to the axis defined by the hadronic debris) is large, we should be able to disentangle it from the rest of the hadron fragments, which will continue mostly along the axis of collision, as shown in Fig. 5.5.2. The transverse momentum of the jet can be inferred from that of the e^+e^- pair, which is why the present process is easier to study experimentally than the former one. The theoretical calculations may be found in Sterman and Libby (1978), Altarelli, Parisi and Petronzio (1978), Parisi and Petronzio (1979), Curci, Greco and Srivastava (1979), Dokshitzer, Dyakonov and Troyan (1980), including subleading corrections (Collins and Soper, 1982; Kodaira and Trentadue, 1983; Davies and Stirling, 1984; Altarelli, Ellis, Greco and Martinelli, 1984; Altarelli, Ellis and Martinelli, 1985).

After these somewhat cursory descriptions, we will consider in more detail the production of jets in purely hadronic collisions, probably the first evidence, together with the observation of three jet events at PETRA, that quarks and gluons, and not merely *currents* made out of quark fields, are real.

Consider, typically, pp scattering at the Intersecting Storage Rings (ISR) at CERN; it is there, both at the ISR and the $\bar{p}p$ Collider (as well as somewhat later in Fermilab) where pointlike structures were first discovered in hadron–hadron collisions. The ISR c.m. energy is $s^{1/2} \sim 60$ GeV. Hadron–hadron scattering had been studied for a long time and it was known that at small momentum transfer t (or, equivalently, small transverse momentum p_t) the scattering is dominated by diffractive and/or Regge phenomena

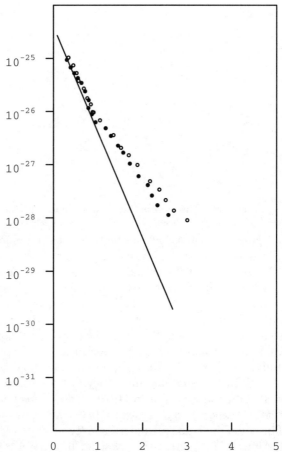

Fig. 5.5.3. The cross section, at large p_t, for various average multiplicities $\langle N/\Delta y \rangle$. The quantity plotted is $Ed^3\sigma/d^3p$ in $\mathrm{cm}^2\mathrm{c}^3\,\mathrm{GeV}^{-2}$, against p_t. Data from the CERN collider, UA1 experiment. The solid line is the extrapolation of the low p_t exponential fit.

(Barger and Cline, 1969). In both cases, the cross section decreases exponentially with p_t at fixed c.m. energy:

$$\frac{d\sigma}{d\langle p_t \rangle} \sim \exp(-bp_t), \quad b \simeq 6 \text{ GeV}. \tag{5.5.1}$$

This behaviour is well followed by the cross section at ISR energies for small average transverse momentum. At large $\langle p_t \rangle$, however, the decrease expected from (5.5.1) stops; the cross section becomes much larger than the

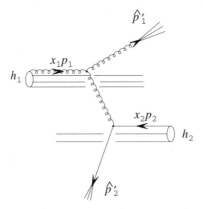

Fig. 5.5.4. Example of a two jet production in a hadron–hadron
collision, via scattering of a gluon and a quark.

value implied by (5.5.1) and in fact the experimentally observed decrease for
large transverse momentum is only power-like, precisely as occurs for scattering of elementary, pointlike particles.[11] This experimental trend is shown
very clearly in Fig. 5.5.3. For large $\langle p_t \rangle$, most of the cross section there can be
interpreted in terms of scattering of the point-like constituents of the hadrons.

At large momentum transfer, the scattered partons – quarks, antiquarks
and gluons – will generate individual jets (Fig. 5.5.4). The cross section for
the process can thus be calculated in terms of the elementary scattering
of the constituent partons, plus jet formation, much as in deep inelastic or
Drell–Yan jet formation.

Let us distinguish the variables for the elementary subprocess by putting
carets over them. With this notation, the c.m. energy squared of the colliding
partons is

$$\hat{s} = (\hat{p}_1 + \hat{p}_2)^2 = (x_1 p_1 + x_2 p_2)^2 = x_1 x_2 s, \qquad (5.5.2a)$$

with s the c.m. energy of the hadron–hadron collision, x_i the energy frac-
tions carried by the partons, and we neglect parton and hadron masses. The
momentum transfer of the subprocess is

$$\hat{t} = (\hat{p}_1 - \hat{p}_1')^2, \qquad (5.5.2b)$$

and we also define the variable $\hat{u}$ by

$$\hat{u} = (\hat{p}_1 - \hat{p}_2')^2 = \sum m - \hat{s} - \hat{t}, \qquad (5.5.2c)$$

---

[11]Of course this pointlike scattering ought to be present also at small $\langle p_t \rangle$ but there,
because $\alpha_s(p_t) \sim 1$, it is masked by rescattering corrections which do, presumably,
generate diffractive and Regge type cross sections.

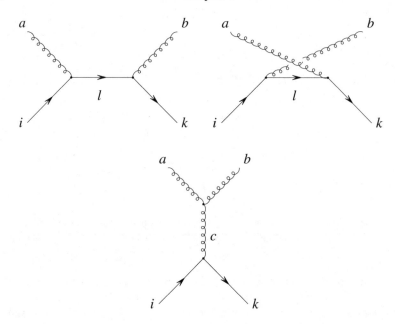

Fig. 5.5.5. Diagrams involved in the calculation of qG scattering.

where $\sum m$ is the sum of the masses of the four partons (two in and two out) intervening; in our approximation, $\sum m = 0$.

Let us denote generically by $q_{hf}(x, \hat{s})$ the parton densities, including the density of gluons, for which we have $q_{hG}(x, \hat{s}) = G(x, \hat{s})$. The cross section for the process

$$h_1 + h_2 \to j(\hat{p}_1') + j(\hat{p}_2') + \text{ anything}, \tag{5.5.3}$$

is obtained from the elementary cross sections

$$\frac{d\hat{\sigma}(f_1 + f_2 \to f_1' + f_2')}{d\hat{t}} \tag{5.5.4}$$

where this last equation is evaluated with the help of diagrams like those in Fig. 5.5.5, which correspond precisely to the process shown previously in Fig. 5.5.4.

One then multiplies these cross sections by the appropriate partonic densities, and sums over possible final states if, as is usually the case, the partons producing the jets are not identified, and averages over the possible initial partons.

Schematically, one evaluates

$$\frac{1}{N_1 N_2} \sum_{f_1 f_2} \sum_{f_1' f_2'} q_{h_1 f_1}(x_1, \hat{s}) q_{h_2 f_2}(x_1, \hat{s}) \frac{d\hat{\sigma}(f_1 + f_2 \to f_1' + f_2')}{d\hat{t}}, \qquad (5.5.5)$$

where N_i are the number of possible partons (quarks, antiquarks and gluons, counting colour) in hadron h_i.

The elementary cross sections $d\hat{\sigma}/d\hat{t}$ are evaluated straightforwardly. If, for example, we have the elementary process Gq as in Fig. 5.5.4, then the Feynman diagrams that have to be considered are those shown in Fig. 5.5.5. Denoting by $\mathcal{A}_1$, $\mathcal{A}_2$, $\mathcal{A}_3$ the amplitudes there ($\mathcal{A}_3$ being that involving the triple gluon coupling), *excluding* the colour factors, then the elementary cross section is proportional to

$$\frac{1}{8 \times 3} \sum_{ik,ab} \left| \sum_l t_{kl}^b t_{li}^a \mathcal{A}_1 + \sum_l t_{kl}^a t_{li}^b \mathcal{A}_2 + \sum_c \mathrm{i} f_{abc} t_{ki}^c \mathcal{A}_3 \right|^2.$$

After tedious but simple colour and Dirac algebra evaluations, the cross section is obtained; it is included among those given in the following formulas. We write[12]

$$d\hat{\sigma}/d\hat{t} = (\pi \alpha_s^2 / \hat{s}^2) \Phi,$$

and then we have

$$\Phi(q\bar{q} \to q'\bar{q}') = \tfrac{4}{9} \frac{\hat{t}^2 + \hat{u}^2}{\hat{s}^2},$$

$$\Phi(qq' \to qq') = \Phi(q\bar{q}' \to q\bar{q}') = \tfrac{4}{9} \frac{\hat{s}^2 + \hat{u}^2}{\hat{t}^2}, \quad q \neq q';$$

$$\Phi(qq \to qq) = \tfrac{4}{9} \left(\frac{\hat{s}^2 + \hat{u}^2}{\hat{t}^2} + \frac{\hat{s}^2 + \hat{t}^2}{\hat{u}^2} \right) - \tfrac{8}{27} \frac{\hat{s}^2}{\hat{u}\hat{t}},$$

$$\Phi(q\bar{q} \to q\bar{q}) = \tfrac{4}{9} \left(\frac{\hat{s}^2 + \hat{u}^2}{\hat{t}^2} + \frac{\hat{u}^2 + \hat{t}^2}{\hat{s}^2} \right) - \tfrac{8}{27} \frac{\hat{u}^2}{\hat{s}\hat{t}};$$

$$\Phi(q\bar{q} \to GG) = \tfrac{32}{27} \left(\frac{\hat{u}^2 + \hat{t}^2}{\hat{u}\hat{t}} \right) - \tfrac{8}{3} \left(\frac{\hat{u}^2 + \hat{t}^2}{\hat{s}^2} \right),$$

$$\Phi(GG \to q\bar{q}) = \tfrac{1}{6} \left(\frac{\hat{u}^2 + \hat{t}^2}{\hat{u}\hat{t}} \right) - \tfrac{3}{8} \left(\frac{\hat{u}^2 + \hat{t}^2}{\hat{s}^2} \right),$$

$$\Phi(Gq \to Gq) = \tfrac{4}{9} \left(\frac{\hat{u}^2 + \hat{s}^2}{\hat{u}\hat{s}} \right) + \frac{\hat{u}^2 + \hat{s}^2}{\hat{t}^2},$$

$$\Phi(GG \to GG) = \tfrac{9}{2} \left(3 - \frac{\hat{u}\hat{t}}{\hat{s}^2} - \frac{\hat{u}\hat{s}}{\hat{t}^2} - \frac{\hat{s}\hat{t}}{\hat{u}^2} \right).$$

$$(5.5.6)$$

---

[12]Combridge, Kripfganz and Ranft (1978); Calahan, Geer, Kogut and Susskind (1975); Cutler and Sivers (1977).

These formulas assume that the quarks are light (with respect to the energies) and their masses are neglected. The production of heavy quarks has features of interest. The tree level cross section may be found in Glück, Owens and Reya (1978) and Combridge (1979). Radiative corrections are evaluated in Altarelli, Diemoz, Martinelli and Nason (1988), Dawson, Ellis and Nason (1989) and Beenakker et al. (1991). Soft gluons may actually be summed to all orders for production of heavy quarks (Fadin, Khoze and Sjöstrand, 1990; Laenen, Smith and van Neerven, 1992). For *massless* quarks, the radiative corrections are not known, so Eq. (5.5.6) is as far as one can go at present. This is unfortunate because we would expect large K-type corrections that will at times enhance (for GG or $q\bar{q}'$, when the quantum numbers of the state are such that it is a colour singlet) and at times suppress the cross section (when the quantum numbers are those of a colour octet). What is known, on the other hand, is the three jet cross section (Sachrajda, 1978). We refer to the review of Jacob and Landshoff (1978) for more information on this subject.

5.6 The SVZ Sum Rules

In this section we will consider a method for obtaining *static* properties of hadrons from perturbative QCD, plus some nonperturbative input. Conversely, the method can be used together with experimental information, to obtain the value of QCD parameters, notably quark masses and quark and gluon condensates $\langle \bar{q}q \rangle$, $\langle G^2 \rangle$ etc. The method is variously known as SVZ sum rules, from its originators (Shifman, Vainshtein and Zakharov, 1979a,b); ITEP sum rules, from the institution where these and a good number of their followers did the earlier work; or simply QCD sum rules. A very comprehensive review is that of Narison (1989). Here we will discuss a few typical examples, leaving for Sect. 7.4 an application to estimates of light quark masses, and for Sect. 10.5 an evaluation of the gluon condensate.

The first example, which will serve to illustrate the philosophy at work, is connected with the ϕ resonance. We consider the two point function

$$\Pi_\phi^{\mu\nu}(q) \equiv \left(-g^{\mu\nu}q^2 + q^\mu q^\nu\right) \Pi_\phi(q^2) = i \int d^4x \, e^{iq\cdot x} \langle T\phi^\mu(x)\phi^\nu(0)\rangle_{\text{vac}}, \quad (5.6.1)$$

where ϕ^μ is a (composite) operator with the quantum numbers of the ϕ; specifically, we take

$$\phi^\mu(x) = C_\phi : \bar{s}(x)\gamma^\mu s(x) : . \quad (5.6.2)$$

The constant C_ϕ is chosen so that $\langle \text{vac}|\phi^\mu(0)|\phi(p,\lambda)\rangle = (2\pi)^{-3/2}\epsilon^\mu(p,\lambda)$; its numerical value may be fixed from the decay $\phi \to e^+e^-$. In the nonrelativistic approximation for the $\bar{s}s$ quarks inside the ϕ resonance we, would have

$$C_\phi = \frac{m_s}{\sqrt{N_c M_\phi}\Psi(\mathbf{0})},$$

with Ψ the ss wave function normalized to $\int d^3\mathbf{r}\,|\Psi(\mathbf{r})|^2 = 1$.

In perturbation theory, the function $\Pi_\phi(q^2)$ grows at most as a logarithm as $|q^2| \to \infty$; hence, any derivative

$$d^N \Pi_\phi(q^2)/(dq^2)^N \equiv \Pi_\phi^{(N)}(q^2)$$

with $N \geq 1$ will satisfy an unsubtracted dispersion relation (Cauchy representation)

$$\frac{d^N \Pi_\phi}{(dq^2)^N} \equiv \Pi_\phi^{(N)}(q^2) = \frac{N!}{\pi} \int ds\, \frac{\operatorname{Im}\Pi_\phi(s)}{(s - q^2)^{N+1}}.$$

For $|q^2|$ near M_ϕ, we expect the representation to be dominated by the ϕ pole. Using the unitarity relation at the ϕ,

$$\operatorname{Im}\Pi_\phi^{\mu\nu}(q) = \tfrac{1}{2}\sum_\lambda \int \frac{d^3\mathbf{P}}{2p_0} \langle 0|\phi^\mu(0)|\phi(p,\lambda)\rangle\langle\phi(p,\lambda)|\phi^\nu(0)|0\rangle(2\pi)^4\delta(p - q),$$

we immediately find the pole value,

$$\operatorname{Im}\Pi_\phi(s) = \frac{\pi}{M_\phi^2}\delta(s - M_\phi^2),$$

and we approximate, for $s \sim q^2$,

$$\Pi_\phi^{(N)}(q^2) \approx \frac{N!}{M_\phi^2(M_\phi^2 - q^2)^{N+1}}. \tag{5.6.3}$$

The mass M_ϕ is of 1 GeV; hence it is not totally absurd to evaluate $\Pi_\phi^{(N)}(q^2)$ using perturbative QCD: as we saw in Sect. 5.1, perturbative QCD describes ϕ decay fairly well. If we took the vacuum to be the perturbative vacuum, we would obtain

$$\Pi_\phi^{(N)}(q^2) \approx \frac{3C_\phi^2}{12\pi^2}(N - 1)!\frac{1}{(-q^2)^N}\left\{1 - \frac{m_s}{q^2} + \cdots\right\}.$$

It turns out that it is impossible to fit (5.6.3) to (5.6.2), with the value $m_s \simeq 200$ MeV.[13] This indicates that nonperturbative effects are important. These can be implemented most easily (at least the leading ones) by replacing, in the perturbative calculation, the perturbative quark and gluon propagators by propagators with the nonperturbative pieces included, as in Sect. 3.9ii; to

---

[13]The discrepancy is even more clear if we take ratios of consecutive derivatives, for here the C_ϕ drop out.

lowest order we only need Eqs. (3.9.12) and (3.9.15). Examples of detailed calculations will be given later; we now only quote the result, which is

$$\Pi_\phi^{(N)}(q^2) \approx \frac{3C_\phi^2}{12\pi^2}(N-1)! \frac{1}{(-q^2)^N} \left\{ 1 - \frac{m_s}{q^2} - \frac{4\pi^2 N(N+1)}{q^4} m_s \langle : \bar{s}s : \rangle \right.$$
$$\left. - \frac{3\pi N(N+1)}{8q^2} \langle \alpha_s : G^2 : \rangle + \cdots \right\}.$$

$$(5.6.4)$$

One now finds that it is possible to fit (5.6.4) to (5.6.2) in a region $|q^2| \sim M_\phi^2$, which indicates that the ϕ receives a good part of its mass not from the perturbative mass of its constituent s quarks, but from the condensates.

We now present two detailed sample calculations of nonperturbative effects. The first is that of the quark condensate $\langle \bar{s}s \rangle$ to Π_ϕ in Eq. (5.6.4). From (5.6.1),

$$\Pi_\phi^{\mu\nu}(q) = iC_\phi^2 \int d^4x\, e^{iq\cdot x} \langle T\bar{s}(x)\gamma^\mu s(x)\bar{s}(0)\gamma^\mu s(0)\rangle_{\text{vac}}. \qquad (5.6.5)$$

Therefore, to zero order in α_s,

$$\Pi_\phi^{\mu\nu}(q) = -iC_\phi^2 \int d^D\hat{k}\, \text{Tr}\, \gamma^\mu S_s(k)\gamma^\nu S_s(k+q). \qquad (5.6.6)$$

If we only considered the perturbative piece of the propagator, $S_s = S_P$, we would have obtained the perturbative piece,

$$\Pi_P^{\mu\nu} = \frac{8N_c C^2}{6} \frac{1}{16\pi^2}(-g^{\mu\nu}q^2 + q^\mu q^\nu)\left[N_\epsilon - \log q^2 + \cdots \right], \qquad (5.6.7)$$

and then (5.6.3) would follow. The nonperturbative correction is obtained by using the full expression, $S_s = S_P + S_{NP}$. The leading term is the mixed term,

$$\Pi_{NP,\text{quark}}^{\mu\nu}(q) \simeq -iC_\phi^2 \int d^D\hat{k}\, \text{Tr}\left\{ \gamma^\mu S_{NP}(k)\gamma^\nu S_P(k+q) \right.$$
$$\left. + \gamma^\mu S_P(k)\gamma^\nu S_{NP}(k+q) \right\}$$

$$(5.6.8)$$

with S_{NP} given by Eq. (3.9.12) and $S_P(k) = i/(\not{k} - m_s)$. We find

$$\Pi_{NP,\text{quark}}^{\mu\nu}(q) \simeq (-g^{\mu\nu}q^2 + q^\mu q^\nu)\frac{-2C_\phi^2 m_s\langle \bar{s}s\rangle}{q^4},$$

from which the $\langle \bar{s}s \rangle$ piece in (5.6.4) follows immediately by differentiation.

As a second example of a detailed calculation, we take the evaluation of the gluon condensate using charmonium spectroscopy.[14] We now consider the current

$$J_c^\mu(x) = \bar{c}(x)\gamma^\mu c(x); \qquad (5.6.9)$$

---

[14]The same calculations would work (in fact, better) for bottomium.

a two point function similar to the one we have been considering will now be

$$\Pi_c^{\mu\nu}(q) = i \int d^4 x \, e^{iq\cdot x} \langle T J_c^{\mu}(x) J_c^{\nu}(0) \rangle_{\text{vac}} = (-g^{\mu\nu} q^2 + q^{\mu} q^{\nu}) \Pi_c(Q^2), \quad (5.6.10)$$

with $Q^2 = -q^2$.

The function $\Pi_c(Q^2)$ is analytic in Q^2 except for a cut running from $-4m_c^2$ to $-\infty$ and poles for $-Q^2 = M_n^2$, with M_n^2 the masses of the bound states $V_n = J/\psi, \psi', \ldots$. Writing a Cauchy representation for the Nth derivative, with $N > 1$ to avoid subtractions, we have

$$\Pi_c^{(N)}(Q^2) = \sum_n \frac{(-1)^N N! r_n}{M_n^2 (M_n^2 + Q^2)^{N+1}} + \frac{1}{\pi} \int dt \, \frac{(-1)^N N! \operatorname{Im} \Pi_c(-t)}{(t + Q^2)^{N+1}}. \quad (5.6.11)$$

The r_n are the residues of the poles, calculable in terms of $V_n \to e^+ e^-$.

In contrast with the previous situation, the quantity $4m_c^2$ is now large enough for us to apply perturbative QCD at that scale. In fact, we will choose an intermediate scale Q_0^2 such that

$$\Lambda^2 \ll Q_0^2 \ll 4m_c^2;$$

for example, we may take $Q_0^2 \simeq 2$ GeV^2. We can still use perturbation theory at Q_0^2; because $Q_0^2 \ll 4m_c^2$, it then follows that

$$\Pi^{(N)}(Q_0^2) \simeq \Pi^{(N)}(0)$$

is *calculable* in perturbative QCD.

The purely perturbative contribution is elementary. The contribution of the quark condensate can be evaluated as in the previous situation; it turns out to be subleading with respect to the contribution of the gluon condensate, which we shall calculate in detail in a moment. Neglecting the quark condensate contribution, the result is

$$\Pi^{(N)}(0) = R_N^{(0)} \left\{ 1 + \frac{B_N \langle \alpha_s G^2 \rangle}{m_c^4} + \cdots \right\}. \quad (5.6.12a)$$

Here

$$R_N^{(0)} = \frac{N_c m_c^{2N}}{2\pi^2} (-1)^N \frac{(N-1)! [(N+1)!]^2}{(2N+3)!}, \quad N_c = 3, \quad (5.6.12b)$$

is the purely perturbative piece and

$$B_N = -\frac{\pi C_F N(N+1)(N+2)(N+3)}{48(2N+5)}. \quad (5.6.12c)$$

This last equation is obtained by replacing, in the two loop expression for Π_c (given by the diagrams in Fig. 4.1.2) the gluon propagator by its nonperturbative piece, Eq. (3.9.15). Thus we have the diagrams of Fig. 5.6.1, where

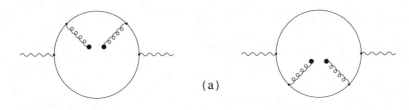

(a)

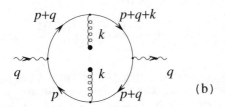

(b)

Fig. 5.6.1. Diagrams for the LO gluon condensate contribution to Π_c.

the blobs represent VEVs. The contribution of e.g. the third diagram there, Fig. 5.6.1b, yields the term, to be included in $\Pi_c^{\mu\nu}(q)$,

$$(5.6.1\text{b}) = -\,g^2 C_F \int \mathrm{d}^D\hat{p} \int \mathrm{d}^D\hat{k}\; D_{NP}^{\alpha\beta}(k)$$

$$\times \mathrm{Tr}\,\gamma_\mu \frac{\mathrm{i}}{\not{p}-m_c}\gamma_\alpha\frac{\mathrm{i}}{\not{p}+\not{k}-m_c}\gamma_\nu\frac{\mathrm{i}}{\not{p}+\not{q}+\not{k}-m_c}\gamma_\beta\frac{\mathrm{i}}{\not{p}+\not{q}-m_c}$$

$$= -\,g^2 C_F \frac{\langle G^2\rangle}{4(N_c^2-1)D(D-1)(D+2)}\left[(D+1)g^{\alpha\beta}\partial^2 - 2\partial^\alpha\partial^\beta\right]$$

$$\times \int \mathrm{d}^D\hat{p}\; \mathrm{Tr}\,\gamma_\mu\frac{\mathrm{i}}{\not{p}-m_c}\gamma_\alpha\frac{\mathrm{i}}{\not{p}+\not{k}-m_c}\gamma_\nu\frac{\mathrm{i}}{\not{p}+\not{q}+\not{k}-m_c}\gamma_\beta\frac{\mathrm{i}}{\not{p}+\not{q}-m_c},$$

$$(5.6.13)$$

where the derivatives are with respect to k, and are evaluated at $k = 0$. The calculation is simplified with the help of two tricks. First, since the integral is convergent, we can set $D = 4$ directly. Secondly, the term $\partial^\alpha\partial^\beta$ comes from the Fourier transform of $x^\alpha x^\beta$; recall Sect. 3.9ii. If we write

$$x^\alpha x^\beta = -\tfrac{1}{2}x^2 g^{\alpha\beta} + \tfrac{1}{8}\frac{\partial^2}{\partial x_\alpha \partial x_\beta}(x^2)^2,$$

it follows that, up to gauge terms proportional to ∂^α, ∂^β (hence, after Fourier transform, to k^α, k^β) that will give zero in the end, we can replace

$$(D+1)g^{\alpha\beta}\partial_k^2 - 2\frac{\partial^2}{\partial k_\alpha \partial k_\beta} \to (D+2)g^{\alpha\beta}\partial_k^2. \qquad (5.6.14)$$

The rest of the calculation is straightforward. Adding the contribution of the other diagrams, Fig. 5.6.1a, the result reported in Eq. (5.6.12c) is obtained.

Let us return to (5.6.11). We can use the experimental values of Im $\Pi_c(-t)$ obtained by subtracting the theoretically known contribution of the u, d, s quarks from the value obtained from the experimental cross section $e^+e^- \to$ hadrons. Likewise, we can take the residues r_n from experiment: so the whole right hand side of (5.6.11) is known. Integrating, we get an experimental value for the left hand side, which we may compare with the theoretical evaluation, Eqs. (5.6.12). In this way we obtain a determination of m_c, $\langle \alpha_s G^2 \rangle$, or of m_b, $\langle \alpha_s G^2 \rangle$ if we apply the same calculation to b quarks. The values found in older determinations are[15]

$$\bar{m}_c(\bar{m}_c^2) = 1.27 \pm 0.05 \text{ GeV}, \quad \bar{m}_b(\bar{m}_b^2) = 4.25 \pm 0.1 \text{ GeV} \qquad (5.6.15)$$

and

$$\langle \alpha_s G^2 \rangle = 0.044^{+0.014}_{-0.010} \text{ GeV}^4 . \qquad (5.6.16)$$

The errors, however, have been lately shown to be overoptimistic. A more recent evaluation of m_b (Jamin and Pich, 1997) gives

$$\bar{m}_b(\bar{m}_b^2) = 4.25 \pm 0.1 \text{ GeV},$$

but the more precise determinations are perhaps those from bottomium spectroscopy (Chap. 6). More recent detemations of the gluon condensate tend to give much larger values than those reported in (5.6.16), up to 0.1 GeV^4; see Narison (1997). We refer to Chapter 10 for a more complete discussion on these quantities, and specifically to Sect. 10.5ii for a detailed evaluation of the gluon condensate using sum rules.

SVZ-type sum rules have been evaluated for a large number of correlators. We will see in Sect. 7.4 applications to determinations of light quark masses. Sum rules for correlators with the quantum numbers of the *proton* are particularly interesting in that they provide a connection between the mass of this particle and quark and gluon condensates (Ioffe, 1981; Reinders, Rubinstein and Yazaki, 1981; Espriu, Pascual and Tarrach, 1983; and, particularly, Chung, Dosch, Kremer and Schall, 1984).

---

[15]See, for example, the reviews of Novikov et al. (1978) and Narison (1989).

5.7 Exclusive Processes

We will present a detailed discussion for the pion form factor; this will, we hope, pave the way for the extension to other processes, for which we only give the results.

One can define the *pion form factor*, F_π by writing (cf. Fig. 5.7.1)

$$\begin{aligned} V^\mu(p_1, p_2) &= (2\pi)^3 \langle \pi(p_2) | J^\mu_{\text{em}}(0) | \pi(p_1) \rangle \\ &= (p_1^\mu + p_2^\mu) F_\pi(q^2), \quad q = p_2 - p_1; \end{aligned} \tag{5.7.1}$$

so defined, F_π is normalized to $F_\pi(0) = 1$.

To calculate this we write, suppressing the index "em" in the current,

$$V^\mu(p_1, p_2) = (2\pi)^3 \langle \pi(p_2) | T J_0^\mu(0) e^{i \int d^4 x \, \mathcal{L}^0_{\text{int}}(x)} | \pi(p_1) \rangle,$$

with the index (or superindex) 0 indicating free fields. To second order,

$$V^\mu(p_1, p_2) = -(2\pi)^3 \frac{g^2}{2!} \sum_{f=u,d} Q_f \int d^4x \, d^4y \, \langle \pi(p_2) | T \bar{q}_{0f}(0) \gamma^\mu q_{0f}(0)$$

$$\times \sum_{a,b} \left\{ \bar{u}_0(x) \gamma^\rho t^a u_0(x) \bar{d}_0(y) \gamma^\sigma t^b d_0(y) + (x \leftrightarrow y) \right\} B_{0\rho}^a B_{0\sigma}^b | \pi(p_1) \rangle + \cdots. \tag{5.7.2}$$

The various combinations give rise to the terms depicted in the diagrams of Figs. 5.7.2A and B. Actually, the diagrams in Fig. 5.6.2A give a zero contribution, as can be checked by explicit calculation, and as is intuitively obvious: in those diagrams there is no exchange of momentum between the struck quark and the rest, so it is impossible that the pion bound state (which implies, in particular, collinear momenta of the quarks travelling together) can be formed again after the collision; for this reason their contribution is omitted. The contribution of the diagrams in Fig. 5.7.2B is now, with i, j, k colour indices and α, β and δ Dirac ones, and dropping the indices 0 for free fields,

$$V^\mu(p_1, p_2) = -(2\pi)^3 g^2 \sum \int d^4x \, d^4y \, \langle \pi(p_2) | \bar{u}_\alpha^i(0) d_{\delta'}^{k'}(y) \gamma^\mu_{\alpha\alpha'} S_{\alpha'\beta}(-x)$$

$$\times t_{ii'}^a t_{kk'}^b \gamma^\rho_{\beta\beta'} \gamma^\sigma_{\delta\delta'} D_{\rho\sigma}(x-y) \delta_{ab} u_{\beta'}^{i'}(x) \bar{d}_\delta^k(y) | \pi(p_1) \rangle + \text{"crossed term"},$$

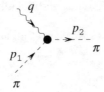

Fig. 5.7.1. Kinematics for the pion form factor.

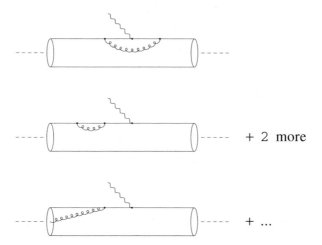

Fig. 5.7.2A. Diagrams *not* contributing to the pion form factor.

where the "crossed term" is that obtained from the other contraction in (5.7.2), and normal ordering of the operators is understood.

Next, we perform a spacetime shift by y, and insert a sum over a complete set of states, $\sum_\Gamma |\Gamma\rangle\langle\Gamma|$. Letting $z = x - y$ and neglecting the quark masses, we find

$$
\begin{aligned}
V^\mu(p_1, p_2) = (2\pi)^3 g^2 \sum \int \mathrm{d}^4 k \int \mathrm{d}^4 p \int \frac{\mathrm{d}^4 z \, \mathrm{e}^{iz\cdot(p-k)}}{(2\pi)^4} \int \frac{\mathrm{d}^4 y \, \mathrm{e}^{iy\cdot(p+p_2-p_1)}}{(2\pi)^4} \\
\times \langle \pi(p_2)|\bar{u}^i_\alpha(-y)d^{k'}_\delta(0)\sum_\Gamma |\Gamma\rangle\langle\Gamma|u^i_{\beta'}(z)\bar{d}^k_\delta(0)|\pi(p_1)\rangle\gamma^\mu_{\alpha\alpha'} \\
\times \frac{-\not p_{\alpha'\beta}}{p^2 k^2} g_{\rho\sigma}\gamma^\rho_{\beta\beta'}\gamma^\sigma_{\delta\delta'} t^c_{ii'} t^c_{kk'} + (p_1 \leftrightarrow p_2).
\end{aligned}
$$

The term $(p_1 \leftrightarrow p_2)$ comes from the "crossed term". We have not written explicitly the contribution of the gauge terms as they give zero, to leading order. In fact, we have evaluated the expression above in the Fermi–Feynman gauge, but, after adding the $(p_1 \leftrightarrow p_2)$ piece the result turns gauge invariant. To leading order we can replace the complete sum of states by only the vacuum state, $\sum_\Gamma |\Gamma\rangle\langle\Gamma| \to |0\rangle\langle 0|$.

Let us next write

$$
u^i_{\beta'}(z)\bar{d}^k_\delta(0) = \frac{\delta_{ik}}{4N_c}(\gamma^\lambda\gamma_5)_{\beta'\delta}\bar{d}(0)\gamma_\lambda\gamma_5 u(z) - \frac{\delta_{ik}}{4N_c}(\gamma_5)_{\beta'\delta}\bar{d}(0)\gamma_5 u(z) + \cdots ;
$$

$$
(5.7.3)
$$

$N_c = 3$. Other terms (the dots) will not contribute to the pion form factor due to the pseudoscalar and colour singlet nature of the pion. Of the two

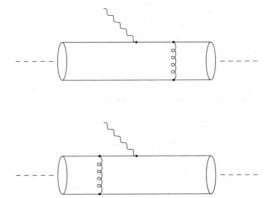

Fig. 5.7.2B. Diagrams contributing to the pion form factor.

terms in (5.7.3), the second contains the twist three operator $\bar{d}\gamma_5 u$, and we will neglect it for the time being. Then we find

$$
V^\mu(p_1, p_2) = (2\pi)^3 \frac{C_F g^2}{48} \int \mathrm{d}^4 k \int \mathrm{d}^4 p \int \frac{\mathrm{d}^4 z \, e^{iz\cdot(p-k)}}{(2\pi)^4} \int \frac{\mathrm{d}^4 y \, e^{iy\cdot(p+p_2-p_1)}}{(2\pi)^4}
$$
$$
\times \frac{\operatorname{Tr} \gamma^\mu \not{p}\, \gamma^\rho \gamma^\lambda \gamma_5 \gamma_\rho \gamma^\tau \gamma_5}{p^2 k^2} \, \langle \pi(p_2)| : \bar{u}(y)\gamma_\tau \gamma_5 d(0) : |0\rangle
$$
$$
\times \langle 0| : \bar{d}(0)\gamma_\lambda \gamma_5 u(z) : |\pi(p_1)\rangle + (p_1 \leftrightarrow p_2),
$$
$$
\tag{5.7.4}
$$

and the normal ordering is now written explicitly.

Let us concentrate on the terms $\langle 0| \ldots |\pi\rangle$, $\langle \pi| \ldots |0\rangle$. We expand them in powers of z and y; for example,

$$
\langle 0| : \bar{d}(0)\gamma_\lambda \gamma_5 u(z) : |\pi(p_1)\rangle
$$
$$
= \sum_n \frac{z^{\mu_1} \ldots z^{\mu_n}}{n!} \mathcal{S}\langle 0| : \bar{d}(0)\gamma_\lambda \gamma_5 D_{\mu_1} \ldots D_{\mu_n} u(0) : |\pi(p_1)\rangle, \tag{5.7.5a}
$$

and, neglecting terms proportional to the pion mass, we can define

$$
(2\pi)^{3/2} \langle 0|\mathcal{S} : \bar{d}(0)\gamma_\lambda \gamma_5 D_{\mu_1} \ldots D_{\mu_n} u(0) : |\pi(p_1)\rangle \equiv i^{n+1} p_{1\lambda} p_{1\mu_1} \ldots p_{1\mu_n} A_n. \tag{5.7.5b}
$$

Furthermore, we introduce the "parton wave function", $\Psi(\xi)$, such that

$$
A_n = \int_0^2 \mathrm{d}\xi \, \xi^n \Psi(\xi); \tag{5.7.5c}
$$

then,

$$
\langle 0| : \bar{d}(0)\gamma_\lambda \gamma_5 u(z) : |\pi(p_1)\rangle = i p_{1\lambda} \int_0^1 \mathrm{d}\xi \, e^{i\xi p_1 \cdot z} \Psi(\xi). \tag{5.7.6}
$$

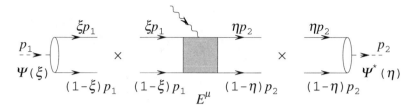

Fig. 5.7.3. Splitting of the pion form factor into the wave functions, and a "hard" piece, E^μ.

All of this has been accomplished formally. When renormalizing, we will have to replace $g \to g(\nu)$ and realize that $A_n = A_n(\nu)$, $\Psi(\xi) = \Psi(\xi, \nu^2)$. To avoid $\log Q^2/\nu^2$ terms we choose $\nu^2 = Q^2 = -(p_1 - p_2)^2$. We then carry over the z, y; k, p integrations in (5.7.4), using (5.7.6), so that we get

$$V^\mu(p_1, p_2) = \frac{C_F g^2(\nu)}{48} \int_0^1 d\xi\, \Psi(\xi, \nu^2) \int_0^1 d\eta\, \Psi^*(\eta, \nu^2) \frac{\mathrm{Tr}\, \gamma^\mu \slashed{p}\, \gamma^\rho \slashed{p}_1 \gamma_5 \gamma_\rho \slashed{p}_2 \gamma_5}{p^2 k^2}$$
$$+ (p_1 \leftrightarrow p_2),$$
$$p = p_1 - (1-\eta)p_2, \quad k = (1-\eta)p_2 - (1-\xi)p_1.$$
$$(5.7.7)$$

We have succeeded in splitting the vertex into a "soft" part, buried in the wave functions Ψ, Ψ^*, and a "hard" piece, E^μ (Fig. 5.7.3; see also the lowest order diagram for E^μ in Fig. 5.7.4). We see the physical interpretation of the variables ξ, η as the fraction of momentum carried by each quark. Evaluating the trace in (5.7.7), we finally find

$$F_\pi(q^2) = \frac{2\pi C_F \alpha_s(Q^2)}{3Q^2} \left| \int_0^1 d\xi\, \frac{\Psi(\xi, Q^2)}{1 - \xi} \right|^2 + O(M_\pi^2/Q^4) + O(\alpha_s^2). \quad (5.7.8)$$

The last task is the evaluation of the Q^2 evolution of the Ψ. The operators that define Ψ via Eqs. (5.7.5) are the same as those for the nonsinglet part of deep inelastic scattering structure functions, cf. Sects. 4.5 and 4.6. However, we have here an extra complication: because the matrix elements are nondiagonal, the total derivatives yield a nonzero contribution. Then, the operators $N_{A,n,j}^{\lambda\mu_1\ldots\mu_n}$, with $j = 0, \ldots, n$, given by

$$N_{A,n,j}^{\lambda\mu_1\ldots\mu_n} = \partial^{\mu_{j+1}} \ldots \partial^{\mu_n} \bar{d}(0) \gamma^\lambda \gamma_5 D^{\mu_1} \ldots D^{\mu_j} u(0), \quad (5.7.9)$$

mix under renormalization. They are thus renormalized by a matrix,

$$N_{A,n,j} \to \sum_{j'} Z_{n+1,j'} N_{A,n,j'}. \quad (5.7.10a)$$

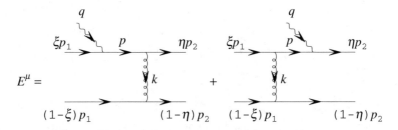

Fig. 5.7.4. The "hard" piece of the pion form factor, E^μ.

For $j = n$, the $Z_{n+1,n}$ coincide with those calculated in Sect. 4.6:

$$Z_{n+1,n} = 1 + \frac{g^2 N_\epsilon}{16\pi^2} C_F \left\{ 4S_1(n+1) - 3 - \frac{2}{(n+1)(n+2)} \right\}; \qquad (5.7.10b)$$

for $j \leq n - 1$ we get, after a similar calculation,

$$Z_{n+1,j} = \frac{g^2 N_\epsilon}{16\pi^2} C_F \left\{ \frac{2}{n+2} - \frac{2}{n-j} \right\}. \qquad (5.7.10c)$$

To obtain the operators with definite behaviour as $Q^2 \to \infty$, we have to diagonalize[16] Z. Let S be the matrix that accomplishes this; we define the $\hat{A}_j$ as the transform of the A under S,

$$A_n(Q^2) = \sum_{j=0}^{n} S_{nj} \hat{A}_j(Q^2). \qquad (5.7.11a)$$

Then the anomalous dimensions of the $\hat{A}_j$ are given by the eigenvalues of Z. But, because Z is triangular, it follows that its eigenvalues are simply its diagonal elements. Therefore,

$$\hat{A}_j(Q^2) \underset{Q^2 \to \infty}{\simeq} \left[\alpha_s(Q^2) \right]^{d_{NS}(j+1)} \hat{A}_j^{(0)},$$

with $\hat{A}_j^{(0)}$ constant. To leading order, and since $d_{NS}(j+1) > d_{NS}(1) = 0$, we need retain only one term in (5.7.11a), so that

$$A_n(Q^2) \underset{Q^2 \to \infty}{\simeq} S_{n0} \hat{A}_0^{(0)},$$

and we then find

$$\int_0^1 d\xi \, \frac{\Psi(\xi, Q^2)}{1 - \xi} \underset{Q^2 \to \infty}{\to} \hat{A}_0^{(0)} \sum_{n=0}^{\infty} S_{n0}.$$

---

[16]An alternative method uses properties of conformal invariance (Ferrara, Gatto and Grillo, 1972).

The values of the S_{n0} are easily verified to be

$$S_{n0} = \frac{1}{n+2} - \frac{1}{n+3}. \tag{5.7.11b}$$

In addition, the value of $\hat{A}_0^{(0)}$ is also known. Because of the PCAC equation (cf. Sect. 7.3, especially Eqs. (7.3.1) later on),

$$(2\pi)^{3/2}\langle 0|\bar{d}(0)\gamma^\lambda\gamma_5 u(0)|\pi(p)\rangle = ip^\lambda\sqrt{2}f_\pi, \quad f_\pi \simeq 93 \text{ MeV}.$$

Hence,

$$A_0 = \int_0^1 d\xi\, \Psi(\xi, Q^2) = \sqrt{2}f_\pi, \quad \text{independent of } Q^2.$$

From this,

$$\hat{A}_0^{(0)} = 6\sqrt{2}f_\pi,$$

so we finally obtain the result[17]

$$F_\pi(t) \underset{Q^2\to\infty}{\simeq} \frac{12\pi C_F f_\pi^2 \alpha_s(-t)}{-t}. \tag{5.7.12}$$

The corrections are $O(\alpha_s^{d_{NS}(3)} \simeq \alpha_s^{0.6})$; even terms actually vanish due to charge conjugation invariance.

An important feature to notice is the following. The "hard" part of the pion form factor appears to be infrared divergent (the term $1/(1-\xi)$ in Eq. (5.7.8)). However, for the leading order we are lucky, as this is cancelled by a zero of the wave functions. In fact, we have found that

$$\int d\xi\, \Psi(\xi, Q^2)\xi^n \underset{Q^2\to\infty}{\to} S_{n0}\hat{A}_0^{(0)},$$

which, given the values of the S_{n0}, Eq. (5.7.11b), implies the behaviour

$$\Psi(\xi, Q^2) \underset{Q^2\to\infty}{\to} \xi(1-\xi)\hat{A}_0^{(0)}. \tag{5.7.13}$$

With the pion form factor we are apparently in an ideal situation: both behaviour and absolute normalization are predicted theoretically. There are, unfortunately, a number of snags.

First of all, the perturbative corrections decrease slowly, only as $\alpha_s^{0.6}$. Worse still, the convergence of the wave function to its asymptotic value, (5.7.13), is also extremely slow. Isgur and Llewellyn Smith (1989) have carefully examined this issue and conclude that huge energies are necessary before (5.7.13) is approximated to some 90%; and the correction, though small

---

[17]Farrar and Jackson (1979); Brodsky, Frishman, Lepage and Sachrajda (1980); Efremov and Radyushin (1980a,b), which we have followed. The same result may be obtained using so-called light cone perturbation theory (Brodsky and Lepage, 1980).

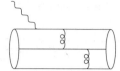

Fig. 5.7.5. Gluon exchanges in the nucleon form factor.

(10%), becomes much amplified by the divergence of the hard piece. Lastly, if we evaluate the next twist (twist three) contributions (Espriu and Ynduráin, 1983) we find that the corresponding wave function *diverges* like $\alpha_s^{-d_m} \log(1-\xi)$, with $d_m = 12/(33-2n_f)$, for $\xi \to 1$. Thus a cut-off becomes necessary and the corresponding contribution decreases no longer as $1/t^2$, but rather goes like $C(\log t)^\nu/|t|^{3/2}$ with unknown C, ν.

It would appear that (5.7.12) only has an asymptotic value, as a *quantitative* prediction, because the corrections are so out of control; and, indeed, if employed at experimentally accessible values of Q^2, say $Q^2 \lesssim 10$ GeV^2, the estimate (5.7.12) is merely *qualitative*: experiment lies well above it.

This quantitative failure should not hide the qualitative success: F_π really decreases proportional to $1/t$. We may then use the example of the pion form factor to infer general *qualitative* rules. To do so, consider the amplitude for an exclusive process. We take it to be of the form

$$\mathcal{A} = \int \Phi^\dagger K \Phi,$$

where Φ is the wave function of the bound state B, made out of n quarks, $\Phi \simeq \langle 0|\mathrm{T}q_1(x_1)\ldots q_n(x_n)|B\rangle$;

K is a hard kernel

$$K \sim \left[\frac{\alpha_s(Q^2)}{Q^2}\right]^{n-1} :$$

the momentum has to be shared among the n constituents, so each time we get a denominator $\sim 1/Q^2$ and we use two powers of the coupling, g^2. This yields the *counting rules* of Brodsky and Farrar (1973). For example, for the nucleon form factor (Fig. 5.7.5), one finds the celebrated dipole form factor

$$F_N \sim \left[\frac{\alpha_s(-t)^2}{-t}\right]^2 ;$$

for the deuteron,

$$F_D \sim \left[\frac{\alpha_s(-t)^2}{-t}\right]^5 ,$$

a behaviour that is in fact seen experimentally. For fixed angle scattering of particles A, B into C, D with form factors F_A, F_B...,

$$\left.\frac{d\sigma(A + B \to C + D)}{dt}\right|_{\theta \text{ fixed}} \sim \frac{\alpha_s(t)}{-t} F_A(t)F_B(t)F_C(t)F_D(t)f(\theta),$$

with $f(\theta)$ an unknown function of the scattering angle. Further details and references may be found in Brodsky and Lepage (1980). Many of these results have been made rigorous in terms of renormalization group analyses by Duncan and Muller (1980b); see also the review of Duncan (1981).

5.8 Other Processes that can be Described with Perturbative QCD

i Deep Inelastic Scattering on π, K, γ Targets

By looking at processes such as

$$e^+e^- \to (\pi, K, \gamma) + \text{ hadrons},$$

we can deduce the properties of deep inelastic scattering on π, K, γ targets,

$$\gamma^* + (\pi, K, \gamma) \to \text{ hadrons};$$

see Fig. 5.8.1a. These processes present the peculiarity of the continuation of the momentum of the γ^* to timelike values; apart from that, π of K targets are not very different (for deep inelastic scattering) from nucleon ones.

The situation is different for photon targets. The cross section

$$\gamma^* + \gamma \to \text{ hadrons},$$

can be deduced, as stated, from $e^+e^- \to \gamma +$ hadrons; it can also be obtained by Weiszäcker–Williams scattering,

$$e^+e^- \to e^+e^- + \underbrace{\gamma + \gamma}_{\hookrightarrow \text{ hadrons}} \quad ,$$

as shown in Fig. 5.8.1b. This last process also provides information on

$$\gamma^*(p_1^2) + \gamma^*(p_2^2) \to \text{ hadrons},$$

with one or both of the photons off shell. The process $\gamma^*(p_1^2) + \gamma^*(p_2^2) \to$ hadrons presents the peculiarity of being *calculable* for $x \to 1$ provided that one of the two momenta is $|p_i|^2 \gg \Lambda^2$. That is to say, not only the evolution but the absolute normalization is known except in a region $x \approx 0$. The reason is that the dominant piece is given by a set of operators built solely from the photon field, whose matrix elements are known. This was first remarked by Witten (1977); see also Kingsley (1973) and Walsh and Zerwas (1973). The

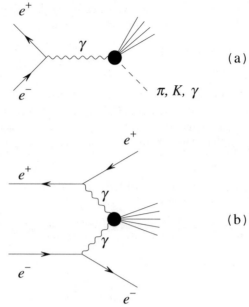

Fig. 5.8.1. (a) Diagrams for deep inelastic scattering on π, K and photon targets. (b) Weiszäcker–Williams scattering of $\gamma\gamma$.

second order calculation is due to Bardeen and Buras (1979); this and other features are reviewed by Buras (1981) and Walsh (1983), where comments on peculiarities of the region $x \simeq 0$ may also be found. A complete set of radiative corrections may be found in Laenen, Riemersma, Smith and van Neerven (1994).

ii Strong Interaction Corrections to Weak and Electromagnetic Decays of Hadrons

The methods employed to study QCD corrections to weak and electromagnetic decays are not very different[18] from those already encountered, so only a brief review for weak decays will be presented here.

Broadly speaking, we can classify the decays into three categories. In the first we have semileptonic decays of particles containing a heavy quark; typical examples are D and B meson decays, such as

$$D^+ \to e^+ + \nu + \text{hadrons}.$$

At the partonic level, the process proceeds via the diagram shown in Fig. 5.8.2. QCD corrections are of three types. There are corrections that can

---

[18]With the exception of complications caused by the bound nature of the decaying quarks.

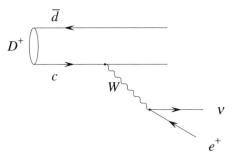

Fig. 5.8.2. Semileptonic decay of D^+.

be incorporated into corrections to the wave function of the decaying particle, or the hadron debris. Then there are corrections to the Wcs coupling, and gluon radiation. The last two can be combined to give an enhancement factor essentially equal to that for τ decay, $1 + 3C_F\alpha_s/4\pi$. Details and references may be found in Altarelli et al. (1982), and the excellent review of the first author (Altarelli, 1983).

A second type of processes that we consider are inclusive nonleptonic decays of heavy particles. At the partonic level, and to lowest order in weak interactions, we find the S-matrix amplitude for $D^+ \to$ hadrons, for example,

$$\langle f|S|i\rangle = (2\pi)^4\delta_4(P_f - P_i)\frac{g_W^2}{2} \int \mathrm{d}^4x\, D_{W\alpha\beta}(x)\langle f|\mathrm{T}J^\alpha_{Lud}(x)J^\beta_{Lsc}(0)|i\rangle$$

$$(5.8.1\mathrm{a})$$

(see Fig. 5.8.3), where we neglect Cabibbo mixing and the contribution of the annihilation channel, proportional to m_s. We have defined

$$J^\mu_{Lff'} = \bar{q}_f\gamma^\mu\frac{1 - \gamma_5}{2}q_{f'};\qquad (5.8.1\mathrm{b})$$

g_W is the weak coupling[19] and D_W is the x-space W propagator. Dropping quark masses as compared to the W mass, M_W, we can write

$$D_{W\alpha\beta}(x) = -\mathrm{i}g_{\alpha\beta} \int \frac{\mathrm{d}^4k}{(2\pi)^4} \frac{e^{\mathrm{i}k\cdot x}}{k^2 - M_W^2} \simeq \mathrm{i}g_{\alpha\beta}\frac{1}{M_W^2}\delta_4(x). \qquad (5.8.1\mathrm{c})$$

It follows that we can replace the T-product of two currents in (5.8.1a) by its short distance expansion. The leading terms will be of dimension six,

---

[19]Cf. Sect. 4.3 for definitions concerning weak interactions.

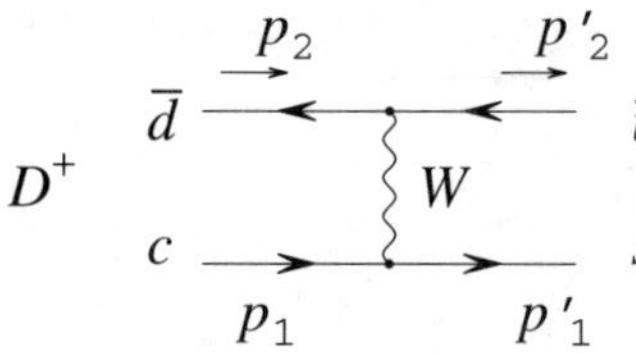

Fig. 5.8.3. Nonleptonic decay of D^+.

with the appropriate quantum numbers. There are four such operators:

$$\begin{aligned}
N_1(0) &= :\bar{s}(0)\gamma_\alpha\frac{1-\gamma_5}{2}c(0)\bar{u}(0)\gamma^\alpha\frac{1-\gamma_5}{2}d(0) :\equiv (\bar{s}c_L)(\bar{u}d_L),\\
N_2(0) &= (\bar{u}c_L)(\bar{s}d_L),\\
N_1'(0) &= \sum_a :\bar{s}(0)\gamma_\alpha\frac{1-\gamma_5}{2}t^a c(0)\bar{u}(0)\gamma^\alpha\frac{1-\gamma_5}{2}t^a d(0) :\equiv (\bar{s}\vec{t}c_L)(\bar{u}\vec{t}d_L),\\
N_2'(0) &= (\bar{u}\vec{t}c_L)(\bar{s}\vec{t}d_L).
\end{aligned} \tag{5.8.2a}$$

The last two may be written as combination of the first two by using the identity

$$\sum_a t^a_{ij}t^a_{lk} = \tfrac{1}{2}\left[-\frac{1}{N_c}\delta_{il}\delta_{jk} + \delta_{ik}\delta_{lj}\right],$$

so that we find

$$(\bar{q}_1\vec{t}q_2)(\bar{q}_3\vec{t}q_4) = \tfrac{1}{2}\left[(\bar{q}_1q_2)(\bar{q}_3q_4) - \frac{1}{N_c}(\bar{q}_1q_4)(\bar{q}_3q_2)\right].$$

The operators N_1 and N_2 mix under renormalization; a set of operators that do not mix, because they behave differently under flavour transformations, are

$$N_\pm = \tfrac{1}{2}\left(N_1 \pm N_2\right), \tag{5.8.2b}$$

so we write

$$\begin{aligned}
\langle f|g_{\alpha\beta}\mathrm{T}J^\alpha_{Lud}(x)J^\beta_{Lsc}(0)|i\rangle =& \,C_+\langle f|N_+(0)|i\rangle + C_-\langle f|N_-(0)|i\rangle\\
&+ \text{higher orders in } x.
\end{aligned} \tag{5.8.3a}$$

The Wilson coefficients, $C_\pm = C_\pm(\alpha_s, \mu^2)$, equal unity in the free field theory. μ^2 is a reference momentum. For the record, we give another usual expression for the $N_\pm$ obtained by reordering:

$$\begin{aligned}
N_\pm =& \frac{N_c \pm 1}{2N_c} :\bar{s}\gamma_\alpha\frac{1-\gamma_5}{2}c\,\bar{u}\gamma^\alpha\frac{1-\gamma_5}{2}d :\\
&\pm\sum_a :\bar{s}\gamma_\alpha\frac{1-\gamma_5}{2}t^a c\,\bar{u}\gamma^\alpha\frac{1-\gamma_5}{2}t^a d : .
\end{aligned} \tag{5.8.3b}$$

The effect of QCD corrections[20] to leading order is to renormalize the operators $N_\pm$. We will follow the custom in this area and *define* the $C_\pm$ including the renormalization constants of the $N_\pm$, $Z_\pm$. Therefore we obtain, to leading order,

$$C_\pm(\alpha_s, \mu^2) = \left[\frac{\alpha_s(\mu^2)}{\alpha_s(M_W^2)}\right]^{d_\pm}, \tag{5.8.4a}$$

i.e., a scaling of the weak interaction strength from where it is defined, on the W propagator pole, to the reference momentum μ^2 which we take to be of the order of magnitude of the mass squared of the decaying particle, in our case $\mu^2 \simeq M_{D^+}^2$. The $d_\pm$ are, up to a factor β_0, the anomalous dimensions of the operators $N_\pm$. A simple calculation gives

$$d_+ = -\frac{9(N_c - 1)}{N_c(11N_c - 2n_f)}, \quad d_- = \frac{9(N_c + 1)}{N_c(11N_c - 2n_f)}. \tag{5.8.4b}$$

Further details may be found in the reports of Altarelli (1982, 1983).

The third class of process consists of the decays $K \to mesons$. The mass of the kaon is $m_K \sim 1/2$ GeV, and hence too low to apply perturbative QCD. Here the philosophy is somewhat different. One calculates as before; but, instead of evaluating QCD corrections at $\mu^2 = m_K^2$, one takes a reference value, say $\mu_0^2 \simeq 1$ GeV^2, sufficiently high for perturbative QCD to be applicable. And then one *hopes* that the value so obtained for the decay amplitude does not change much between this μ_0^2 and the $\mu^2 \simeq m_K^2$ where the process really takes place. At times, this is refined by the use of SVZ sum rule methods for the extrapolation, or inclusion of low energy estimates from chiral dynamics. An enormous amount of work has been done on these processes, from the pioneering evaluations that helped to pin down the predictions for the c quark mass[21] to more recent studies, in particular in connection with the $\Delta I = 1/2$ rule; the reviews of Gaillard (1978), Altarelli (1982, 1983), Pich and de Rafael (1991) and de Rafael (1995) may be consulted for this. Here we will leave the subject, remarking only that, generally speaking, the results are quantitatively good when the *second order* QCD corrections are small; and fall short of experiment when large coefficients appear: not a surprising situation.

---

[20]Gaillard and Lee (1974a,b); Altarelli and Maiani (1974); Shifman, Vainshtein and Zakharov (1977a,b).

[21]Glashow, Iliopoulos and Maiani (1970); Gaillard and Lee (1974a,b).

6 Hadrons as Bound States of Quarks

"You boil it in sawdust: you salt it in glue:
You condense it with locusts and tape:
Still keeping one principal object in view –
To preserve its symmetrical shape"

LEWIS CARROLL, 1897

6.1 Generalities. The Quark Model of Hadrons

As stated at the very beginning of this text, the first evidence in the direction of QCD came from the quark model of hadrons; that is to say, from the fact that hadrons can be classified as colour singlet bound states $\bar{q}q'$, $qq'q''$. These states, including radial and angular excitations, do indeed accommodate the vast majority of the hundreds of hadrons known today (see the Particle Data Group tables). There are only a few dubious cases, and two or three hadrons that can be interpreted as being made mostly of gluons, called *glueballs*. Not only this, but some of the quantitative properties of these hadrons, in particular mass differences, were roughly understood in simple potential models well before the advent of QCD. In the present chapter we will review the situation, of course (whenever possible) within the context of the fullfledged theory of quark and gluon interactions. From this point of view, it is convenient to split the subject into three broad areas.

First of all, we have the lowest-lying bound states of $\bar{c}c$ and, especially, $\bar{b}b$. Here a Coulombic approximation is valid to first order, and we may estimate with it various quantities. In general we have, besides the QCD parameter Λ and the confinement radius $R \sim \Lambda^{-1}$, two more scales: the size of the bound state, which for heavy quarks and the ground state or lowest lying ones is of the order of the equivalent to the Bohr radius, $a \sim 1/mC_F\alpha_s$; and the inverse of the binding energy, $T_q \sim 1/mC_F^2\alpha_s^2$; note that $B_E \sim \frac{1}{4}mC_F^2\alpha_s^2$ (more precise formulas will be given later). In the case of the lowest lying $\bar{c}c$ and $\bar{b}b$ bound states, we have $a \ll R$, so we may neglect confinement as a first approximation, and treat its effects as a perturbation; furthermore, radiative corrections, which involve $\alpha_s(B_E^2)$, are small and we can evaluate the potential using perturbation theory. Finally, the average velocity of the quarks, proportional to $a/T_q \sim C_F\alpha_s$ is small, so the nonrelativistic approximation may be used, with eventual inclusion of relativistic effects as first order corrections. Under these circumstances we have what may be described as an *ab initio*, rigorous QCD evaluation of the properties of the corresponding quarkonium states; the quality of the approximations being estimated from the size of the higher order effects calculated – radiative, nonperturbative and relativistic.

The second type of situation occurs for *excited* bound states of heavy quarks. Here the velocity is small (in fact, even smaller than in the previous case) but the perturbative calculations are of little use: not only are the radiative corrections large, involving $\alpha_s(B_E^2/n^2)$ for the nth excited state; but the system, with a size of order na, extends to the confinement radius R. Because of this, and although, as results from general properties of Galilean invariance, the interaction can be described in terms of a potential, the derivation of this potential involves nonperturbative evaluations which entail assumptions that are more or less reasonable, but unfortunately not unique. All calculations produce a Coulomb-type potential at short distances, plus a linear potential at long distances; but the corrections to both are somewhat less clear. In spite of this, it is possible to present a reasonable description of these states, hiding one's ignorance in a few phenomenological terms.

The third type of situation arises when we have bound states involving light quarks. Here, and except for a few general results that may be obtained for states involving one heavy quark using effective field theories,[1] we are in a difficult situation. The light quarks move ultra-relativistically inside the hadrons; so a potential picture is not appropriate. To study these states rigorously, we would have to perform a full nonperturbative calculation, as in lattice QCD. Alternatively, one may invent phenomenological models incorporating features suggested by QCD. One such model is the constituent quark model. Here one assumes that the net effect of having the quarks travelling in a sea of gluons and light $q\bar{q}$ pairs inside the hadron can be approximated by giving light quarks an effective, *constituent* mass common to all of them, $\mu_0 \simeq 330$ MeV. A potential is then used, possessing the features suggested by the study of the heavy quarks case. Another type of models are *bag* models: one keeps quarks massless, but confines them in a sphere of radius of the order of the confinement radius, R. Needless to say, both models present striking successes together with serious drawbacks.

Glueballs are a case apart. Because the experimental situation is unclear, and the theoretical understanding so poor (any nonrelativistic model for glueballs will be largely arbitrary) we will not discuss them here, although something will be said about them in the chapter dedicated to lattice QCD.

---

[1] See, for example, the reviews of Lepage and Thacker (1988) and Grinstein (1991).

6.2 Pole Masses and the Schrödinger Equation. Corrections

i Confinement. Pole Mass. Relation with the $\overline{\text{MS}}$ Mass

We will, in this and the next sections, consider bound states of heavy quarks, specifically $\bar{q}q$ (*quarkonium*), under the assumption that its characteristic sizes, a and T_q, are much smaller[2] than the confinement radius $R \sim \Lambda^{-1}$; we are thus in the situation depicted in Fig. 6.2.1. Under these circumstances it would appear that the fact that quarks are confined would have little influence on their motion. Otherwise stated, although the forces between quarks grow with the distance between them (something that we will make plausible in Sect. 6.4 and Chap. 9) this growth is only supposed to be important for distances $r \sim R$ and should have little bearing on bound states with an average size $a \ll R$.

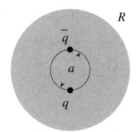

Fig. 6.2.1. The confinement region, and the region of motion of the $\bar{q}q$ pair.

Actually, for no quark states among those observed does one really have a, $T_q \ll R$ by a wide margin; this only occurs clearly for the lowest states of *toponium*. For $\bar{b}b$ in the ground state the inequality is reasonably fulfilled; for $\bar{c}c$ in the ground state and for the first radial excitations of $\bar{b}b$ we are on the borderline. Only detailed calculation can then tell for which states, and for which observables, is the approximation of neglecting confinement (or at least treating it as a small perturbation) actually valid; but we will in this and the next section assume validity of the approximation, leaving the detailed evaluations for Sects. 6.3, 4.

According to this we will, as a first approximation, take the limit $R \to \infty$, i.e., neglect confinement altogether. We can therefore treat quarks as if they could become free particles, and in particular as if asymptotic quark states could exist. To make these assumptions more quantitative, we should consider the *corrections* to our approximation. Obviously, these will be of the order of $1/R \sim \Lambda$ to some power; thus, in particular, they cannot be seen in perturbation theory because $\Lambda \sim e^{-2\pi/\beta_0 \alpha_s}$. One should thus look for these corrections in the appearance of *nonperturbative* effects which, as we shall see,

---

[2] The condition $T_q^{-1} \ll \Lambda$ is equivalent to $B \gg \Lambda$, with B the binding energy.

can be parametrized in terms of contributions of the *condensates* to binding
energies and wave functions. Not surprisingly, the leading contribution will
be given by the gluon condensate and in fact will turn out to be proportional
to $a^{-4}\langle \alpha_s G^2 \rangle$, a being the average size of the state. The approximation of
neglecting confinement will be reasonable when these corrections are small.

If one neglects confinement, a concept that becomes useful is that of what
is called the *on-shell*, or, more appropriately, *pole mass*, m^{pole}, of quarks
(Coquereaux, 1981; Tarrach, 1981). We define it as the location of the pole of
the quark propagator, in *perturbation theory*, and thus through the equation

$$S_P^{-1}(\not{p} = m^{\text{pole}}) = 0, \qquad (6.2.1)$$

where the label P in S_P emphasizes that it is to be evaluated to a *finite* order
in perturbation theory.

This mass may be easily related to the $\overline{\text{MS}}$ mass. To one loop, and working
in the Landau gauge to get rid of the inessential wave function renormalization, we have, for the bare propagator, the expression reported in Sect. 3.1
with $\xi = 1$:

$$S_D(p) = \frac{i}{\not{p} - m} - \frac{1}{\not{p} - m} C_F g^2 \Sigma_D(p) \frac{i}{\not{p} - m},$$

$$\Sigma_D(p) = (\not{p} - m)A + m B_D,$$

with

$$A = \frac{1}{16\pi^2} \left\{ -1 - \int_0^1 dx\, (1 - 2x) \log\left[x m^2 - x(1-x)p^2 \right] \right.$$

$$\left. - (p^2 - m^2) \int_0^1 dx\, \frac{x}{m^2 - xp^2} \right\};$$

and

$$B_D = \frac{1}{16\pi^2} \left\{ -3N_\epsilon + 1 + 2 \int_0^1 dx\, (1+x) \log \frac{x m^2 - x(1-x)p^2}{\nu_0^2} \right.$$

$$\left. - (p^2 - m^2) \int_0^1 dx\, \frac{x}{m^2 - xp^2} \right\}.$$

In the $\overline{\text{MS}}$ scheme,

$$S_{\overline{\text{MS}}}(p) = \frac{i}{\not{p} - \bar{m}(\nu^2)} - \frac{1}{\not{p} - m} C_F g^2 \Sigma_{\overline{\text{MS}}}(p) \frac{i}{\not{p} - m},$$

$$\Sigma_{\overline{\text{MS}}}(p) = (\not{p} - m)A + m B_{\overline{\text{MS}}},$$

$$B_{\overline{\text{MS}}} = \frac{1}{16\pi^2} \left\{ 1 + 2 \int_0^1 dx\, (1+x) \log \frac{x m^2 - x(1-x)p^2}{\nu^2} \right. \qquad (6.2.2a)$$

$$\left. - (p^2 - m^2) \int_0^1 dx\, \frac{x}{m^2 - xp^2} \right\}.$$

(Being finite in this gauge, A is independent of the scheme of renormalization so its value is as before). For the propagator in terms of the pole mass, on the other hand, we have to determine the counterterm by requiring (6.2.1) and thus the condition $B_{\text{pole}}(\not{p} = m^{\text{pole}}) = 0$ so that

$$S_{\text{pole}}(p) = \frac{i}{\not{p} - m^{\text{pole}}} - \frac{1}{\not{p} - m} C_F g^2 \Sigma_{\text{pole}}(p) \frac{i}{\not{p} - m},$$

$$\Sigma_{\text{pole}}(p) = (\not{p} - m)A + mB_{\text{pole}},$$

$$B_{\text{pole}} = \frac{1}{16\pi^2} \left\{ 2 \int_0^1 dx\, (1+x) \log \frac{xm^2 - x(1-x)p^2}{x^2 m^2} \right.$$

$$\left. - (p^2 - m^2) \int_0^1 dx\, \frac{x}{m^2 - xp^2} \right\}. \tag{6.2.2b}$$

One then requires equality of $S_{\overline{\text{MS}}}$ and S_{pole}. Writing

$$m^{\text{pole}} = \bar{m}(\nu^2) \left\{ 1 + C_m \frac{\alpha_s}{\pi} \right\},$$

expanding to second order and comparing (6.2.2a,b) we find (Coquereaux, 1981; Tarrach, 1981)

$$C_m = -\frac{C_F}{4\pi} \left\{ 1 + 2 \int_0^1 dx\, (1+x) \log \frac{x^2 m^2}{\nu^2} \right\}.$$

In particular, for $\nu^2 = m^2$,

$$m^{\text{pole}} = \bar{m}(\bar{m}^2) \left\{ 1 + \frac{C_F \alpha_s}{\pi} \right\}.$$

To two loops,

$$m^{\text{pole}} = \bar{m}(\bar{m}^2) \left\{ 1 + \frac{C_F \alpha_s((m^{\text{pole}})^2)}{\pi} + (K - 2C_F) \left[\frac{\alpha_s((m^{\text{pole}})^2)}{\pi} \right]^2 \right\},$$
$$\tag{6.2.3a}$$

where, denoting by n_f the number of quark flavours with mass less than or equal to m,

$$K = K_0 + \sum_{i=1}^{n_f - 1} \Delta \left(\frac{m_i^{\text{pole}}}{m^{\text{pole}}} \right),$$

$$K_0 = \tfrac{1}{9}\pi^2 \log 2 - \tfrac{7}{18}\pi^2 - \tfrac{1}{6}\zeta(3) + \tfrac{3673}{288} - \left(\tfrac{1}{18}\pi^2 + \tfrac{71}{144} \right) n_f \tag{6.2.3b}$$

$$\simeq 17.15 - 1.04\, n_f,$$

with

$$\Delta(\rho) = \tfrac{1}{8}\pi^2 \rho - \tfrac{3}{4}\rho^2 + \cdots.$$

The full expression for K may be found in Gray, Broadhurst, Grafe and Schilcher (1990), to whom the two loop calculation is due.

ii The Schrödinger Equation. Ladders

To any order in perturbation theory, the S-matrix for $\bar{q}q$ scattering is free of bound state poles. However, bound states may be generated from perturbation theory using any of the two methods to be described presently, and which can (and will) be shown to be equivalent. We will work in the *nonrelativistic* (NR) limit; later on relativistic corrections will be evaluated. Radiative corrections will also be considered at a later stage.

In the NR limit one can show quite generally that Galilean invariance implies that the interaction between particles can be implemented by a potential which is local (i.e., depending only on the relative distance[3]), that we denote by $V(r)$, with r the relative coordinate of $\bar{q}$ and q. The energy levels and wave functions may then be found by solving the Schrödinger equation

$$\left\{ 2m + \frac{-1}{m}\Delta + V(r) \right\} \Psi(\mathbf{r}) = E\Psi(\mathbf{r}); \qquad (6.2.4)$$

note that for the $\bar{q}q$ system the reduced mass is $m_{\text{red}} = m/2$.

The form of the potential may be found using the following trick. In the NR limit, $V(r)$ is given by the Fourier transform of the transition amplitude evaluated in the Born approximation:

$$T_{NR}^{\text{Born}}(\mathbf{p} \to \mathbf{p}') = -\frac{1}{4\pi^2} \int d^3\mathbf{r}\, e^{i\mathbf{r}(\mathbf{p}-\mathbf{p}')} V(r), \qquad (6.2.5)$$

where $\mathbf{p}, \mathbf{p}'$ are the initial and final momenta in the centre of mass reference system. On the other hand, T_{NR}^{Born} may be calculated as the nonrelativistic limit of the relativistic scattering amplitude. With the normalization used in this text (Appendix G),

$$T_{NR}^{\text{Born}} = \lim_{m \to \infty} \frac{1}{4\sqrt{p_{10}p_{20}p'_{10}p'_{20}}} F(p_1 + p_2 \to p'_1 + p'_2); \qquad (6.2.6)$$

formally the NR limit is equivalent to taking the limit of infinite quark masses, keeping the three-momenta fixed. One can thus calculate the Born approximation to F, F^{Born} using the familiar Feynman rules (actually at tree level) for $\bar{q}q$ scattering, take the NR limit and hence obtain T_{NR}^{Born}. From it, by inverting (6.2.5) one finds V, and solving then (6.2.4) we get the energy spectrum and wave functions.

An interesting point to clear is that the results that one obtains for these are *gauge independent* (in the case of the wave function, for the measurable modulus). It is true that, in general, F and hence T_{NR}^{Born} and *a fortiori* V are gauge dependent; but the (time dependent) Schrödinger equation

$$i\frac{\partial}{\partial t}\Psi(\mathbf{r}, t) = \left\{ 2m + \frac{-1}{m}\Delta + V(r) \right\} \Psi(\mathbf{r}, t)$$

---

[3] We neglect spin for the time being.

is gauge invariant: the alteration of V in a change of gauge may be compensated by a change in the phase of the wave function. We fix the gauge by requiring

$$V(r) \underset{r \to \infty}{\to} 0, \qquad (6.2.7)$$

and no vector potential for static colour charges.

Note that (6.2.7) is less harmless than it looks at first sight. In fact, and because of confinement, one can only require

$$V(r) \underset{r \simeq R}{\simeq} 0,$$

so we always have a nonperturbative indeterminacy of a constant. One can fix it so that, as $R \to \infty$, the leading nonperturbative correction to observables such as energy levels is that given by the condensates, i.e., that there is no renormalon of order Λ^2; see Sect. 10.2ii.

In QED, because one can renormalize on the mass shell for photons as well as for electrons, the procedure described above to find the potential would be *exact*: radiative corrections do indeed vanish in the strict NR limit (see below). In QCD one can only obtain the static potential in a power series in the coupling. The way this works is most clearly seen in the second method for getting static properties, to which we now turn. We will discuss it first in QED to show clearly the mechanism at hand; then we will extend the results, with due modifications, to QCD.

Consider the QED analogue of quarkonium, viz., positronium. The interaction Hamiltonian is now

$$H_{\text{int}} = e \int d^3\mathbf{x} \, : \bar{\psi}(x)\gamma_\mu A^\mu(x)\psi(x) :, \qquad (6.2.8)$$

if we quantize the theory in a covariant gauge. However, in the Coulomb gauge we have to add to this an extra piece,[4]

$$H'_{\text{int}} = \frac{e^2}{8\pi} \int d^3\mathbf{x} \, d^3\mathbf{y} \, \frac{\psi^\dagger(x)\psi(x)\psi^\dagger(y)\psi(y)}{|\mathbf{x} - \mathbf{y}|}. \qquad (6.2.9)$$

As is easily seen, this corresponds to an instantaneous Coulomb interaction. It appears because, in the Coulomb gauge, the scalar potential is not an independent variable, and it has to be expressed in terms of the ψ:

$$A_0(x) = \frac{e}{4\pi} \int d^3\mathbf{y} \, \frac{\psi^\dagger(y)\psi(y)}{|\mathbf{x} - \mathbf{y}|}, \qquad (6.2.10)$$

and then (6.2.9) corresponds precisely to the longitudinal piece of the electric field contribution to the radiation Hamiltonian,

$$\tfrac{1}{2} \int d^3\mathbf{r} \, \mathcal{E}_l^2,$$

---

[4] See, e.g., Bjorken and Drell (1965), Sect. 15.2.

Fig. 6.2.2. Ladder giving bound states in the NR approximation.

with $\mathcal{E}_l$ the longitudinal part of the electric field.

The photon propagator in this gauge is, in p-space,

$$
D_{\text{Coul.}}^{\mu\nu}(k) = \frac{1}{k^2}\left\{-g^{\mu\nu} - \frac{k^\mu k^\nu}{\mathbf{k}^2} + \frac{k^0(k^\nu g^{0\mu} + k^\mu g^{0\nu})}{\mathbf{k}^2} - \frac{k^2 g^{0\mu} g^{0\nu}}{\mathbf{k}^2}\right\}.
$$
(6.2.11)

Now, this propagator has the important property of vanishing in the static limit (formally as the fermion's mass, m, goes to infinity). To show this, let us take it between physical electron–positron states with momenta p, p' so that $k = p - p'$; the more general situation can be reduced to this with some effort. Of the terms in (6.2.11), those proportional to k^μ or k^ν give zero contribution because of current conservation. The terms with μ or ν different from zero give, when contracted with the spinors, terms proportional to the velocity of the particles, thus vanishing in the static limit. Finally, the remaining term with $\mu = \nu = 0$ is equivalent to

$$
D_{\text{Coul.}}^{00} \overset{\text{equiv.}}{\simeq} \frac{1}{k^2}\left\{-g^{00} - \frac{k^2 g^{00} g^{00}}{\mathbf{k}^2}\right\} = \frac{-1}{k^2}\frac{k_0^2}{\mathbf{k}^2},
$$
(6.2.12)

and this vanishes in the static limit because

$$
k_0 = p_0 - p_0' = \sqrt{m^2 + \mathbf{p}^2} - \sqrt{m^2 + \mathbf{p}'^2} \simeq \tfrac{1}{2}\frac{\mathbf{p}^2 - \mathbf{p}'^2}{m},
$$

which indeed goes to zero as $m \to \infty$.

In view of this, we find that the piece of the interaction (6.2.8) that involves A^μ will give a vanishing contribution in the NR limit, and the whole of the S-matrix for $e^+ e^-$ scattering will be obtained with the static interaction (6.2.9): in diagrammatic language, by the sum of a ladder like that in Fig. 6.2.2 with the rungs replaced by the instantaneous interaction (6.2.9). This may be summed easily, in the NR limit; not surprisingly, we obtain the well-known nonrelativistic Coulombic amplitude, which is the same as the result that we could have found by solving the Schrödinger equation with the Coulomb potential.

An important remark is that we need *not* go through a Coulomb gauge, for actual calculations. In fact, in a covariant gauge (say, the Fermi–Feynman one) the propagator is

$$
D_{\text{F}}^{\mu\nu}(k) = \mathrm{i}\frac{-g^{\mu\nu}}{k^2}.
$$
(6.2.13)

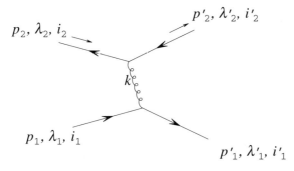

Fig. 6.2.3. Tree level (Born) scattering of $\bar{q}q$.

Apart from "gauge terms" (the terms proportional to k^μ, k^ν that may be neglected due to gauge invariance) the difference between this and (6.2.11) is

$$D_{\mathrm{F}}^{\mu\nu}(k) - D_{\mathrm{Coul.}}^{\mu\nu}(k) = +\frac{g^{0\mu}g^{0\nu}}{\mathbf{k}^2} + \text{gauge terms.}$$

The inverse Fourier transform of this is precisely the Coulomb potential and, as was to be expected, reproduces the effect of the term (6.2.9) in the Coulomb gauge. We can therefore calculate in a covariant gauge, which simplifies the job enormously, especially for QCD.

iii The Static Potential; Radiative Corrections

The calculation of bound states in QCD requires extra refinements for the following reason. Due to the non-Abelian character of the interaction, the expressions equivalent to both (6.2.9) and (6.2.10) involve higher order interactions. The separation of the interaction into an instantaneous interaction equivalent to a simple Coulombic potential and an interaction that involves a propagator that vanishes in the static limit is therefore less straightforward: the potential must include, even in the static limit, radiative corrections. Nevertheless, the equivalence of sums of ladders and solving the Schrödinger equation suggests the strategy to follow, which for the moment we only describe for the static limit, $m \to \infty$. Consider the scattering amplitude for slowly moving quarks, say a $\bar{q}q$ pair, that we expand in powers of α_s:

$$F = \alpha_s F^{(0)} + \alpha_s^2 F^{(1)} + \alpha_s^3 F^{(2)} + \cdots ; \tag{6.2.14}$$

the superscript in $F^{(n)}$ indicates the number of loops. So, $\alpha_s F^{(0)}$ is given by the tree diagram of Fig. 6.2.3, and $\alpha_s^2 F^{(1)}$ involves the one loop radiative corrections, some of which are shown in Fig. 6.2.4.

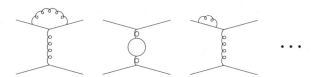

Fig. 6.2.4. Some diagrams for one loop corrections to $\bar{q}q$ scattering.

In the strict nonrelativistic limit, we then invert the Fourier transform and find the static potential,

$$V_{\text{stat}}(r) = \alpha_s U^{(0)}(r) + \alpha_s^2 U^{(1)}(r) + \alpha_s^3 U^{(2)}(r) + \cdots. \qquad (6.2.15)$$

One then solves the Schrödinger equation to increasing orders with the potential

$$V_{\text{stat}}^N \equiv \alpha_s \sum_0^N \alpha_s^n U^{(n)}.$$

At each order, and to avoid double counting, one has to subtract the iteration of the lowest orders. This is equivalent to a rearrangement of the (infinite) perturbative series in which one evaluates a kernel for $\bar{q}q$ scattering, given by (6.2.14) to a fixed order,

$$F^N = \alpha_s \sum_0^N \alpha_s^n F^{(n)},$$

and then adds to it infinite ladders at both ends of the external legs, as in Fig. 6.2.5. Note that this is valid in the nonrelativistic limit; otherwise, crossed diagrams have to be included in the ladder to ensure gauge invariance. Fortunately, we do not have this problem, as only lowest order relativistic corrections, *to tree level*, will be considered.

A totally equivalent method for calculating bound states based on the Bethe–Salpeter equation may be found in Itzykson and Zuber (1980).

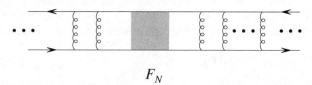

$$F_N$$

Fig. 6.2.5. Infinite ladders at both sides of the kernel F in $\bar{q}q$ scattering.

6.3 Relativistic, Radiative and Nonperturbative Corrections to Heavy Quarkonium.
Evaluation of Lowest Lying $\bar{c}c$ and $\bar{b}b$ States

i Coulomb Potential and Relativistic Corrections

Let us consider for definiteness a state $\bar{q}_i q_j$, with i, j colour indices. The evaluations we are going to make can be extended to states with different flavours, to quark–quark states or to states of three quarks without excessive effort; we refer to the pertinent literature for the details.[5] At tree level the scattering amplitude, with the conventions of Fig. 6.2.3, is

$$F\Big(q(p_1, \lambda_1, i_1) + \bar{q}(p_2, \lambda_2, i_2) \to q(p_1', \lambda_1', i_1') + \bar{q}(p_2', \lambda_2', i_2')\Big)$$

$$= \sum_a t^a_{i_2 i_2'} t^a_{i_1' i_1} \frac{g^2}{4\pi^2} \bar{u}(p_1', \lambda_1') \gamma^\mu u(p_1, \lambda_1) \frac{g_{\mu\nu}}{k^2} \bar{v}(p_2, \lambda_2) \gamma^\nu v(p_2', \lambda_2')$$

$$= \sum_a t^a_{i_2 i_2'} t^a_{i_1' i_1} \frac{-g^2}{4\pi^2} \bar{u}(p_1', \lambda_1') \gamma^\mu u(p_1, \lambda_1) \frac{g_{\mu\nu}}{k^2} \bar{u}(p_2', \lambda_2') \gamma^\nu u(p_2, \lambda_2).$$

$$(6.3.1)$$

The nonrelativistic amplitude is connected to F by

$$T_{NR} = \frac{1}{4\sqrt{p_{10} p_{10}' p_{20} p_{20}'}} F, \qquad (6.3.2a)$$

and, including spin now, T is related to the potential by

$$T_{NR}^{\text{Born}} = -\frac{1}{4\pi^2} \int d^3 \mathbf{r}\, e^{i\mathbf{k}\cdot\mathbf{r}} \chi^\dagger(\lambda_1') \chi^\dagger(\lambda_2') V(\mathbf{r}) \chi(\lambda_1) \chi(\lambda_2), \qquad (6.3.2b)$$

with the χ two-component Pauli spinors, and the potential $V(\mathbf{r})$ is a matrix in spin space. In the strict static limit, as we will see, this matrix is diagonal and V only depends on $r = |\mathbf{r}|$, but already the lowest relativistic corrections introduce a dependence on spin and angular momentum, and on the angular components of $\mathbf{r}$.

We will take the nonrelativistic limit, including the lowest order relativistic corrections. For this, we write

$$p_0 = \sqrt{\mathbf{p}^2 + m^2} \simeq m + \frac{\mathbf{p}^2}{2m} + \frac{\mathbf{p}^4}{8m^3},$$

$$k^2 = (p_{10} - p_{20})^2 - \mathbf{k}^2 \simeq -\mathbf{k}^2 + \frac{\mathbf{p}^2 - \mathbf{p}'^2}{4m^2} \qquad (6.3.3a)$$

and

$$\frac{1}{\sqrt{2p_0}} u(p, \lambda) \simeq \begin{pmatrix} (1 - \mathbf{p}^2/4m^2)\chi(\lambda) \\ (1/2m)\mathbf{p}\boldsymbol{\sigma}\chi(\lambda) \end{pmatrix}. \qquad (6.3.3b)$$

---

[5] See e.g., Gupta and Radford (1981); Brambilla, Consoli and Prosperi (1994).

In the static limit, we neglect all but the first terms in (6.3.3) and find

$$T_{NR}^{\text{Born}} \underset{\text{stat.}}{\simeq} \sum_a t^a_{i_2 i'_2} t^a_{i'_1 i_1} \chi^\dagger(\lambda'_1)\chi^\dagger(\lambda'_2) \frac{-g^2}{4\pi^2\mathbf{k}^2} \chi(\lambda_1)\chi(\lambda_2)$$

$$= \sum_a t^a_{i_2 i'_2} t^a_{i'_1 i_1} \delta_{\lambda_1 \lambda'_1}\delta_{\lambda_2 \lambda'_2} \frac{-g^2}{4\pi^2\mathbf{k}^2},$$

so, as announced, the transition amplitude and thus the potential are spin-independent.

For the colour algebra it is convenient to decompose the colour representation of the $\bar{q}q$ state, $\bar{3} \otimes 3$, into a singlet S and an octet, 8 state. After a simple calculation using formulas from Appendix C, we find

$$\sum_a t^a_{i_2 i'_2} t^a_{i'_1 i_1} \rightarrow \begin{cases} -C_F, & \text{singlet, } S, \\ \dfrac{1}{2N_c} = \dfrac{1}{6}, & \text{octet, 8.} \end{cases}$$

Inverting the Fourier transform we therefore find the Coulombic potentials in the static limit:

$$V_{\text{stat}}^{(0)}(r) = -\frac{C_F \alpha_s}{r} \quad \text{(singlet)} \tag{6.3.4a}$$

and

$$V_{\text{stat}}^{8(0)}(r) = \frac{\alpha_s}{2N_c r} \quad \text{(octet).} \tag{6.3.4b}$$

We will find uses for the octet potential later on; for the moment, we remark that a physical bound state of $\bar{q}q$ is necessarily singlet, so (6.3.4a) applies.

Let us now pause to solve what will be our zero order approximation, viz., the Schrödinger equation with the potential (6.3.4a)

$$H^{(0)}\Psi_{nlM}^{(0)}(\mathbf{r}) \equiv \left(2m + \frac{-1}{m}\Delta - \frac{C_F \alpha_s}{r} \right) \Psi_{nlM}^{(0)}(\mathbf{r}) = E_n^{(0)}\Psi_{nlM}^{(0)}(\mathbf{r}). \tag{6.3.5}$$

This is a standard hydrogen-like problem and we have, also with standard notation,

$$E_n^{(0)} = 2m - \frac{C_F^2 \alpha_s^2}{4n^2}m; \tag{6.3.6}$$

$$\Psi_{nlM}^{(0)}(\mathbf{r}) = Y_M^l(\theta, \phi) R_{nl}^{(0)}(r),$$

$$R_{nl}^{(0)}(r) = \frac{2}{n^2 a^{3/2}} \sqrt{\frac{(n-l-1)!}{(n+l)!}} \left(\frac{2r}{na} \right)^l e^{-r/na} L_{n-l-1}^{(2l+1)}\left(\frac{2r}{na} \right), \tag{6.3.7}$$

where we define, quite generally, the analogue of the Bohr radius,

$$a \equiv \frac{2}{mC_F\alpha_s}.$$

Relativistic corrections of first order (the second order corrections are not known) are evaluated keeping an extra term in (6.3.3). They are identical to those for positronium[6] and we only give the result for the singlet states. We write

$$V^{(0)}(r) \simeq V_{\text{stat}}^{(0)}(r) + V_{1,\text{rel}}^{(0)}, \tag{6.3.8}$$

where the superscript zero indicates that the potential is still obtained from the tree level (zero loop) amplitude. The relativistic corrections, which are to be treated as first order corrections to the unperturbed equation (6.3.5), are

$$V_{1,\text{rel}}^{(0)} = V_{\text{si}}^{(0)} + V_{\text{tens}}^{(0)} + V_{LS}^{(0)} + V_{\text{hf}}^{(0)}, \tag{6.3.9a}$$

where the various pieces, spin-independent (in which we also include the correction to the kinetic energy), tensor, LS and hyperfine, are

$$
\begin{aligned}
V_{\text{si}}^{(0)} &= -\frac{1}{4m^3}\Delta^2 + \frac{C_F\alpha_s}{m^2}\frac{1}{r}\Delta, \\
V_{\text{tens}}^{(0)} &= \frac{C_F\alpha_s}{4m^2}\frac{1}{r^3}S_{12}, \\
V_{LS}^{(0)} &= \frac{3C_F\alpha_s}{2m^2}\frac{1}{r^3}\mathbf{LS}, \\
V_{\text{hf}}^{(0)} &= \frac{4\pi C_F\alpha_s}{3m^2}\mathbf{S}^2\delta(\mathbf{r}).
\end{aligned}
\tag{6.3.9b}
$$

Here $\mathbf{L}$ is the orbital angular momentum operator, $\mathbf{S}$ the total spin operator, and S_{12} the tensor operator:

$$\mathbf{L} = -i\mathbf{r}\times\nabla, \quad \mathbf{S} = \frac{\boldsymbol{\sigma}_1+\boldsymbol{\sigma}_2}{2}, \quad S_{12} = 2\sum_{ij}\left(\frac{2r_ir_j}{r^2} - \delta_{ij}\right)S_iS_j.$$

The Pauli matrices $\boldsymbol{\sigma}_a$ act on spinor $\chi(\lambda_a)$, $a = 1, 2$. The operator $r^{-1}\Delta$ in (6.3.9b) is not well defined, as r^{-1} and Δ do not commute. We refer to the quoted literature for details; in the present text we will only consider diagonal matrix elements of $r^{-1}\Delta$ between states $\Psi_{nlM}^{(0)}$ for which the problem does not matter. In fact, one can write

$$\Delta = -m\left\{H^{(0)} - 2m\right\}$$

and, acting on the right or left on the $\Psi_{nlM}^{(0)}$, one can thus replace

$$\Delta = -m\left\{H^{(0)} - 2m\right\} \to -m\left\{E_n^{(0)} - 2m\right\} = \text{constant},$$

so $r^{-1}\Delta$ and Δr^{-1} will produce the same result.

Another peculiarity of (6.3.9b), for whose detailed explanation we again refer to the quoted literature, is that one has to take the expectation values of the terms containing $\mathbf{LS}$ and S_{12} to be zero between states with angular

---

[6] See standard textbooks on relativistic quantum mechanics: Akhiezer and Berestetskii (1963); Berestetskii, Lifshitz and Pitaevskii (1971); Ynduráin (1996).

momentum equal to zero, because their angular average vanishes, and this in spite of the singularity of the factor $1/r^3$ at the origin.

It is important to realize the size of the relativistic corrections. They are proportional to the squared velocities of the quarks, $\mathbf{k}^2/m^2$. In the Coulombic approximation, the average value of this quantity is

$$\langle \mathbf{k}^2/m^2 \rangle_{nl} = \left(\frac{C_F \alpha_s}{2n} \right)^2. \tag{6.3.10}$$

Therefore, relativistic corrections are of relative order α_s^2. In fact, for e.g. the spin-independent energy levels, one gets the shifts

$$E_n^{(0)} \rightarrow E_n^{(0)} + \delta_{\mathrm{rel}} E_{nl}, \tag{6.3.11}$$

where

$$\begin{aligned} \delta_{\mathrm{rel}} E_{nl} &= \langle V_{\mathrm{si}}^{(0)} \rangle_{nl} \\ &= \frac{C_F \alpha_s}{m^2 a^3} \frac{2l+1-4n}{(2l+1)n^4} - \frac{2}{m^3 a^4} \left[\frac{1}{(2l+1)n^4} - \frac{3}{8n^5} \right]. \end{aligned} \tag{6.3.12}$$

Besides this, relativistic corrections induce LS, tensor and hyperfine splittings, obtained taking expectation values between unperturbed states. For example, the hyperfine splittings are obtained evaluating $\langle V_{\mathrm{hf}}^{(0)} \rangle_{n0}$. In the particularly important case $n = 1$, we get the mass difference between states Υ, η_b (or J/ψ, η_c):

$$M_\Upsilon - M_{\eta_b} = \delta_{\mathrm{hf}} E_{10} = \langle V_{\mathrm{hf}}^{(0)} \rangle_{10} = \frac{8 C_F \alpha_s}{3 m^2 a^3}. \tag{6.3.13}$$

ii One and Two Loop Radiative Corrections to the Coulombic Potential. Mixed Radiative–Relativistic Corrections

Before embarking on the discussion of the radiative corrections, we have to consider the matter of the renormalization scheme. Because we want to use the Schrödinger equation, and we are for the moment neglecting confinement, we will renormalize the mass in the on-shell scheme; that is to say, throughout the following equations m will represent the pole mass. However, we renormalize the coupling constant, the wave function and so on, in the $\overline{\mathrm{MS}}$ scheme. We can recover the $\overline{\mathrm{MS}}$ mass using the formulas of Sect. 6.2.

We consider first the spin-independent corrections, that is to say, corrections for states where the spins of the quarks compose to total spin $s = 0$. To one loop we have the radiative corrections to the static potential that can be written, in p-space

$$V(\mathbf{r}) = (2\pi)^{-3} \int \mathrm{d}^3 \mathbf{k}\, e^{i\mathbf{k}\mathbf{r}} \hat{V}(\mathbf{k}),$$

and to one loop as (Fischler, 1977; Billoire, 1980)

$$\hat{V}_{\text{si}}^{(1)}(\mathbf{k}) = -\frac{4C_F \alpha_s^2(\mu^2)}{\mathbf{k}^2}\left[-\frac{\beta_0}{2}\log\frac{|\mathbf{k}|}{\mu} + a_1\right],$$

$$a_1 = \frac{31C_A - 20T_F n_f}{36}.$$

(6.3.14a)

Here and in what follows, n_f will denote the number of flavours *lighter* than m. μ is the renormalization point. If we now add the two loop correction we find

$$\hat{V}_{\text{si}}^{(1)}(\mathbf{k}) + \hat{V}_{\text{si}}^{(2)}(\mathbf{k}) = \frac{4C_F \alpha_s^2(\mu^2)}{\mathbf{k}^2}\left\{\left(a_1 + b_1\frac{\alpha_s}{\pi}\right)\right.$$

$$\left. -\left[\frac{\beta_0}{2} + \left(a_1\beta_0 + \frac{\beta_1}{8}\right)\frac{\alpha_s}{\pi}\right]\log\frac{k}{\mu} + \beta_0^2\frac{\alpha_s}{4\pi}\log^2\frac{k}{\mu}\right\},$$

$$b_1 = \frac{1}{16}\left\{\left[\frac{4343}{162} + 4\pi^2 - \frac{1}{4}\pi^4 + \frac{22}{3}\zeta(3)\right]C_A^2\right.$$

$$\left. -\left[\frac{1798}{81} + \frac{56}{3}\zeta(3)\right]C_A T_F n_f - \left[\frac{55}{3} - 16\zeta(3)\right]C_F T_F n_f + \frac{400}{81}T_F^2 n_f^2\right\}$$

$$\simeq 13.20.$$

(6.3.14b)

The coefficient b_1 has been evaluated by Peter (1997). This calculation has been recently repeated by Schröder (1998), who has checked the value of all the pieces in Peter's expression except one where, instead of the term $\frac{1}{16}6\pi^2 C_A^2$, he finds $\frac{1}{16}4\pi^2 C_A^2$, as shown in (6.3.14b). Although this result has not yet been verified by an independent calculation we will give numerical results with Schröder's value for b_1.

For the same order of accuracy as two loop static corrections, we have to consider one loop, semi-relativistic $O(|\mathbf{k}|)$ corrections. They are (Titard and Ynduráin, 1994)

$$\hat{V}_{\text{s.rel}}(\mathbf{k}) = \frac{a_2\pi^2 C_F \alpha_s}{m|\mathbf{k}|},$$

$$a_2 = \frac{C_F - 2C_A}{2}.$$

(6.3.14c)

We have to add (6.3.14), transform to x-space and include the spin independent part of the relativistic corrections given in (6.3.9)[7]. When adding the Fourier transforms of (6.3.14) it is convenient to separate the pieces proportional to $1/r$: these can be combined with the Coulombic potential (6.3.4a) to form a Hamiltonian that can be solved exactly, and which can be taken as the basis for perturbation theory. We thus write

$$H = \tilde{H}^{(0)} + H_1,$$

(6.3.15)

---

[7] The Fourier transforms of the various terms may be easily evaluated with the help of the table in the appendix of Titard and Ynduráin (1994).

where $\widetilde{H}^{(0)}$ is given by

$$\widetilde{H}^{(0)} = 2m + \frac{-1}{m}\Delta - \frac{C_F \widetilde{\alpha}_s(\mu^2)}{r}, \qquad (6.3.16a)$$

$$\begin{aligned}
\widetilde{\alpha}_s(\mu^2) = \alpha_s(\mu^2) &\left\{ 1 + \left(a_1 + \frac{\gamma_E \beta_0}{2} \right) \frac{\alpha_s(\mu^2)}{\pi} \right. \\
&\left. + \left[\gamma_E \left(a_1 \beta_0 + \frac{\beta_1}{8} \right) + \left(\frac{\pi^2}{12} + \gamma_E^2 \right) \frac{\beta_0^2}{4} + b_1 \right] \frac{\alpha_s^2}{\pi^2} \right\}
\end{aligned} \qquad (6.3.16b)$$

and will be solved exactly. H_1 may be split as

$$H_1 = V_{\text{tree}} + V_1^{(L)} + V_2^{(L)} + V^{(LL)} + V_{\text{s.rel}} + V_{\text{hf}}^{(0)} \qquad (6.3.17a)$$

and

$$\begin{aligned}
V_{\text{tree}} &= \frac{-1}{4m^3}\Delta^2 + \frac{C_F \alpha_s}{m^2 r}\Delta, \\
V_1^{(L)} &= \frac{-C_F \beta_0 \alpha_s(\mu^2)^2}{2\pi} \frac{\log r\mu}{r}, \\
V_2^{(L)} &= \frac{-C_F \alpha_s^3}{\pi^2} \left(a_1 \beta_0 + \frac{\beta_1}{8} + \frac{\gamma_E \beta_0^2}{2} \right) \frac{\log r\mu}{r}, \\
V^{(LL)} &= \frac{-C_F \beta_0^2 \alpha_s^3}{4\pi^2} \frac{\log^2 r\mu}{r}, \\
V_{\text{s.rel}} &= \frac{C_F a_2 \alpha_s^2}{2mr^2}, \\
V_{\text{hf}}^{(0)} &= \frac{4\pi C_F \alpha_s}{3m^2} s(s+1)\delta(\mathbf{r}).
\end{aligned} \qquad (6.3.17b)$$

Here the running coupling constant has to be taken to three loops. Note that, to the present level of accuracy, we have to include the (lowest order) hyperfine interaction that produces energy splittings of order α_s^4; s is the combined spin of the $\bar{q}q$ pair, $s = 0, 1$. Because we solve $\widetilde{H}^{(0)}$ exactly, we redefine the analogue of the Bohr radius as

$$a = \frac{2}{mC_F \widetilde{\alpha}_s(\mu^2)}.$$

The solution of the unperturbed Schrödinger equation,

$$\widetilde{H}^{(0)}\Psi_{nlM} = E_n^{(0)}\Psi_{nlM}, \qquad (6.3.18)$$

is identical to (6.3.6) and (6.3.7) replacing $\alpha_s \to \widetilde{\alpha}_s$. The remaining terms in (6.3.17) are to be treated as *first order* perturbations to this, *except for* $V_1^{(L)}$ for which second order corrections should be included. We have

$$E_{nl}^P = 2m - m\frac{C_F^2 \widetilde{\alpha}_s^2}{4n^2} + \sum_V \delta_V^{(1)} E_{nl} + \delta_{V_1^{(L)}}^{(2)} E_{nl} + \delta_{\text{NP}} E_{nl}, \qquad (6.3.19)$$

where the superscript "P" in E_{nl}^P is a reminder that we have only taken into account effects generated by perturbation theory, and where

$$\delta_{V_{\text{tree}}^{(1)}} E_{nl} = -\frac{2}{n^4 m^3 a^4}\left[\frac{1}{2l+1} - \frac{3}{8n}\right] + \frac{C_F \alpha_s}{m^2}\frac{2l+1-4n}{n^4(2l+1)a^3}, \quad (6.3.20a)$$

$$\delta_{V_1^{(L)}}^{(1)} E_{nl} = -\frac{\beta_0 C_F \alpha_s^2(\mu^2)}{2\pi n^2 a}\left[\log\frac{na\mu}{2} + \psi(n+l+1)\right], \quad (6.3.20b)$$

$$\delta_{V_2^{(L)}}^{(1)} E_{nl} = -\frac{C_F c_2^{(L)}\alpha_s^3}{\pi^2 n^2 a}\left[\log\frac{na\mu}{2} + \psi(n+l+1)\right], \quad (6.3.20c)$$

$$\delta_{V^{(LL)}}^{(1)} E_{nl} = -\frac{C_F \beta_0^2 \alpha_s^3}{4\pi^2 n^2 a}\left\{\log^2\frac{na\mu}{2} + 2\psi(n+l+1)\log\frac{na\mu}{2}\right.$$
$$+ \psi(n+l+1)^2 + \psi'(n+l+1) \quad (6.3.20d)$$
$$\left. + \theta(n-l-2)\frac{2\Gamma(n-l)}{\Gamma(n+l+1)}\sum_{j=0}^{n-l-2}\frac{\Gamma(2l+2+j)}{j!(n-l-j-1)^2}\right\},$$

$$\delta_{V_{\text{s.rel}}}^{(1)} E_{nl} = \frac{C_F a_2 \alpha_s^2}{m}\frac{1}{n^3(2l+1)a^2}. \quad (6.3.20e)$$

The calculation of the second order contribution of $V_1^{(L)}$, denoted by $\delta_{V_1^{(L)}}^{(2)} E_{nl}$, is nontrivial. We define

$$\delta_{V_1^{(L)}}^{(2)} E_{nl} \equiv -m\frac{C_F^2 \beta_0^2 \alpha_s^4}{4n^2\pi^2}\left\{N_0^{(n,l)} + N_1^{(n,l)}\log\frac{na\mu}{2} + \tfrac{1}{4}\log^2\frac{na\mu}{2}\right\} \quad (6.3.20f)$$

and one has, for the lowest states,

$$N_1^{(1,0)} = -\frac{\gamma_E}{2} \simeq -0.288608,$$

$$N_1^{(2,0)} = \frac{1-2\gamma_E}{4} \simeq -0.0386078,$$

$$N_1^{(2,1)} = \frac{5-6\gamma_E}{12} \simeq 0.128059,$$

$$N_0^{(1,0)} = \frac{3+3\gamma_E^2 - \pi^2 + 6\zeta(3)}{12} \simeq 0.111856,$$

$$N_0^{(2,0)} = -\frac{5}{16} - \frac{\gamma_E}{4} + \frac{\gamma_E^2}{4} - \frac{\pi^2}{12} + \zeta(3) \simeq 0.00608043,$$

$$N_0^{(2,1)} = -\frac{865}{432} - \frac{5\gamma_E}{12} + \frac{\gamma_E^2}{4} - \frac{11\pi^2}{36} + \zeta(3) \simeq 0.0314472.$$

The values of the coefficients $N_i^{(n,l)}$ for $i = 0$, 1, and arbitrary n, l, may be found, together with the details of the calculation, in Pineda and Ynduráin (1998).

These formulas give the masses of the pseudoscalar states. To obtain the masses of the vector states (Υ, Υ', Υ''; J/ψ, ψ', ...) one has to add the hyperfine shift, at tree level,

$$\delta^{(1)}_{V_{\text{spin}}} E_{nl} = \delta_{s1}\delta_{l0}\frac{8C_F\alpha_s}{3n^3m^2a^3}. \tag{6.3.20g}$$

Substituting the expression for a we see that, indeed, these formulas produce the energy spectrum correct up to and including order α_s^4 effects.

For the spin-dependent potential and energy shifts, the radiative corrections have been evaluated by Buchmüller, Ng and Tye (1981) and, especially, by Gupta and Radford (1981); see also Titard and Ynduráin (1994), where the formulas are checked and collected in detail. For e.g. the hyperfine splitting, we have the potential

$$V^{(0)}_{\text{hf}} + V^{(1)}_{\text{hf}} = \frac{4\pi C_F\alpha_s^2(\mu^2)}{3m^2}\mathbf{S}^2$$

$$\times\left\{\delta(\mathbf{r}) + \left[\frac{\beta_0}{2}\left(\frac{1}{4\pi}\text{reg}\frac{1}{r^3} + (\log\mu)\delta(\mathbf{r})\right)\right.\right. \tag{6.3.21}$$

$$\left.\left. + b_{\text{hf}}\delta(\mathbf{r}) - \frac{21}{4}\left(\frac{1}{4\pi}\text{reg}\frac{1}{r^3} + (\log m)\delta(\mathbf{r})\right)\right]\frac{\alpha_s}{\pi}\right\},$$

$$b_{\text{hf}} = \frac{3}{2}(1 - 2\log 2)T_F - \frac{5}{9}T_F n_f = \frac{11C_A - 9C_F}{18}.$$

The function reg r^{-3} is defined by its integral with any φ. In n dimensions,

$$\int \mathrm{d}^n\mathbf{r}\,\varphi(\mathbf{r})\text{reg}\frac{1}{r^n} \equiv \lim_{\epsilon\to 0}\left\{\int \mathrm{d}^n\mathbf{r}\,\varphi(\mathbf{r})\frac{r^\epsilon}{r^n} - A(n,\epsilon)\varphi(0)\right\},$$

$$A(n,\epsilon) = \frac{2\pi^{n/2}}{\Gamma(n/2)}\left\{\frac{1}{\epsilon} + \log 2 + \frac{\psi(n/2) - \gamma_{\text{E}}}{2}\right\}.$$

iii Nonperturbative Corrections

In all the preceding evaluations we have neglected nonperturbative corrections (which, we believe, are responsible, among other effects, for the confinement of quarks). We will now take them into account, to lowest order.
These nonperturbative corrections, to be denoted by NP from now on, are
relativistic in that they are proportional to inverse powers of m. Nevertheless,
the coefficients are large and they turn out to be important even for lowest
lying $\bar{b}b$ states; indeed, it is their size that limits the validity of the present
approach to quarkonium bound states.

We may understand the NP effects as being due to the complicated structure of the QCD vacuum, full of soft gluons and light quark–antiquark pairs.
It is not difficult to become convinced, following the methods to be described,
that for *heavy* quark bound states the contribution of quark condensates is
subleading, leading effects being provided by the gluon condensates.

A first approach to the effects of gluon condensates may be obtained by
considering the NP gluon propagator corrections to the scattering of a $\bar{q}q$ pair;
this is shown diagrammatically in Fig. 6.3.1. Among those diagrams there,
and because to leading order in relativistic corrections we have to take $k \to 0$
(k being the momentum of the exchanged gluon) we will obtain contributions
that diverge every time a nonperturbative gluon is attached to a quark. In
fact, the quark propagator becomes

$$\mathrm{i}\frac{\not{p} + m}{p^2 - m^2}.$$

The denominator here is different from zero only because, being bound, the
quarks are off-shell, so that this denominator is proportional to the binding
energy:

$$p^2 - m^2 \sim m^2\alpha_s^2.$$

Therefore, the dominating diagrams will be those in which both nonperturbative gluons are attached to the quarks.[8]

Under these circumstances, we may view the NP interactions as being
due to interactions of the quarks with a gluonic medium (the NP vacuum).
We will evaluate this following Leutwyler (1981) and Voloshin (1979, 1982).

For heavy quarks the region where they move, of order $\langle \mathbf{k}^2 \rangle^{1/2} \sim a \sim 2/mC_F\tilde{\alpha}_s$, is much smaller than $1/\Lambda$. Therefore, we can assume that fluctuations of the field strengths $G_a^{\mu\nu}$, proportional to $\partial G_a^{\mu\nu}$, can be neglected,
as they will produce extra powers of m in the denominator (we will consider
size corrections in Sect. 6.4). So we may imagine that the quarks move in
a medium filled with nonzero, constant colour fields. Because of rotational
invariance we can take these fields to be random, i.e., we average over their

---

[8] We will show this later in an explicit calculation of the diagram where both nonperturbative gluons are attached to the exchanged one. A more rigorous discussion
will be presented in Sect. 6.4, following the methods of Dosch and Simonov.

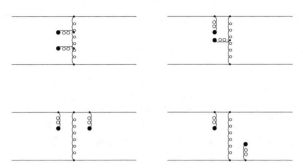

Fig. 6.3.1. Diagrams for gluon condensate nonperturbative contributions to $\bar{q}q$ scattering.

orientations. It is convenient to split $G_a^{\mu\nu}$ into a chromomagnetic piece, $\boldsymbol{\mathcal{B}}_a$, and a chromoelectric one, $\mathcal{E}_a^i = G_a^{0i}$. Since the interaction of $\boldsymbol{\mathcal{B}}_a$ is proportional to the velocity of the quarks, we can neglect it for the spin-independent spectrum in the NR limit. Relativistic corrections, however, are important for the spin-dependent splittings, which will involve $\boldsymbol{\mathcal{B}}_a$; we will take them into account later.

The interaction Hamiltonian can now be written at once by analogy with the usual electromagnetic case: we have

$$H_\mathcal{E} = -g\mathbf{r}\sum_a t^a \boldsymbol{\mathcal{E}}_a. \qquad (6.3.22)$$

Since we consider the fields $\boldsymbol{\mathcal{E}}_a$ to be oriented at random, only the second order perturbation in $H_\mathcal{E}$ will be different from zero: apart from colour complications, the problem is like the familiar one of the second order Stark effect. To solve it we take the second order expectation value in the functions we obtained solving the NR Coulombic Schrödinger equation (6.3.5), or, if we want to be more precise, (6.3.18). We then find

$$\delta_{\mathrm{NP}} E_{nl} = -\left\langle \Psi_{nlM}, H_\mathcal{E} \frac{1}{H^{(8)} - E_n^{(0)}} H_\mathcal{E}\Psi_{nlM} \right\rangle. \qquad (6.3.23)$$

There are a few points that we have to clarify regarding this equation. First of all, we may, since we average over directions, neglect the magnetic quantum number, M. Secondly, we have used in the denominator of (6.3.23) the *octet* Hamiltonian, $H^{(8)}$ which, from (6.3.4), is

$$H^{(8)} = -\frac{1}{m}\Delta + \frac{1}{2N_c}\frac{\alpha_s}{r}.$$

This happens because the perturbed state,

$$H_{\mathcal{E}}|\Psi\rangle = -g\mathbf{r}\sum_a t^a \mathcal{E}_a|\Psi\rangle,$$

is manifestly an octet one, as $\mathcal{E}_a$ creates a gluon on top of the singlet $|\Psi\rangle$.

Next, we write the gluon radiation Hamiltonian as

$$H_{\text{rad}} = \frac{1}{8\pi} \int d^3\mathbf{r} : \mathcal{E}^2 + \mathcal{B}^2 :,$$

with sums over omitted colour indices understood. Its expectation value in the physical vacuum should vanish, so we conclude

$$\langle \text{vac}| \int d^3\mathbf{r} : \mathcal{E}^2 : |\text{vac}\rangle = -\langle \text{vac}| \int d^3\mathbf{r} : \mathcal{B}^2 : |\text{vac}\rangle.$$

Because we assume the field intensities to be constant, we may replace the integrals by the volume times the integrands at $x = 0$. Canceling then the volume and recalling that $G^2 = -2(\mathcal{E}^2 - \mathcal{B}^2)$, we find

$$\langle \text{vac}| : \mathcal{E}^2 : |\text{vac}\rangle = -\tfrac{1}{4}\langle \text{vac}| : G^2(0) : |\text{vac}\rangle.$$

Finally, using Lorentz and colour invariance of the physical vacuum,

$$g^2 \langle : \mathcal{E}_a^i(0)\mathcal{E}_b^j(0) : \rangle = \frac{4\pi\alpha_s \delta_{ij}\delta_{ab}}{(D-1)(N_c^2-1)}\langle : \mathcal{EE} : \rangle = -\frac{\pi\delta_{ij}\delta_{ab}}{24}\langle \alpha_s G^2 \rangle.$$

With this, we return to (6.2.23) and get

$$\delta_{\text{NP}}E_{nl} = \left\langle \Psi_{nl}, \left(\mathbf{r}\sum_a t^a \mathcal{E}_a(0)\right) \frac{1}{H^{(8)} - E_n^{(0)}} H_{\mathcal{E}} \left(\mathbf{r}\sum_b t^b \mathcal{E}_b(0)\right) \Psi_{nl} \right\rangle$$

$$= \frac{\pi\langle \alpha_s G^2 \rangle}{18} \sum_i \left\langle \Psi_{nl}, r_i \left(-\frac{1}{m}\Delta + \frac{\alpha_s}{6r} - E_n^{(0)}\right)^{-1} r_i\Psi_{nl} \right\rangle.$$

$$(6.3.24)$$

To finish the calculation, we need to invert the operator $\left(-m^{-1}\Delta + \alpha_s/6r\right)$. For the simple case above, the method may be found in Schiff (1968); in more complicated situations, see Titard and Ynduráin (1994). We may write the final result as (Voloshin, 1979, 1982; Leutwyler, 1981)

$$\delta_{\text{NP}}E_{nl} = m \frac{\pi\epsilon_{nl}n^6 \langle \alpha_s G^2 \rangle}{(mC_F\tilde{\alpha}_s)^4}. \qquad (6.3.25a)$$

The (complicated) explicit expression for the numbers ϵ_{nl}, of order unity, may be found in Leutwyler (1981); for the lowest states,

$$\epsilon_{10} = \tfrac{624}{425}, \quad \epsilon_{20} = \tfrac{1051}{663}, \quad \epsilon_{21} = \tfrac{9929}{9945}. \qquad (6.3.25b)$$

As promised, the correction is relativistic in that it is of order $1/m^4$; but the coefficient is very large. The powers of α_s and n can be understood easily. Two come from the energy denominators, and four from the expectation value $\langle r_i r_j \rangle_{nl} \sim n^4/(mC_F\alpha_s)^2$. The extra powers of α_s^{-1} are due to the energy denominator in the quark propagators (cf. Fig. 6.3.1). In this, they are similar to the Bethe logarithm in the ordinary Lamb shift. The right hand side of (6.3.25) thus grows as the sixth power of the radial quantum number, n. It is in fact this very fast growth with n, a fourth power, that leads to the breakdown of the method as soon as n exceeds, or in some cases even equals, the value 2.

The NP corrections to the wave function may be obtained with the same methods. For $n = 1$, $l = 0$, we have

$$\Psi_{10}(r) \rightarrow (1 + \delta_{\mathrm{NP}}(r))\,\Psi_{10}(r), \qquad\qquad (6.3.26a)$$

where the correction is

$$\delta_{\mathrm{NP}}(r) = \left\{ \tfrac{2968}{425} - \tfrac{104}{425}\rho^2 - \tfrac{52}{1275}\rho^3 - \tfrac{1}{225}\rho^4 \right\} \frac{\pi\langle \alpha_s G^2 \rangle}{m^4(C_F\alpha_s)^6}, \qquad \rho = \frac{2r}{a}.$$
$$(6.3.26b)$$

In obtaining this, one has to take into account a subtle point concerning a change in normalization, because the perturbed state contains a gluon. We refer to Voloshin (1982) for the details. It turns out that the coefficient of the correction is larger than for the energy shifts, both in powers of α_s^{-1} and of n: the effects of confinement are larger for the wave function than for the energy levels.

In the leading relativistic order, new nonperturbative interactions are generated. They add to the dipole interaction (6.3.22) two terms:

$$\frac{g}{2m^2}(\mathbf{S} \times \mathbf{p})\sum_a \boldsymbol{\mathcal{E}}_a t^a, \qquad -\frac{g}{m}(\mathbf{S}_1 - \mathbf{S}_2)\sum_a \boldsymbol{\mathcal{B}}_a t^a, \qquad\qquad (6.3.27)$$

where the $\mathbf{S}_i$ are the spin operators of each quark, and $\mathbf{p} = -i\nabla$ acts on the relative coordinate $\mathbf{r}$. They introduce spin dependent nonperturbative shifts; their effects may be found in Titard and Ynduráin (1994).[9]

The effects of higher order condensates, or the radiative corrections to the contribution of the gluon condensate, are not known. Some of the first, nominally the more important (but renormalization may alter this), are evaluated in Pineda (1997b) to where we send for details. Among the second we have the contribution of the diagrams where one (or two) nonperturbative gluon is (are) attached to the exchanged gluon. For example, the diagram

---

[9] With correction for one case by Pineda (1997a).

in which both nonperturbative gluons are attached to the exchanged gluon modify the propagator of the latter by (in the Fermi–Feynman gauge)

$$\frac{ig^{\mu\nu}}{k^2} \rightarrow \frac{ig^{\mu\nu}}{k^2} + \frac{ig^{\mu\nu}}{k^2}\frac{11\pi\langle\alpha_sG^2\rangle}{6k^4},$$

plus terms proportional to k^{μ}, k^{ν}. The corresponding alteration of the Coulombic potential is, for $\bar{q}q$ systems,

$$-\frac{C_F\alpha_s}{r} \rightarrow -\frac{C_F\alpha_s}{r} - \frac{11C_F\alpha_s\langle\alpha_sG^2\rangle}{12\pi}\int d^3\mathbf{k}\frac{e^{i\mathbf{k}\mathbf{r}}}{|\mathbf{k}|^6}$$

$$= -\frac{C_F\alpha_s}{r} - \frac{11\pi\langle\alpha_sG^2\rangle}{144}C_F\alpha_s r^3.$$

This produces energy shifts

$$\delta_{\text{prop}}E_{nl} = -\frac{11\pi\langle\alpha_sG^2\rangle}{144}C_F\alpha_s\langle r^3\rangle_{nl} \sim \frac{\pi\langle\alpha_sG^2\rangle}{(mC_F\alpha_s)^4}(C_F\alpha_s)^2,$$

as stated of higher order (α_s^2) than the piece $\delta_{\text{NP}}E_{nl}$ of Eq. (6.3.25).

The shape of this piece raises two questions. First, would it also be possible to represent the NP corrections (6.3.25) by a potential? Secondly, what is the connection with the confining potential?

As to the first, the answer is *no*. As shown by Leutwyler (1981), one cannot obtain the shifts $\delta_{\text{NP}}E_{nl}$ from a *local* potential; although one can get them, approximately, with a cubic potential proportional to $\langle\alpha_sG^2\rangle r^3$. With regard to the second question, the NP corrections we have discussed can be described as the *short distance* part of the confining forces, very different from the *long distance* piece which, as we will see, grows as r.

iv QCD Analysis of Lowest Lying $\bar{c}c$ and $\bar{b}b$ States

When comparing the *ab initio* calculations of quarkonium states of the previous subsections with experiment, we have two different situations. For the energy and, to a lesser extent, the wave function of the $\bar{b}b$ states with $n=1$, radiative corrections are small and so are nonperturbative corrections. Thus we have reasonably accurate theoretical calculations, which do not depend excessively on the renormalization point μ provided that it is not too far from its natural value, $\mu \sim 2/a$ (see below). The second type of situation occurs for the energy levels of bottomium with $n=2$, and for the energy of the ground state of $\bar{c}c$. Here radiative corrections and especially NP corrections, while still smaller than the leading term, are not so by a wide margin. The calculations are thus less reliable and depend much more on the renormalization point μ. The *errors* of the calculations are now large; but it is still possible to get *global* agreement with experiment for the $n=2$ system, including tensor and LS splittings, at the price of taking the renormalization point μ as a free parameter, and fitting it.

Let us first consider $\bar{b}b$ states with $n = 1$. Here one could take Λ, $\langle\alpha_s G^2\rangle$ and m_b from other sources and then predict the mass of the Υ resonance; but it is preferable to take M_Υ from experiment and then obtain a determination of m_b: in principle at least, this should be the more precise approach because it will be accurate including $O(\alpha_s^4)$ corrections. Of course, we still have to take Λ and $\langle\alpha_s G^2\rangle$ from other sources. Specifically, we will use the values

$$\Lambda(n_f = 4,\ \text{three loops}) = 0.23^{+0.08}_{-0.05}\ \text{GeV}, \quad \langle\alpha_s G^2\rangle = 0.06 \pm 0.02\ \text{GeV}^2\,.$$

The first corresponds to the world average for processes with spacelike momenta; see Sect. 10.3 for details. We next choose the renormalization point. From our calculations of radiative corrections we see that in the argument of the logarithms the combination $na\mu/2$ appears systematically, as is clear from Eqs. (6.3.20). So we choose $\mu = 2/na$; for $n = 1$ this gives $\mu^2 = 6.6\ \text{GeV}^2$. Specializing (6.3.19) for $n = 1$, $l = 0$, adding (6.3.20g) for the Υ state and also including the NP contribution (6.3.25), we find (Pineda and Ynduráin, 1998)

$$m_b = 5\,001^{+104}_{-66}\ \text{MeV}, \quad \bar{m}_b(\bar{m}_b^2) = 4\,440^{+43}_{-28}\ \text{MeV} \tag{6.3.28}$$

where, to find the error, we have composed quadratically the errors produced by the errors in the input, the error due to a variation of 25% in μ^2 and, as an extra error, the value of estimated $O(\alpha_s^5)$ contributions.

To get an independent idea of the dependence of the result on μ, we remark that even varying μ^2 by a factor of *two* only produces a shift of less than a hundred MeV in m_b: the resulting value of m_b is thus stable against variations of the renormalization point.

The hyperfine splitting is obtained by adding to the tree level value (6.3.13), with $n = 1$, the radiative correction stemming from the correction (6.3.21) to the hyperfine potential and the NP correction given in (6.3.26). Likewise, to obtain the rate for $\Upsilon \to e^+ e^-$ we include radiative (δ_{wf}) and NP (δ_{NP}) corrections to the wave function, as well as the one loop "hard" radiative corrections already considered in Sect. 5.1, δ_{rad}, calculated by Barbieri, Gatto, Kögerler and Kunszt (1975). Thus we obtain the results

$$M_\Upsilon - M_{\eta_b} = m\,\frac{C_F^4 \alpha_s(\mu^2)\tilde{\alpha}_s(\mu^2)^3}{3}\,[1 + \delta_{\text{wf}} + \delta_{\text{NP}}]^2$$

$$\times\left\{1 + \left[\frac{\beta_0}{2}\left(\log\frac{a\mu}{2} - 1\right) + \tfrac{21}{4}\left(\log C_F\tilde{\alpha}_s + 1\right) + b_{\text{hf}}\right]\frac{\alpha_s}{\pi} + \tfrac{1\,161}{8\,704}\frac{\pi\langle\alpha_s G^2\rangle}{m^4\tilde{\alpha}_s^6}\right\}$$
$$\tag{6.3.29}$$

and

$$\Gamma(\Upsilon \to e^+ e^-) = \Gamma^{(0)} \times [1 + \delta_{\text{wf}} + \delta_{\text{NP}}]^2\,(1 + \delta_{\text{rad}}),$$

$$\Gamma^{(0)} = 2\left[\frac{Q_b\alpha_{\text{QED}}}{M_\Upsilon}\right]^2\,\left(mC_F\tilde{\alpha}_s(\mu^2)\right)^3, \tag{6.3.30}$$

where

$$\delta_{\text{rad}} = -\frac{4C_F \alpha_s}{\pi}, \quad \delta_{\text{wf}} = \frac{3\beta_0}{4}\left(\log\frac{a\mu}{2} - \gamma_{\text{E}}\right)\frac{\alpha_s}{\pi},$$

$$\delta_{\text{NP}} = \frac{1}{2}\left[\frac{270\,459}{108\,800} + \frac{1\,838\,781}{2\,890\,000}\right]\frac{\pi\langle\alpha_s G^2\rangle}{m^4\tilde{\alpha}_s^6}.$$

Numerically, we have

$$\Gamma(\Upsilon \to e^+ e^-) = 1.07 \pm 0.28 \text{ keV},$$

in nice agreement with the experimental figure of 1.32 ± 0.04 keV, and the hyperfine splitting is predicted to be

$$M(\Upsilon) - M(\eta) = 47^{+15}_{-10} \text{ MeV}.$$

As for the states with $n = 2$, the energy levels can be evaluated using the values found for m_b and taking now $\mu = 1/a$. However, since (as stated) radiative and nonperturbative corrections are large, the results are very sensitive to the value of μ chosen. For this reason it is more profitable to fit μ. This is the procedure followed in Titard and Ynduráin (1994), from where the following table for the mass splittings of the shown states is taken:

States	Theory	Experiment
$2^3 P_2 - 2^3 P_1$	21 ± 7	21 ± 1 MeV
$2^3 P_1 - 2^3 P_0$	29 ± 9	32 ± 2 MeV
$2^3 S_1 - \overline{2^3 P}$	181 ± 60	123 ± 1 MeV
$2^3 S_1 - 1^3 S_1$	428 ± 105	563 ± 0.4 MeV
$\overline{2^3 P} - 2^1 P_1$	1.5 ± 1	$-$

Here we use standard spectroscopic notation; the common, fitted value of μ is $\mu \sim 1$ GeV, and the errors are those generated by the errors in Λ, $\langle\alpha_s G^2\rangle$ given above.

The *wave functions* for states with $n = 2$ present such large errors that the calculation using the methods described become meaningless for them. For the $\bar{c}c$ state, only the calculation of the mass of the J/ψ particle, or equivalently m_c, is reliable. One finds

$$m_c = 1\,866^{+190}_{-136} \text{ MeV}, \quad \bar{m}_c(\bar{m}_c^2) = 1\,531^{+132}_{-127} \text{ MeV}.$$

6.4 Higher Excited $\bar{c}c$ and $\bar{b}b$ States. Confinement Forces. Effective Potentials

As we have shown in the previous section, the *nonperturbative* corrections to the energy levels and wave functions grow very quickly for large n, thus making the approach followed there useless for excited states; certainly when $n > 2$ for $\bar{b}b$, and for all observables except the ground state energy for $\bar{c}c$. As noted by, for example, Campostrini, Di Giacomo and Olejnik (1986) this growth can be traced to the fact that we have taken the correlators $\langle G_{\mu\nu}(x)G_{\alpha\beta}(y)\rangle$ as being independent of $x - y$, while (as we will see in Chap. 9), they should decrease exponentially for large spacelike $x - y$.

In the nonrelativistic limit, which is the only case we will treat explicitly in this section, we expect nonperturbative correlators to involve the chromo-electric fields,

$$\langle g^2 \mathcal{E}_i(x)\mathcal{E}_j(0)\rangle_{\text{vac}},$$

with sum over omitted colour indices understood. On invariance grounds, this quantity can be written with all generality as

$$\langle g^2 \mathcal{E}_i(x)\mathcal{E}_j(0)\rangle_{\text{vac}} = \tfrac{1}{12}\left\{\delta_{ij}\Delta(x) + x_i x_j \,\Box D_1(x^2)\right\},$$
$$\Delta(x) \equiv D(x^2) + D_1(x^2) + x^2\,\Box D_1(x^2). \tag{6.4.1}$$

In the approximation of neglecting the x-dependence of the correlator, only the piece $(\tfrac{1}{12}\delta_{ij}\Delta(0)$ survives and it can be related to the gluon condensate,

$$\Delta(0) = 2\pi\langle\alpha_s G^2\rangle. \tag{6.4.2}$$

This suggests that one could rederive both the Leutwyler–Voloshin approximation for small n as well as a potential valid for large n, taking into account in particular (6.4.1), keeping the x-dependence. This forms the basis of the so-called stochastic vacuum model, developed by Dosch and Simonov,[10] which we will review briefly in this section.

Note that the Dosch–Simonov approach is not the only one available. In some other papers the approach is, rather, to *input* a long distance potential (namely, the linear potential that, as we will see in Sect. 9.4 is suggested by lattice QCD) and derive from it the effective potential, including relativistic and spin corrections. This method can be found in Eichten and Feinberg (1981) and, more recently, Brambilla, Consoli and Prosperi (1994), from where the relevant literature may be traced. We prefer however the Dosch–Simonov approach because here the linear potential at long distances is *deduced* and, moreover, their short distance nonperturbative potential agrees qualitatively with what one would expect from renormalon considerations.

---

[10]Dosch (1987); Simonov (1988, 1989b); Dosch and Simonov (1988); Bertmann, Dosch and Krämer (1989). See also Simonov, Titard and Ynduráin (1995).

Let us start by considering the Green's function G for a quark and an antiquark to move from $\mathbf{x}_1$, $\bar{\mathbf{y}}_1$ respectively, at time $t = 0$, to $\mathbf{x}_2$, $\bar{\mathbf{y}}_2$ after the long period of time, T. We will consider the static limit and hence take

$$|\mathbf{x}_1 - \bar{\mathbf{y}}_1| \simeq |\mathbf{x}_2 - \bar{\mathbf{y}}_2| \simeq R,$$

with $R \ll T$. The quarks will be treated nonrelativistically and in first quantization (so in particular we will neglect quark loops) but the gluons will be second-quantized. In the path integral formalism we may write G as (recall Sect. 1.3)

$$G(x_1, \bar{y}_1; x_2, \bar{y}_2) = \langle \mathbf{x}_2, \bar{\mathbf{y}}_2 | e^{-iT\hat{H}} | \mathbf{x}_1, \bar{\mathbf{y}}_1 \rangle = N \int \mathrm{d}\mathbf{z}\, \mathrm{d}\bar{\mathbf{z}} \int \mathcal{D}B\, \mathrm{e}^{i \int_0^T \mathrm{d}t\, L},$$
(6.4.3)

where we integrate over the trajectories of $\mathbf{z}$, $\bar{\mathbf{z}}$ of quark and antiquark. N is a normalization constant, and the Lagrangian L is the sum of the nonrelativistic kinetic energies for the quarks, the nonrelativistic quark–gluon interaction and the gluon radiation Lagrangian:[11]

$$L = L_{\mathrm{kin}} + L_{\mathrm{int}} + L_{\mathrm{rad}};$$
$$L_{\mathrm{kin}} = \tfrac{1}{2}m\dot{\mathbf{z}}^2 + \tfrac{1}{2}m\dot{\bar{\mathbf{z}}}^2,$$
$$L_{\mathrm{int}} = \int \mathrm{d}^3\mathbf{w}\, g j^\mu(w) B_\mu(w) = g\left(B_0(z) - B_0(\bar{z})\right),$$
(6.4.4)
$$L_{\mathrm{rad}} = \frac{1}{4} \int \mathrm{d}^3\mathbf{w}\, G^2(w).$$

We assume the $\bar{q}q$ to be in a singlet state, and do not write colour factors explicitly. The second expression for L_{int} above was obtained using the expression of the current for static quarks,

$$j^\mu(w) = g^{\mu 0}\left(\delta(\mathbf{w} - \mathbf{z}) - \delta(\mathbf{w} - \bar{\mathbf{z}})\right).$$

The stochastic vacuum model may be formulated as follows. We work in the background field formalism and thus write

$$B_\mu = A_\mu + \phi_\mu,$$
(6.4.5)

where the background field ϕ is fixed so that it reproduces the nonperturbative gluon correlator. Specifically, we require

$$\langle : G_{A\mu\nu}(x) G_{\phi\alpha\beta}(0) : \rangle_{\mathrm{vac}} = 0,$$
$$\langle : G_{\mu\nu}(x) G_{\alpha\beta}(0) : \rangle_{\mathrm{vac}} = \langle G_{\phi\mu\nu}(x) G_{\phi\alpha\beta}(0) \rangle_{\mathrm{vac}},$$
(6.4.6)

---

[11]By "Lagrangians" we mean here the integrated Lagrangian densities.

where G_A is constructed with A, and G_ϕ with ϕ. Because of these equalities it follows that the VEV constructed with the field A vanishes:

$$\langle : G_{A\mu\nu}(x)G_{A\alpha\beta}(0) : \rangle_{\text{vac}} = 0. \tag{6.4.7}$$

Let us consider the so-called *Wilson loop*, defined as

$$W\left(C(R,T)\right) \equiv \int \mathcal{D}B e^{i \int_0^T \mathrm{d}t\,\{L_{\text{int}}+L_{\text{rad}}\}}, \tag{6.4.8}$$

with $C(R,T)$ the contour limited by the trajectories of the quarks. It can be shown that the potential $V(R)$ is related to the Wilson loop by

$$W\left(C(R,T)\right) \sim \exp\left(iTV(R)\right). \tag{6.4.9}$$

Both the Wilson loop and its connection with the potential are described in more detail later in this text, in Sect. 9.4.

We now expand W in powers of the background field ϕ_μ. Odd powers vanish, so that we have

$$W = W_0 + W_2 + \cdots. \tag{6.4.10}$$

In the first term only A_μ appears, and its VEV is zero. We can than apply ordinary perturbation theory to it, so that

$$W_0 = \int \mathcal{D}A \Big\{ 1 + \frac{(ig)^2}{2!} \int_0^T \mathrm{d}t \int_0^T \mathrm{d}t' \left(A_0(z) - A_0(\bar{z})\right)\left(A_0(z') - A_0(\bar{z}')\right)$$
$$+ \cdots \Big\} e^{\int_0^T \mathrm{d}\tau\, L_{\text{rad}}^0} + \cdots.$$

After a simple evaluation it is seen that we can identify this with the expansion of the 00 component of the gluon propagator and thus, to order g^2, we find the Coulombic potential, $-C_F \alpha_s / r$.

The interesting piece is W_2 in (6.4.10). At very short distances it provides the leading nonperturbative corrections because, on dimensional grounds, it is clear that higher order terms $W_4, \ldots$ would involve higher dimensional condensates, which are thus suppressed by inverse powers of the quark mass. In the stochastic vacuum model one *assumes* that this dominance also holds at long distances. That is to say, here we approximate W by $W_0 + W_2$. The form of the potential that W_2 implies may be found by the trick of considering it as a perturbation on the Coulombic potential, and calculating the energy shifts δE_{nl} that this induces. Note that this does *not* mean that we suppose W_2 to be small; the expansion is purely formal. The details of the calculation, which is somewhat involved technically in particular because of the non-Abelian character of the interaction, may be found in the quoted paper of Simonov, Titard and Ynduráin (1994). One finds

$$\delta E_{nl} = \frac{1}{36} \int \frac{d^3\mathbf{p}\, dp_0}{(2\pi)^4} \int d\beta \int d\beta'\, \widetilde{\Delta}(p) \int d^3\mathbf{k} \sum_j \langle nl|r_j e^{i\mathbf{p}(\beta-\frac{1}{2})\mathbf{r}}|8;\mathbf{k}\rangle$$

$$\times \frac{1}{E_k^{(8)} - E_n^{(0)} + p_0} \langle 8;\mathbf{k}|r'_j e^{i\mathbf{p}(\beta-\frac{1}{2})\mathbf{r}'}|n,l\rangle$$

$$(6.4.11)$$

where the $|nl\rangle$ are the states which are solution of the singlet Coulombic Hamiltonian,

$$\left\{ -\frac{1}{m}\Delta - \frac{C_F\alpha_s}{r} \right\} |nl\rangle = E_n^{(0)}|nl\rangle, \quad E_n^{(0)} = -m\frac{(C_F\alpha_s)^2}{4n^2},$$

and the $|8;\mathbf{k}\rangle$ those of the octet one:

$$\left\{ -\frac{1}{m}\Delta + \frac{\alpha_s}{6r} \right\} |8;\mathbf{k}\rangle = E_k^{(8)}|8;\mathbf{k}\rangle.$$

$\widetilde{\Delta}(p)$ is the Fourier transform of $\Delta(x)$,

$$\widetilde{\Delta}(p) = \int d^4x\, e^{ip\cdot x}\Delta(x). \qquad (6.4.12)$$

Equation (6.4.11) is our basic equation. The function $\Delta(x)$ can, on dimensional grounds, be written as

$$\Delta(x) = f\left(\frac{x^2}{T_g^2} \right),$$

where the quantity T_g has dimensions of time (or length) and may be interpreted as a correlation time/length. We now have two well-defined regimes.[12] For very heavy quarks and small n, we have

$$T_g^{-1} \ll |E_n^{(0)}|.$$

We may then approximate $\Delta(x)$ by $\Delta(0)$; hence $\widetilde{\Delta}(p) = 2\pi\delta_4(p)\langle\alpha_s G^2\rangle$ and (6.4.11) becomes

$$\delta E_{nl} = \frac{\pi\langle\alpha_s G^2\rangle}{18} \int d^3\mathbf{k} \sum_i \frac{\langle nl|r_i|8;\mathbf{k}\rangle \langle 8;\mathbf{k}|r_i|nl\rangle}{E_k^{(8)} - E_n^{(S)}}$$

$$= \frac{\pi\langle\alpha_s G^2\rangle}{18} \sum_i \langle nl|r_i \frac{1}{H^{(8)} - E_n^{(S)}} r_i|nl\rangle. \qquad (6.4.13)$$

---

[12]In actual quarkonium states, none of the regimes to be described is fully operative; the first regime would be certainly applicable with very good approximation only for toponium with n up to $n = 4$.

This is identical to the expression (6.3.24) that we found with the Leutwyler–Voloshin model which is thus rigorously justified.

The opposite regime is when, even for heavy quarks (so that the nonrelativistic approximation is still justified) we have

$$T_g^{-1} \gg |E_n^{(0)}|.$$

In this case the velocity tends to zero, the nonlocality of the interaction tends to zero as compared to the quark rotation period, which in the Coulombic approximation would be $T_q = 1/|E_n^{(0)}|$, and the interaction may therefore be described by a local potential. In fact, considering Eq. (6.4.11), it now turns out that we can neglect both $E_n^{(S)}$ and the kinetic energy term in $E_k^{(8)}$ (indeed, all of it) as compared to p_0. Then one gets

$$\delta E_{nl} = \tfrac{1}{36} \int \int \frac{\mathrm{d}^3\mathbf{p}\,\mathrm{d}p_0}{(2\pi)^4} \int \mathrm{d}\beta \int \mathrm{d}\beta'\ \tilde{\Delta}(p) \sum_i \langle nl|r_i \mathrm{e}^{\mathrm{i}\mathbf{p}(\beta-\frac{1}{2})\mathbf{r}} r_i' \mathrm{e}^{\mathrm{i}\mathbf{p}(\beta-\frac{1}{2})\mathbf{r}'}|nl\rangle$$

$$= \langle nl|U(r)|nl\rangle,$$

where the local potential U is[13]

$$U(r) = \tfrac{1}{36}\Big\{ 2r \int_0^r \mathrm{d}\lambda \int_0^\infty \mathrm{d}\nu\, D(\nu^2 - \lambda^2)$$
$$+ \int_0^r \mathrm{d}\lambda\,\lambda \int_0^\infty \mathrm{d}\nu\, [-2D(\nu^2 - \lambda^2) + D_1(\nu^2 - \lambda^2)] \Big\}. \tag{6.4.14}$$

This is the nonperturbative potential that follows from the stochastic vacuum model, so that the full nonrelativistic potential is

$$V(r) = -\frac{C_F \alpha_s}{r} + U(r).$$

Radiative corrections can be included for the Coulombic piece.

$U(r)$ has a number of desirable properties. As Eq. (6.4.14) shows, one has

$$U(r) \underset{r\to\infty}{\simeq} Kr + \text{constant}, \quad K = 2 \int_0^\infty \mathrm{d}\lambda \int_0^\infty \mathrm{d}\nu\, D(\nu^2 - \lambda^2). \tag{6.4.15}$$

Therefore a linear potential at long distances is a consequence of the model. For small distances, (6.4.14) gives

$$U(r) \underset{r\to 0}{\simeq} r^2 \int_0^\infty \mathrm{d}\nu\, \{D(\nu^2) + \tfrac{1}{2}D_1(\nu^2)\} + \text{constant}. \tag{6.4.16}$$

---

[13]Our derivation of these formulas is not rigorous. A rigorous derivation would require to make the calculations in Euclidean QCD, to be described in detail in Sects. 9.1 ff, and go back to Minkowski space at the end.

This is the behaviour also suggested on the basis of renormalon calculations, as we will discuss in Sect. 10.2ii. To be precise, however, we should add a few extra words specifying further the region of validity of the last equation. Eq. (6.4.16) is only valid in the regime $T_g^{-1} \gg |E_n^{(0)}|$. That is to say, strictly speaking, (6.4.16) is only valid in an intermediate region in which r is small, but the *state* is located, on the average, at a large distance of the center of mass.

An equation such as (6.4.14) still does *not* fix the potential; to obtain it further approximations have to be made. In particular, one has to assume a functional form for $\Delta(x)$. An exponential one is usually taken, mostly because of its simplicity. Thus Campostrini, Di Giacomo and Olejnik (1986) take

$$\widetilde{\Delta}(p) = \frac{3(2\pi)^3 T_g^{-1}}{p^2 + T_g^{-2}} \langle \alpha_s G^2 \rangle.$$

A fit to the $\bar{b}b$ and $\bar{c}c$ with the ensuing potential, including spin and relativistic corrections, but *not* radiative ones, has been carried out successfully by Badalian and Yurov (1990), who were able in particular to *predict* correctly the P-wave hyperfine splitting in quarkonium $\delta_{\mathrm{hf}} E(P)$ as

$$-1.76 \le \delta_{\mathrm{hf}} E(P) \le 0.32 \text{ MeV}$$

(the range induced by variation of the parameters inside their accepted domains), against the experimental figure of

$$\delta_{\mathrm{hf}} E(P) = -0.9 \pm 0.23 \text{ MeV}$$

(Armstrong et al., 1992).

An alternative method of calculation is based on postulating a specific long distance behaviour for the confining forces and evaluating the relativistic, and at least partially, short distance corrections, by various methods. For example, we may write the propagator for a fictitious scalar particle such that its NR limit agrees with Kr. The inverse Fourier transform of this will be $\sim 1/(\mathbf{k}^2)^2$, so we may postulate the relativistic propagator $1/(\mathbf{k}^2)^2$. Then, the standard analysis of retardation effects (Akhiezer and Berestetskii, 1963) applied to this propagator produces the potential[14]

$$U_W(\mathbf{r}) = Kr\gamma_1^0\gamma_2^0 + \frac{K}{2}\left\{3r\gamma_1^0\boldsymbol{\gamma}_1\gamma_2^0\boldsymbol{\gamma}_2 + \frac{1}{r}(\mathbf{r}\gamma_1^0\boldsymbol{\gamma}_1)(\mathbf{r}\gamma_2^0\boldsymbol{\gamma}_2)\right\}. \qquad (6.4.17)$$

This potential is to be used in a two-particle Dirac equation acting on wave functions $\Psi_{A_1 A_2}$ with two Dirac indices A_1, A_2. The gamma matrices γ_a^μ act on the index A_a. The last terms in (6.4.17) are relativistic corrections; they will produce fine and hyperfine splittings (in addition to the Coulombic

---

[14]Buchmüller (1982). See also Eichten and Feinberg (1981) and Brambilla, Consoli and Prosperi (1994) for more refined derivations.

ones) and even an alteration of the Coulomb static interaction. For example, to lowest order in the relativistic corrections the term $\frac{3}{2}Kr\gamma_1^0\boldsymbol{\gamma}_1\gamma_2^0\boldsymbol{\gamma}_2$ gives a perturbation, to be used with nonrelativistic wave functions, of

$$\frac{3K}{2}r\gamma_1^0\boldsymbol{\gamma}_1\gamma_2^0\boldsymbol{\gamma}_2 \rightarrow \frac{K}{4m^2}\left\{4r\Delta + 4\frac{\partial}{\partial r} + \frac{2}{r}\right.$$
$$\left. - \frac{4}{r}\mathbf{SL} - \frac{4r}{3}\boldsymbol{\sigma}_1\boldsymbol{\sigma}_2 - \frac{1}{r^3}\left((\mathbf{r}\boldsymbol{\sigma}_1)(\mathbf{r}\boldsymbol{\sigma}_2) - \frac{r^3}{3}\boldsymbol{\sigma}_1\boldsymbol{\sigma}_2\right)\right\}. \tag{6.4.18}$$

In some of the analyses, radiative corrections are included (Gupta, Radford and Repko, 1982; Pantaleone,Tye and Ng, 1986; Halzen, Olson, Olsson and Stong, 1993). Generally speaking, the agreement between theory and experiment is good for the spin-independent splittings, and less so for the spin-dependent ones, particularly tensor splittings. This probably implies that the models include genuine features of the nonperturbative potentials, but the fact that approximations are made shows up more in quantities that are sensitive to relativistic effects.

6.5 The Constituent Quark Model

In the previous sections we have shown how one can use QCD to obtain reliable calculations of static hadronic quantities. In some cases (particularly in the last section) reasonable, but unproven assumptions were needed to obtain results; but even in these cases, rigorous QCD formed the backbone of the methods. In this and the following section, the situation will be somewhat reversed. We will make assumptions, some of which are actually incompatible with some features of QCD (notably chiral invariance) but which may, under certain circumstances, imitate a real situation, and we will supplement these assumptions with features borrowed from QCD. The ensuing models, the bag models to be discussed in next section, and the constituent quark model which is the object of the present one, although not tremendously reliable, allow us to get a handle on otherwise intractable areas of hadron physics, providing an understanding of some of its features.

We first discuss the constituent quark model. Here, we assume that the fact that quarks inside hadrons move through a medium made up of gluons and quark–antiquark pairs can, under certain circumstances, be represented by ascribing an effective mass, called the *constituent* mass even to light quarks; a qualitative way to implement this will be discussed in Sect. 9.5iii. As stated, this mass represents the inertia acquired by quarks due to their having to drag in their motion the gluon–quark soup; so we expect it to be universal, and to add to the *mechanical* mass, which is the one that appears in the Lagrangian. Concentrating on light quarks, we then assume masses

$$m_u(\text{const}) = m_u + \mu_0, \quad m_d(\text{const}) = m_d + \mu_0, \quad m_s(\text{const}) = m_s + \mu_0,$$

$$(6.5.1)$$

where

$$\mu_0 \sim \left(\Lambda, \ \langle \bar{q}q \rangle^{1/3}, \ \langle \alpha_s G^2 \rangle^{1/4} \right) \sim 330 \text{ MeV}. \qquad (6.5.2)$$

In these formulas we take the mechanical masses m_q renormalized at 1 GeV.

Equations (6.5.1) break chiral invariance, and therefore pions and kaons (in particular) will be very poorly described by the constituent quark model: for these particles we have to use other methods (see Chap. 7). But one can use the constituent quark model to describe with success other hadrons (ρ, K^*, Σ, Λ, nucleons, Δ,...). In fact, and as already noted, most of the evidence in favour of the quark picture of hadrons was gathered by considerations of such models (notably by Dalitz and coworkers) and the extension to include QCD features ranks among the first successes of the theory (Appelquist and Politzer, 1975; De Rújula, Georgi and Glashow, 1975; see Hey and Kelly, 1983 for a review and references).

To implement the interactions among quarks, we introduce two potentials: a confining potential, linear in r,

$$U_{\text{conf}}(r) = \lambda r, \quad \lambda \sim K, \qquad (6.5.3)$$

and a Coulombic-type interaction,

$$U_{\text{Coul}}(r) = \frac{-\kappa^2}{r}, \qquad (6.5.4\text{a})$$

together with corresponding QCD-type hyperfine interactions. For quarks with indices i, j, we take

$$U_{\text{hyp}}(r) = -\kappa^2 \sum_{i \neq j} \frac{1}{m_i m_j} \sum_a t_i^a t_j^a \boldsymbol{\sigma}_i \boldsymbol{\sigma}_j, \qquad (6.5.4\text{b})$$

and t_i, $\boldsymbol{\sigma}_i$ act on the wave function of quark i. κ^2 may be connected with the running coupling at, say, the reference momentum of 1 GeV:

$$\kappa^2 \sim C_F \alpha_s (1 \text{ GeV}^2).$$

Because the model is in any case not terribly precise, one at times replaces the linear potential by a quadratic potential, which can be solved explicitly.

The model provides quantitatively correct predictions, to within some 20%, for the bulk of the masses of the particles, the magnetic moments of the baryons and – a triumph of the QCD inspired interaction (6.5.4b) – an explanation of the sign and size of the mass splittings of the nucleons and Δ resonance, and even of that between Λ and Σ. The wave functions at the origin, as measured experimentally in, for example, the e^+e^- decays of ρ, ω and ϕ, are also well reproduced.

These successes should not hide the shortcomings. As already stated, pions and kaons are very poorly described. Worse, for some particles the velocities of the quarks are *above* the velocity of light. Finally, the status of the tensor and L-S couplings is unclear. We leave the model here; the interested reader may consult, besides the articles already quoted, the papers of Isgur and Karl (1979), Close and Dalitz (1981) and the monographs of Alvarez-Estrada et al. (1986) and Flamm and Schoberl (1981).

6.6 Bag Models

i Introduction. Bogoliubov's Model

As we know, the strength of the interaction among quarks and gluons decreases at short distances; while at long ones it will grow to the point that confinement is produced. These features are reproduced, albeit in a gross manner, by bag models.[15] In these models the hadron is envisaged as a bag, i.e., an infinite spherical well of radius $R \sim \Lambda^{-1}$ assumed to contain the quarks (and gluons) that make up the hadron. The model is certainly crude (we will discuss refinements in the third subsection here) but it has the important property of being solvable and it thus affords a method to obtain, easily, at least estimates of quantities such as bound state energies or wave functions of light quarks, that cannot be easily obtained by other methods. By comparison with the constituent model of the previous section, the bag model wins in that it is relativistic: it is surely worth devoting some time to it.

It should, however, be borne in mind that the bag model presents serious drawbacks. An important one is that the introduction of a fixed hadron radius R violates chiral invariance; thus we expect, and it so happens, that the bag model will poorly describe pions and kaons. Another drawback is that the vacuum inside the bag is empty, in which it differs radically from the real vacuum inside hadrons, chock full of gluons and the light quark–antiquark sea. Thus, one can perform a bag model calculation of, say, nonsinglet structure functions, as in the work of Jaffe and Ross (1980); but singlet ones are beyond the reach of the model.

We start by considering what may be called a quantum mechanical bag (as opposed to a field-theoretic, or MIT, bag), first discussed by Bogoliubov (1967). In this approach we consider a fermion in a potential, $U(r)$,

$$U(r) = \begin{cases} 0, & r < R, \\ v_0, & r \geq R. \end{cases} \tag{6.6.1}$$

---

[15]Chodos et al. (1974); Chodos, Jaffe, Johnson and Thorn (1974). See Hasenfratz and Kuti (1978), Johnson (1975) and Alvarez-Estrada et al. (1988) for reviews.

Later, we will let $v_0 \to \infty$. We will assume the potential to be scalar, both because of indications from lattice evaluations, as well as from the potential calculations of previous sections, and to avoid the Klein paradox. The Dirac equation in this potential is solved as follows.[16] The Dirac equation may be written as

$$i\frac{\partial}{\partial t}\Psi(\mathbf{r},t) = (-i\gamma_0\boldsymbol{\gamma}\nabla + m\gamma_0)\,\Psi(\mathbf{r},t) = E\Psi(\mathbf{r},t). \tag{6.6.2}$$

We then form the *spherical harmonics with spin*, $\mathcal{Y}$, by composing ordinary spherical harmonics $Y_M^l(\theta,\phi)$ with Pauli spinors $\chi(s_3)$ corresponding to the third component of spin s_3. If $(l, M; \frac{1}{2}, s_3|j)$ are the Clebsch–Gordan coefficients corresponding to coupling of angular momentum l and spin $1/2$ to total angular momentum j, then

$$\mathcal{Y}_\lambda^{l\pm}(\theta,\phi) = \sum_{\lambda=M+s_3} (l, M; \tfrac{1}{2}, s_3|l \pm \tfrac{1}{2})Y_M^l(\theta,\phi)\chi(s_3). \tag{6.6.3a}$$

We then form the wave functions, in the Pauli realization of the gamma matrices (with $\gamma_0 = \mathrm{diag}(1,1,-1,1)$)

$$\Psi_\lambda^{j+}(\mathbf{r}) = \begin{pmatrix} f_{l+}(r)\mathcal{Y}_\lambda^{l+}(\theta,\phi) \\ g_{l+}(r)\mathcal{Y}_\lambda^{l+1,-}(\theta,\phi) \end{pmatrix}, \quad \Psi_\lambda^{j-}(\mathbf{r}) = \begin{pmatrix} f_{l-}(r)\mathcal{Y}_\lambda^{l+1,-}(\theta,\phi) \\ g_{l-}(r)\mathcal{Y}_\lambda^{l+}(\theta,\phi) \end{pmatrix}. \tag{6.6.3b}$$

In both cases $j = l = 1/2$ and f, g are scalar functions. Substituting into the time independent Dirac equation, and after simple manipulations, we obtain two coupled scalar equations for the f, g:

$$i\left(\frac{\partial}{\partial r}g_\kappa + \frac{1}{r}g_\kappa + \frac{\kappa}{r}g_\kappa\right) + mf_\kappa + U(r)f_\kappa = Ef_\kappa,$$
$$i\left(\frac{\partial}{\partial r}f_\kappa + \frac{1}{r}f_\kappa - \frac{\kappa}{r}f_\kappa\right) - mg_\kappa + U(r)g_\kappa = Eg_\kappa. \tag{6.6.4}$$

Here m is the mass of the quark, $\kappa = \omega(j + 1/2)$, $\omega = \pm$ and we have introduced the notation, to be used interchangeably,

$$f_{+(j+1/2)} \leftrightarrow f_{l+}, \quad l = j - 1/2,$$
$$f_{-(j+1/2)} \leftrightarrow f_{l-}, \quad l = j - 1/2,$$

and identical ones for the g. In terms of the f, g, the normalization is given by

$$\langle\Psi_\lambda^{j\omega}|\Psi_{\lambda'}^{j'\omega'}\rangle = \delta_{jj'}\delta_{\lambda\lambda'}\delta_{\omega\omega'}\int_0^\infty dr\, r^2\,(f^*f' + g^*g'). \tag{6.6.5}$$

---

[16]More details of this solution may be found in standard textbooks: Akhiezer and Berestetskii (1963); Greiner, Müller and Rafelski (1985); Ynduráin (1996).

The bound state solutions to (6.6.4) are easily found. Because the potential is constant they are like free waves. We distinguish two regions. For $r < R$, we have exactly free waves,

$$g_{l+}(r) = \frac{-ik}{m + E} f_{l+1,+}(r), \quad f_{l+}(r) = N_+ j_l(kr);$$

$$g_{l-}(r) = \frac{ik}{mc^2 + E} f_{l-1,-}(r), \quad f_{l-}(r) = N_- j_{l+1}(kr), \tag{6.6.6}$$

where

$$k = \sqrt{E^2 - m^2}, \quad l = j - 1/2, \quad \kappa = \pm(j + 1/2);$$

the $N_\pm$ are normalization constants and the j_l spherical Bessel functions:

$$j_l(x) = \sqrt{\frac{\pi}{2x}} J_{l+1/2}(x) = (-x)^l \left(\frac{d}{xdx} \right)^l \frac{\sin x}{x}.$$

For $r > R$, we define $\bar{k} = \sqrt{(m + v_0)^2 - E^2}$. Equations (6.6.4) with $U \equiv v_0$ are formally equal to the free ones if we replace m by $m + v_0$, but now we have to impose the condition of decrease at infinity. Thus,

$$g_{l+}(r) = \frac{-i\bar{k}}{mc^2 + v_0 + E} f_{l+1,+}(r),$$

$$g_{l-}(r) = \frac{i\bar{k}}{mc^2 + v_0 + E} f_{l-1,-}(r), \tag{6.6.7}$$

with

$$f_{l+}(r) = C \sqrt{\frac{\pi}{2r\bar{k}}} K_{l+1/2}(\bar{k}r),$$

$$f_{l-}(r) = C' \sqrt{\frac{\pi}{2r\bar{k}}} K_{l+3/2}(\bar{k}r).$$

Here K is the Bessel function of the second kind, given by

$$K_{n+1/2}(x) = \left(\frac{\pi}{2x} \right)^{1/2} x^{n+1} \left(-\frac{1}{x} \frac{d}{dx} \right)^n \frac{e^{-x}}{x}.$$

Explicit formulas for the $g_{l\pm}$ can be found by using the differentiation properties for $k_n(x) \equiv (\pi/2x)^{1/2} K_{n+1/2}(x)$,

$$k_n' = \frac{n}{z} k_n - k_{n+1}.$$

Matching $f(r)$, $\partial f(r)/\partial r$ at $r = R$, one finds the constants $N_\pm$, C, C' and the energy values. For the simple case of the S wave, $l = 0$, the quantization condition is

$$\tan kR = -k/\bar{k}, \tag{6.6.8}$$

quite analogous to the nonrelativistic one. In the limit $v_0 \to \infty$, and if we neglect the mass of the particles (a case of practical interest for bound states of light quarks, u, d), the quantization condition becomes $k_n R = n\pi$, $n = 1, 2, \ldots$, so we find the energies

$$E_n = n\pi R^{-1}. \tag{6.6.9}$$

This reproduces *qualitatively* the corresponding hadronic spectrum, with the already noted exception of the π, K. *Quantitatively*, we have to take $R \simeq 1.7\,\mathrm{fm}$ to get $E_1 \simeq 370\,\mathrm{GeV}$ and then the correct masses for a two quark system such as the ρ, or a three quark one (the p, n). The model as it stands does not, of course, feature fine or hyperfine splittings which have to be introduced with extra interactions like (6.5.4b). The value of R is of the right order of magnitude, albeit on the largish side. The mean radius of a hadron would be $\langle r \rangle \simeq 3R/4 \simeq 1.28\,\mathrm{fm}$, to be compared with the size of nucleons from e.g. nuclear physics, $R_N \sim 0.6\,\mathrm{fm}$.

ii The MIT Bag

We will only give the briefest of descriptions; the interested reader may find details in the reviews of Johnson (1975) and Alvarez-Estrada et al. (1986).

The MIT bag may be described as ordinary QCD *plus* the boundary condition that no quark current or energy-momentum may leave a prescribed bag, which simulates nonperturbative (confinement) effects. Let $n_\mu(x)$ be a spacelike vector orthogonal to the sphere $|\mathbf{x}| = R$ (the bag). We *cannot* postulate the condition $q(x) = 0$ for $|\mathbf{x}| = R$ for the quark fields, as this is incompatible with the Dirac equation for free quarks, $i\slashed{\partial}\, q(x) = mq(x)$. A condition which *is* compatible is

$$i\slashed{n}\, q_j(x) = q_j(x), \quad |\mathbf{x}| = R, \text{ all } j, \tag{6.6.10}$$

where j is a colour index and (6.6.10) is assumed to be valid for all flavours, q. This is sufficient to ensure that the flow of any current $\bar{q}_j \gamma^\mu q'_{j'}$ vanishes on the sphere, $|\mathbf{x}| = R$, and hence no colour or flavour leaves the bag. The proof is elementary: on one hand, on the surface of the bag, $n^\mu \bar{q}_j \gamma_\mu q'_{j'} = \bar{q}_j \slashed{n}\, q'_{j'} = -i\bar{q}_j q'_{j'}$; on the other, and in the same region,

$$n^\mu \bar{q}_j \gamma_\mu q'_{j'} = (\slashed{n}\, q)^\dagger \gamma_0 q'_{j'} = (-iq_j)^\dagger \gamma_0 q'_{j'} = i\bar{q}_j q'_{j'};$$

hence $n^\mu \bar{q}_j \gamma_\mu q'_{j'} = 0$ on the bag surface, as desired. A similar condition, involving an adjustable parameter (the bag pressure) ensures no energy-momentum flow.

The condition (6.6.10) is incompatible with chiral invariance, even for massless quarks.[17] Moreover, introduction of ordinary QCD interactions and a bag confinement certainly leads to double counting, as QCD is supposed to be confining by itself. One should not neglect the insights gained by the study of the field-theoretic (MIT) bag, butthe the above problems justify our leaving its detailed study to specialized articles.

---

[17]It has been proposed to at least partially repair this by introducing an independent pion field, defined to be zero inside the bag, and an elementary pseudoscalar field outside; this is the so-called "little bag", which has enjoyed some success, particularly in the realm of nuclear physics. We refer the reader to the original papers: Brown, Rho and Vento (1979); Vento et al., (1980).

7 Light Quarks; PCAC;
Chiral Dynamics;
the QCD Vacuum

*"If the Lord Almighty had consulted me before embarking upon creation,
I should have recommended something simpler"*

ALPHONSE X *"The Wise"* (1211–1284), King of Castille and León,
on having the Ptolemaic system of epicycles explained to him

7.1 Mass Terms and Invariances: Chiral Invariance

In this section we will consider quarks with masses $m \ll \Lambda$, to be referred to as *light quarks*.[1] Because the only dimensional parameter intrinsic to QCD is, we believe, Λ, we may expect that to some approximation we may neglect the masses of such quarks, which will yield only contributions of order m^2/Λ^2 or m^2/Q^2.

To study this, we retake the discussion of Sect. 2.8. Consider the QCD Lagrangian,

$$\mathcal{L} = -\sum_{l=1}^{n} m_l \bar{q}_l q_l + i \sum_{l=1}^{n} \bar{q}_l \not{D} q_l - \tfrac{1}{4}(D \times B)^2 + \text{gauge fixing} + \text{ghost terms}.$$

$$(7.1.1)$$

The sum runs only over *light* quarks; the presence of heavy quarks will have no effect in what follows and consequently we neglect them. We may then split the quark fields into left-handed and right-handed components:

$$q_l = q_{L,l} + q_{R,l}; \quad q_{L,l} \equiv q_{-,l} = \frac{1 - \gamma_5}{2} q_l, \quad q_{R,l} \equiv q_{+,l} = \frac{1 + \gamma_5}{2} q_l.$$

In terms of these, the quark part of the Lagrangian may be written as

$$\mathcal{L} = -\sum_{l=1}^{n} m_l \left(\bar{q}_{R,l} q_{L,l} + \bar{q}_{L,l} q_{R,l} \right) + i \sum_{l=1}^{n} \left(\bar{q}_{L,l} \not{D} q_{L,l} + \bar{q}_{R,l} \not{D} q_{R,l} \right) + \cdots .$$

---

[1] It is, of course, unclear whether the meaningful parameter in this respect is Λ or Λ_0 defined by $\alpha_s(\Lambda_0) \approx 1$. From considerations of chiral dynamics (see later), it would appear that the scale for smallness of quark masses is $4\pi f_\pi \sim 1$ GeV, where f_π is the pion decay constant; but even if we accept this, it is not obvious at which scale m has to be computed. We will see that m_u, $m_d \sim 4$ to 10 MeV so there is little doubt that u, d quarks should be classed as "light" quarks with any reasonable definition; but the situation is less definite for the s quark with $m_s \sim 200$ MeV.

We then consider the set of transformations $W^{\pm}$ in $U_L \times U_R$ (left-handed times right-handed) given by the independent transformations of the $q_{R,l}$, $q_{L,l}$:

$$q_{R,l} \to \sum_{l'} W^+_{ll'} q_{R,l'}, \quad q_{L,l} \to \sum_{l'} W^-_{ll'} q_{L,l'}; \quad W^{\pm} \text{ unitary.} \tag{7.1.2}$$

Clearly, the only term in $\mathcal{L}$ that is not invariant under all the transformations (7.1.2) is the mass term,

$$\mathcal{M} = \sum_{l=1}^{n} m_l \bar{q}_l q_l = \sum_{l=1}^{n} m_l \left(\bar{q}_{R,l} q_{L,l} + \bar{q}_{L,l} q_{R,l} \right). \tag{7.1.3}$$

When written in this form, the mass term is invariant under the set of transformations $[U(1)]^n$,

$$q_l \to e^{i\theta_l} q_l, \tag{7.1.4}$$

but this would not have been the case if we had allowed for nondiagonal terms in the mass matrix. To resolve this question of which are the general invariance properties of a mass term, we will prove two theorems.[2]

THEOREM 1. *Any general mass matrix can be written in the form (7.1.3) by appropriate redefinition of the quark fields. Moreover, we may assume that $m \geq 0$. Thus, (7.1.3) is actually the most general mass term possible.*

For the proof we consider that the most general mass term compatible with hermiticity is

$$\mathcal{M}' = \sum_{ll'} \left\{ \bar{q}_{L,l} M_{ll'} q_{R,l'} + \bar{q}_{R,l} M^*_{ll'} q_{L,l'} \right\}. \tag{7.1.5}$$

Let us temporarily denote matrices in flavour space by putting a tilde under them. If $\underset{\sim}{M}$ is the matrix with components $M_{ll'}$, then the well-known polar decomposition, valid for any matrix, allows us to write

$$\underset{\sim}{M} = \underset{\sim}{m}\underset{\sim}{U},$$

where $\underset{\sim}{m}$ is positive-semidefinite, so all its eigenvalues are ≥ 0, and $\underset{\sim}{U}$ is unitary. Eq. (7.1.5) may then be rewritten as

$$\mathcal{M}' = \sum_{ll'} \left\{ \bar{q}_{L,l} m_{ll'} q'_{R,l'} + \bar{q}'_{R,l} m_{ll'} q_{L,l'} \right\}, \quad q'_{R,l} = \sum_{l'} U_{ll'} q_{R,l'}, \tag{7.1.6}$$

and we have used the fact that $\underset{\sim}{m}$ is Hermitian. Define $q' = q_L + q'_R$; because $\bar{q}_R q_R = \bar{q}_L q_L = 0$, (7.1.6) becomes, in terms of q',

$$\mathcal{M}' = \sum \bar{q}'_l m_{ll'} q'_{l'}.$$

---

[2] The theorems are valid for *any* quark mass matrix, i.e., also including heavy flavours.

It then suffices to transform q' by the matrix that diagonalizes $\underset{\sim}{m}$ to obtain (7.1.3) with positive m_l. The term $\bar{q}\not{D}q$ in the Lagrangian is left invariant by all these transformations, so the theorem is proved.

THEOREM 2. *If all the m_l are nonzero and different, then the only invariance left is the $[U(1)]^n$ of (7.1.4).*

Let us consider the $W_\pm$ of (7.1.2), and assume that $W_+ = W_- \equiv W$; to show that this must actually be the case is left as an exercise. The condition of invariance of $\mathcal{M}$ yields the relation

$$W^\dagger \underset{\sim}{m} W = \underset{\sim}{m}, \quad \text{i.e.,} \quad [\underset{\sim}{m}, W] = 0. \tag{7.1.7}$$

It is known that any $n \times n$ diagonal matrix can be written as $\sum_{k=0}^{n-1} c_k \underset{\sim}{m}^k$ if, as occurs in our case, all the eigenvalues of $\underset{\sim}{m}$ are different and nonzero. Because of (7.1.7), it then follows that W commutes with all diagonal matrices, and hence it must itself be diagonal: because it is also unitary, it consists of diagonal phases, i.e., it may be written as a product of transformations (7.1.4), as was to be proved. We leave it to the reader to check that the conserved quantity corresponding to the $U(1)$ that acts on flavour q_f is the corresponding flavour number.

In the preceding theorems, we have not worried whether the masses $\underset{\sim}{m}$ were bare, running or invariant masses. This is because, in the $\overline{\text{MS}}$ scheme, the mass matrix becomes renormalized as a whole:

$$\underset{\sim}{M} = Z_m^{-1} \underset{\sim}{M}_u,$$

where Z_m is a *number*. The proof is easy: all we have to do is to repeat the analysis of Sects. 3.1, 2, 3 and 3.7, allowing for the matrix character of M, Z_m. We find, for the divergent part and in an arbitrary covariant gauge,

$$\underset{\sim}{S}_R^\xi(p) = \frac{\mathrm{i}}{\not{p} - \underset{\sim}{M}} + \frac{1}{\not{p} - \underset{\sim}{M}} \left\{ -[\underset{\sim}{\Delta}_F(\not{p} - \underset{\sim}{M}) + (\not{p} - \underset{\sim}{M})\underset{\sim}{\Delta}_F^\dagger] - \delta\underset{\sim}{M} \right.$$
$$\left. - (1 - \xi)(\not{p} - \underset{\sim}{M})N_\epsilon C_F \frac{g^2}{16\pi^2} + 3N_\epsilon C_F \frac{g^2}{16\pi^2} \underset{\sim}{M} \right\} \frac{\mathrm{i}}{\not{p} - \underset{\sim}{M}},$$

and we have defined

$$\underset{\sim}{M} = \underset{\sim}{M}_u + \delta\underset{\sim}{M}, \quad \underset{\sim}{Z}_F = 1 - \underset{\sim}{\Delta}_F.$$

The renormalization conditions then yield

$$\underset{\sim}{\Delta}_F^\dagger + \underset{\sim}{\Delta}_F = -(1 - \xi)N_\epsilon C_F \frac{g^2}{16\pi^2} = \text{diagonal},$$
$$[\underset{\sim}{\Delta}_F, \underset{\sim}{M}] = 0, \quad [\underset{\sim}{M}, \delta\underset{\sim}{M}] = 0,$$
$$\delta\underset{\sim}{M} = 3N_\epsilon C_F \frac{g^2}{16\pi^2} \underset{\sim}{M}.$$

Thus, the set of fermion fields and the mass matrix get renormalized as a whole:

$$Z_F^{-1} = 1 + N_\epsilon C_F \frac{g^2}{16\pi^2}, \quad Z_m = 1 - 3N_\epsilon C_F \frac{g^2}{16\pi^2}, \tag{7.1.8a}$$

i.e.,

$$Z_F = Z_F \, 1, \quad Z_m = Z_m \, 1. \tag{7.1.8b}$$

We have proved this to lowest order, but the renormalization group equations guarantee the result to leading order in α_s.

This result can be understood in yet another way. The invariance of $\mathcal{L}$ under the transformations (7.1.4) implies that we may choose the counterterms to satisfy the same invariance, so the mass matrix will remain diagonal after renormalization. In fact, this proof shows that in mass independent renormalization schemes (such as the $\overline{MS}$), Eqs. (7.1.8b) actually hold to all orders.

The results we have derived show that, if all the m_i are different and nonvanishing,[3] the only global symmetries of the Lagrangian are those associated with flavour conservation, (7.1.4). As stated above, however, under certain conditions it may be a good approximation to neglect the m_l. In this case, all the transformations of Eq. (7.1.2) become symmetries of the Lagrangian. The measure of the accuracy of the symmetry is given by, for example, the divergences of the corresponding currents or, equivalently, the conservation of the charges. This has been discussed in Sect. 2.8, and we now present some extra details.

Let us parametrize the W as $\exp\{(i/2)\sum \theta_a \lambda^a\}$, where the λ are the Gell-Mann matrices. (We consider the case $n = 3$; for $n = 2$, replace the λ by the σ of Pauli.) We may denote by $U_\pm(\theta)$ the operators that implement (7.1.2):

$$U_\pm(\theta) \frac{1 \pm \gamma_5}{2} q_l U_\pm^{-1}(\theta) = \sum_{l'} \left(e^{(i/2) \sum \theta_a \lambda^a} \right)_{ll'} \frac{1 \pm \gamma_5}{2} q_{l'}. \tag{7.1.9}$$

For infinitesimal θ, we write

$$U_\pm(\theta) \simeq 1 - \frac{i}{2} \sum \theta_a L_\pm^a, \quad (L_\pm^a)^\dagger = L_\pm^a,$$

so that (7.1.9) yields

$$[L_\pm^a, q_{\pm,l}(x)] = -\sum_{l'} \lambda_{ll'}^a q_{\pm,l'}(x), \quad q_{\pm,l} \equiv \frac{1 \pm \gamma_5}{2} q_l. \tag{7.1.10}$$

---

[3] As seems to be the case in nature. As we will see, one finds $\hat{m}_d/\hat{m}_u \sim 2$, $\hat{m}_s/\hat{m}_d \sim 20$, $\hat{m}_u \sim 6$ MeV.

Because the U leave the interaction part of the Lagrangian invariant, and since QCD is a free field theory at zero distance, we may solve (7.1.10) using free-field commutation relations. The result is

$$L_\pm^a(t) =: \int \mathrm{d}^3\mathbf{x} \sum_{ll'} \bar{q}_{\pm,l}(x)\gamma^0\lambda_{ll'}^a q_{\pm,l'}(x) :, \quad t = x^0. \tag{7.1.11}$$

These will be recognized as the charges corresponding to the currents

$$J_\pm^{a\mu}(x) =: \sum_{ll'} \bar{q}_l(x)\lambda_{ll'}^a\gamma^\mu\frac{1 \pm \gamma_5}{2}q_{l'}(x) : . \tag{7.1.12}$$

If the symmetry is exact, $\partial_\mu J_\pm^{a\mu} = 0$, and a standard calculation shows that the $L_\pm^a(t)$ are actually independent of t. Otherwise, we have to define *equal time* transformations and modify (7.1.9, 10) writing, for example,

$$[L_\pm^a(t), q_{\pm,l}(x)] = -\sum_{l'}\lambda_{ll'}^a q_{\pm,l'}(x), \quad t = x^0. \tag{7.1.13}$$

The set of transformations

$$U_\pm(\theta,t) = \exp\left\{-\frac{\mathrm{i}}{2}\sum L_\pm^a(t)\theta_a\right\},$$

builds up the group of *chiral transformations* generated by the currents (7.1.12). In our case we find the chiral $SU_F^+(3) \times SU_F^-(3)$ group. Its generators may be rearranged in terms of the set of vector and axial currents $V_{ll'}^\mu(x)$, $A_{ll'}^\mu(x)$ introduced in Sect. 2.8. (Actually, not all diagonal elements are in $SU_F^+(3) \times SU_F^-(3)$, but they are in the group $U_F^+(3) \times U_F^-(3)$.) An important subgroup of $SU_F^+(3) \times SU_F^-(3)$ is that generated by the vector currents, which is simply the flavour $SU(3)$ of Gell-Mann and Ne'eman.

The exactness of the symmetries is related to the time independence of the charges $L_\pm$, which in turn is linked to the divergence of the currents. These divergences are proportional to differences of masses, $m_l - m_{l'}$ for the vector, and sums $m_l + m_{l'}$ for the axial currents[4] (cf. Eq. (2.8.5)). Thus, we conjecture that $SU_F(3)$ will be good to the extent that $|m_l - m_{l'}|^2 \ll \Lambda^2$ and chiral $SU_F^+(3) \times SU_F^-(3)$ to the extent that $m_l \ll \Lambda$. In the real world, it appears that mass differences are of the same order as the masses themselves, so we expect chiral symmetries to be almost as good as flavour symmetries. This seems to be the case experimentally.[5]

---

[4] The *diagonal* axial currents have peculiar extra terms in the divergence that we will discuss in detail in coming sections.

[5] Chiral symmetry and chiral dynamics is a subject in itself. Here we only touch upon some of its aspects. This omits many important applications. The interested reader may consult the review of Pagels (1975), the excellent text of Georgi (1984) and, more recently, the basic paper of Gasser and Leutwyler (1984) and the reviews of Pich (1995) and Ecker (1995).

7.2 Wigner–Weyl and Nambu–Goldstone Realizations
of Symmetries

The fact that flavour and chiral $SU(3)$ (or $SU(2)$) appear to be valid to similar orders of approximation does not mean that these symmetries are realized in the same manner. In fact, we will see that there are good theoretical and experimental reasons why they are very different.

Let us begin by introducing the charges with definite parity,

$$Q^a = L^a_+ + L^a_-, \quad Q^a_5 = L^a_+ - L^a_-. \tag{7.2.1}$$

Their *equal time commutation relations* are

$$
\begin{aligned}
[Q^a(t), Q^b(t)] &= 2\mathrm{i} \sum f^{abc} Q^c(t), \\
[Q^a(t), Q^b_5(t)] &= 2\mathrm{i} \sum f^{abc} Q^c_5(t), \\
[Q^a_5(t), Q^b_5(t)] &= 2\mathrm{i} \sum f^{abc} Q^c(t).
\end{aligned}
\tag{7.2.2}
$$

The set Q^a builds the group $SU_F(3)$. In the limit $m_l \to 0$, all Q, Q_5 are t-independent and

$$[Q^a, \mathcal{L}] = [Q^a_5, \mathcal{L}] = 0. \tag{7.2.3}$$

The difference between Q^a, Q^a_5, however, lies in the vacuum. In general, given a set of generators L^j of symmetry transformations of $\mathcal{L}$, we have two possibilities:

$$L^j |0\rangle = 0, \tag{7.2.4}$$

which is called a *Wigner–Weyl* symmetry, or

$$L^j |0\rangle \neq 0, \tag{7.2.5}$$

or *Nambu–Goldstone* symmetry. Obviously, we will in general have a mixture of the two symmetries, with some L^i, $i = 1, \ldots, r$, verifying (7.2.4) and the rest, L^k, $k = r + 1, \ldots, n$, satisfying (7.2.5). Since the commutator of two operators that annihilate the vacuum also annihilates the vacuum, it follows that the subset of Wigner–Weyl symmetries forms a subgroup.

Two theorems are especially relevant with respect to these questions. The first, due to Coleman (1966), asserts that "the invariance of the vacuum is the invariance of the world", or, in more transparent terms, that the physical states (including bound states) are invariant under the transformations of a Wigner–Weyl group of symmetries. It follows that, if we assumed that chiral symmetry was all of it realized in the Wigner–Weyl mode, we could conclude that the masses of all mesons in a flavour multiplet would be degenerate, up to corrections of order m_q^2/M_h^2, with M_h the (average) hadron mass. This is true of the ω, ρ, K^*, ϕ, but if we include parity doublets this is no longer the case. Thus, for example, there is no scalar meson with a mass anywhere near that of the pion, and the axial vector meson masses are more than half a GeV larger than the masses of ω or ρ. Thus it is strongly suggested that

$SU_F(3)$ is a Wigner–Weyl symmetry, but chiral $SU_F^+(3) \times SU_F^-(3)$ contains generators of the Goldstone–Nambu type. We assume, therefore,

$$Q^a(t)|0\rangle = 0, \quad Q_5^a(t)|0\rangle \neq 0. \tag{7.2.6}$$

The second relevant theorem is Goldstone's (1961). It states that, for each generator that fails to annihilate the vacuum, there must exist a massless boson with the quantum numbers of that generator. Therefore, we "understand" the smallness of the masses of the pion or kaon[6] because, in the limit m_u, m_d, $m_s \to 0$, we would also have $m_\pi \to 0$, $m_K \to 0$. Indeed, we will later show that

$$m_\pi^2 \sim m_u + m_d, \quad m_K^2 \sim m_{u,d} + m_s. \tag{7.2.7}$$

We will not prove either theorem here, but we note that (7.2.7) affords a quantitative criterion for the validity of chiral symmetries; they hold to corrections of order m_π^2/m_ρ^2 for $SU(2)$ and of m_K^2/m_{K*}^2 for $SU(3)$.

We also note that a Nambu–Goldstone realization (Nambu, 1960; Nambu and Jona–Lasinio, 1961a,b) is never possible in perturbation theory. Since the symmetry generators are Wick-ordered products of field operators, it is clear that to all orders of perturbation theory $Q_5^a(t)|0\rangle = 0$. This means that the physical vacuum is different from the vacuum of perturbation theory in the limit $m \to 0$. We emphasize this by writing $|0\rangle$ for the perturbation-theoretic vacuum and $|\text{vac}\rangle$ for the physical one when there is danger of confusion, a practice that we have already followed. So we rewrite (7.2.6) as

$$Q^a(t)|\text{vac}\rangle = 0, \quad Q_5^a(t)|\text{vac}\rangle \neq 0. \tag{7.2.8}$$

It is not difficult to see how this comes about in QCD. Let $a_P^\dagger(\mathbf{k})$ be the creation operator for a particle P. The states

$$a_P^\dagger(\mathbf{0}) \overbrace{\ldots}^{n} a_P^\dagger(\mathbf{0})|0\rangle = |n\rangle$$

are all degenerate in the limit $m_P \to 0$. Therefore, the physical vacuum will be, in this limit,

$$|\text{vac}\rangle = \sum_n c_n |n\rangle,$$

i.e., it will contain zero-frequency massless particles. In QCD we have the gluons which are massless, and so will the light quarks be, to a good approximation, in the chiral limit.

---

[6] The particles with zero flavour quantum numbers present problems of their own (the so-called $U(1)$ problem) that will be discussed later.

7.3 PCAC and Light Quark Mass Ratios

We are now in a position to obtain quantitative results on the masses of the light quarks. To do so, consider the current

$$A_{ud}^\mu(x) = \bar{u}\gamma^\mu\gamma_5 d(x),$$

and its divergence

$$\partial_\mu A_{ud}^\mu(x) = i(m_u + m_d)\bar{u}\gamma_5 d(x).$$

The latter has the quantum numbers of the π^+, so we can use it as a composite pion field operator. We thus write

$$\partial_\mu A_{ud}^\mu(x) = \sqrt{2}f_\pi m_\pi^2 \phi_\pi(x). \tag{7.3.1a}$$

The factors in (7.3.1a) are chosen for historical reasons (our convention is not universal, however). $\phi_\pi(x)$ is the pion field normalized to

$$\langle 0|\phi_\pi(x)|\pi(p)\rangle = \frac{1}{(2\pi)^{3/2}}e^{-ip\cdot x}, \tag{7.3.1b}$$

with $|\pi(p)\rangle$ the state of a pion with momentum p. The constant f_π may be obtained from experiment as follows. Consider the weak decay $\pi^+ \to \mu^+\nu$. With the effective Fermi Lagrangian for weak interactions (see, e.g., Marshak, Riazzudin and Ryan, 1969)

$$\mathcal{L}_{int}^{Fermi} = \frac{G_F}{\sqrt{2}}\bar{\mu}\gamma_\lambda(1-\gamma_5)\nu_\mu\bar{u}\gamma^\lambda(1-\gamma_5)d + \cdots,$$

we find

$$F(\mu \to \mu\nu) = \frac{2\pi G_F}{\sqrt{2}}\bar{u}_{(\nu)}(p_2)\gamma_\lambda(1-\gamma_5)v_{(\mu)}(p_1,\sigma)\langle 0|A_{ud}^\lambda(0)|\pi(p)\rangle. \tag{7.3.2a}$$

Now, on invariance grounds,

$$\langle 0|A_{ud}^\lambda(0)|\pi(p)\rangle = ip^\lambda C_\pi; \tag{7.3.2b}$$

contracting with p_μ we find $C_\pi = f_\pi\sqrt{2}/(2\pi)^{3/2}$ and hence

$$m_\pi^2 C_\pi = \langle 0|\partial_\lambda A_{ud}^\lambda(0)|\pi(p)\rangle = \sqrt{2}f_\pi m_\pi^2 \frac{1}{(2\pi)^{3/2}}. \tag{7.3.2c}$$

Therefore

$$\tau(\pi \to \mu\nu) = \frac{4\pi}{(1 - m_\mu^2/m_\pi^2)^2 G_F^2 f_\pi^2 m_\pi m_\mu^2},$$

and we obtain f_π from the decay rate. Experimentally, one has $f_\pi \approx$ 93.3 MeV. A remarkable fact is that, if we repeat the analysis for kaons,

$$\partial_\mu A_{us}^\mu(x) = \sqrt{2}f_K m_K^2 \phi_K(x), \tag{7.3.3}$$

we find that, experimentally, $f_K \approx 110$ MeV: it agrees with f_π to 20%. Actually, this is to be expected because, in the limit $m_{u,d,s} \to 0$, there is no difference between pions and kaons, and we would find strict equality. That f_π, f_K are so similar in the real world is a good point in favour of $SU_F(3)$ chiral ideas.

The relations (7.3.1) and (7.3.3) are at times called PCAC[7] but this is not very meaningful, for these equations are really *identities*. One may use any pion field operator one wishes, in particular (7.3.1), provided that it has the right quantum numbers and its vacuum-one pion matrix element is not zero. The nontrivial part of PCAC will be described below.

The next step is to consider the two-point function (we drop the ud index from A_{ud})

$$F^{\mu\nu}(q) = \mathrm{i} \int \mathrm{d}^4 x\, \mathrm{e}^{\mathrm{i}q\cdot x} \langle \mathrm{T} A^\mu(x) A^\nu(0)^\dagger \rangle_{\mathrm{vac}},$$

and contract with q_μ, q_ν:

$$
\begin{aligned}
q_\nu q_\mu F^{\mu\nu}(q) &= - q_\nu \int \mathrm{d}^4 x\, \mathrm{e}^{\mathrm{i}q\cdot x} \partial_\mu \langle \mathrm{T}\, A^\mu(x) A^\nu(0)^\dagger \rangle_{\mathrm{vac}} \\
&= - q_\nu \int \mathrm{d}^4 x\, \mathrm{e}^{\mathrm{i}q\cdot x} \delta(x^0) \langle [A^0(x), A^\nu(0)^\dagger] \rangle_{\mathrm{vac}} \\
&\quad - q_\nu \int \mathrm{d}^4 x\, \mathrm{e}^{\mathrm{i}q\cdot x} \langle \mathrm{T}\, \partial \cdot A(x) A^\nu(0)^\dagger \rangle_{\mathrm{vac}} \\
&= 2\mathrm{i} \int \mathrm{d}^4 x\, \mathrm{e}^{\mathrm{i}q\cdot x} \delta(x^0) \langle [A^0(x), \partial \cdot A(0)^\dagger] \rangle_{\mathrm{vac}} \\
&\quad + \mathrm{i} \int \mathrm{d}^4 x\, \mathrm{e}^{\mathrm{i}q\cdot x} \langle \mathrm{T}\, \partial \cdot A(x) \partial \cdot A(0)^\dagger \rangle_{\mathrm{vac}}.
\end{aligned}
$$

Using Eqs. (7.3.1, 2) and evaluating the commutator, we find

$$
\begin{aligned}
q_\nu q_\mu F^{\mu\nu}(q) = {}& 2(m_u + m_d) \int \mathrm{d}^4 x\, \mathrm{e}^{\mathrm{i}q\cdot x} \delta(x) \langle \bar{u}(x) u(x) + \bar{d}(x) d(x) \rangle_{\mathrm{vac}} \\
&+ 2\mathrm{i} f_\pi^2 m_\pi^4 \int \mathrm{d}^4 x\, \mathrm{e}^{\mathrm{i}q\cdot x} \langle \mathrm{T}\, \phi_\pi(x) \phi_\pi(0)^\dagger \rangle_{\mathrm{vac}},
\end{aligned}
$$

or, in the limit $q \to 0$,

$$
\begin{aligned}
&2(m_u + m_d) \langle : \bar{u}(0) u(0) + \bar{d}(0) d(0) : \rangle_{\mathrm{vac}} \\
&= -2\mathrm{i} f_\pi^2 m_\pi^4 \int \mathrm{d}^4 x\, \mathrm{e}^{\mathrm{i}q\cdot x} \langle \mathrm{T}\, \phi_\pi(x) \phi_\pi(0)^\dagger \rangle_{\mathrm{vac}} \big|_{q\to 0},
\end{aligned}
$$

and we have reinstated explicitly the colons of normal ordering. The right hand side of this equality has contributions from the pion pole and from the

---

[7] Partially conserved axial current. In fact, in the limit $m_\pi^2 \to 0$, the right hand side of (7.3.1a) vanishes.

continuum; by writing a dispersion relation (Cauchy representation) for $\Pi(t)$, they can be expressed as[8]

$$\mathrm{i}\int \mathrm{d}^4 x\, \mathrm{e}^{\mathrm{i}q\cdot x}\langle \mathrm{T}\,\phi_\pi(x)\phi_\pi(0)^\dagger\rangle_{\mathrm{vac}}\Big|_{q\to 0} = \left\{\frac{1}{m_\pi^2 - q^2} + \frac{1}{\pi}\int \mathrm{d}t\,\frac{\mathrm{Im}\,\Pi(t)}{t - q^2}\right\}_{q\to 0}$$

$$= \frac{1}{m_\pi^2} + \frac{1}{\pi}\int \mathrm{d}t\,\frac{\mathrm{Im}\,\Pi(t)}{t};$$

$$\Pi(q^2) = \mathrm{i}\int \mathrm{d}^4 x\, \mathrm{e}^{\mathrm{i}q\cdot x}\langle \mathrm{T}\,\phi_\pi(x)\phi_\pi(0)^\dagger\rangle_{\mathrm{vac}}.$$

The order of the limits is essential; we first must take $q \to 0$ and the chiral limit afterwards. In the limit $m_\pi^2 \to 0$, the first term on the right hand side above *diverges*, and the second remains finite.[9] We then get

$$(m_u + m_d)\langle \bar{u}u + \bar{d}d\rangle = -2f_\pi^2 m_\pi^2 \left\{1 + O(m_\pi^2)\right\},$$

$$\langle \bar{q}q\rangle \equiv \langle : \bar{q}(0)q(0) :\rangle_{\mathrm{vac}}, \quad q = u, d, s, \ldots. \tag{7.3.4}$$

This is a strong indication that $\langle \bar{q}q\rangle \neq 0$ because, in order to ensure that it vanishes, we would require $f_\pi = 0$. We also note that we have not distinguished in e.g. (7.3.4), between bare or renormalized quark masses and operators; the distinction is not necessary because, as we know, m_q and $\langle \bar{q}q\rangle$ acquire opposite renormalization, so that $m_u\langle \bar{q}q\rangle_u = m_R\langle \bar{q}q\rangle_R$.

We may repeat the derivation of (7.3.4) for kaons. We find, to leading order in m_K^2,

$$(m_s + m_u)\langle \bar{s}s + \bar{u}u\rangle \simeq -2f_{K^+}^2 m_{K^+}^2,$$

$$(m_s + m_d)\langle \bar{s}s + \bar{d}d\rangle \simeq -2f_{K^0}^2 m_{K^0}^2. \tag{7.3.5}$$

We may assume $f_{K^+} = f_{K^0}$ since, in the limit $m_{u,d}^2 \ll \Lambda^2$ they should be strictly equal. For the same reason, one can take it that the VEVs $\langle \bar{q}q\rangle$ are equal for all light quarks. Under these circumstances, we may eliminate the VEVs and obtain

$$\frac{m_s + m_u}{m_d + m_u} \simeq \frac{f_K^2 m_{K^+}^2}{f_\pi^2 m_\pi^2}, \quad \frac{m_d - m_u}{m_d + m_u} \simeq \frac{f_K^2(m_{K^0}^2 - m_{K^+}^2)}{f_\pi^2 m_\pi^2}.$$

A more careful evaluation requires consideration of electromagnetic contributions to the observed π, K masses (Bijnens, 1993; Donoghue, Holsten and

---

[8] The equation below should have been written with subtractions, to compensate for the growth of $\Pi(q^2)$ for large q^2; but these do not alter the conclusions.

[9] Properly speaking, this is the PCAC limit, for in this limit the axial current is conserved.

Wyler, 1993) and higher order chiral corrections (Kaplan and Manohar, 1986; Bijnens, Prades and de Rafael, 1995).[10] In this way we find

$$\frac{m_s}{m_d} = 18 \pm 5, \quad \frac{m_d}{m_u} = 2.0 \pm 0.4. \tag{7.3.6}$$

If we couple this with the phenomenological estimate (from meson and baryon spectroscopy) $m_s - m_d \approx 100$ to 200 MeV, $m_d - m_u \approx 4$ MeV, we obtain the masses (in MeV)

$$\bar{m}_u(Q^2 \sim m_\rho^2) \approx 5, \quad \bar{m}_d(Q^2 \sim m_\rho^2) \approx 9, \quad \bar{m}_s(Q^2 \sim m_\rho^2) \approx 190, \tag{7.3.7}$$

where the symbol $\approx$ here means that a 50% error would not be very surprising.

This method for obtaining light quark masses is admittedly very rough; in the next section we will describe more sophisticated ones.

To conclude this section we make a few comments concerning light quark condensates, $\langle \bar{q}q \rangle$. The fact that these do not vanish implies spontaneous breaking of chiral symmetry because, under $q \to \gamma_5 q$, $\langle \bar{q}q \rangle \to -\langle \bar{q}q \rangle$. One may thus wonder whether chiral symmetry would not be restored in the limit $m_q \to 0$, which would imply

$$\langle \bar{q}q \rangle \underset{m_q \to 0}{\to} 0. \tag{7.3.8}$$

This possibility is discussed for example by Gasser and Leutwyler (1982). The equation (7.3.8) is highly unlikely. If it held, one would expect in particular the ratios,

$$\langle \bar{s}s \rangle : \langle \bar{d}d \rangle : \langle \bar{u}u \rangle \sim 190 : 9 : 5,$$

which runs contrary to all evidence, from hadron spectroscopy to SVZ sum rules which suggest

$$\langle \bar{s}s \rangle \sim \langle \bar{d}d \rangle \sim \langle \bar{u}u \rangle$$

to a few percent. Thus we obtain an extra indication that chiral symmetry is indeed spontaneously broken in QCD.

---

[10]The method originates in the work of Glashow and Weinberg (1968) and Gell-Mann, Oakes and Renner (1968). In QCD, see Weinberg (1978a), Domínguez (1978) and Zepeda (1978). Estimates of the quark masses essentially agreeing with (7.3.6, 7) below had been obtained even before QCD by e.g. Okubo (1969), but nobody knew what to do with them. The first evaluation in the context of QCD is due to Leutwyler (1974).

7.4 Bounds and Estimates of Light Quark Masses

In this section we describe a method for obtaining bounds and estimates of light quark masses. The method was first used (to get rough estimates) by Vainshtein et al. (1978) and further refined by Becchi, Narison, de Rafael and Ynduráin (1981), Gasser and Leutwyler (1982), etc. One starts with the correlator,

$$
\begin{aligned}
\Psi_{ij}^5(q^2) &= \mathrm{i} \int \mathrm{d}^4x\, \mathrm{e}^{\mathrm{i}q\cdot x} \langle \mathrm{T}\partial \cdot A_{ij}(x)\partial \cdot A_{ij}(0)^\dagger \rangle_{\mathrm{vac}} \\
&= \mathrm{i}(m_i + m_j)^2 \int \mathrm{d}^4x\, \mathrm{e}^{\mathrm{i}q\cdot x} \langle \mathrm{T}J_{ij}^5(x)J_{ij}^5(0)^\dagger \rangle_{\mathrm{vac}},
\end{aligned}
\tag{7.4.1}
$$

where $A_{ij}^\mu = \bar{q}_i\gamma^\mu\gamma_5 q_j$, $J_{ij}^5 = \bar{q}_i\gamma_5 q_j$, $i,j = u, d, s$.

To all orders of perturbation theory, the function

$$
F_{ij}(Q^2) = \frac{\partial^2}{\partial(q^2)^2}\Psi_{ij}^5(q^2), \quad Q^2 = -q^2,
$$

vanishes as $Q^2 \to \infty$. Hence, we may write a dispersion relation of the form

$$
F_{ij}(Q^2) = \frac{2}{\pi}\int_{m_P^2}^\infty \mathrm{d}t\, \frac{\mathrm{Im}\,\Psi_{ij}^5(t)}{(t + Q^2)^3}, \quad P = \pi, K.
\tag{7.4.2}
$$

For large values of Q^2, t we may calculate $F_{ij}(Q^2)$, $\mathrm{Im}\,\Psi_{ij}^5(t)$. The calculation has been improved along the years due to increasing precision of the QCD evaluations of these quantities.[11] Here, however, we will consider only leading effects and first order subleading corrections. We then have,

$$
\begin{aligned}
F_{ij}(Q^2) = \frac{3}{8\pi^2}\, &\frac{[\bar{m}_i(Q^2) + \bar{m}_j(Q^2)]^2}{Q^2} \\
\times \Bigg\{ 1 &+ \frac{11}{3}\frac{\alpha_s(Q^2)}{\pi} + \frac{m_i^2 + m_j^2 + (m_i - m_j)^2}{Q^2} + \frac{2\pi}{3}\frac{\langle \alpha_s G^2 \rangle}{Q^4} \\
&- \frac{16\pi^2}{3Q^4}\left[\left(m_j - \frac{m_i}{2}\right)\langle \bar{q}_i q_i \rangle + \left(m_i - \frac{m_j}{2}\right)\langle \bar{q}_j q_j \rangle\right] \Bigg\}
\end{aligned}
\tag{7.4.3a}
$$

and

$$
\mathrm{Im}\,\Psi_{ij}^5(t) = \frac{3[\bar{m}_i(t) + \bar{m}_j(t)]^2}{8\pi}\left\{ \left[1 + \frac{17}{3}\frac{\alpha_s(t)}{\pi}\right]t - (m_i - m_j)^2 \right\}.
\tag{7.4.3b}
$$

---

[11]Broadhurst (1981) and Chetyrkin et al. (1995) for subleading mass corrections; Becchi, Narison, de Rafael and Ynduráin (1981), Generalis (1990), Sugurladze and Tkachov (1990), Chetyrkin, Groshny and Tkachov (1982), Groshny, Kataev, Larin and Sugurladze (1991) and Pascual and de Rafael (1982) for radiative corrections to various terms.

The contributions containing the condensates are easily evaluated taking into account the nonperturbative parts of the quark and gluon propagators (Sect. 3.9). The quantities $m_i \langle \bar{q}_j q_j \rangle$ may be reexpressed in terms of experimentally known quantities, $f_{K,\pi}$, $m_{K,\pi}$ as in (7.3.4, 5). For the case $ij = ud$, which is the one we will consider in more detail, their contribution is negligible, as are the terms of order m^2/Q^2 in Eqs. (7.4.3). We will henceforth neglect these quantities. Because one can write the imaginary part of the spectral function as

$$\operatorname{Im} \Psi_{ij}^5(t) = \tfrac{1}{2} \sum_\Gamma \left| \langle \mathrm{vac} | \partial^\mu A_\mu^{ij}(0) | \Gamma \rangle \right|^2 (2\pi)^4 \delta_4(q - p_\Gamma)$$

it follows that $\operatorname{Im} \Psi_{ij}^5(t) \geq 0$: it is this positivity that will allow us to derive quite general bounds. To obtain tight ones it is important to use the information contained in both Eqs. (7.4.3a,b); to this end, we define the function

$$
\begin{aligned}
\varphi_{ij}(Q^2) &= F_{ij}^5(t) - \int_{Q^2}^\infty dt \, \frac{1}{(t + Q^2)^3} \frac{2 \operatorname{Im} \Psi_{ij}^5(t)}{\pi} \\
&= \int_{m_P^2}^{Q^2} dt \, \frac{1}{(t + Q^2)^3} \frac{2 \operatorname{Im} \Psi_{ij}^5(t)}{\pi}.
\end{aligned}
\tag{7.4.4}
$$

For sufficiently large Q^2 we may use (7.4.3) and integrate the imaginary part to obtain, for $ij = ud$,

$$
\begin{aligned}
\varphi_{ud}(Q^2) = \frac{3}{8\pi^2} \Bigg\{ & \frac{[\bar{m}_u(Q^2) + \bar{m}_d(Q^2)]^2}{Q^2} \left[\tfrac{1}{4} + \left(\tfrac{5}{12} + 2 \log 2 \right) \frac{\alpha_s}{\pi} \right] \\
& + \frac{1}{3Q^6} \left[8\pi^2 f_\pi^2 m_\pi^2 + 2\pi \langle \alpha_s G^2 \rangle \right] \Bigg\},
\end{aligned}
\tag{7.4.5a}
$$

and to this accuracy the two loop expression (3.7.5c) for the running masses is to be used. For the $ij = us, ds$ cases, and neglecting $m_{u,d}/m_s$,

$$
\begin{aligned}
\varphi_{us,ds}(Q^2) = \frac{3}{8\pi^2} \Bigg\{ & \frac{\bar{m}_s^2}{Q^2} \left[\tfrac{1}{4} + \left(\tfrac{5}{12} + 2 \log 2 \right) \frac{\alpha_s}{\pi} \right] \\
& - \frac{2\bar{m}_s^4}{Q^4} \left[\tfrac{3}{4} + (6 + 4 \log 2) \frac{\alpha_s}{\pi} \right] + \frac{1}{3Q^6} \left[8\pi^2 f_K^2 m_K^2 + 2\pi \langle \alpha_s G^2 \rangle \right] \Bigg\}.
\end{aligned}
\tag{7.4.5b}
$$

We can extract the pion (or kaon, as the case may be) pole explicitly from the low energy dispersive integral in (7.4.4) thus getting for e.g., φ_{ud}

$$
\varphi_{ud}(Q^2) = \frac{4 f_\pi^2 m_\pi^4}{(m_\pi^2 + Q^2)^3} + \int_{t_0}^{Q^2} dt \, \frac{1}{(t + Q^2)^3} \frac{2 \operatorname{Im} \Psi_{ij}^5(t)}{\pi};
\tag{7.4.6}
$$

the continuum threshold t_0 is $3m_\pi^2$ for $ij = ud$ or $(m_K + 2m_\pi)^2$ for $ij = (u, d)s$. Because of the positivity of $\operatorname{Im} \Psi$ this immediately gives bounds on

$m_i(Q^2) + m_j(Q^2)$ as soon as $Q^2 \geq Q_0^2$, where Q_0^2 is a momentum large enough for the QCD estimates (7.4.5) to be valid: thus to leading order,

$$\bar{m}_u(Q_0^2) + \bar{m}_d(Q_0^2) \geq \left\{ \frac{2^7 \pi^2 f_\pi^2 m_\pi^4}{3} \frac{Q_0^2}{(Q_0^2 + m_\pi^2)^3} \right\}^{\frac{1}{2}} ; \tag{7.4.7a}$$

for the combination us,

$$\bar{m}_s(Q_0^2) \geq \left\{ \frac{2^7 \pi^2 f_K^2 m_K^4}{3} \frac{Q_0^2}{(Q_0^2 + m_K^2)^3} \right\}^{\frac{1}{2}} . \tag{7.4.7b}$$

The bound depends a lot on the value of Q_0^2. We find, for example, the bounds

$$\begin{aligned}
\bar{m}_u(1 \text{ GeV}^2) + \bar{m}_d(1 \text{ GeV}^2) &\geq 13 \text{ MeV}, \quad Q_0^2 = 1.75 \text{ GeV}^2, \\
\bar{m}_u(1 \text{ GeV}^2) + \bar{m}_d(1 \text{ GeV}^2) &\geq 7 \text{ MeV}, \quad Q_0^2 = 3.5 \text{ GeV}^2
\end{aligned} \tag{7.4.8a}$$

and

$$\begin{aligned}
\bar{m}_s(1 \text{ GeV}^2) &\geq 245 \text{ MeV}, \quad Q_0^2 = 1.75 \text{ GeV}^2, \\
\bar{m}_s(1 \text{ GeV}^2) &\geq 150 \text{ MeV}, \quad Q_0^2 = 3.5 \text{ GeV}^2 .
\end{aligned} \tag{7.4.8b}$$

As is customary, we have translated the bounds (as we will also do for the estimates later on) to bound on the running masses defined at 1 GeV. The bounds can be stabilized somewhat by considering derivatives of F_{ij}^5, but (7.4.8) do not change much.

To get *estimates* for the masses, a model is necessary for the low energy piece of the dispersive integral (7.4.6). At very low energy, one can calculate $\text{Im}\, \Psi^5$ using chiral perturbation theory (see for example Pagels and Zepeda, 1972; Gasser and Leutwyler, 1982); the contribution is minute. The important region is that where the quasi-two body channels are open, the $(\rho, \omega) - \pi$ channels for the ud case. This is expected to be dominated by the π' resonance, with a mass of 1.2 GeV. One can take the residue of the resonance as a free parameter, and fit the QCD expression (7.4.5). This is the procedure followed by Narison and de Rafael (1981), Hubschmid and Mallik (1981), Gasser and Leutwyler (1982), Kataev, Krasnikov and Pivovarov (1983), Domínguez and de Rafael (1987), Chetyrkin, Pirjol and Schilcher (1997), etc. The errors one finds in the literature are many times *overoptimistic* because they do not take into account the important matter of the value Q_0^2 at which the perturbative QCD evaluation is supposed to be valid (Ynduráin, 1998). Now, as is clear from Eq. (7.4.5), the radiative corrections feature a large coefficient, so it is difficult to estimate reliably a figure for Q_0^2. Both bounds (as shown above) and estimates will depend on this. As reasonably safe estimates we may quote the values

$$\begin{aligned}
\bar{m}_u(Q^2 = 1 \text{ GeV}^2) &= 4.2 \pm 2 \text{ MeV}, \\
\bar{m}_d(Q^2 = 1 \text{ GeV}^2) &= 8.9 \pm 4.3 \text{ MeV}, \\
\bar{m}_s(Q^2 = 1 \text{ GeV}^2) &= 200 \pm 50 \text{ MeV},
\end{aligned} \tag{7.4.9}$$

and we have, to reduce the errors a bit, taken also into account the chiral theory estimates of the mass ratios given in the previous section, Eq. (7.3.6).

7.5 The Decay $\pi^0 \to \gamma\gamma$; the Axial Anomaly

Historically, one of the first motivations for the colour degree of freedom came from the study of the decay $\pi^0 \to \gamma\gamma$, which we now consider in some detail.

The amplitude for the process $\pi^0 \to \gamma\gamma$ may be written, using the reduction formulas, as

$$\langle\gamma(k_1,\lambda_1),\gamma(k_2,\lambda_2)|S|\pi^0(q)\rangle = \frac{-ie^2}{(2\pi)^{9/2}}\,\epsilon_\mu^*(k_1,\lambda_1)\epsilon_\nu^*(k_2,\lambda_2)$$

$$\times \int d^4x_1\,d^4x_2\,d^4z\,e^{i(x_1\cdot k_1 + x_2\cdot k_2 - z\cdot q)}(\Box+m_\pi^2)\langle TJ_{em}^\mu(x_1)J_{em}^\nu(x_2)\phi_{\pi^0}(z)\rangle_0,$$

$$(7.5.1)$$

and we have used the relation $\Box A_{ph}^\mu(x) = J_{em}^\mu(x)$, with A_{ph}^μ the photon field. We leave it as an exercise for the reader to check this, as well as to verify that, in our particular case, one can replace

$$\Box_{x_1}\Box_{x_2}TA_{ph}^\mu(x_1)A_{ph}^\nu(x_2)\phi_{\pi^0}(z) \to T(\Box A_{ph}^\mu(x_1))(\Box A_{ph}^\nu(x_2))\phi_{\pi^0}(z),$$

i.e., that potential delta function terms that appear when the derivatives in the d'Alembertians act on the theta functions $\theta(x_1 - z), \ldots$ implicit in the T-product make no contribution. Separating off the delta of four-momentum conservation, we then find

$$F\left(\pi^0 \to \gamma(k_1,\lambda_1),\gamma(k_2,\lambda_2)\right) = \frac{e^2(q^2 - m_\pi^2)}{\sqrt{2\pi}}\epsilon_\mu^*(k_1,\lambda_1)\epsilon_\nu^*(k_2,\lambda_2)F^{\mu\nu}(k_1,k_2),$$

$$(7.5.2a)$$

where we have defined the VEV

$$F^{\mu\nu}(k_1,k_2) = \int d^4x\,d^4y\,e^{i(x\cdot k_1 + y\cdot k_2)}\langle TJ^\mu(x)J^\nu(y)\phi(0)\rangle_0, \quad q = k_1 + k_2.$$

$$(7.5.2b)$$

We henceforth suppress the indices "em" and "π^0" in J and ϕ respectively.

We next use the equation (7.3.1), generalized to include the π^0:

$$\partial_\mu A_0^\mu(x) = \sqrt{2}f_\pi m_\pi^2\phi(x), \quad \phi \equiv \phi_{\pi^0},$$

$$A_0^\mu(x) = \frac{1}{\sqrt{2}}\left\{\bar{u}(x)\gamma^\mu\gamma_5 u(x) - \bar{d}(x)\gamma^\mu\gamma_5 d(x)\right\}. \quad (7.5.3a)$$

It will prove convenient to use, instead of A_0, the current A_3, defined as

$$A_3^\mu(x) = \left\{\bar{u}(x)\gamma^\mu\gamma_5 u(x) - \bar{d}(x)\gamma^\mu\gamma_5 d(x)\right\}; \quad (7.5.3b)$$

with it, we write

$$F^{\mu\nu}(k_1,k_2) = \frac{1}{f_\pi m_\pi^2}T^{\mu\nu}(k_1,k_2),$$

$$T^{\mu\nu}(k_1,k_2) = \frac{1}{2}\int d^4x\,d^4y\,e^{i(x\cdot k_1 + y\cdot k_2)}\langle TJ^\mu(x)J^\nu(y)\partial\cdot A_3(0)\rangle_0.$$

$$(7.5.4)$$

Up to this point, everything has been exact. The next step involves using the PCAC hypothesis in the following form: we assume that $F(\pi \to \gamma\gamma)$ can be approximated by its leading term in the limit $q \to 0$. On purely kinematic grounds, this is seen to imply that also $k_1, k_2 \to 0$. One may write

$$T^{\mu\nu}(k_1, k_2) = \epsilon^{\mu\nu\alpha\beta} k_{1\alpha} k_{2\beta} \Phi + O(k^3). \qquad (7.5.5)$$

The PCAC hypothesis means that we retain only the first term in Eq. (7.5.5). As will be seen presently, this will lead us to a contradiction, the resolution of which will involve introducing the so-called *axial*, or *triangle anomaly*, and will allow us actually to calculate $T^{\mu\nu}$ exactly to all orders of perturbation theory (in the PCAC approximation).

The first step is to consider the quantity

$$R^{\lambda\mu\nu}(k_1, k_2) = i \int d^4x \, d^4y \, e^{i(x \cdot k_1 + y \cdot k_2)} \langle T J^\mu(x) J^\nu(y) A_3^\lambda(0) \rangle_0. \qquad (7.5.6)$$

On invariance grounds, we may write the general decomposition,

$$R^{\lambda\mu\nu}(k_1, k_2) = \epsilon^{\mu\nu\lambda\alpha} k_{1\alpha} \Phi_1 + \epsilon^{\mu\nu\lambda\alpha} k_{2\alpha} \Phi_2 + O(k^3), \qquad (7.5.7)$$

where the $O(k^3)$ terms are of the form

$$\epsilon^{\mu\lambda\alpha\beta} k_{i\alpha} k_{j\beta} k_{l\lambda} \Phi_{ijl} + \text{three permutations of } i, j, l = 1, 2, 3,$$

and, for quarks with nonzero mass, the Φ are regular as $k_i \to 0$.

The conservation of the e.m. current, $\partial \cdot J = 0$, yields two equations:

$$k_{1\mu} R^{\mu\nu\lambda} = k_{2\nu} R^{\mu\nu\lambda} = 0. \qquad (7.5.8)$$

The first implies

$$\Phi_2 = O(k^2); \qquad (7.5.9a)$$

the second gives

$$\Phi_1 = O(k^2). \qquad (7.5.9b)$$

Now we have, from (7.5.4) and (7.5.6),

$$q_\lambda R^{\lambda\mu\nu}(k_1, k_2) = T^{\mu\nu}(k_1, k_2), \quad \text{i.e.,} \quad \Phi = \Phi_2 - \Phi_1, \qquad (7.5.10)$$

and hence we find the result of Veltman (1967) and Sutherland (1967),

$$\Phi = O(k^2). \qquad (7.5.11)$$

Because the scale for k is m_π, this means that Φ should be of order m_π^2/M^2, where M is a typical hadronic mass. Thus, we expect that Φ would be vanishing in the chiral limit, and hence very small in the real world. Now, this is in disagreement with experiment, as the decay $\pi^0 \to 2\gamma$ is in no way suppressed;

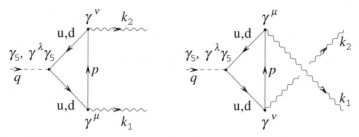

Fig. 7.5.1. Diagrams connected with the anomaly ($\pi^0 \to \gamma\gamma$ decay).

but worse still, (7.5.11) contradicts a direct calculation. In fact, we may use the equations of motion and write

$$\partial_\mu A_3^\mu(x) = 2\mathrm{i}\left\{ m_u \bar{u}(x)\gamma_5 u(x) - m_d \bar{d}(x)\gamma_5 d(x) \right\}. \tag{7.5.12}$$

We will calculate first neglecting strong interactions; (7.5.11) should certainly be valid in this approximation. This involves the diagrams of Fig. 7.5.1 with a γ_5 vertex. The result, as first obtained by Steinberger (1949) is, in the limit $k_1, k_2 \to 0$, and defining $\delta_u = 1$, $\delta_d = -1$,

$$
\begin{aligned}
T^{\mu\nu}(k_1, k_2) &= 2N_c \sum_{f=u,d} \delta_f Q_f^2 m_f \\
&\times \int \frac{\mathrm{d}^4 p}{(2\pi)^4} \frac{\operatorname{Tr}\gamma_5(\not{p} + \not{k}_1 + m_f)\gamma^\mu(\not{p} + m_f)\gamma^\nu(\not{p} - \not{k}_2 + m_f)}{[(p+k_1)^2 - m_f^2][(p-k_2)^2 - m_f^2](p^2 - m_f^2)} \\
&= -\frac{1}{4\pi^2}\epsilon^{\mu\nu\alpha\beta}k_{1\alpha}k_{2\beta}\left\{ 3(Q_u^2 - Q_d^2) \right\} + O(k^4) \\
&= -\frac{1}{4\pi^2}\epsilon^{\mu\nu\alpha\beta}k_{1\alpha}k_{2\beta} + O(k^4).
\end{aligned}
$$

The factor $N_c = 3$ comes from the sum over the three colours of the quarks and the factor 2 from the two diagrams in Fig. 7.5.1 (which in fact contribute equally to the amplitude). We thus find that

$$\Phi = -\frac{1}{4\pi}, \tag{7.5.13}$$

which contradicts (7.5.11). This is the *triangle anomaly* (Bell and Jackiw, 1969; Adler, 1969).

What is wrong here? Clearly, we cannot maintain (7.5.12), which was obtained with free-field equations of motion, $\mathrm{i}\not{\partial} q = m_q q$; we must admit that in the presence of interactions with vector fields (the photon field in our case),

Eq. (7.5.12) is no longer valid. To obtain agreement with (7.5.13) we have to write (Adler, 1969)

$$\partial_\mu A_3^\mu(x) = 2\mathrm{i}\left\{m_u \bar{u}(x)\gamma_5 u(x) - m_d \bar{d}(x)\gamma_5 d(x)\right\}$$
$$+ N_c(Q_u^2 - Q_d^2)\frac{e^2}{16\pi^2}F_{\mu\nu}(x)\widetilde{F}^{\mu\nu}(x), \tag{7.5.14}$$

where the *dual* $\widetilde{F}$ has been defined as

$$\widetilde{F}^{\mu\nu} = \tfrac{1}{2}\epsilon^{\mu\nu\alpha\beta}F_{\alpha\beta}, \quad F^{\mu\nu} = \partial^\mu A_{\mathrm{ph}}^\nu - \partial^\nu A_{\mathrm{ph}}^\mu.$$

More generally, for fermion fields interacting with vector fields with strength h, we find

$$\partial_\mu \bar{f}\gamma^\mu\gamma_5 f = 2\mathrm{i}m_f \bar{f}\gamma_5 f + \frac{T_F h^2}{8\pi^2}H^{\mu\nu}\widetilde{H}_{\mu\nu}; \tag{7.5.15}$$

$H^{\mu\nu}$ is the vector field strength tensor.

Let us return to the decay $\pi^0 \to 2\gamma$. From (7.5.13) we calculate the amplitude, in the PCAC limit $m_\pi \sim 0$,

$$F(\pi^0 \to 2\gamma) = \frac{\alpha}{\pi}\frac{\epsilon^{\mu\nu\alpha\beta}k_{1\alpha}k_{2\beta}\epsilon_\mu^*(k_1,\lambda_1)\epsilon_\mu^*(k_2,\lambda_2)}{\sqrt{2\pi}}, \tag{7.5.16}$$

and the decay rate

$$\Gamma(\pi^0 \to 2\gamma) = \left(\frac{\alpha}{\pi}\right)^2\frac{m_\pi^3}{64\pi f_\pi^2} \approx 7.25 \times 10^{-6} \text{ MeV},$$

to be compared with the experimental figure,

$$\Gamma_{\exp}(\pi^0 \to 2\gamma) = 7.95 \times 10^{-6} \text{ MeV}.$$

Actually, the sign of the decay amplitude can also be measured (from the Primakoff effect) and it agrees with the theory. It is important to note that, if we had no colour, our result would have decreased by a factor $1/N_c^2$, i.e., it would have been off experiment by a full order of magnitude.

One may wonder what credibility to attach to this calculation: after all, it was made to zero order in α_s. In fact, the calculation is exact to all orders in QCD;[12] the only approximation is the PCAC one $m_\pi \approx 0$. To show this we will give an alternate derivation of the basic result, Eq. (7.5.13). Let us then return to (7.5.6). To zero order in α_s,

$$R^{\mu\nu\lambda} = \sum \delta_f Q_f^2$$
$$\times \int \frac{\mathrm{d}^4 p}{(2\pi)^4}\frac{\mathrm{Tr}\,\gamma^\lambda\gamma_5(\not{p} + \not{k}_1 + m_f)\gamma^\mu(\not{p} + m_f)\gamma^\nu(\not{p} - \not{k}_2 + m_f)}{[(p+k_1)^2 - m_f^2][(p-k_2)^2 - m_f^2](p^2 - m_f^2)}$$
$$+ \text{ crossed term}$$

---

[12]The proof is essentially contained in the original paper of Adler and Bardeen (1969). See also Wilson (1969), Crewther (1972) and Bardeen (1974).

(Fig. 7.5.1 with the $\gamma^\lambda \gamma_5$ vertices). More generally, we regulate the integral by working in dimension D, and consider an arbitrary axial triangle with

$$R_{ijl}^{\mu\nu\lambda} = 2 \int \frac{d^D p}{(2\pi)^D} \; \text{Tr} \, \gamma^\lambda \gamma_5 \frac{1}{\not{p} + \not{k}_1 - m_i} \gamma^\mu \frac{1}{\not{p} - m_j} \gamma^\nu \frac{1}{\not{p} - \not{k}_2 - m_l}. \tag{7.5.17}$$

We would like to calculate $q_\lambda R_{ijl}^{\mu\nu\lambda}$. Writing identically

$$(\not{k}_1 + \not{k}_2)\gamma_5 = -(\not{p} - \not{k}_2 - m_l)\gamma_5 + (\not{p} + \not{k}_1 + m_i)\gamma_5 - (m_i + m_l)\gamma_5,$$

we have

$$q_\lambda R_{ijl}^{\mu\nu\lambda} = -2(m_i + m_l)$$
$$\times \int \frac{d^D p}{(2\pi)^D} \frac{\text{Tr} \, \gamma_5 (\not{p} + \not{k}_1 + m_i)\gamma^\mu (\not{p} + m_j)\gamma^\nu (\not{p} - \not{k}_2 + m_l)}{[(p+k_1)^2 - m_i^2][(p-k_2)^2 - m_l^2](p^2 - m_j^2)}$$
$$+ a_{ijl}^{\mu\nu\lambda}, \tag{7.5.18a}$$

$$a_{ijl}^{\mu\nu\lambda} = -2 \int d^D \hat{p} \; \text{Tr} \{ (\not{p} - \not{k}_2 - m_l)\gamma_5 - (\not{p} + \not{k}_1 + m_i)\gamma_5 \} \tag{7.5.18b}$$
$$\times \frac{1}{\not{p} + \not{k}_1 - m_i} \gamma^\mu \frac{1}{\not{p} - m_j} \gamma^\nu \frac{1}{\not{p} - \not{k}_2 - m_l}.$$

The first term on the right hand side of (7.5.18a) is what we would have obtained by naive use of the equations of motion, $\partial_\mu \bar{q}_i \gamma^\mu \gamma_5 q_l = i(m_i + m_l) \bar{q}_i \gamma_5 q_l$; $a_{ijl}^{\mu\nu\lambda}$ is the anomaly. If we accepted the commutation relations $\{\gamma^\mu, \gamma_5\} = 0$ also for dimension $D \neq 4$, we could rewrite it as

$$a_{ijl}^{\mu\nu\lambda} = -2 \int d^D \hat{p} \left\{ \text{Tr} \, \gamma_5 \frac{1}{\not{p} + \not{k}_1 - m_i} \gamma^\mu \frac{1}{\not{p} - m_l} \gamma^\nu \right.$$
$$\left. + \text{Tr} \, \gamma_5 \gamma^\mu \frac{1}{\not{p} - m_j} \gamma^\mu \frac{1}{\not{p} - \not{k}_2 - m_l} \right\}. \tag{7.5.18c}$$

Then we could conclude that $a_{ijl}^{\mu\nu\lambda}$ vanishes because each of the terms in (7.5.18c) consists of an antisymmetric tensor that depends on a single vector (k_1 for the first term, k_2 for the second) and this is zero. It is thus clear that the nonvanishing of $a_{ijl}^{\mu\nu\lambda}$ is due to the fact that it is given by an ultraviolet divergent integral: if it was convergent, one could take $D \to 4$ and $a_{ijl}^{\mu\nu\lambda}$ would vanish. Incidentally, this shows that $a_{ijl}^{\mu\nu\lambda}$ is actually independent of the masses because $(\partial/\partial m)a_{ijl}^{\mu\nu\lambda}$ is convergent, and thus the former argument applies. We may therefore write $a_{ijl}^{\mu\nu\lambda} = a^{\mu\nu}$, where $a^{\mu\nu}$ is obtained by setting all masses to zero. A similar argument shows that $a^{\mu\nu}$ has to be of the form

$$a^{\mu\nu}(k_1, k_2) = a\epsilon^{\mu\nu\alpha\beta} k_{1\alpha} k_{2\beta}, \quad a = \text{constant}, \tag{7.5.19a}$$

and thus we may obtain a as

$$a\epsilon^{\mu\nu\alpha\beta} = \frac{\partial^2}{\partial k_{1\alpha}\partial k_{2\beta}}a^{\mu\nu}(k_1, k_2)\bigg|_{k_i=0}. \tag{7.5.19b}$$

If we could write the formula (7.5.18c) for a, we would immediately conclude from (7.5.19b) that $a = 0$, in contradiction with the Veltman–Sutherland theorem. But this is easily seen to be inconsistent: if we would have shifted variables in (7.5.18c), say $p \to p - \xi k_2$, we would have found a finite but nonzero value, actually ξ-dependent for a, $a = -\xi/2\pi^2$. This shows that the commutation relations[13] $\{\gamma^\mu, \gamma_5\} = 0$ cannot be accepted for $D \neq 0$, for they lead to an undefined value for the anomaly. If, however, we start from (7.5.18b) and refrain from commuting γ_5 and γ^μs,

$$a\epsilon^{\mu\nu\alpha\beta} = -2\int d^D\hat{p}\,\text{Tr}\,\gamma_5\left\{\frac{1}{\not{p}}\gamma^\alpha\frac{1}{\not{p}}\gamma^\mu\frac{1}{\not{p}}\gamma^\nu\frac{1}{\not{p}}\gamma^\beta - \frac{1}{\not{p}}\gamma^\mu\frac{1}{\not{p}}\gamma^\nu\frac{1}{\not{p}}\gamma^\beta\frac{1}{\not{p}}\gamma^\alpha\right\}.$$

Performing symmetric integration (Appendix B) and using only the rules of Appendix A for $D \neq 4$, we obtain an unambiguous result:

$$a\epsilon^{\mu\nu\alpha\beta} = \frac{8(D-1)(4-D)}{D(D+2)}\frac{i}{16\pi^2}\frac{2}{4-D}\,\text{Tr}\,\gamma_5\gamma^\mu\gamma^\nu\gamma^\alpha\gamma^\beta + O(4-D)$$
$$\xrightarrow[D\to 4]{} \frac{1}{2\pi^2}\epsilon^{\mu\nu\alpha\beta}.$$

This is one of the peculiarities of the anomaly: a *finite* Feynman integral whose value depends on the regularization prescription. Fortunately, we may eschew the problem by using the Veltman–Sutherland theorem to conclude that, at any rate, there is a *unique* value of $a^{\mu\nu}$ compatible with gauge invariance for the e.m. current, viz.,

$$a^{\mu\nu}_{ijl} = a^{\mu\nu} = -\frac{1}{2\pi^2}\epsilon^{\mu\nu\alpha\beta}k_{1\alpha}k_{2\beta}. \tag{7.5.20}$$

We have explicitly checked that our regularization leads to precisely this value; to verify that it also respects gauge invariance is left as as simple exercise.

---

[13]These commutation relations are actually self-contradictory. For example, using only the relations of Appendix A for $D \neq 4$, we have

$$\text{Tr}\,\gamma_5\gamma^\alpha\gamma^\mu\gamma^\nu\gamma^\rho\gamma_\alpha\gamma^\sigma = (6 - D)\,\text{Tr}\,\gamma_5\gamma^\mu\gamma^\nu\gamma^\rho\gamma^\sigma,$$

while, if we allow γ_5 anticommutation, we can obtain

$$\text{Tr}\,\gamma_5\gamma^\alpha\gamma^\mu\gamma^\nu\gamma^\rho\gamma_\alpha\gamma^\sigma = -\,\text{Tr}\,\gamma_5\gamma^\mu\gamma^\nu\gamma^\rho\gamma_\alpha\gamma^\sigma\gamma^\alpha = (D-2)\,\text{Tr}\,\gamma_5\gamma^\mu\gamma^\nu\gamma^\rho\gamma^\sigma,$$

which differs from the former by a term $O(D-4)$. These problems, however, only arise for arrays with an odd number of γ_5 and at least four other gammas.

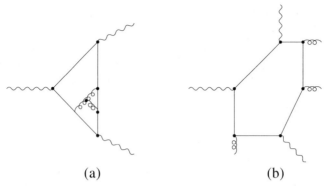

(a) (b)

Fig. 7.5.2. (a) A nonanomalous diagram. (b) "Opened" diagram corresponding to (a).

Before continuing, a few words on the Veltman–Sutherland theorem for zero quark masses are necessary. In this case, the first term on the right hand side of (7.5.18a) is absent: it would appear that we could not maintain our result for the anomaly, Eq. (7.5.20), because this would imply

$$q_\lambda R^{\mu\nu\lambda}_{ijl} = -\frac{1}{2\pi^2}\epsilon^{\mu\nu\alpha\beta}k_{1\alpha}k_{2\beta} \neq 0,$$

thus contradicting the Veltman–Sutherland conclusion, $q_\lambda R^{\mu\nu\lambda}_{ijl} = 0$. This is not so. The relation $q_\lambda R^{\mu\nu\lambda}_{ijl} = a^{\mu\nu}$ and the value of $a^{\mu\nu}$ are correct. What occurs is that for vanishing masses the functions Φ_i in (7.5.7) possess singularities of the type $1/k_1 \cdot k_2$, singularities coming from the denominators in, for example, Eq. (7.5.17) when $m_i = 0$. Therefore, the Veltman–Sutherland theorem is *not* applicable. This is yet another peculiarity of the anomalous triangle: we have the relation

$$\lim_{m \to 0} q_\lambda R^{\mu\nu\lambda}_{ijl} = 0$$

but, if we begin with $m = 0$,

$$q_\lambda R^{\mu\nu\lambda}_{m\equiv 0} = a^{\mu\nu} \neq 0.$$

Let us return to our original discussion, in particular for $m \neq 0$. The present method shows how one can prove that the result does not get renormalized. The Veltman–Sutherland theorem is exact; so we have actually shown that it is sufficient to prove that (7.5.20) is not altered by higher orders in α_s. Now, consider a typical higher order contribution (Fig. 7.5.2a). It may be written as an integral over the gluon momenta and an integral over the quark momenta. But for the latter, the triangle has become an hexagon (Fig. 7.5.2b) for which the quark integral is convergent and here the limit $D \to 4$ may be taken: it vanishes identically. In addition, the above arguments have shown that the anomaly is in fact related to the large momentum behaviour of the theory and thus we expect that the exactness of (7.5.13) will not be spoiled by nonperturbative effects.

We will not make the proof more precise, but refer to the literature.[14] However, we will present an alternative derivation (Wilson, 1969) that will clearly reveal the short-distance character of the anomaly. An axial current involves products of two fields at the same space–time point, so it should be properly defined as

$$A_q^\mu(x) = \lim_{\xi \to 0} A_{q,\mathrm{gn}}^\mu(x, \xi),$$
$$A_{q,\mathrm{gn}}^\mu(x, \xi) \equiv \bar{q}\,(x + \xi/2)\,\gamma^\mu \gamma_5 q\,(x - \xi/2)\,.$$

$$(7.5.21\text{a})$$

For $\xi \neq 0$ this is, however, not gauge invariant. To restore gauge invariance we have to replace $A_{q,\mathrm{gn}}^\mu$ in (7.5.21a) by (cf. Appendix I)

$$A_{q,\mathrm{gn}}^\mu \to A_{q,\mathrm{gi}}^\mu(x, \xi) \equiv \bar{q}\,(x + \xi/2)\,\gamma^\mu \gamma_5 \mathrm{e}^{\,\mathrm{i}e \int_{x-\xi/2}^{x+\xi/2} \mathrm{d}y_\alpha\, A_{\mathrm{ph}}^\alpha(y)}\, q\,(x - \xi/2)\,.$$
$$(7.5.21\text{b})$$

Thus,

$$\partial_\mu A_{q,\mathrm{gi}}^\mu(x, \xi) = \lim_{\xi \to 0}\left\{ 2\mathrm{i}m_q \bar{q}(x)\gamma_5 q(x) + \mathrm{i}e A_{q,\mathrm{gi}}^\mu(x, \xi) F_{\mu\alpha}\xi^\alpha + O(\xi^2)\right\}.$$

Because $A_{q,\mathrm{gi}}^\mu(x, \xi)$ diverges as $1/\xi$ for $\xi \to 0$, the second term on the right hand side does not vanish in this limit. The explicit calculations (Wilson, 1969; Crewther, 1972) show that, as could be expected, Eq. (7.5.14) is reproduced.

The axial current is not the only one that possesses anomalies. The trace of the energy–momentum tensor Θ_μ^μ is also anomalous, due to the fact that renormalization breaks scale invariance. This is discussed in some detail by Callan, Coleman and Jackiw (1970) and, in the context of QCD, by Collins, Duncan and Joglekar (1977). This anomaly is rather harmless; indeed, its analysis is closely related to that of the renormalization group.

---

[14]For a detailed discussion, see the reviews of Adler (1971) and Ellis (1976). The triangle graph is the only one that has *primitive* anomalies; it does, however, induce secondary anomalies in square and pentagon graphs. The triangle with three axial currents has an anomaly closely related to the one we have discussed, cf. the text of Taylor (1976). An elegant discussion of currents with anomalies for arbitrary interaction may be found in Wess and Zumino (1971). The derivation of the anomaly in the context of the path integral formulation of field theory, where it is connected with the *divergence* of the measure, may be found in Fujikawa (1980, 1984, 1985).

7.6 The $U(1)$ Problem. The Gluon Anomaly

In the previous section, we discussed the triangle anomaly in connection with the decay $\pi^0 \to \gamma\gamma$. As remarked there, the anomaly is not restricted to photons; in particular, we have a gluon anomaly. Defining the current

$$A_0^\mu = \sum_{f=1}^n \bar{q}_f \gamma^\mu \gamma_5 q_f, \qquad (7.6.1)$$

we find that it has an anomaly

$$\partial^\mu A_0^\mu = i \sum_{f=1}^n 2m_f \bar{q}_f \gamma_5 q_f + \frac{ng^2}{16\pi^2} \tilde{G}G, \qquad (7.6.2)$$

where

$$\tilde{G}_a^{\mu\nu} \equiv \tfrac{1}{2}\epsilon^{\mu\nu\alpha\beta} G_{a\alpha\beta}, \quad \tilde{G}G = \sum_a \tilde{G}_a^{\mu\nu} G_{a\mu\nu}.$$

The current (7.6.1) is the so-called $U(1)$ current (pure flavour singlet) and is atypical in more respects than one. In particular, it is associated with the $U(1)$ problem, to which we now turn.

Assume that we have n light quarks; we only consider these and will neglect (as irrelevant to the problem at hand) the existence of heavy flavours. We may take $n = 2$ (u, d) and then we speak of "the $U(1)$ problem of $SU(2)$" or $n = 3$ (u, d, s), which is the $SU(3)$ $U(1)$ problem. Consider now the $n^2 - 1$ matrices in flavour space $\lambda_1, \ldots, \lambda_{n^2-1}$; for $SU(3)$ they coincide with the Gell-Mann matrices, and for $SU(2)$ with the Pauli matrices. Define further $\lambda_0 \equiv 1$. Any $n \times n$ Hermitian matrix may be written as a linear combination of the n^2 matrices λ_α, $\alpha = 0, 1, \ldots, n^2 - 1$. Because of this completeness, it is sufficient to consider the currents

$$A_\alpha^\mu = \sum_{ff'} \bar{q}_f \gamma^\mu \gamma_5 \lambda_{ff'}^\alpha q_{f'}; \quad \alpha = 0, 1, \ldots, n^2 - 1.$$

Of course, only A_0 has an anomaly. We will consistently let the indices $a, b, \ldots$ for the currents run from 1 to $n^2 - 1$, and we will let Greek indices $\alpha, \beta, \ldots$ also include the value 0.

Now let $N_1(x), \ldots, N_k(x)$ denote local operators (simple or composite) and consider the quantity

$$\langle \text{vac}|\text{T}A_\alpha^\mu(x) \prod_j N_j(x_j)|\text{vac}\rangle. \qquad (7.6.3)$$

For $\alpha = a \neq 0$, the Goldstone theorem implies that the masses of the pseudoscalar particles P_a with the quantum numbers of the A_a vanish in the chiral limit; introducing a common parameter ϵ for all the quark masses by letting $m_f = \epsilon r_f$, $f = 1, \ldots, n$, where the r_f remain fixed in the chiral limit, we have

$$m_{P_a}^2 \approx \epsilon. \tag{7.6.4}$$

This was shown in Sect. 7.3, Eqs. 7.3.4, 5. Therefore, in this limit, the quantity (7.6.3) develops a pole at $q^2 = 0$, for $\alpha = a \neq 0$. To be precise, what this means is that in the chiral limit (zero quark masses),

$$\lim_{q \to 0} \int d^4 x \, e^{iq \cdot x} \langle \text{vac} | T A_\alpha^\mu(x) \prod_j N_j(x_j) | \text{vac} \rangle \approx (\text{const.}) \times q^\mu \frac{1}{q^2}. \tag{7.6.5}$$

If we neglect anomalies, the derivation of (7.6.4) can be repeated for the case $\alpha = 0$ and we would thus find that the $U(1)$ (flavour singlet) particle P_0 would also have vanishing mass in the chiral limit (Glashow, 1968). This statement was made more precise by Weinberg (1975) who proved the bound $m_{P_0} \leq \sqrt{n} \times$ (average m_{P_a}). Now, this is a catastrophe since, for the $SU(2)$ case, $m_\eta \gg \sqrt{2}\, m_\pi$ and, for $SU(3)$, the mass of the η' particle also violates the bound. This is the $U(1)$ problem. In addition, Brandt and Preparata (1970) proved that under these conditions the decay $\eta \to 3\pi$ is forbidden, which is also in contradiction with experiment. We are thus led to *assume* that (7.6.3) remains regular as $\epsilon \to 0$ for $\alpha = 0$. If we could *prove* that this is so, we would have solved the $U(1)$ problem. This will be discussed later on; for the moment we shall assume that there are no massless $U(1)$ bosons, without asking for a proof. It is quite clear that, if there was no anomaly, this assumption would be inconsistent, so it looks a good strategy to see what we can obtain from the interplay of the absence of P_0 Goldstone bosons and the existence of an anomaly for the A_0 current. We will proceed to do this, following the excellent review of Crewther (1979b).

The current A_0, as defined in (7.6.1), is gauge invariant but not $U(1)$ invariant: its divergence does not vanish due to the anomaly, Eq. (7.6.2). We may construct another current which is $U(1)$ invariant in the chiral limit (but is not gauge invariant), as shown by Adler (1969) for the Abelian case and by Bardeen (1974) in general. We define

$$\hat{A}_0^\mu = A_0^\mu - 2nK^\mu, \tag{7.6.6}$$

where K^μ is the purely gluonic current,

$$K^\mu(x) = \frac{g^2}{16\pi^2} \epsilon^{\mu\nu\rho\sigma} \sum B_{a\nu}(x) \left\{ \partial_\rho B_{a\sigma}(x) + \tfrac{1}{3} f_{abc} B_{b\rho}(x) B_{c\sigma}(x) \right\}. \tag{7.6.7}$$

That this is a correct construction may be easily checked by noting that

$$\partial_\mu K^\mu = \frac{g^2}{32\pi^2} \tilde{G} G, \tag{7.6.8}$$

so that using (7.6.2) we obtain, in the chiral limit,

$$\partial_\mu \hat{A}_0^\mu = 0. \tag{7.6.9}$$

It should be remarked that K is not unique, even requiring (7.6.8): indeed, it is gauge dependent. Another useful remark is that, in principle, Eq. (7.6.6) is defined for the bare quantities; but we may always renormalize in such a way that it remains valid after renormalization. The reason, of course, is that the anomaly does not get renormalized.

The generator of $U(1)$ transformations must be the current that is conserved, viz., $\hat{A}_0$. We therefore define the *chiralities* χ, quantum numbers associated with the $U(1)$ symmetry,

$$\delta(x^0 - y^0)[\hat{A}_0^0(x), N_j(y)] = -\chi_j \delta(x - y)N_j(y), \qquad (7.6.10a)$$

or, in integrated form,

$$[\hat{Q}_0, N_j] = -\chi_j N_j, \qquad (7.6.10b)$$

where we have defined the $U(1)$ chiral charge operator

$$\hat{Q} = \int \mathrm{d}^3\mathbf{x}\, \hat{A}_0^0(x). \qquad (7.6.11)$$

Since $\hat{A}$ verifies (7.6.9), $\hat{Q}$ is time independent and hence we will expect not only that (7.6.10) makes sense, but that the numbers χ_j will not become renormalized. To prove this more formally, consider the VEV

$$\langle \mathrm{vac}|\mathrm{T}\hat{A}_0^\mu(x) \prod_j N_j(x_j)|\mathrm{vac}\rangle,$$

and apply ∂_μ to it. We obtain the Ward identity,

$$\partial_\mu \langle \mathrm{vac}|\mathrm{T}\hat{A}_0^\mu(x) \prod_j N_j(x_j)|\mathrm{vac}\rangle$$

$$= -\left\{\sum_l \chi_l \delta(x - y)\right\} \langle \mathrm{vac}|\mathrm{T} \prod_j N_j(x_j)|\mathrm{vac}\rangle; \qquad (7.6.12)$$

we have used (7.6.9) and (7.6.10a). Since $\hat{A}$ is (partially) conserved, we know that it is not renormalized, and so the χ must share this property. In the following section we will see that (7.6.12), plus the absence of massless $U(1)$ bosons, leads to peculiar properties of the QCD vacuum.

7.7 The θ Parameter; the QCD Vacuum; the Effect of Massless Quarks; Solution to the $U(1)$ Problem

So far, we have been working with the QCD Lagrangian (omitting gauge fixing and ghost terms)

$$\mathcal{L} = \sum_q \bar{q}(i\not{D} - m_q)q - \tfrac{1}{4}GG; \qquad (7.7.1)$$

we now ask what would be the modifications introduced by adding a term

$$\mathcal{L}_{1\theta} = -\frac{\theta g^2}{32\pi^2}\widetilde{G}G, \qquad (7.7.2a)$$

obtaining

$$\mathcal{L}_\theta = \mathcal{L} + \mathcal{L}_{1\theta}. \qquad (7.7.2b)$$

In fact, $\mathcal{L}_{1\theta}$ is the only extra term that can be added to $\mathcal{L}$ and which is allowed by gauge invariance and renormalizability. Moreover, as shown in the previous section, Eq. (7.6.8), it is a four-divergence and thus leaves the equations of motion unchanged. Certainly we can dispose of it by setting $\theta = 0$; but, although there are indications that θ is very small indeed, there are also reasons why it may be nonzero. At any rate, it is of interest to find the implications of choosing the more general form (7.7.2).

First, because we are adding a new interaction, we expect the physical vacuum to depend on it, so we write $|\theta\rangle$ for it. Our next task is to explore the θ dependence of the Green's functions. To do this, consider the *topological charge operator*[15]

$$Q_K = \frac{g^2}{32\pi^2} \int \mathrm{d}^4x\, \widetilde{G}G; \qquad (7.7.3)$$

we may use (7.6.8) and Gauss's theorem to write it as a surface integral:

$$Q_K = \int \mathrm{d}s_\mu K^\mu.$$

We will choose as the surface of integration that of a cylinder oriented along the time axis, with bases at $t_+ \to +\infty$ and $t_- \to -\infty$ (Fig. 7.7.1). When the sides approach infinity, we find

$$Q_K = \int \mathrm{d}^3\mathbf{x}\, K^0(t_+ \to +\infty, \mathbf{x}) - \int \mathrm{d}^3\mathbf{x}\, K^0(t_- \to -\infty, \mathbf{x}) \equiv K_+ - K_-. \qquad (7.7.4)$$

---

[15]More about the θ-vacua and the topics of this section will be found in Sect. 8.4 where, in particular, the reasons for some seemingly peculiar names will become apparent.

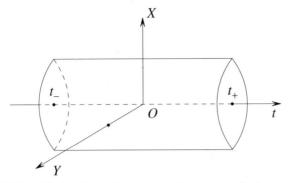

Fig. 7.7.1. The region of integration for the topological charge.

The operators $K_\pm$ are Hermitian and related one to another by time reversal, so their spectra coincide. We label their eigenstates as $|n_\pm\rangle \equiv |n, t_\pm \to \pm\infty\rangle$, such that

$$K_\pm|n_\pm\rangle = n|n_\pm\rangle. \tag{7.7.5}$$

Because of the hermiticity of the $K_\pm$, the $|n_\pm\rangle$ form complete bases: we may expand the physical vacuum as

$$|\theta\rangle = \sum c_n(\theta)|n_+\rangle = \sum c_n(\theta)|n_-\rangle; \tag{7.7.6}$$

the c_n are the same in both bases. Indeed, the vacuum is invariant under time translations and hence we may take it at any time, in particular $t = 0$: so, applying time reversal, we find (7.7.6) with equal c_n.

We then have to determine the c_n. To do so, apply the operation $i\partial/\partial\theta$ to a Green's function. Recalling the formalism of Sect. 1.2,

$$i\frac{\partial}{\partial\theta}\langle\theta|T\prod N_j(x_j)|\theta\rangle$$

$$= i\frac{\partial}{\partial\theta}\langle 0|T\prod N_j^0(x_j)e^{i\int d^4x\,\{\mathcal{L}_{\text{int}}^0(x)+\mathcal{L}_{1\theta}^0(x)\}}|0\rangle$$

$$= \frac{g^2}{32\pi^2}\int d^4x\,\langle 0|T\tilde{G}^0(x)G^0(x)\prod N_j^0(x_j)e^{i\int d^4x\,\{\mathcal{L}_{\text{int}}^0(x)+\mathcal{L}_{1\theta}^0(x)\}}|0\rangle$$

$$= \frac{g^2}{32\pi^2}\int d^4x\,\langle\theta|T\tilde{G}(x)G(x)\prod N_j(x_j)|\theta\rangle : \tag{7.7.7}$$

we find that the operation $i\partial/\partial\theta$ is equivalent to the insertion of Q_K. Using (7.7.3) and (7.7.4) and because $+\infty$ is later, and $-\infty$ earlier than any time, (7.7.7) becomes

$$i\frac{\partial}{\partial\theta}\langle\theta|T\prod N_j(x_j)|\theta\rangle = \langle\theta|K_+T\prod N_j(x_j)|\theta\rangle - \langle\theta|T\prod N_j(x_j)K_-|\theta\rangle.$$

Expanding as in (7.7.6), we obtain the equation

$$\mathrm{i}\frac{\partial}{\partial\theta}\sum_{n,k}c_n^*(\theta)c_k(\theta) = \sum_{n,k}(n-k)c_n^*(\theta)c_k(\theta)$$

with the solution

$$c_n(\theta) = Ce^{in\theta}. \tag{7.7.8}$$

The constant C is arbitrary, and may be taken to be unity.

A more rigorous derivation may be found in the review of Crewther (1979a); later, in Sect. 8.4, we will present an alternative derivation of these results.

A first consequence of (7.7.8) is that different θ-vacua are orthogonal:

$$\langle\theta|\theta'\rangle = \delta(\theta - \theta'), \tag{7.7.9}$$

so each value of θ (up to periodicity) characterizes a different world.

Until now we have not taken into account the existence of fermions. We will now describe how the analysis is modified if we introduce n fermions of vanishing mass. We begin by rewriting our familiar Ward identity (7.6.12) as

$$\partial_\mu\langle\theta|\mathrm{T}\hat{A}_0^\mu(x)\prod_j N_j(x_j)|\theta\rangle = -\left\{\sum_l \chi_l\delta(x-y)\right\}\langle\theta|\mathrm{T}\prod_j N_j(x_j)|\theta\rangle,$$

and integrate it with d^4x:

$$\int \mathrm{d}^4x\,\partial_\mu\langle\theta|\mathrm{T}\hat{A}_0^\mu(x)\prod_j N_j(x_j)|\theta\rangle = -\left(\sum_l \chi_l\right)\langle\theta|\mathrm{T}\prod_j N_j(x_j)|\theta\rangle.$$

Using (7.6.6) and (7.6.8), we find

$$\int \mathrm{d}^4x\,\partial_\mu\langle\theta|\mathrm{T}\sum_f \bar{q}_f(x)\gamma^\mu\gamma_5 q_f(x)\prod_j N_j(x_j)|\theta\rangle$$

$$= 2n\frac{g^2}{32\pi^2}\int \mathrm{d}^4x\,\langle\theta|\mathrm{T}\tilde{G}(x)G(x)\prod_j N_j(x_j)|\theta\rangle - \left(\sum_l \chi_l\right)\langle\theta|\mathrm{T}\prod_j N_j(x_j)|\theta\rangle. \tag{7.7.10}$$

Two remarks are in order. Clearly,

$$\int \mathrm{d}^4x\,\partial_\mu\langle\theta|\mathrm{T}\sum_f \bar{q}_f(x)\gamma^\mu\gamma_5 q_f(x)\prod_j N_j(x_j)|\theta\rangle$$

$$= -\lim_{q\to 0}\mathrm{i}q_\mu\int \mathrm{d}^4x\,e^{iq\cdot x}\langle\theta|\mathrm{T}\sum_f \bar{q}_f(x)\gamma^\mu\gamma_5 q_f(x)\prod_j N_j(x_j)|\theta\rangle.$$

Now, if there is no massless $U(1)$ boson, this VEV above has no pole at $q^2 = 0$, so the limit as $q \to 0$ vanishes. But, as we saw earlier, insertion of Q_K is equivalent to $i\partial/\partial\theta$: so (7.7.10) becomes

$$2ni\frac{\partial}{\partial\theta}\langle\theta|\mathrm{T}\prod N_j(x_j)|\theta\rangle = \left(\sum \chi_l\right)\langle\theta|\mathrm{T}\prod N_j(x_j)|\theta\rangle. \tag{7.7.11}$$

For massless quarks, the vacuum is invariant under chiral rotations:

$$|\theta\rangle = U_\varphi|\theta\rangle, \quad U_\varphi = \mathrm{e}^{-i\varphi\hat{Q}_0}; \tag{7.7.12}$$

using (7.6.10b), on the other hand, we have

$$i\frac{\partial}{\partial\varphi}U_\varphi^{-1}\prod N_j U_\varphi = \left(\sum \chi_l\right)U_\varphi^{-1}\prod N_j U_\varphi, \tag{7.7.13}$$

so the right hand side of (7.7.11) may be rewritten as

$$i\frac{\partial}{\partial\varphi}\langle\theta|\mathrm{T}\prod_j N_j(x_j)|\theta\rangle :$$

we find that the operation

$$2ni\frac{\partial}{\partial\theta} - i\frac{\partial}{\partial\varphi}$$

annihilates all Green's functions. This means that a charge in θ may be compensated for by a change in φ. Therefore, the theory is equivalent to one with $\theta = 0$, because it is certainly chiral invariant. Thus it follows that, in the special case where the quarks are massless, the θ parameter may be taken to be zero,[16] and the old QCD Lagrangian $\mathcal{L}$ of (7.7.1) is actually the most general one.

One may argue that the quark masses are of weak origin, generated in the manner discussed by Higgs and Weinberg so, for pure QCD, quarks should be assumed to be massless. However, we are interested in the real world, and thus the effects of perturbing QCD by weak and electromagnetic interactions (at least to first order) cannot be eschewed.[17] It would also seem that, since $\mathcal{L}_{1\theta}$ violates time reversal and parity invariance, we could put it to zero by requiring invariance under P, T. Again, this view cannot be maintained. Weak interactions violate P and T, and some of this may seep into strong interactions. If this is the origin of θ, however, there are reasonable

---

[16]A more detailed analysis shows that it is enough that *one* quark is massless. This result was first obtained by Peccei and Quinn (1977).

[17]Another possibility to obtain $\theta = 0$ is to use a system of Higgs fields which is nonminimal (Peccei and Quinn, 1977). This can be shown to lead to the existence of a new pseudoscalar boson, the "axion" (Weinberg, 1978b; Wilczek, 1978). There is not enough experimental evidence to rule out completely the existence of this particle.

arguments (Ellis and Gaillard, 1979) that the effect is small, provided that θ_{QCD} is originally zero.

Perhaps it is more profitable to discuss experimental bounds on θ. As will be argued later (Sect. 8.4), the effects of $\mathcal{L}_{1\theta}$ on processes such as deep inelastic scattering are quite negligible; the only place where one can obtain a substantial effect is in T and P violating effects. The best such quantity is the neutron dipole moment, d_n. This was noted first by Baluni (1979) and the calculation was refined by Crewther, Di Vecchia, Veneziano and Witten (1980). One finds

$$d_n \approx 4 \times 10^{-16} |\theta| \text{ (in e-cm)}.$$

Experimentally (Smith et al., 1990) $d_n^{\text{exp}} \leq 8 \times 10^{-26}$, so we obtain $|\theta| < 10^{-9}$, a very small value indeed.

Let us return to the vacuum problem. We have discussed the effect of massless quarks; now we need to study the influence of chiral symmetry breaking by "small" mass terms. That is to say, what happens after introducing the perturbation

$$\mathcal{M} = \sum m_f \bar{q}_f q_f$$

at least to first order in ϵ; recall that we take $m_f = \epsilon r_f$, r_f fixed. We will not enter into the details here; the interested reader is referred to the lectures of Crewther (1979b). We merely summarize the results. Consider the inequality

$$m_u^{-1} > \sum_{f=2}^{n} m_f^{-1}; \qquad (7.7.14)$$

note that the results of Sects. 7.3, 4 imply that it is probably satisfied in the real world. Then we have the following situation. (i) If (7.7.14) holds, the topological charge is quantized in integer units; that is to say, the difference ν between two eigenvalues of the $K_\pm$ is an integer. (ii) If (7.7.14) does *not* hold, then there are at least fractional values of ν. In fact, for some particular values of the masses, ν must take irrational values.

We end this section with two comments. First, we have obtained constraints on the spectrum of the $K_\pm$, and the expression of the vacuum in terms of the eigenvectors $|n_\pm\rangle$; but we have not proved that the spectrum is nontrivial. One could imagine that all the n coincided, and thus the contents of the last sections would be much ado about nothing. Luckily (or unluckily, according one's the point of view), the existence of instantons implies that at least there exists a denumerable infinity, $\ldots, -1, 0, 1, 2, \ldots$ of different values of n. This will be shown in subsequent sections (particularly Sect. 8.4).

Secondly, we have assumed that massless $U(1)$ bosons do not exist. The mass of a pseudoscalar meson may be evaluated as in Sect. 7.3. If we repeat the calculation for the singlet current A_0^μ, we find that, because of the anomaly, Eq. (7.3.5) is modified by the appearance of a term

$$n_f^2 \left(\frac{g^2}{32\pi^2}\right)^2 \int \mathrm{d}^4 x \, \langle \mathrm{T} G(x) \widetilde{G}(x) G(0) \widetilde{G}(0) \rangle_{\text{vac}}. \qquad (7.7.15)$$

This would vanish to all orders in perturbation theory because $G\tilde{G}$ is a four-divergence; but the existence of instantons (see the next chapter) shows that, at least in the semiclassical approximation, (7.7.15) remains nonzero in the chiral limit ('t Hooft, 1976a,b). One may question the validity of this approximation. Alternatively, the same result is obtained in the large N_c (= number of colours) limit (Witten, 1979a). Thus, and although we do not have a totally rigorous proof, it appears extremely likely that the structure of the vacuum solves the $U(1)$ problem.

8 Instantons

8.1 The WKB Approximation
in the Path Integral Formalism; Tunnelling

In ordinary quantum mechanics, the WKB approximation is obtained by expanding in powers of Planck's constant, $\hbar$. To zero order we have the classical trajectory; higher orders yield the quantum fluctuations around this trajectory. The path integral formulation lends itself particularly well to the extension of the method to the field-theoretic case. To accomplish this, we reintroduce $\hbar$ into the expression for the generating functions of a field theory that, to simplify, we start by taking to be scalar. From (1.3.11) we then have,

$$Z[\eta] = \int \prod_x d\phi(x) \, \exp \frac{i}{\hbar} A_\eta[\phi]. \tag{8.1.1}$$

We next write

$$\phi(x) = \phi_{cl}(x) + \hbar^{1/2}\widetilde{\phi}(x) + \cdots, \quad \pi(x) = \partial_0\phi_{cl}(x) + \hbar^{1/2}\widetilde{\pi}(x) + \cdots \tag{8.1.2}$$

and match the powers of $\hbar$. The field ϕ_{cl} is the solution of the *classical* equation of motion,

$$\Box\phi_{cl} + m^2\phi_{cl} = \left.\frac{\mathcal{L}_{int}}{\partial\phi}\right|_{\phi=\phi_{cl}}, \tag{8.1.3a}$$

or, equivalently,

$$\phi_{cl}(x) = \phi_0(x) + i\int d^4y\, \Delta(x-y) \left.\frac{\mathcal{L}_{int}}{\partial\phi}\right|_{\phi=\phi_{cl}}, \tag{8.1.3b}$$

where ϕ_0 is the free classical field, $(\Box+m^2)\phi_0 = 0$. Because ϕ_{cl} satisfies the equation of motion, we have that $A[\phi_{cl}]$ is stationary: we are expanding (8.1.1) around the stationary phase. The zero order approximation yields the tree approximation; higher orders correspond to an expansion in the number of loops. The usefulness of the method lies in the fact that, to each order, the functional integral is of Gaussian type and can therefore be evaluated.

Let us show this for the first correction. To order $\hbar$,

$$\mathcal{A} = \mathcal{A}[\phi_{\mathrm{cl}}] - \frac{1}{2} \int \mathrm{d}^4 x \left\{ \widetilde{\phi}(x)(\Box + m^2)\widetilde{\phi}(x) - \left. \frac{\partial^2 \mathcal{L}_{\mathrm{int}}(\phi)}{\partial \phi^2} \right|_{\phi = \phi_{\mathrm{cl}}} \widetilde{\phi}(x)\widetilde{\phi}(x) \right\}.$$

Next we perform a change of variables,

$$\widetilde{\phi}(x) \to \phi'(x) = \left\{ \Box + m^2 - \frac{\partial^2 \mathcal{L}_{\mathrm{int}}(\phi)}{\partial \phi^2} \right\}^{1/2} \widetilde{\phi}(x),$$

and thus find

$$Z = (\text{const.}) \exp \left\{ -\frac{1}{2} \operatorname{Tr} \log \left[1 - (\Box + m^2)^{-1} \left. \frac{\partial^2 \mathcal{L}_{\mathrm{int}}(\phi)}{\partial \phi^2} \right|_{\phi = \phi_{\mathrm{cl}}} \right] \right\} Z_{\mathrm{tree}}, \tag{8.1.4a}$$

where, using (8.1.1) and the relation $\mathrm{i}(\Box + m^2)\Delta(x) = \delta(x)$, we can write

$$Z_{\mathrm{tree}} = N \exp \frac{\mathrm{i}}{\hbar} \left\{ \int \mathrm{d}^4 x\, \mathcal{L}_{\mathrm{int}}(\phi_{\mathrm{cl}}) \right. \tag{8.1.4b}$$
$$\left. - \frac{\mathrm{i}}{2} \int \mathrm{d}^4 x\, \mathrm{d}^4 y\, \left. \frac{\partial \mathcal{L}_{\mathrm{int}}}{\partial \phi(x)} \right|_{\phi = \phi_{\mathrm{cl}}} \Delta(x - y) \left. \frac{\partial \mathcal{L}_{\mathrm{int}}}{\partial \phi(y)} \right|_{\phi = \phi_{\mathrm{cl}}} \right\}.$$

The constant in (8.1.4a) contains the term

$$\int \mathcal{D}\phi' e^{-(\mathrm{i}/2) \int \mathrm{d}^4 x \phi'(x)^2} \det(\Box + m^2)^{1/2},$$

and we have used the identity, valid for any A,

$$\det(A^{-1/2}) = \exp \left\{ -\frac{1}{2} \operatorname{Tr} \log A \right\}.$$

It is known that there are quantum mechanical situations for which no classical trajectory exists. This occurs when there is tunnelling through a potential barrier. However, one can still adapt the WKB method to cope with this situation. We will exemplify this with the typical case of a particle in one dimension, subject to a potential $V(x)$. The wave function is (see, e.g., Landau and Lifshitz, 1958)

$$\psi(x) = C e^{\mathrm{i}\mathcal{A}_{\mathrm{cl}}}, \tag{8.1.5}$$

where $\mathcal{A}_{\mathrm{cl}}$ is now the action calculated along the classical trajectory

$$\tfrac{1}{2} m \ddot{x} + V(x) = E. \tag{8.1.6}$$

Take a potential with two minima, both corresponding to $V = 0$, and located at $x = x_0$, x_1 (Fig. 8.1.1a). If $E > \max V$, the motion from x_0 to x_1 is possible, and (8.1.5) yields the "transition" or "diffusion" amplitude. However, if

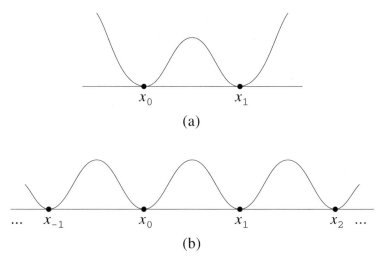

Fig. 8.1.1. (a) Potential with two minima. (b) Periodic potential.

$E < \max V$, the correct WKB analysis gives a result in which the transition amplitude

$$\langle x_1 | x_0 \rangle = C e^{i \mathcal{A}_{\text{cl}}(x_1, x_0)}, \qquad (8.1.7)$$

is to be replaced by the *tunnelling amplitude*,

$$\langle x_1 | x_0 \rangle = C e^{-\underline{\mathcal{A}}(x_1, x_0)}, \qquad (8.1.8)$$

where $\underline{\mathcal{A}}$ is not calculated along the solution of (8.1.6), but for

$$-\tfrac{1}{2} m \ddot{x} + V(x) = E. \qquad (8.1.9)$$

We see that to obtain a tunnelling amplitude we can use the same formula as that for a transition, making only the formal replacement of t by it, both in the expression for the action,

$$\mathcal{A} = \int_{t(\xi_0)}^{t(\xi_1)} dt \, L \to i\underline{\mathcal{A}},$$

with ξ_i the turning points, and in the equations of motion, (8.1.6) and (8.1.9).

Equations (8.1.5) and (8.1.8) do not give the normalization, which may, however, be readily obtained by dividing by $\langle x_0 | x_0 \rangle$. We thus infer that, in quantum field theory, the leading tunnelling amplitude will be

$$\langle \Psi_1, t = +\infty | \Psi_0, t = -\infty \rangle \approx C \exp \left\{ - \int \mathrm{d}^4 \underline{x} \, \mathcal{L}(\underline{\phi}_{\mathrm{cl}}) \right\}, \qquad (8.1.10)$$

where $\underline{\phi}_{\mathrm{cl}}$ is the classical solution to the *Euclidean* equations of motion, i.e., with x^0 replaced by $\pm i x_4$, x_4 real. (The sign $\pm$ depends on the boundary conditions; the reason for the name Euclidean is that, under the transformation $x^0 \to i x_4$ the *Minkowski* metric becomes Euclidean, up to a global sign).

According to the discussion at the beginning of this section, we may consider this to be the leading order of the exact expression,

$$\langle \Psi_1, t = +\infty | \Psi_0, t = -\infty \rangle = N \exp \int \mathcal{D}\underline{\phi} \left\{ - \int \mathrm{d}^4 \underline{x} \, \mathcal{L}(\underline{\phi}) \right\}, \qquad (8.1.11)$$

when expanding the field $\underline{\phi}$ in powers of $\hbar$ around $\underline{\phi}_{\mathrm{cl}}$.

An important property of the states of a system in a situation when tunnelling is possible is that the stationary states (in particular, the ground state, to be identified with the vacuum in quantum field theory) are not those in which the system is localized in one minimum of the potential, but is shared by all minima. The situation is familiar in solid state theory, where the potentials are periodic (like that in Fig. 8.1.1b). We will see an example of this in QCD soon.

8.2 Euclidean QCD

Consider the energy–momentum tensor of the pure Yang–Mills QCD, given in Eq. (2.8.2), leaving quarks aside, as they are irrelevant for the considerations of this and the next section. We can rewrite it as

$$\Theta^{\mu\nu} = -\tfrac{1}{2} g_{\alpha\beta} \sum_a G_a^{\mu\alpha} G_a^{\nu\beta} - \tfrac{1}{2} g_{\alpha\beta} \sum_a \widetilde{G}_a^{\mu\alpha} \widetilde{G}_a^{\nu\beta}. \qquad (8.2.1)$$

It follows that Θ^{00} is positive for *real* gluon fields:

$$\Theta^{00} = \tfrac{1}{2} \sum_{k,a} \left\{ (G_a^{0k})^2 + (\widetilde{G}_a^{0k})^2 \right\}. \qquad (8.2.2)$$

Therefore $\Theta^{\mu\nu} = 0$ requires $G \equiv 0$, and thus only the zero-field configurations may be identified with the vacuum. However, (8.2.2) no longer has a definite sign if we allow for complex $G^{\mu\nu}$. Particularly important is the case where a complex Minkowskian $G^{\mu\nu}$ corresponds to a real $\underline{G}^{\mu\nu}$ in Euclidean space; for, according to the discussion at the end of Sect. 8.1, this will indicate a tunnelling situation. This is the rationale for seeking solutions to the QCD equations in Euclidean space.[1]

Another point is that in Minkowski space,

$$\widetilde{\widetilde{G}}_a^{\mu\nu} = -G_a^{\mu\nu},$$

so only the trivial $G = 0$ may be *dual*,

$$\widetilde{G} = \pm G. \tag{8.2.3}$$

(If the sign is $(+)$ we say G is *self-dual*, if $(-)$ *anti-dual*.) However, in Euclidean space,

$$\widetilde{\widetilde{\underline{G}}}_a^{\mu\nu} = \underline{G}_a^{\mu\nu},$$

so nontrivial dual values of G may, and indeed do, exist. In addition, Euclidean dual G automatically satisfy the equations of motion.

This last property comes about as follows: the equations of motion for G read (Eq. (2.1.6))

$$D_\mu G_a^{\mu\nu} \equiv \partial_\mu G_a^{\mu\nu} + g \sum f_{abc} B_{b\mu} G_c^{\mu\nu} = 0; \tag{8.2.4}$$

the condition

$$D_\mu \widetilde{G}_a^{\mu\nu} = 0 \tag{8.2.5}$$

is the Bianchi identity, identically satisfied by any $G = D \times B$ whether or not B solves the equations of motion. However, if $\underline{G}$ is *dual*, then (8.2.5) implies (8.2.4), as was to be shown (Polyakov, 1977).

The connection with the problem of the vacuum occurs because, in the Euclidean case, (8.2.1) is replaced by

$$\underline{\Theta}_{\mu\nu} = -\tfrac{1}{2} \sum_\lambda \left\{ \underline{G}_{\mu\lambda} \underline{G}_{\nu\lambda} - \widetilde{\underline{G}}_{\mu\lambda} \widetilde{\underline{G}}_{\nu\lambda} \right\}, \tag{8.2.6}$$

so for dual fields $\underline{\Theta}_{\mu\nu} = 0$: dual $\underline{G}$ may represent nontrivial vacuum states.

Another property of dual fields has to do with a condition of minimum of the Euclidean action. We can write

$$\begin{aligned}
\underline{A} &= \tfrac{1}{4} \int \mathrm{d}^4\underline{x} \sum \underline{G}_{\mu\nu} \underline{G}_{\mu\nu} \\
&= \tfrac{1}{4} \sum \int \mathrm{d}^4\underline{x} \left\{ \tfrac{1}{2} \left(\underline{G}_{\mu\nu} \pm \widetilde{\underline{G}}_{\mu\nu} \right)^2 \mp \underline{G}_{\mu\nu} \widetilde{\underline{G}}_{\mu\nu} \right\} \geq \tfrac{1}{4} \left| \int \mathrm{d}^4\underline{x} \sum \underline{G}\,\widetilde{\underline{G}} \right|.
\end{aligned} \tag{8.2.7}$$

---

[1] This is usually referred to as Euclidean QCD or, more generally, Euclidean field theory. We will distinguish Euclidean quantities from the corresponding Minkowskian ones by underlining the first. Also, sums over repeated space–time indices will be written explicitly, while sums over implicit colour indices will continue to be understood.

Thus the action is positive-definite and reaches its minimum for dual fields, where one has the equality

$$\underline{A} = \tfrac{1}{4} \left| \int d^4 \underline{x} \sum \underline{G} \tilde{\underline{G}} \right| = \tfrac{1}{4} \int d^4 \underline{x} \sum_{a,\mu\nu} (\underline{G}_a^{\mu\nu})^2. \qquad (8.2.8)$$

Now, and at least in situations where the semi-classical approximation WKB holds, we know that the tunnelling amplitude is given by $\exp(-\underline{A})$, so the leading tunnelling effect, if it exists, will be provided by dual configurations.

We have been talking about "nontrivial vacuum states". It is not difficult to see that nonzero values of B exist for which $G = 0$. In fact, the general form of such B is what is called a *pure gauge*, and may be obtained from $B = 0$ by a gauge transformation. To see this, write a finite gauge transformation as

$$B_a^\mu(x) \to B'^\mu_a(x) = 2 \operatorname{Tr} t^a U^{-1}(x) t^b U(x) B_b^\mu(x) - \frac{2}{ig} \operatorname{Tr} t^a U^{-1}(x) \partial^\mu U(x), \qquad (8.2.9)$$

cf. Eq. (2.1.1).[2] Here U is any x-dependent matrix with $U^\dagger(x) = U^{-1}(x)$, $\det U(x) = 1$. Now, if $B = 0$,

$$B'^\mu_a(x) = -\frac{2}{ig} \operatorname{Tr} t^a U^{-1}(x) \partial^\mu U(x) : \qquad (8.2.10)$$

the gauge covariance of G ensures that $G'^{\mu\nu} = G^{\mu\nu} = 0$. Nontrivial solutions of the equations will be such that $G \neq 0$.

8.3 Instantons

We now seek Euclidean field configurations that lead to a dual field strength tensor, G. We are interested in fields with finite action. This means that we require, in particular,

$$\lim_{x \to \infty} |x|^2 \underline{G}_{\mu\nu}(x) = 0, \qquad (8.3.1)$$

where the Euclidean length is

$$|x| \equiv + \left\{ \sum_{\mu=1}^{4} (x_\mu)^2 \right\}^{1/2}.$$

Note that, to lighten the notation, we still write x for the Euclidean four-vector which, if we had been strictly consistent, should have been denoted by $\underline{x}$.

---

[2] To check with (2.1.1) we have to identify $U(x) = \exp(+i \sum \theta_a t^a)$, i.e., the U of (8.2.9) is the inverse of the U defined in Sect. 2.1.

Let $U(x)$ be a gauge transformation. The condition (8.3.1) will be satisfied provided that, at large x, B is the gauge transform of a null field, i.e., that it is asymptotically *pure gauge*. Thus,

$$\underline{B}_{a\mu}(x) \underset{x\to\infty}{\to} -\frac{2}{ig} \operatorname{Tr} t^a U^{-1}(x)\partial_\mu U(x),$$
$$\underline{G}_{a\mu\nu}(x) \underset{x\to\infty}{\to} 0,$$
(8.3.2)

and we try the ansatz

$$\underline{B}_\mu^a(x) = \varphi(|x|^2)\hat{\underline{B}}_\mu^a(x), \quad \hat{\underline{B}}_\mu^a(x) = -\frac{2}{ig} \operatorname{Tr} t^a U^{-1}\partial_\mu U, \quad \varphi(|x|^2) \underset{x\to\infty}{\to} 1.$$
(8.3.3)

It is instructive to check that the $\hat{\underline{G}}$ corresponding to $\hat{\underline{B}}$ is zero: for this, we define the matrices

$$\underline{\mathcal{B}}_\mu \equiv \sum_a t^a \underline{B}_\mu^a, \quad \underline{\mathcal{G}}_{\mu\nu} \equiv \sum_a t^a \underline{G}_{\mu\nu}^a.$$
(8.3.4a)

Clearly,

$$\underline{B}_\mu^a = 2\operatorname{Tr} t^a \underline{\mathcal{B}}_\mu, \quad \underline{G}_{\mu\nu}^a = 2\operatorname{Tr} t^a \underline{\mathcal{G}}_{\mu\nu},$$
(8.3.4b)

and

$$\underline{\mathcal{G}}_{\mu\nu} = \partial_\mu \underline{\mathcal{B}}_\nu - \partial_\nu \underline{\mathcal{B}}_\mu - ig[\underline{\mathcal{B}}_\mu, \underline{\mathcal{B}}_\nu].$$
(8.3.4c)

Of course Eqs. (8.3.4) also hold in the Minkowskian case. Now, if $\underline{B}$ is given by (8.3.3),

$$\underline{\mathcal{B}}_\mu \underset{x\to\infty}{\simeq} -\frac{1}{ig}U^{-1}\partial_\mu U,$$
(8.3.5)

so that

$$\underline{\mathcal{G}}_{\mu\nu}^a \underset{x\to\infty}{\simeq} -\frac{1}{ig}\left\{\partial_\mu(U^{-1}\partial_\nu U) - \partial_\nu(U^{-1}\partial_\mu U)\right\}$$
$$-ig\left(\frac{-1}{ig}\right)^2 [U^{-1}\partial_\mu U, U^{-1}\partial_\nu U]$$
$$= \frac{-1}{ig}\left\{-U^{-1}(\partial_\mu U)U^{-1}\partial_\nu U + U^{-1}(\partial_\nu U)U^{-1}\partial_\mu U\right\}$$
$$+\frac{-1}{ig}[U^{-1}\partial_\mu U, U^{-1}\partial_\nu U] = 0.$$

Note that the terms generated by the trilinear and quadrilinear ones cancel one another: the factor $1/g$ is essential because of the nonlinear character of $\underline{G}$. Its appearance heralds the nonperturbative character of the solutions.

If U is a group element that can be continuously connected to the identity, then $\mathcal{G}$ vanishes not only asymptotically, but identically. So we need U to couple space–time and colour indices. This can be managed because the dimension of space–time is four. Its group of invariance (in the Euclidean version) is $SO(4)$, whose (complex) Lie algebra is isomorphic to the product of the Lie algebra of $SU(2)$ by itself. Thus, we may couple $SO(4)$ with an $SU(2)$ subgroup of colour $SU(3)$. In view of this, we try a matrix of the form

$$U = \begin{pmatrix} u & 0 \\ 0 & 1 \end{pmatrix},$$

where u is a 2×2 matrix in $SU(2)$.

Let $\sigma_4 = 1$, and let σ_i be the Pauli matrices. Any 2×2 matrix A may be expanded in the σ_4, σ_i: $A = \sum a_\mu \sigma_\mu$. If we let $\tilde{a}_i = -a_i$, $\tilde{a}_4 = a_4$, then

$$\left(\sum a_\mu \sigma_\mu\right)\left(\sum \tilde{a}_\mu \sigma_\mu\right) = \sum a_\mu \tilde{a}_\mu,$$

and

$$\det A = \sum a_\mu \tilde{a}_\mu;$$

we find that the most general u may be written as

$$u_f = \frac{1}{|f(x)|}\left\{\sigma_4 f_4(x) + i\boldsymbol{\sigma}\mathbf{f}(x)\right\}, \quad f(x) = \text{real}. \tag{8.3.6}$$

The simplest choice is to take $f_\mu(x) = x_\mu$, so that

$$u(x) = \frac{1}{|x|}(\sigma_4 x_4 + i\boldsymbol{\sigma}\mathbf{x}). \tag{8.3.7a}$$

The space–time and colour indices are coupled in a nontrivial way. One then tries, as stated,[3]

$$\underline{\mathcal{B}}_\mu(x) = \varphi(|x|^2)\hat{\underline{\mathcal{B}}}_\mu(x), \quad \hat{\underline{\mathcal{B}}}_\mu(x) = -\frac{1}{ig}U^{-1}(x)\partial_\mu U(x),$$

$$U(x) = \begin{pmatrix} u(x) & 0 \\ 0 & 1 \end{pmatrix}. \tag{8.3.7b}$$

For the subsequent calculation it is useful to remember that, because $\hat{\underline{\mathcal{B}}}$ is pure gauge, the corresponding $\hat{\mathcal{G}}$ vanishes. We have

$$\begin{aligned}
\underline{\mathcal{G}}_{\mu\nu} &= \partial_\mu\underline{\mathcal{B}}_\nu - \partial_\nu\underline{\mathcal{B}}_\mu - ig[\underline{\mathcal{B}}_\mu, \underline{\mathcal{B}}_\nu] \\
&= (\partial_\mu\varphi)\hat{\underline{\mathcal{B}}}_\nu - (\partial_\nu\varphi)\hat{\underline{\mathcal{B}}}_\mu + \varphi(\partial_\mu\hat{\underline{\mathcal{B}}}_\nu - \partial_\nu\hat{\underline{\mathcal{B}}}_\mu) \\
&\quad - ig\varphi^2[\hat{\underline{\mathcal{B}}}_\mu, \hat{\underline{\mathcal{B}}}_\nu] \\
&= 2\varphi'\left\{x_\mu\hat{\underline{\mathcal{B}}}_\nu - x_\nu\hat{\underline{\mathcal{B}}}_\mu\right\} + (\varphi - \varphi^2)\left\{\partial_\mu\hat{\underline{\mathcal{B}}}_\nu - \partial_\nu\hat{\underline{\mathcal{B}}}_\mu\right\}, \\
\varphi' &= \frac{d\varphi(|x|^2)}{d|x|^2}.
\end{aligned}$$

---

[3] More general ansätze have been described by Corrigan and Fairlie (1977) and Wilczek (1977).

This is most easily calculated by defining 't Hooft's mixed colour and space–time tensor η by

$$\eta_{\mu\nu}^a = \begin{cases} \epsilon_{a\mu\nu4} + \delta_{\mu4}\delta_{a\nu} - \delta_{\nu4}\delta_{a\mu}, & a = 1, 2, 3, \\ 0, & a = 4, \ldots, 8, \end{cases} \tag{8.3.8}$$

so that $\hat{\underline{B}}_{\mu}^a = -(2/g|x|^2)\sum \eta_{\rho\mu}^a x_\rho$. We then find

$$\underline{G}_{\mu\nu}^a = \frac{4\mathrm{i}^2}{|x|^2 g}\left(\varphi' - \frac{\varphi - \varphi^2}{|x|^2}\right)\sum_\rho(\eta_{\rho\nu}^a x_\rho x_\mu - \eta_{\rho\mu}^a x_\rho x_\nu) + \frac{4\mathrm{i}^2}{|x|^2 g}(\varphi - \varphi^2)\eta_{\mu\nu}^a.$$

We note that η is self-dual, $\eta_{\mu\nu} = \tilde{\eta}_{\mu\nu}$; therefore, the condition of self-duality for $\underline{G}$ is met if φ satisfies the equation

$$\varphi' - \frac{\varphi - \varphi^2}{|x|^2} = 0.$$

Solving this, we finally have

$$\underline{B}_\mu(x) = \frac{|x|^2}{|x|^2 + \lambda^2}\frac{\mathrm{i}}{g}U^{-1}(x)\partial_\mu U(x), \quad \lambda \text{ arbitrary.} \tag{8.3.9}$$

This can be made more explicit by substituting U so that we write

$$\underline{B}_\mu^a(x) = \frac{1}{g}\frac{-2}{|x|^2 + \lambda^2}\sum_\rho \eta_{\rho\mu}^a x_\rho, \tag{8.3.10}$$

and the coupling of space–time and colour is obvious from the form of η. This is the original *instanton* solution found by Belavin, Polyakov, Schwartz and Tyupkin (1975). We note that it is concentrated around $x \approx 0$, i.e., in space *and* time (hence the name instanton). Solutions concentrated around $x \approx y$, any y, are obtained from (8.3.9, 10) by displacing $x \to x - y$; this will be useful later.

The field strength tensor may be readily calculated from e.g. (8.3.10) to get

$$\underline{G}_{\mu\nu}^a = \frac{1}{g}\frac{-4\lambda^2\eta_{\mu\nu}^a}{(|x|^2 + \lambda^2)^2}. \tag{8.3.11}$$

It turns out that there is perfect symmetry between self-dual and anti-dual solutions: the anti-dual solutions may be obtained from (8.3.10) by replacing η by $\bar{\eta}$,

$$\bar{\eta}_{\mu\nu}^a = \eta_{\mu\nu}^a, \quad \mu, \nu = 1, 2, 3, \quad \bar{\eta}_{\mu\nu}^a = -\eta_{\mu\nu}^a \text{ for } \mu \text{ or } \nu = 4. \tag{8.3.12}$$

They may be called *anti-instantons*.

A remarkable property of instantons is that, whereas $\underline{B} \sim 1/|x|$ for large x, a sufficient number of cancellations occur when forming $\underline{G}$, so that $\underline{G} \sim 1/|x|^4$, well within the requirements of (8.3.1). Also, and as was to be expected, both $\underline{B}$ and $\underline{G}$ become singular (and complex!) when continued to

Minkowski space, because then $|x|^2$ is replaced by $x \cdot x$, which is no longer positive, and hence $x^2 + \lambda^2$ may vanish.

The solution (8.3.9, 10) is all that we will use here; but other solutions have been found by De Alfaro, Fubini and Furlan (1976, 1977) and by Cerveró, Jacobs and Nohl (1977) with finite Minkowski action, but infinite Euclidean action.

Let us next compute the action corresponding to the instanton. Using $\sum \eta_{\mu\nu}^a \eta_{\mu\nu}^a = 12$ and the formulas of Appendix B,

$$\mathcal{A} = \tfrac{1}{4} \int \mathrm{d}^4 x \sum \underline{G}_{\mu\nu}^a \underline{G}_{\mu\nu}^a = \frac{48\lambda^2}{g^2} \int \mathrm{d}^4 x \, \frac{1}{(|x|^2 + \lambda^2)^4} = \frac{8\pi}{g^2}. \qquad (8.3.13)$$

In the following section we will show that instantons provide tunnelling between states $|n_\pm\rangle$ and $|n_\pm + \nu\rangle$, where ν is an integer. In this sense, they provide the "existence proof" for the reality of the complicated vacuum structure discussed in Sect. 7.7. One may thus wonder about the necessity of the sophisticated discussion there, since we have found explicit solutions. The answer lies in the requirement of finite action under which instantons were found. As discussed in Sect. 8.1, the observable tunnelling amplitude between two states $|a\rangle$ and $|b\rangle$ is

$$\langle a|b\rangle_{\mathrm{phys.}} = \frac{\langle a|\mathrm{e}^{-\mathcal{A}}|b\rangle}{\langle b|\mathrm{e}^{-\mathcal{A}}|b\rangle}, \qquad (8.3.14)$$

so even configurations with infinite action may yield finite tunnelling probability, provided that the infinities in numerator and denominator of (8.3.14) cancel. The requirement of finite action may be appealing, but is not compelling. In fact, we will see in Sect. 8.4 that instantons lead to *integer* values of ν, while we know from the work of Crewther (1979b), discussed in Sect. 7.7, that some patterns of quark masses lead to noninteger values of ν.[4] The importance of instantons lies in the fact that they provide explicit tunnelling effects, and they thus give indications on how to estimate these, but it is unlikely that they exhaust all the possibilities. With this proviso in mind, we continue the study of instantons, keeping the requirement of finite action.

---

[4] "Semi-instantons" with finite Euclidean action and half-integer topological charge seem to have been found by Forgács, Horváth and Palla (1981).

8.4 Connection with the Topological Quantum Number and the QCD Vacuum

Consider the quantity

$$Q_K = \frac{g^2}{32\pi^2} \int d^4x \sum \widetilde{\underline{G}}^a_{\mu\nu} \underline{G}^a_{\mu\nu} \qquad (8.4.1)$$

(cf. Eq. (7.7.3)). The gluon fields that approach zero at infinity are, as we discussed, of the form

$$\underline{\mathcal{B}}_\mu \underset{x\to\infty}{\simeq} \frac{-1}{ig} T_B^{-1}(x)\partial_\mu T_B(x), \qquad (8.4.2)$$

where T_B is a general matrix in $SU(3)$. Consider x varying on the boundary of a four-dimensional sphere, ∂S_4. The gauge fields map each point x into a $T_B(x)$ in the gauge group: so we have a mapping of ∂S_4 into $SU(3)$. We say that two fields are *homotopic*, written $\underline{\mathcal{B}} \approx \underline{\mathcal{B}}'$, if they can be continuously deformed one into another. Clearly, this relation is an equivalence relation and thus we may split the set of gauge fields into homotopy classes. The number of homotopy classes[5] is a *countable* infinity, so we may label fields with an integer n according to their homotopy class. Our next task is to show that n coincides with Q_K as given in (8.4.1). The quantity Q_K is called the *topological*, *winding* or *Pontryagin* quantum number; the second name refers to the number of times the mapping wraps the sphere around the group.

To see this, we first remark that (8.4.1) is invariant under continuous gauge transformations, as can be seen by direct computation. Next, we note that the integrand there is actually a four divergence. In fact, as shown in Sect. 7.7,

$$\frac{g^2}{32\pi^2} \sum \widetilde{\underline{G}}^a_{\mu\nu} \underline{G}^a_{\mu\nu} = \sum \partial_\mu \underline{K}_\mu, \qquad (8.4.3a)$$

where $\underline{K}$ is the "chiral current",

$$\underline{K}_\mu = \frac{g^2}{16\pi^2} \sum \epsilon_{\mu\nu\rho\sigma} \left\{ (\partial_\rho \underline{B}^a_\sigma)\underline{B}^a_\nu + \tfrac{1}{3} g f_{abc} \underline{B}^a_\rho \underline{B}^b_\sigma \underline{B}^c_\nu \right\}. \qquad (8.4.3b)$$

Because of Gauss's theorem,

$$Q_K = \frac{g^2}{32\pi^2} \int d^4x \sum \widetilde{\underline{G}}^a_{\mu\nu} \underline{G}^a_{\mu\nu} = \int_{\partial S_4} \sum ds_\mu \underline{K}_\mu, \qquad (8.4.4)$$

where ds is the surface element in ∂S_4. Using (8.4.3b), we then find

$$Q_K = \frac{g^3}{48\pi^2} \sum \epsilon_{\mu\nu\rho\lambda} f_{abc} \int_{\partial S_4} ds_\mu \underline{B}^a_\rho \underline{B}^b_\lambda \underline{B}^c_\nu.$$

---

[5] This holds for any gauge group that is simple and contains an $SU(2)$ subgroup.

The calculation simplifies if we assume $\underline{B}^a = 0$ except for $a = 1, 2, 3$; this is possible because the homotopy relation is dependent only on an $SU(2)$ subgroup. In this case, we let

$$\underline{B}_\mu = \tfrac{1}{2}\sigma_k \underline{B}_\mu^k,$$

and (8.4.2) holds with T in $SU(2)$. We thus have

$$Q_K = \frac{1}{12\pi^2} \sum \epsilon_{\mu\nu\rho\lambda} \int_{\partial S_4} \mathrm{d}s_\mu \, \mathrm{Tr}\left\{(T^{-1}\partial_\rho T)(T^{-1}\partial_\lambda T)(T^{-1}\partial_\nu T)\right\}. \quad (8.4.5)$$

Let us parametrize the elements of $SU(2)$ by the three Euler angles ξ_i; the invariant measure over the group is

$$\mathrm{d}\mu = \mathrm{Tr}\left\{T^{-1}\frac{\partial T}{\partial \xi_1}T^{-1}\frac{\partial T}{\partial \xi_2}T^{-1}\frac{\partial T}{\partial \xi_3}\right\} \mathrm{d}\xi_1\mathrm{d}\xi_2\mathrm{d}\xi_3, \quad \int_{SU(2)} \mathrm{d}\mu = 12\pi^2.$$

We see that (8.4.5) indeed gives the number of times the surface of the sphere is wrapped around $SU(2)$. Our instanton/anti-instanton solution has $Q_K = \pm 1$, as is clear from (8.3.13) using the self-dual/anti-dual property. It is also not difficult to construct solutions for any ν. Suppose ν positive, and consider the dilute gas of ν instantons,

$$\underline{B}_\mu^{a(\nu)}(x) = \sum_{k=1}^{\nu} \underline{B}_\mu^a(x - y_k), \quad (8.4.6a)$$

with $\underline{B}$ given by (8.3.10), and let then $|y_j - y_k| \to \infty$. Clearly, the overlap between two different terms in (8.4.6a) when building $\underline{G}^{(\nu)}$ tends towards zero as $|y_j - y_k| \to \infty$; hence, in this limit,

$$\frac{g^2}{32\pi^2} \int \mathrm{d}^4x \, \underline{G}^{(\nu)} \widetilde{\underline{G}}^{(\nu)} = \nu. \quad (8.4.6b)$$

We have succeeded in finding a representative in each homotopy class. What is more interesting, the multi-instanton field configurations are dual; hence, the corresponding energy–momentum tensors vanish, $\Theta^{(\nu)} = 0$. This means that in QCD (at least, in the Euclidean version) there is not a single vacuum, but an infinity of vacuum configurations, $|\nu\rangle$, $\nu = \ldots, -1, 0, 1, 2, \ldots$, that are topologically inequivalent: the situation is like that of Fig. 8.1.1b.

To explore this in greater detail, let us use a different hypersurface for integration; to be precise, we take a cylinder along the time axis, as in Fig. 8.4.1a. First, we choose a Coulomb-like gauge so that $\underline{B}_4 = 0$ for $x \to \infty$. Thus, only the integrals along the bases of the cylinder remain (Fig. 8.4.1b), so that

$$\nu = \left\{\int_{t''} - \int_{t'}\right\} \mathrm{d}x_1 \, \mathrm{d}x_2 \, \mathrm{d}x_3 \, K_4^{(\nu)}.$$

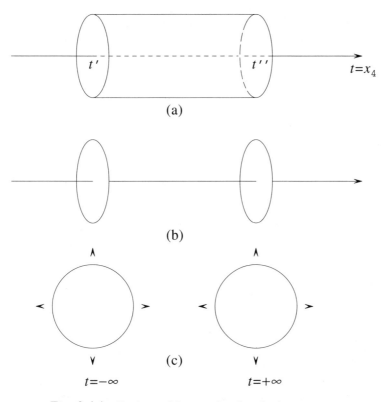

Fig. 8.4.1. Regions of integration for the instantons.

Since the field vanishes at infinity, we may identify the points at spatial infinity of the bases of the cylinder, so we obtain integrals over large three-dimensional spheres, one at $t = -\infty$ and the other at $t = +\infty$ as in Fig. 8.4.1c.

We can further select the gauge so that

$$\int_{t'\to-\infty} dx_1\, dx_2\, dx_3\, K_4^{(\nu)} = n(-\infty) = \text{integer}.$$

The proof that these properties may be achieved by a gauge choice, continuously connected to the identity, may be found, for example, in the lectures of Sciuto (1979). In view of (8.4.6b), we see that this implies

$$\int_{t''} dx_1\, dx_2\, dx_3\, K_4^{(\nu)} = n(t''), \quad n(+\infty) - n(-\infty) = \nu. \tag{8.4.7}$$

A multi-instanton $\underline{B}^{(\nu)}$ connects vacua separated by ν units of the topological quantum number between $-\infty$ and $+\infty$. So, in the quantum case, and according to the discussion of Sect. 8.1, we expect that these vacua will be connected by the tunnel effect, the leading amplitude for tunnelling being

$$\langle n(+\infty)|n(-\infty)\rangle = (\text{const.}) \times \exp(-\mathcal{A}).$$

As we discussed earlier, the minimum of the action is reached for self-dual/anti-dual solutions, i.e., for the instanton/anti-instanton for the case where $|\nu| = 1$. Thus, to leading order,

$$\langle n(+\infty)|n(-\infty)\rangle \simeq (\text{const.}) \times \exp\left\{-\frac{8\pi^2|\nu|}{g^2}\right\}. \tag{8.4.8}$$

The corrections to this may be calculated ('t Hooft, 1976a,b) by expanding the exact action not around $\underline{B}_{\mathrm{cl}} = 0$, but around $\underline{B}_{\mathrm{cl}} = \underline{B}_{\mathrm{cl}}^{(\nu)} = \underline{B}^{(\nu)}$. They are important in that they yield the constant in (8.4.8). Indeed,

$$\exp\left\{-\frac{8\pi^2|\nu|}{g^2}\left(1 + \frac{g^2}{16\pi^2}a\right)\right\} = e^{-a/2}\exp\left\{-\frac{8\pi^2|\nu|}{g^2}\right\},$$

but they do not substantially alter the result. What occurs is that, in order to believe the calculation (8.4.8), one has to consider situations where g is small, and then the exponential $\exp(-2\pi/\alpha_s)$ overwhelms any constant.

Let us now turn to the vacuum. The generating functional was defined in Sects. 1.3, 8.1 and 8.2. Neglecting gauge-fixing and ghost terms, and converted to Euclidean theory,

$$_+\langle 0|0\rangle_- = \underline{Z} = \int \mathcal{D}\underline{B} \exp\left\{-\int d^4x\,\mathcal{L}(\underline{B})\right\}. \tag{8.4.9a}$$

Now, however, we have to decide which homotopy classes to integrate. We may recall (Sect. 1.3) that in the equation (8.4.9a) the left hand side was, really, $\langle 0, t = +\infty|0, t = -\infty\rangle$; so it appears that we should reinterpret (8.4.9a) as

$$\langle n(+\infty)|m(-\infty)\rangle = \int \mathcal{D}\underline{B}_{n-m} \exp\left\{-\int d^4x\,\mathcal{L}(\underline{B})\right\}. \tag{8.4.9b}$$

In perturbation theory only the vacuum $|n = 0\rangle$ is considered; but, because of tunnelling, it is clear that all the $|n\rangle$ are connected ('t Hooft, 1976a,b; Callan, Dashen and Gross, 1976; Jackiw, Nohl and Rebbi, 1977), so none of them is the true vacuum state, in that it is not stationary. Stationary states are formed like the Bloch states in solids, by considering the superpositions

$$\sum_n e^{in\theta}|n\rangle \equiv |\theta\rangle.$$

These states are certainly invariant under changes of topological charge for, if we let Γ_k be the operator that changes n by k units,

$$\Gamma_k|\theta\rangle = \sum_n e^{in\theta}|n+k\rangle = \sum_m e^{i(m-k)\theta}|m\rangle = e^{-ik\theta}|\theta\rangle,$$

i.e., the vacuum only undergoes a change of phase. The generating functional is now, in terms of the θ-vacua,

$$\langle\theta(+\infty)|\theta'(-\infty)\rangle = N\delta(\theta-\theta')\sum_\nu e^{-i\nu\theta}\int \mathcal{D}\underline{B}^{(\nu)}\, e^{-\int d^4x\,\underline{\mathcal{L}}(\underline{B}^{(\nu)})}. \quad (8.4.10)$$

We may drop the $\delta(\theta-\theta')$, which only expresses the fact that worlds corresponding to different values of θ are unconnected. Moreover, we can extend the integral over $\underline{B}$ to all field configurations by introducing a factor of

$$\delta\left(\nu - \frac{g^2}{32\pi^2}\int d^4x \sum \underline{G}\widetilde{\underline{G}}\right);$$

then the sum over ν may be carried over trivially and we obtain the result

$$\underline{Z} = N\int \mathcal{D}\underline{B}\, e^{-\int d^4x\,\underline{\mathcal{L}}_\theta}, \quad (8.4.11a)$$

where

$$\underline{\mathcal{L}}_\theta = -\tfrac{1}{4}\sum\underline{G}\,\underline{G} + \frac{i\theta g^2}{32\pi^2}\sum\underline{G}\,\widetilde{\underline{G}}. \quad (8.4.11b)$$

Now we can finally return to Minkowski space and conclude that the existence of instantons suggests that the true Lagrangian is actually

$$\mathcal{L}_\theta = -\frac{1}{4}\sum_a G_a^{\mu\nu}G_{a\mu\nu} - \frac{\theta g^2}{32\pi^2}\sum_a G_a^{\mu\nu}\widetilde{G}_{a\mu\nu}, \quad (8.4.12)$$

thereby justifying the necessity of the introduction, in the general case, of the $\mathcal{L}_{1\theta}$ term (recall the discussion at the beginning of Sect. 7.7).

One may wonder to what extent the phenomena we have discussed will modify the results found previous to Sect. 7.6. First, the phenomenological bounds obtained for the value of θ (Sect. 7.7) force it to be so small that $\mathcal{L}_{1\theta}$ by itself should have practically no effect. In addition, instanton and related effects are *long-distance*; field configurations that vanish sufficiently rapidly as $x \to \infty$ have $Q_K = 0$. Since in this text we have discussed mostly short distance effects (for $\pi^0 \to 2\gamma$, deep inelastic scattering, etc.), we would think that the perturbative regime should continue to be relevant there. This may be seen clearly if one considers the tunnelling effect due to an instanton:

$$\langle 0|\pm 1\rangle \sim \exp\left(-\frac{8\pi^2}{g^2}\right).$$

After renormalization g should be replaced by $\bar{g}$ so that, up to logarithmic corrections,

$$\langle 0| \pm 1 \rangle \sim \left(\frac{\Lambda^2}{Q^2} \right)^{(33-2n_f)/6} . \tag{8.4.13}$$

This shows that at large momenta Q^2, tunnelling should be negligible and we may work with $|0\rangle$ as if it were the true vacuum; the error induced by (8.4.13) is much smaller than, for example, twist four or twist six effects in deep inelastic scattering. In fact, estimates by Baulieu, Ellis, Gaillard and Zakrewski (1979) show that the instanton correction for e^+e^- annihilations or deep inelastic scattering are utterly negligible for $Q^2 \gtrsim 1$ GeV^2. Thus, when instanton effects are important, the calculational tools do not work; when they work, the instanton effects are unobservable. In this, the instantons resemble that mythical animal, the basilisk, whose sight was supposed to cause the death of the beholder.

9 Lattice QCD

Alice laughed. "There is no use trying", she said, "One cannot believe impossible things".

"I daresay you haven't had much practice", said the Queen. . .,

"Why, sometimes I have believed as many as six impossible things before breakfast!"

LEWIS CARROLL, 1896

9.1 Quarks (and Gluons) on an Euclidean Lattice

The functional formalism apparently fulfils the theorist's dream: field theory reduced to quadratures! This and the following sections of the present chapter will be devoted to an introduction to field theory (especially, QCD) on a lattice, precisely the tool to implement such a programme.[1]

The basic equations for the functional formalism, say Eqs. (2.5.11), present, as they stand, a number of difficulties. First of all, we have there a *continuous infinity* of integrals. To evaluate them numerically we have to replace the spacetime continuum by a finite lattice of points. However, even with a finite lattice the integrals in (2.5.11) are not suitable for numerical treatment. The reason is that the exponential $\exp i \int d^4 x \, \mathcal{L}$ oscillates violently: the integrals are not convergent.

The device usually employed to deal with this is to work in Euclidean space, i.e., continue analytically to imaginary time. Temporarily denoting Euclidean quantities as in the last chapter by underlining them, we write

$$\underline{x}_4 \equiv i x_0, \quad \underline{x}_j \equiv x^j, \tag{9.1.1a}$$

and, for the gamma matrices,

$$\underline{\gamma}_4 \equiv \gamma_0, \quad \underline{\gamma}_j \equiv i \gamma^j, \tag{9.1.1b}$$

so that

$$\{\underline{\gamma}_\mu, \underline{\gamma}_\mu\} = \delta_{\mu\nu}.$$

The scalar product in Euclidean space becomes

$$\sum_\mu \underline{x}_\mu \underline{y}_\mu = -x \cdot y. \tag{9.1.1c}$$

---

[1] The formulation of field theory, and specifically QCD on a lattice, was given by Wilson (1975), who first proved confinement in the strong coupling limit. Application to actual calculations followed the pioneering work of Creutz. In our presentation we will follow mostly Wilson's (1975) paper and Creutz's (1983) text. Summaries of results of recent calculations may be found in the proceedings of specialized conferences; some are presented in Sect. 9.5.

Equations (9.1.1) permit the passage from Euclidean to Minkowski space and conversely. The free fermion Euclidean Lagrangian is defined as[2]

$$\mathcal{L}_q \equiv \bar{q}(x)(\underline{\slashed{\partial}} + m)\underline{q}(x), \quad \underline{\slashed{\partial}} \equiv \sum_\mu \underline{\gamma}_\mu \partial_\mu. \tag{9.1.2a}$$

and it goes into minus the Lagrangian in Minkowski space: under (9.1.1),

$$\mathcal{L}_q \to -\mathcal{L}_q.$$

For the gluon fields we use the matrix notation of Sect. 8.3, so we may write the Lagrangian as

$$\mathcal{L}_{\text{YM}} = \tfrac{1}{2} \operatorname{Tr} \sum_{\mu\nu} \mathcal{G}^2_{\mu\nu}(\underline{x}), \tag{9.1.2b}$$

cf. Eq. (8.3.4c). Also in matrix notation, the quark–gluon interaction term is, for one quark flavour and with q a vertical matrix in colour space,

$$\mathcal{L}_{qG} = -\mathrm{i}g\bar{q}(\underline{x})\slashed{B}(\underline{x})\underline{q}(\underline{x}). \tag{9.1.2c}$$

In this chapter we will work in Euclidean space, and use the matrix formalism for the gluon fields. We will accordingly simplify the notation by *removing the underlining of Euclidean quantities and representing matrices by ordinary italics*. So we write the full Lagrangian as, simply,

$$\mathcal{L} = \bar{q}(x)(\slashed{\partial} + m)q(x) + \tfrac{1}{2} \operatorname{Tr} \sum_{\mu\nu} G^2_{\mu\nu}(x) - \mathrm{i}g\bar{q}(x)\slashed{B}(x)q(x). \tag{9.1.3}$$

With this notation, gauge transformations can be written as

$$\begin{aligned}
B_\mu(x) &\to U^{-1}(x)B_\mu(x)U(x) + \frac{\mathrm{i}}{g}U^{-1}(x)\partial_\mu U(x), \\
G_{\mu\nu}(x) &\to U^{-1}(x)G_{\mu\nu}(x)U(x), \\
q(x) &\to U^{-1}(x)q(x),
\end{aligned} \tag{9.1.4}$$

with U an arbitrary $SU_c(3)$ matrix.[3]

Finally, the action and Euclidean generating functional are defined as

$$S \equiv \int \mathrm{d}^4x\,\mathcal{L}, \quad Z \equiv \int \mathcal{D}q\mathcal{D}\bar{q}\mathcal{D}B\,\mathrm{e}^{-S}, \tag{9.1.5}$$

and we do not write for the moment sources or gauge terms explicitly. We have used the letter S for the Euclidean action to follow current practice.

---

[2] We write the Lagrangian for a single quark flavour. For several flavours, replace m by m_q, and sum over the flavours q.

[3] Note that (9.1.4) uses a convention *different* from that of Sect. 2.1; now we set $U(x) \equiv \exp(+\mathrm{i}\sum \theta_a(x)t^a)$. The notation (9.1.4) is forced by the fact that $\mathcal{L}_q$ corresponds to $-\mathcal{L}$ (see above).

Equation (9.1.5) is further clarified if we return momentarily to the underlining notation for Euclidean quantities:

$$\underline{\mathcal{A}} \equiv \underline{S} = \int \mathrm{d}^4\underline{x}\,\underline{\mathcal{L}} = \frac{1}{\mathrm{i}} \int \mathrm{d}^4x\,\mathcal{L}; \quad \mathcal{A} = \mathrm{i}\underline{S},$$

because $\mathrm{d}^4\underline{x} = \mathrm{i}\mathrm{d}^4x$. Thus, the exponential in the generating functional $\exp \mathrm{i}\mathcal{A}$ becomes $\exp(-S)$ in Euclidean space with positive S, thus providing the desired convergence factor.

We will now define the lattice. We will take a cubic lattice, with periodic boundary conditions. Other types of lattices, and boundary conditions, have been considered in the literature, but, in the author's opinion, to little advantage; the interested reader may find references in Creutz (1983). We let a be the spacing, so the sites of the lattice are the points na, with n an (Euclidean) four-vector with components n_μ,

$$n_\mu = 0, \pm 1, \pm 2, \ldots, \pm N.$$

We thus have $(2N+1)^4$ points. Because of the periodic boundary conditions, quantities defined on the lattice $Q(n_\mu)$ are extended for arbitrary n by

$$Q(n_\mu + 2N + 1) = Q(n_\mu).$$

The physical limit is $N \to \infty$, $a \to 0$, in this order: first $N \to \infty$, then $a \to 0$.

Not only does the lattice give a meaning to (9.1.5), so that we have now a finite number of convergent integrals; but it also provides a *regularization*. Because a is finite, ultraviolet divergences do not occur; and as long as N stays bounded, infrared ones are prevented. With respect to the first, they will reappear as $a \to 0$ in the form of terms proportional to $1/a$ and to $\log a$: the limit of the continuum will have to take this into account.

The lattice formulation does *not* simply consist of writing (9.1.5) replacing integrals by sums, and derivatives by finite differences. Some elaboration is necessary both for gluons and quarks. For the first, the naive replacement would violate gauge invariance, thus rendering the theory meaningless. This will be discussed in detail in the coming section; here we start by presenting the formulation for fermions.

We define the quark variables

$$q_n \equiv q(an) \underset{an \to x}{\to} q(x).$$

Colour and flavour indices are implicitly understood. To keep hermiticity of the $\mathrm{i}\partial$ operators we replace derivatives by symmetric finite differences:

$$\frac{1}{2a}\left(q_{n+\hat{\mu}} - q_{n-\hat{\mu}}\right) = \frac{1}{2a}[q(an + a\hat{\mu}) - q(an - a\hat{\mu})] \underset{\substack{a \to 0 \\ an \to x}}{\to} \partial_\mu q(x).$$

Here, and in all that follows, we denote by $\hat{\mu}$ a unit vector along the μth axis. The action for free quarks may then be written as

$$S_q = a^4 \sum_n \left\{ m\bar{q}_n q_n + \frac{1}{2a} \sum_\mu \bar{q}_n \gamma_\mu \left(q_{n+\hat{\mu}} - q_{n-\hat{\mu}} \right) \right\},$$

$$Z_q = \int \prod_n dq_n \prod_k d\bar{q}_k e^{-S_q}. \qquad (9.1.6)$$

These harmless-looking expressions will reveal their shortcomings when we calculate the quark propagator, to which we now turn. The calculation follows closely that of Sect. 1.3, with appropriate alterations. We write the action as

$$S_q = \sum_{n,k} \bar{q}_n D_{nk} q_k, \qquad (9.1.7a)$$

where the matrix D has elements

$$D_{nk} = a^4 m \delta_{nk} + \frac{a^3}{2} \sum_\mu \gamma_\mu \left(\delta_{k,n+\hat{\mu}} - \delta_{k,n-\hat{\mu}} \right), \qquad (9.1.7b)$$

and we note that Z_q is proportional to the determinant of D (Sect. 2.5),

$$Z_q = (\det D)\frac{(-1)^{2N+1}}{(2N+1)!}. \qquad (9.1.8)$$

This will be of use later on.

To obtain the propagator we have to invert D. We do this with the help of a (finite) Fourier transform. We write

$$(D^{-1})_{nk} = a^{-4}(2N+1)^{-4} \sum_j \tilde{D}_j^{-1} \exp\left\{ \frac{2\pi i}{2N+1} \sum_\mu j_\mu (n-k)_\mu \right\}, \qquad (9.1.9)$$

and use the relation

$$\sum_{j_\mu=-N}^{N} \exp \frac{2\pi i}{2N+1} j_\mu(n-k)_\mu = (2N+1)\delta_{n_\mu k_\mu}$$

to find

$$\tilde{D}_j = m + \frac{i}{a} \sum_\mu \gamma_\mu \sin \frac{2\pi j_\mu}{2N+1}. \qquad (9.1.10a)$$

The p-space propagator is thus

$$S(j) \equiv \tilde{D}_j^{-1} = \left(m + \frac{i}{a} \sum_\mu \gamma_\mu \sin \frac{2\pi j_\mu}{2N+1} \right)^{-1}. \qquad (9.1.10b)$$

This expression becomes more transparent in the limit of a large lattice. We define new variables p_μ by

$$\frac{2\pi j_\mu}{2N+1} \equiv a p_\mu. \tag{9.1.11}$$

The p-space propagator then becomes

$$S(p) = \left(m + \frac{i}{a} \sum_\mu \gamma_\mu \sin a p_\mu \right)^{-1}; \tag{9.1.12}$$

sums over j are replaced by integrals,

$$\frac{1}{2N+1} \sum_{j_\mu=-N}^{N} \rightarrow a \int_{-\pi/a}^{+\pi/a} \frac{\mathrm{d}p_\mu}{2\pi}. \tag{9.1.13}$$

Thus, (9.1.9) now reads

$$(D^{-1})_{nk} = \int_{-\pi/a}^{+\pi/a} \frac{\mathrm{d}^4 p}{(2\pi)^4} \frac{\exp i \sum_\mu p_\mu (an - ak)_\mu}{m + (i/a) \sum_\mu \gamma_\mu \sin a p_\mu}. \tag{9.1.14}$$

Everything would appear to be above board: if we take the limit $a \to 0$, $an \to x$, $an \to y$ we get that, for example, (9.1.14) becomes

$$(D^{-1})_{nk} \underset{a \to 0}{\to} S(x-y) = \int_{-\infty}^{+\infty} \frac{\mathrm{d}^4 p}{(2\pi)^4} \frac{\exp i \sum_\mu p_\mu (x-y)_\mu}{m + i \sum_\mu \gamma_\mu p_\mu}:$$

as one should wish, the Euclidean expression for the propagator.

The problem, however, is that, for *finite* a, (9.1.12) has too many poles. To see this, consider for simplicity the case $m = 0$. Then, the denominator in (9.1.12) does not only vanish for all $p_\mu = 0$, but also for $p_\mu = \pi/a$, or any combination thereof. In all, there are $2^4 = 16$ poles: each flavour gets multiplied by sixteen on the lattice.

This is a catastrophe in more respects than one. Asymptotic freedom is lost. The π^0 refuses to decay, and the $U(1)$ anomaly disappears. (This last phenomenon occurs because the sixteen fermions alternate in sign in their contribution to the anomalous triangle; cf. Karsten and Smit, 1981).[4] In fact, one may doubt the connection between the lattice theory and the continuum one.

Several solutions have been proposed for this problem of *fermion doubling*, some of which may be found in Creutz (1983). Here we will present

---

[4] That something like this had to happen is obvious if one realizes that the lattice regularization preserves dimension and gauge invariance (as will be seen).

the one due to Wilson (1977). It consists of adding to the Lagrangian a new quadratic term; we then obtain $\mathcal{L}_q^r$, with r an arbitrary parameter, and where

$$\mathcal{L}_q^r \equiv m\bar{q}_n q_n + \frac{4r}{a}\bar{q}_n q_n + \frac{1}{2a}\sum_\mu \{(r+\gamma_\mu)q_{n+\hat{\mu}} + (r-\gamma_\mu)q_{n-\hat{\mu}}\}. \quad (9.1.15)$$

The corresponding p-space propagator, in the large lattice limit, is now

$$S^r(p) = \left\{ m + \frac{1}{a}\sum_\mu \left[i\gamma_\mu \sin a p_\mu + \frac{r}{a}(1 - \cos a p_\mu) \right] \right\}^{-1}. \quad (9.1.16)$$

The extra particles still are present, but their masses are

$$m + ra^{-1}\sum_{\mu'}(1 - \cos q_{\mu'}) = m + \frac{2rn_\pi}{a},$$

where the sum over μ' runs over the $q_{\mu'} = \pi$ ($q_\mu \equiv ap_\mu$), and n_π is the number of these. Therefore, in the continuum limit, the unwanted particles decouple as their masses are of order $1/a$. However, Eq. (9.1.16) has the great drawback that it breaks chiral invariance. In fact, this is unavoidable, as shown by Nielsen and Ninomiya (1981); unwanted fermions must accompany any chiral invariant lattice formulation. A way to see this is to recall the comment made before on the disappearance of the anomaly. One then works with Wilson fermions, and hopes that chiral symmetry will be restored in the continuum limit.

9.2 Gluons (and Quarks) on the Lattice.
Paths and Loops. The Wilson Action

i Abelian Gauge Theories

We will start by considering Abelian, Euclidean fields, to be denoted by $A_\mu(x)$. We take them to be interacting with fermion fields $\psi(x)$, with intensity e: we have in mind the important example of electrodynamics.

The elements of the gauge group may now be parametrized as

$$U(x) = e^{ief(x)}, \quad (9.2.1)$$

f arbitrary. A gauge transformation is

$$A_\mu(x) \to U^{-1}(x)A_\mu(x)U(x) + \frac{i}{e}U^{-1}(x)\partial_\mu U(x) \quad (9.2.2a)$$

(cf. (9.1.4)); because the group is Abelian, this agrees with the usual expression, $A_\mu(x) \to A_\mu(x) - \partial_\mu f(x)$. The fermion field will transform as

$$\psi(x) \to U^{-1}(x)\psi(x). \quad (9.2.2b)$$

Consider an expression such as $\bar{\psi}(x)\psi(y)$. It is not gauge invariant. To obtain a gauge invariant expression we have to consider

$$\bar{\psi}(x)\left\{\exp \text{ ie} \int_{P(y\to x)} \sum_\mu dz_\mu \, A_\mu(z)\right\}\psi(y), \qquad (9.2.3)$$

where $P(y \to x)$ is a path from y to x.

We prove gauge invariance for infinitesimal $x - y = \delta$. Under (9.2.2), we have

$$\bar{\psi}(y+\delta)\left\{\exp \text{ ie} \int_y^{y+\delta} \sum_\mu dz_\mu \, A_\mu(z)\right\}\psi(y)$$

$$\to \bar{\psi}(y+\delta)e^{ief(y+\delta)}\left\{\exp \text{ ie} \int_y^{y+\delta} \sum_\mu dz_\mu \, (A_\mu(z) - \partial_\mu f(z))\right\}e^{-ief(y)}\psi(y)$$

$$= \bar{\psi}(y+\delta)\left\{\exp \text{ ie} \int_y^{y+\delta} \sum_\mu dz_\mu \, A_\mu(z)\right\}\psi(y),$$

the last step because

$$-\text{ie} \int_y^{y+\delta} \sum_\mu dz_\mu \, \partial_\mu f(z) = -ief(z)\big|_y^{y+\delta} = ief(y) - ief(y+\delta).$$

Note that in this simple Abelian case the proof does not depend on $y-x$ being infinitesimal; but we want to give methods that can be easily generalized to the non-Abelian case.

Let us now put the theory on the lattice, with spacing a, $(2N + 1)^4$ points and periodic boundary conditions. We must, to do so, define the field variables and the gauge transformations. For this, we start by constructing matrices associated to an infinitesimal link, from lattice site n to lattice site $n + \hat{\mu}$, $\hat{\mu}$ being a unit vector along axis μ. Writing $A_\mu(n)$ instead of $A_\mu(an)$ when convenient to lighten notation, we define

$$U(n,\mu) \equiv \exp \frac{iea}{2}\left\{A_\mu(n) + A_\mu(n+\hat{\mu})\right\}, \qquad (9.2.4a)$$

and we give a meaning to the link in the opposite direction by setting

$$U(n,-\mu) \equiv U^{-1}(n - \hat{\mu}, \mu). \qquad (9.2.4b)$$

One can then rewrite (9.2.3) as the limit $a \to 0$ of

$$\bar{\psi}_k \prod_{\ell(n\to k)} U(j, \mu_j)\psi_n, \qquad (9.2.5)$$

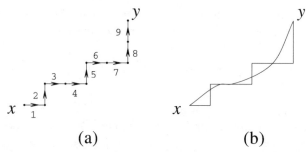

Fig. 9.2.1. (a) A polygonal $\ell(x \to y)$ joining x and y. (b) Approximation of the curved path P by the polygonal.

where the product runs over all the links necessary to join, along a given polygonal path $\ell(n \to k)$, the points $na = x$ to $ka = y$; see Fig. 9.2.1a. Note that the quantity

$$\prod_{\ell(n \to k)} U(j, \mu_j),$$

depends on the particular polygon chosen to join n and k, as indeed the integral

$$\int_{P(y \to x)} \sum_\mu dz_\mu \, A_\mu(z),$$

depends on the path P from y to x which is approximated by the polygonal ℓ (Fig. 9.2.1b).

We assume the product (9.2.5) to be ordered along the path, from right to left. This is irrelevant for Abelian theories, but basic for non-Abelian ones, since there the U will be noncommutative matrices. For example, for the polygonal and numbering of Fig. 9.2.1a,

$$\prod_{\ell(n \to k)} U(j, \mu_j) = U_9 U_8 \ldots U_2 U_1,$$

with U_j associated to link j.

Under a finite gauge transformation, and iterating the method of the proof given before for the infinitesimal case, we obtain in the continuum limit, the mapping

$$\lim_{a \to 0} \prod_{\ell(n \to k)} U(j, \mu_j) \to \lim_{a \to 0} U^{-1}(k) \left\{ \prod_{\ell(n \to k)} U(j, \mu_j) \right\} U(n) \qquad (9.2.6)$$

(for e.g. the path of Fig. 9.2.1a). That is to say: the path ordering picks the transformations given by the group elements associated with the end points.

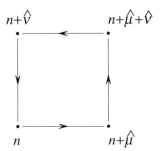

$n+\hat{\nu}$ $\qquad\qquad$ $n+\hat{\mu}+\hat{\nu}$

n $\qquad\qquad\qquad$ $n+\hat{\mu}$

Fig. 9.2.2. A plaquette.

We will see that it is possible to formulate the theory entirely in terms of the $U(n,\mu)$, without reference to the fields A_μ. To do so, we first define the gauge transformation for the link matrices by

$$U(n,\mu) \to U^{-1}(n+\hat{\mu})U(n,\mu)U(n). \qquad (9.2.7a)$$

With this, it follows automatically that

$$\prod_{\ell(n\to k)} U(j,\mu_j) \to U^{-1}(k) \prod_{\ell(n\to k)} U(j,\mu_j)U(n), \qquad (9.2.7b)$$

from which the continuum limit (9.2.6) is straightforward. Reversing the calculation, one easily verifies that the lattice definition (9.2.7a) for the $U(n,\mu)$ becomes the continuum one (9.2.2a) for the $A_\mu(x)$ in the limit $a \to 0$. For fermion fields, the transformation (9.2.2b) is taken over into the lattice, with only the replacement $x \to n$. This completes the definition of gauge transformations on the lattice. An important property of these transformations is that, if $\ell(n \to n)$ is a *closed* line, then

$$\prod_{\ell(n\to n)} U(j,\mu_j) = \text{gauge covariant} \qquad (9.2.8)$$

(invariant for Abelian fields).

We can now introduce a gauge invariant fermion–vector field Lagrangian, and a vector field Lagrangian, also gauge invariant. For the first, we write

$$\mathcal{L}_\psi + \mathcal{L}_{\psi A}$$
$$= m\bar{\psi}_n\psi_n + \frac{1}{2a}\bar{\psi}_n \sum_\mu \gamma_\mu \left\{ U(n+\hat{\mu},-\mu)\psi_{n+\hat{\mu}} - U(n-\hat{\mu},\mu)\psi_{n-\hat{\mu}} \right\}, \qquad (9.2.9)$$

clearly gauge invariant using (9.2.7). It is easy to see, by expanding on a, that this gives the correct continuum limit:

$$\mathcal{L}_\psi + \mathcal{L}_{\psi A} \underset{a\to 0}{\to} m\bar{\psi}(x)\psi(x) + \bar{\psi}(x)\slashed{\partial}\psi(x) - \mathrm{i}e\bar{\psi}(x)\slashed{A}(x)\psi(x).$$

As for the pure gauge field action, we start by considering a *plaquette*. This is denoted by the symbol □ and defined as an elementary square in the lattice with side a (Fig. 9.2.2); the gauge element associated to it is given by the expression

$$U_\square \equiv U(n + \hat{\nu}, -\nu)U(n + \hat{\nu} + \hat{\mu}, -\mu)U(n + \hat{\mu}, \nu)U(n, \mu), \qquad (9.2.10\text{a})$$

the idea being that the circulation around the plaquette will produce a curl of the field, as indeed does happen in the continuum limit. Note that the plaquette is characterized uniquely by one of the corners, n, and the vectors $\hat{\mu}$, $\hat{\nu}$, so we may write (9.2.10a) as

$$U_\square \equiv U_{\mu\nu}(n). \qquad (9.2.10\text{b})$$

Under local gauge transformations, and because the plaquette is closed, we have

$$U_{\mu\nu}(n) \to U^{-1}(n)U_{\mu\nu}(n)U(n).$$

Therefore, the trace[5] is gauge invariant, and a possible action for the gauge field is

$$S'(A) = \lambda \sum_\square \operatorname{Tr} U_\square = \frac{\lambda}{2} \sum_{\mu\nu} \sum_n \operatorname{Tr} U_{\mu\nu}(n); \qquad (9.2.11)$$

the sum over $\mu\nu$ gets a factor $1/2$ because $\mu\nu$ and $\nu\mu$ define the same plaquette.

It turns out that, for an appropriate choice of the constant λ, $S'(A)$ tends to the continuum action for $a \to 0$; but, because we want the action to provide a convergence factor, we will modify (9.4.11) slightly and define

$$S(A) = \lambda \sum_\square \operatorname{Re} \operatorname{Tr} U_\square. \qquad (9.2.12)$$

This expression presents the supplementary advantage over (9.2.11) that it is invariant under CP, which invariance is thus respected on the lattice (and not merely in the continuum limit). Note that taking the real part is necessary because we consider only plaquettes with a given, counter-clockwise orientation. Because the U are unitary, and a clockwise oriented plaquette is the inverse of a counter-clockwise one, we have

$$\operatorname{Tr} U_\square(\text{clock.}) = \operatorname{Tr} U_\square^{-1}(\text{c. clock.}) = \operatorname{Tr} U_\square^{\dagger}(\text{c. clock.}) = (\operatorname{Tr} U_\square(\text{c. clock.}))^* :$$

if we summed over both orientations independently we need not take the real part.

---

[5] The trace is of course irrelevant for Abelian fields, but we write it to ease the transition to the non-Abelian case.

The limit $a \to 0$ is straightforward. We have, using (9.2.4), (9.2.10),

$$U_{\mu\nu}(n) = \exp \frac{iea}{2} \Big\{ A_\mu(n) + A_\mu(n + \hat{\mu}) + A_\nu(n + \hat{\mu}) + A_\nu(n + \hat{\mu} + \hat{\nu})$$
$$- A_\mu(n + \hat{\mu} + \hat{\nu}) - A_\mu(n + \hat{\nu}) - A_\nu(n + \hat{\nu}) - A_\nu(n) \Big\}$$
$$\underset{\substack{a \to 0 \\ an \to x}}{\simeq} \exp iea^2 \{\partial_\mu A_\nu(x) - \partial_\nu A_\mu(x)\},$$

$$(9.2.13a)$$

so that

$$\operatorname{Re} \operatorname{Tr} U_{\mu\nu}(n) \simeq 1 - \frac{e^2 a^4}{2!} F_{\mu\nu}^2(x), \qquad (9.2.13b)$$

and, by choosing $\lambda = -1/e^2$ we find that, up to an irrelevant constant, we recover the correct continuum limit:

$$-\frac{1}{e^2} \sum_\square \operatorname{Re} \operatorname{Tr} U_\square \underset{a \to 0}{\simeq} -\tfrac{1}{4} \int d^4 x \sum_{\mu\nu} F_{\mu\nu}^2(x) + \text{constant}. \qquad (9.2.14)$$

Finally, we have to write the generating functional. Because the fields $A_\mu(na)$ always appear as group elements of the form $U(n, \mu)$, the integration over all $A_\mu(an)$ is redundant: we may limit the range of integration to the interval between $\pm\pi/ea$ or, simpler still, replace integrals over $dA_\mu(an)$ by integrals over the group $dU(n, \mu)$, one for each lattice link. Therefore, we obtain the generating functional,

$$Z = \int \prod_k d\psi_k \int \prod_j d\bar{\psi}_j \int_G \prod_{n,\mu} dU(n, \mu) e^{-S(\psi, A)}, \qquad (9.2.15a)$$

$$S(\psi, A) = -\frac{1}{e^2} \sum_\square \operatorname{Re} \operatorname{Tr} U_\square$$

$$+ a^4 \sum_n \bar{\psi}_n \left\{ m + \frac{1}{2a} \sum_\mu \gamma_\mu \big[U(n + \hat{\mu}, -\mu)\psi_{n+\hat{\mu}} - U(n - \hat{\mu}, \mu)\psi_{n-\hat{\mu}} \big] \right\},$$

$$(9.2.15b)$$

and we have used (9.2.9), (9.2.12) with $\lambda = -1/e^2$. Note that (9.2.15) or, for that matter, (9.2.9), do not take into account fermion doubling, to cope with which a modification like that of the previous section would be necessary. The integral over dU in (9.2.15a) runs over the Abelian $U(1)$ group.

ii QCD. The Wilson Action

The previous discussion was tailored so that it can be carried over, with obvious replacements, to the case of QCD. We define the group element associated with a link,

$$U(n,\mu) \equiv \exp\frac{iag}{2}\{B_\mu(n) + B_\mu(n+\hat\mu)\}; \quad U(n,-\mu) \equiv U^{-1}(n, n-\hat\mu, \mu),$$
(9.2.16a)

and the product of such elements around a plaquette characterized by n, $\hat\mu$, $\hat\nu$:

$$U_\square \equiv U_{\mu\nu}(n) \equiv U(n+\hat\nu, -\nu)U(n+\hat\nu+\hat\mu, -\mu)U(n+\hat\mu, \nu)U(n, \mu). \quad (9.2.16b)$$

Formally, these equations are like (9.2.4), (9.2.10); but now, and unlike for (9.2.10) where the ordering was superfluous, the order in (9.2.16b) is essential: the B and hence the $U(n,\nu)$ are noncommuting matrices, the last in $SU(3)$.

We define gauge transformations on the lattice to act directly on the link elements:

$$U_{\mu\nu}(n) \to U^{-1}(n)U_{\mu\nu}(n)U(n). \quad (9.2.17)$$

With this we get, for any line ordered product,

$$\prod_{\ell(n\to k)} U(j,\mu_j) \to U^{-1}(k) \prod_{\ell(n\to k)} U(j,\mu_j)U(n), \quad (9.2.18)$$

(cf. (9.2.7b)) and, for a closed loop,

$$\prod_{\ell(n\to n)} U(j,\mu_j) \to U^{-1}(n) \prod_{\ell(n\to n)} U(j,\mu_j)U(n). \quad (9.2.19a)$$

In particular, for a plaquette,

$$U_{\mu\nu}(n) \to U^{-1}(n)U_{\mu\nu}(n)U(n). \quad (9.2.19b)$$

Because of these transformation properties (9.2.19) we find that the quantity

$$S_\square \equiv S_{\mu\nu}(n) \equiv -\frac{1}{N_c}\beta\,\mathrm{Re}\,\mathrm{Tr}\,U_{\mu\nu}(n) \quad (9.2.20a)$$

is gauge invariant. $S_\square$ is to be identified as the action on a plaquette (the *Wilson action*[6]). The definition (9.2.20a) is the traditional one; as is customary, we have extracted explicitly a sign and the number of colours, $N_c = 3$. The full action is

$$S_G = \sum_\square S_\square = \tfrac{1}{2}\sum_{\mu\nu}\sum_n S_{\mu\nu}(n). \quad (9.2.20b)$$

If we choose β as

$$\beta = \frac{2N_c}{g^2}, \quad (9.2.20c)$$

---

[6] Other definitions of action, with the same continuum limit, are possible and have been used in the literature; see the treatise of Creutz (1983) and references therein.

then, in the continuum limit $a \to 0$, $an \to x$,

$$S_G \to \tfrac{1}{2} \operatorname{Tr} \int d^4x \sum_{\mu\nu} G^2_{\mu\nu}(x) + \text{constant}. \tag{9.2.21}$$

The proof is similar to that of the Abelian case; the only difference is that an expression such as (9.2.13a) will be modified by commutators according to the rule (a particular case of the familiar Campbell–Hausdorff relation)

$$e^{F_1} e^{F_2} = \exp\left(F_1 + F_2 + \tfrac{1}{2}[F_1, F_2]\right) + O(F^3).$$

These commutators precisely complete the field tensor $G_{\mu\nu}$.

The gluon generating functional can be obtained by integrating over the links (for group integration, cf. Appendix C),

$$Z_G = \prod_{n,\mu} \int_{SU_c(3)} dU(n,\mu) e^{-S_G}. \tag{9.2.22}$$

Interactions between quarks and gluons may be introduced as in the Abelian case. Writing the action directly for Wilson fermions, we have

$$S_{qG} = a^4 \sum_n \bar{q}_n \left\{ \left(m + \frac{4r}{a}\right) q_n \right.$$

$$\left. + \frac{1}{2a} \sum_\mu [(r + \gamma_\mu) U(n + \hat{\mu}, -\mu) q_{n+\hat{\mu}} + (r - \gamma_\mu) U(n - \hat{\mu}, \mu) q_{n-\hat{\mu}}] \right\}, \tag{9.2.23}$$

so that the full generating functional is

$$Z = \prod_{n,\mu} \int dU(n,\mu) \int \prod dq_k \int \prod d\bar{q}_j \, e^{-(S_G + S_{qG})}. \tag{9.2.24}$$

A word of explanation is needed about (9.2.24) in connection with gauge invariance. To obtain vacuum expectation values, sources and functional differentiations have to be introduced just as in the continuum case; but, unless one wants to find the gluon propagator (for example, in perturbation theory) or any other similar gauge dependent quantity, a gauge fixing term is not necessary: (9.2.24) averages over all gauges.

9.3 Feynman Rules on the Lattice.
Renormalization Group.
Connection with the Continuum Parameters

i Feynman Rules

To obtain the propagators on the lattice we have to identify the quadratic terms for the various fields in the action, at zero interaction strength. For quarks, and from (9.2.23), we write $U(n, \mu) \simeq 1$ and find

$$a^4 \sum_n \bar{q}_n \left\{ (m + 4r/a)q_n + \frac{1}{2a} \sum [(r + \gamma_\mu)q_{n+\hat{\mu}} + (r - \gamma_\mu)q_{n-\hat{\mu}}] \right\},$$

from which we find, in the limit of large lattice size $(N \to \infty)$,

$$S_{lj}^r(p) = \left\{ m + \frac{i}{a} \sum_\mu \gamma_\mu \sin ap_\mu + \frac{r}{a} \sum_\mu (1 - \cos ap_\mu) \right\}^{-1} \delta_{lj}, \qquad (9.3.1)$$

(cf. Eq. (9.1.16)). Here l, j, are colour indices. The graph of Fig. 9.3.1a corresponds to (9.3.1).

For the gluon propagator we expand the $U(n, \mu)$ in terms of the fields B_μ. As we are only interested in quadratic terms, the commutators may be neglected. We thus obtain the piece

$$\sum_{n,\mu\nu} \frac{a^2}{8} \operatorname{Tr} \left\{ \sum_c [B_\mu^c(an) + B_\mu^c(an + \hat{\mu}) + B_\nu^c(an + \hat{\mu}) + B_\nu^c(an + \hat{\mu} + \hat{\nu}) \right.$$

$$\left. - B_\mu^c(an + \hat{\mu} + \hat{\nu}) - B_\mu^c(an + \hat{\nu}) - B_\nu^c(an + \hat{\nu}) - B_\nu^c(an)]t^c \right\}^2.$$

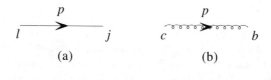

(a)   (b)

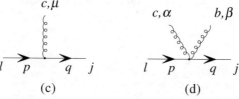

(c)   (d)

Fig. 9.3.1. (a) Quark propagator. (b) Gluon propagator. (c,d) Quark–gluon vertices.

A propagator defined in the continuum requires introduction of a gauge-fixing term. For a large lattice, and in the Fermi–Feynman gauge, a calculation analogous to that of Sect. 9.1 for the quark propagator gives

$$D_{\mu\nu}^{cb}(p) = \delta_{cb} \frac{\delta_{\mu\nu}}{2a^{-2}\sum_\alpha(1 - \cos ap_\alpha)}, \qquad (9.3.2)$$

associated with the graph of Fig. 9.3.1b.

Vertices are more involved. Because in terms of the fields q, B_μ the interactions are nonpolynomial, we have an infinity of vertices, associated with higher and higher powers of the lattice spacing, a. Since loop diagrams will diverge as $a \to 0$, we may not keep only the lowest ones, even in the continuum limit.

We will only give the rules necessary for *one loop* quark self-energy renormalization calculations. Trilinear and quartic couplings, as well as ghost ones, may be found in Kawai, Nakayama and Seo (1981). So we only consider here the quark–gluon vertices. Expanding in (9.2.23), we obtain the interaction Lagrangian, for $r = 0$,

$$\mathcal{L}_{qG\,\text{int}} \simeq -\frac{i}{2}g \sum_c \sum_{\mu\nu} \bar{q}_n \gamma_\mu (q_{n+\hat{\mu}} + q_{n-\hat{\mu}}) B_\mu^c(an)$$

$$-\frac{ag^2}{4} \sum_{cd} \sum_\mu \bar{q}_n \gamma_\mu t^c t^d (q_{n+\hat{\mu}} - q_{n-\hat{\mu}}) B_\mu^c(an) B_\mu^d(an).$$

It is necessary to keep the second term: it will produce a seagull vertex (Fig. 9.3.1d) that will induce quadratic divergences at one loop, and hence one power of a^{-1} over other graphs (which only produce $\log a$ divergences). The vertices, associated respectively with Fig. 9.3.1c,d, are thus, for Wilson fermions with $r = 0$,

$$-igt_{jl}^c \gamma_\mu \cos \frac{a}{2}(p_\mu + q_\mu), \qquad (9.3.3)$$

$$\frac{i}{2}\delta_{\alpha\beta}g^2 \{t^c, t^b\}_{jl} \sum_\mu \gamma_\mu \sin \frac{a}{2}(p_\mu + q_\mu). \qquad (9.3.4)$$

The complicated structure of the laatice Feynman rules explains why only the first orders in perturbation theory are known in this case.

ii Renormalization, and Renormalization Group

As long as a is kept finite, QCD on the lattice is ultraviolet finite, so we have not bothered to distinguish between bare or renormalized quantities. However, as $a \to 0$, loop diagrams (say, in a weak coupling expansion) become divergent. We may consider the lattice as a regularization procedure, and investigate what happens when the cut-off, $1/a$, is allowed to go to infinity. In particular, we can consider the cut-off dependence of the coupling g. To do so we will only have to repeat, with due changes, the analysis of Sects. 3.3, 3.4, 3.7.

We denote the running coupling constant by $\bar{g}$; the renormalization group equation for the coupling in (3.5.6) may be written in terms of a length r instead of a momentum. So we have

$$\frac{r\mathrm{d}\bar{g}}{\mathrm{d}r} = -\beta(\bar{g}(r))\bar{g}(r). \tag{9.3.5a}$$

Expanding β as in (3.7.1),

$$\beta(\bar{g}) = -\beta_0 \frac{\bar{g}^2}{16\pi^2} - \beta_1 \left(\frac{\bar{g}^2}{16\pi^2} \right)^2 - \cdots, \tag{9.3.5b}$$

and the values of the β_n are given in (3.7.2). Integrating (9.3.5) we find the equations corresponding to (3.7.4). For example, to leading order,

$$\alpha_s(r) = \frac{4\pi}{-\beta_0 \log \Lambda^2 r^2}; \tag{9.3.6}$$

Note that we write $\bar{g}(r)$, $\alpha_s(r)$ for simplicity, instead of the more precise expressions $\bar{g}(r^{-1})$, $\alpha_s(r^{-2})$.

All this is valid in the continuum limit, $a \to 0$. To connect with the lattice formulation, we recall that $\bar{g}(r)$ is specified so that it equals the renormalized coupling defined at a fixed $r = r_0$. This in turn is related to the unrenormalized, lattice coupling g_u by the analogue of (3.4.6), say

$$g_u = Z_g(r_0, a)\bar{g}(r_0), \tag{9.3.7}$$

and we have explicitly written the cut-off (a) dependence of the renormalization constant. Because $\bar{g}(r)$ is cut-off independent, it follows that g_u must depend on a: $g_u = g_u(a)$. Its dependence can be obtained from simple dimensional considerations. Because Z_g is dimensionless, it can only depend on the ratio r/a. Therefore,

$$\frac{a\mathrm{d}g_u(a)}{\mathrm{d}a} = -\widetilde{\beta}(g_u)g_u(a),$$

where $\widetilde{\beta}$ is defined as

$$\widetilde{\beta} = Z_g^{-1}\frac{r\mathrm{d}}{\mathrm{d}r}Z_g.$$

Comparing with (3.5.4) we see that, to second order, $\widetilde{\beta}$ has the same functional form as β:

$$\widetilde{\beta} \simeq -\beta_0 \frac{\bar{g}_u^2(a)}{16\pi^2} - \beta_1 \left(\frac{\bar{g}_u^2(a)}{16\pi^2}\right)^2,$$

with the same coefficients as in (9.3.5b). We thus obtain the cut-off dependence of $g_u(a)$; to lowest order,

$$g_u^2(a) = \frac{16\pi^2}{-\beta_0 \log \Lambda^2 a^2}. \tag{9.3.8}$$

Up to now, we have not specified the parameter Λ. As remarked before, and as will be discussed in detail in Sect. 10.2, the value of Λ depends on the regularization and renormalization scheme used. Thus we should really write Λ_{Latt} for Λ. One can relate Λ_{Latt} to the $\overline{\text{MS}}$ value of Λ to first order by simply evaluating one-loop renormalization for α_s, including constant terms, and equating. The calculation was first carried out by Hasenfratz and Hasenfratz (1980).[7] These lattice evaluations are cumbersome because Lorentz invariance is butchered by the lattice regularization, which only respects the subgroup of permutations and finite rotations among the four dimensions of spacetime. In particular, even one loop calculations have to be finished numerically. The ensuing relation between Λ_{Latt} and μ-renormalized Λ, Λ_{MOM} is

$$\Lambda_{\text{Latt}} \simeq \frac{1}{83.5}\Lambda_{\text{MOM}}, \tag{9.3.9}$$

or, in terms of the Λ in the $\overline{\text{MS}}$ scheme *without fermions* (as indeed Λ_{MOM} was defined when obtaining (9.3.9)),

$$\Lambda_{\text{Latt}} \simeq \frac{1}{39}\Lambda_{\overline{\text{MS}}}(n_f = 0). \tag{9.3.10}$$

With the currently accepted values of $\Lambda_{\overline{\text{MS}}}(n_f = 0) \sim 400$ MeV, obtained from deep inelastic scattering, τ decays, ... (see Sect. 10.3 for a summary), we get the surprisingly small value

$$\Lambda_{\text{Latt}} \simeq 10 \pm 4 \text{ MeV}, \tag{9.3.11}$$

and we emphasize that (9.3.11) is obtained from perturbation theory analysis of short distance phenomena.

---

[7] Simplified and extended to actions other than Wilson's by Dashen and Gross (1981) and González-Arroyo and Korthals Altes (1982), using the background field formalism. See also Kawai, Nakayama and Seo (1981) for the introduction of fermions. The values of the quark masses also differ between the lattice and the continuum; see González-Arroyo, Martinelli and Ynduráin (1982).

For a finite lattice, a small but nonvanishing, the theory depends on the two constants a, Λ_{Latt}. It is convenient to work instead in terms of Λ_{Latt} and $g_u(a)$, that we henceforth write simply as $g(a)$, so we convert (9.3.8) into

$$a^2 = \Lambda_{\text{Latt}}^{-2} \exp \frac{-16\pi^2}{\beta_0 g^2(a)}, \qquad (9.3.12a)$$

or, taking into account the second order correction,

$$a^2 = \Lambda_{\text{Latt}}^{-2} e^{-16\pi^2/\beta_0 g^2(a)} \left(\frac{\beta_0 g^2(a)}{16\pi^2} \right)^{-\beta_1/\beta_0^2}, \qquad (9.3.12b)$$

a highly non-analytic result. We recall that (9.3.12a,b) should hold in the limit of a small and for weak coupling:

$$a\Lambda_{\text{Latt}} \ll 1, \quad g^2(a) \ll 1. \qquad (9.3.12c)$$

9.4 The Wilson Loop. Strong Coupling. Confinement

We will start this section by obtaining an expression, in the path integral formalism and in particular on the lattice, for the potential between slowly moving particles, by a method similar to that already used in Sect. 6.4.

In order that the notion of potential be meaningful, we assume the particles to be very heavy (and of equal mass, m). Thus we may treat them with first quantized nonrelativistic formalism. Let Ψ_0 be the ground state of a quark–antiquark pair; the S matrix is

$$\hat{S} = \lim_{\substack{t' \to -\infty \\ t'' \to +\infty}} \exp i(t' - t'')\hat{H},$$

where $\hat{H}$ is the Hamiltonian (in this section we use carets to denote operators). The quantity that will correspond to the generating functional in field theory is the expectation value $\langle \Psi_0 | \hat{S} | \Psi_0 \rangle$. Denoting the position of the particles by $\mathbf{x}$, $\mathbf{y}$, and using the formalism developed in Sect. 1.3,

$$\begin{aligned}
\langle \Psi_0 | \hat{S} | \Psi_0 \rangle &= \int \mathcal{D}\mathbf{x}\, \mathcal{D}\mathbf{y}\, e^{-\int d\tau\, L_{\text{cl}}} \\
&= \int \mathcal{D}\mathbf{x}\, \mathcal{D}\mathbf{y}\, e^{-\int d\tau\, (m/2)(\dot{\mathbf{x}}^2 + \dot{\mathbf{y}}^2)} e^{-\int d\tau\, V(R)}.
\end{aligned} \qquad (9.4.1)$$

We have written (9.4.1) directly for imaginary time, $t \to \tau = t/i$, and have used the corresponding expression for the classical Lagrangian; V is the potential, and $R = |\mathbf{x} - \mathbf{y}|$. Note that, for consistency with the sign choices in these sections about lattice QCD, we define the Euclidean Lagrangian with a sign opposite to the usual one.

Suppose that we now add a second-quantized vector field, say the electromagnetic field $\hat{A}_\mu$. The Lagrangian will thus be

$$L = \frac{m}{2}(\dot{\mathbf{x}}^2 + \dot{\mathbf{y}}^2) + L_{\text{int}} + L_{\text{rad}}.$$

L_{rad} is the pure radiation Lagrangian, $L_{\text{rad}} = \frac{1}{4}\int d^3\mathbf{r}\sum F_{\mu\nu}^2$. The matter–radiation interaction Lagrangian, L_{int}, can be written as

$$L_{\text{int}} = \int d^3\mathbf{z}\sum_\mu j_\mu(z)A_\mu(z) \simeq ieA_4(x) - ieA_4(y).$$

This last expression is valid because, for slowly moving particles, we can neglect $\mathbf{j}$, which is proportional to the velocity, and approximate the fourth component of the current as for static particles with charges $\pm e$:

$$j_4(z) \simeq ie\delta(\mathbf{z} - \mathbf{x}) - ie\delta(\mathbf{z} - \mathbf{y}).$$

This is of course consistent with the nonrelativistic formalism and concepts, and justifies the approximation of *not* quantizing the matter field. We thus have

$$\langle \Psi_0|\hat{S}|\Psi_0\rangle = \int \mathcal{D}\mathbf{x}\,\mathcal{D}\mathbf{y}\,\mathcal{D}A_\mu\, e^{-\int d\tau\, L}$$

$$\simeq \int \mathcal{D}\mathbf{x}\,\mathcal{D}\mathbf{y}\, e^{-\int d\tau\,(m/2)(\dot{\mathbf{x}}^2 + \dot{\mathbf{y}}^2)}\int \mathcal{D}A_\mu\, e^{-S(A) - ie\int d\tau\,(A_4(x) - A_4(y))}$$

$$= \int \mathcal{D}\mathbf{x}\,\mathcal{D}\mathbf{y}\, e^{-\int d\tau\,(m/2)(\dot{\mathbf{x}}^2 + \dot{\mathbf{y}}^2)}\langle e^{-ie\int d\tau\,(A_4(x) - A_4(y))}\rangle;$$

$$(9.4.2)$$

we have defined the average

$$\left\langle \exp\left\{-ie\int d\tau\,(A_4(x) - A_4(y))\right\}\right\rangle$$

$$\equiv \int \mathcal{D}A_\mu\, e^{-S(A)}\exp\left\{-ie\int d\tau\,(A_4(x) - A_4(y))\right\},$$

$$(9.4.3)$$

and $S(A) = \int dt\, L_{\text{rad}}$ is the pure radiation action.

We consider that, because the quarks move slowly,

$$R = |\mathbf{x} - \mathbf{y}| \simeq \text{constant},$$

and, moreover, we will take the infinite time interval to be replaced by a finite, but large one,

$$\int_{-\infty}^{+\infty} d\tau \to \int_0^T d\tau, \quad T \gg R,$$

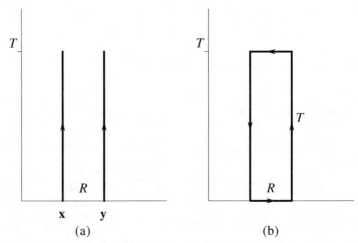

Fig. 9.4.1. (a) Original integration path. (b) The rectangle, ℓ. (The lines along which one integrates are shown thicker).

and thus we also replace the S matrix by the finite time evolution operator, $\hat{U}(T) \equiv \mathrm{e}^{-\mathrm{i}T\hat{H}}$. Under these circumstances we may complete a rectangle (Fig. 9.4.1) for the $\mathrm{d}\tau$ integration so that

$$\int_0^T \mathrm{d}\tau \left(A_4(\mathbf{x}, \tau) - A_4(\mathbf{y}, \tau) \right) \simeq \oint_\ell \sum_\mu \mathrm{d}x_\mu \, A_\mu(x),$$

and the error will be $O(R/T)$, negligible. Alternatively, we may keep terms of order R/T by choosing a gauge where the *horizontal* integrals vanish; for example, the Coulomb gauge. We then define the *Wilson loop* by

$$W(R,T) \equiv \langle \mathrm{e}^{\mathrm{i}e \oint \sum \mathrm{d}x_\mu \, A_\mu(x)} \rangle, \qquad (9.4.4)$$

so that, in terms of it, (9.4.2) becomes

$$\langle \Psi_0 | \hat{U}(T) | \Psi_0 \rangle = \int \mathcal{D}\mathbf{x} \, \mathcal{D}\mathbf{y} \, \mathrm{e}^{-\int \mathrm{d}\tau \, (m/2)(\dot{\mathbf{x}}^2 + \dot{\mathbf{y}}^2)} W(R,T). \qquad (9.4.5)$$

On comparing with (9.4.1) we see that we can interpret W in terms of the potential,

$$W(R,T) = \mathrm{e}^{-\int \mathrm{d}\tau \, V(R)} = \mathrm{e}^{-TV(R)}, \qquad (9.4.6)$$

for $R, T \to \infty$, $R \ll T$. (For R/T fixed, we get the potential in the Coulomb gauge). This completes the discussion for Abelian fields.

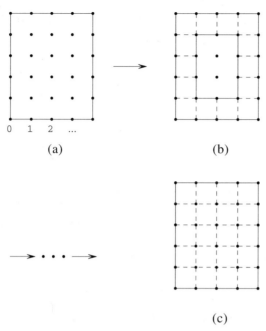

(a)   (b)

(c)

Fig. 9.4.2. The tiling of the rectangle $\ell(I,J)$.

In QCD, (9.4.4) is replaced by (cf. (9.2.22))

$$W(R,T) \equiv \int \mathcal{D}U \, e^{-S_G} \prod_{j \in \ell(R,T)} U(j,\mu_j), \qquad (9.4.7)$$

where $\prod U(j,\mu_j)$ is the path-ordered product of the elements associated with the links in $\ell(R,T)$, $0 \to 1 \to 2 \to 3 \to \ldots$, as in Fig. 9.4.2a. S_G is the gluonic action (9.2.20).

Equation (9.4.7) is written without taking into account the second quantization of the quark fields: in a perturbation-theoretic language, neglecting quark loops. This approximation, called the *quenched* approximation, appears to be reasonable, both from analytical evaluations in the weak coupling regime (the dependence on n_f is usually slight) and in numerical calculations on the lattice, and we will adopt it here.

In general $W(R,T)$ can only be evaluated numerically; but there are two situations when an analytic expression is possible. First, we have a weak coupling expansion, i.e., ordinary perturbation theory. This will be discussed in the next section. The second situation is the *strong coupling* limit, $g^2 \to \infty$, and therefore lowest powers in β. We turn to this now.

Let $R = Ia, T = Ja$, i.e., $I \times J$ is the size of the rectangle $\ell(R,T) \equiv \ell(I,J)$ in lattice units. Writing (9.4.3) explicitly we have,

$$W(I,J) \equiv W(R,T) = \int \prod_n dU(n) \left\{ \prod_{j \in \ell(I,J)} U(j, \mu_j) \right\} \exp \frac{-\beta}{N_c} \sum_\square \text{Tr Re } U_\square .$$
$$(9.4.8)$$

The strong coupling expansion corresponds to the expansion of the exponent in powers of β: recall that $\beta = 2N_c/g^2$; see Eq. (9.2.20c).

Using the formulas of group integration (Appendix C),

$$\int dU \, U_{ik} = 0 \qquad \int dU \, U_{ik} U_{jl}^* = \frac{1}{N_c} \delta_{ij} \delta_{kl},$$

we see that, to get a nonzero result for (9.4.8) we need at least as many U coming from the expansion of S_G as there are in the product $\prod U(j, \mu_j)$. However, because S_G only contains full plaquettes, every time one $U^{-1}(j, \mu_j)$ is brought down from S_G, a plaquette is added: so we get the perimeter $\ell(I,J)$ replaced by a smaller one (Fig. 9.4.2a,b). This goes on until the whole interior of $\ell(I,J)$ is *tiled* with plaquettes (Fig. 9.4.2c). In all, we require $I \times J$ plaquettes, so the first nonzero contribution in the expansion of $\exp[(-\beta/N_c) \sum_\square \text{Re Tr } U_\square]$ is the IJth one. Therefore,

$$W(I,J) \simeq \int \prod_n dU(n) \left\{ \prod_{\ell(I,J)} U(j, \mu_j) \right\}$$
$$\times \left(\frac{-\beta}{N_c} \right)^{IJ} \frac{1}{(IJ)!} \left\{ \sum_{\text{til.}} \text{Re Tr } U_\square \right\}^{IJ} + O(\beta^{IJ+1}).$$
$$(9.4.9)$$

The sum in (9.4.9), $\sum_{\text{til.}}$, runs over the plaquettes in the tiling. We will complete the calculation of (9.4.9) for an Abelian theory. The evaluation for a non-Abelian one requires some extra algebraic machinery (that the interested reader may find in the text of Creutz, 1983) and we will only indicate the result for it.

For an Abelian theory, we may replace $\prod U(j, \mu_j)$ by $\prod U_\square$, because the links traversed in opposite directions cancel one another. Then, since all and every one of the plaquettes is present in $\prod U_\square$, it follows that the only nonzero element of (9.4.9) will be that in the expansion of the product

$$\left\{ \sum_{\text{til.}} \text{Re Tr } U_\square \right\}^{IJ}$$

in which each plaquette appears one, and only one, time. As the order is now irrelevant, it follows that we have $(IJ)!$ of these, a factor that precisely cancels the $1/(IJ)!$ in (9.4.9). Assuming $IJ = $ even, we have found the leading term in the strong coupling expansion for W:

$$W(I,J) \simeq e^{IJ \log \beta}. \qquad (9.4.10)$$

In the non-Abelian case (QCD) one finds a similar result:

$$W(I, J) \simeq e^{IJ \log(\beta/2N_c)} \equiv e^{-KRT},$$

$$K \equiv \frac{1}{a^2} \log \frac{2N_c}{\beta} = \frac{1}{a^2} \log(N_c g^2). \tag{9.4.11}$$

On comparing with (9.4.6) we find the long-distance potential between heavy sources,

$$V(R) = KR, \quad K = \frac{1}{a^2} \log(N_c g^2), \quad g^2 \gg 1, \tag{9.4.12}$$

for QCD – and a similar equation with a slightly different value of K for an Abelian theory.

The linear potential in (9.4.12) strongly suggests confinement; but the situation is not totally clear for the two following reasons. First of all, (9.4.12) has only been derived in the *large* coupling limit, and the quenched approximation. We would like a proof that the result also holds in the unquenched case, and either an evaluation of the potential for arbitrary g^2, or a proof that long distances really imply an exploding g^2. The second snag is that we have got more than we had bargained for: we find confinement for Abelian theories, in particular QED. Thus, it might appear that the linear potential could be an artifact of the lattice, instead of a true feature of the theory.[8]

There are indications that *QED* is an inconsistent theory; and numerical evaluations on the lattice seem to indicate that there really is a phase transition in QED between the weak, $e^2 \ll 1$, and strong, $e^2 \gg 1$, coupling regimes. But then one should worry that the same be not the case for QCD: we would like that the asymptotically free theory, and the confining one, were the same in this case. This hope seems to be fulfilled in that numerical evaluations indicate that there is no phase transition; or, if there is one, that it is of high enough order that the passage from $g^2 \ll 1$ to $g^2 \gg 1$ is smooth.

---

[8] Of course, for QCD we have other reasons than the strong coupling lattice evaluation for believing in a linear potential and/or confinement.

9.5 Observable Consequences of Lattice QCD

i Wilson Loop; String Tension; Connection Between Long and Short Distances

Let us consider the Wilson loop $W(I, J)$. We can extract the quantity K (*string tension*) defined by (9.4.11),

$$K^{1/2} \equiv -\frac{1}{IJa^2} \log W(I, J), \qquad (9.5.1)$$

from the function

$$\chi(I, J) = -\log \frac{W(I, J)W(I - 1, J - 1)}{W(I, J - 1)W(I - 1, J)}. \qquad (9.5.2)$$

We then have, using (9.4.11),

$$\chi(I, J) \simeq a^2 K \underset{\substack{g^2 \to \infty \\ I, J \to \infty \\ I \ll J}}{\simeq} \log(N_c g^2) \simeq \log g^2(a). \qquad (9.5.3)$$

The reason one calculates with the quantity χ is that most perimeter effects cancel out for it (Creutz, 1980).

For $g^2 \to 0$, on the other hand, (9.3.12) implies

$$\chi(I, J) \simeq a^2 K \underset{\substack{g^2 \to 0 \\ I, J \to \infty \\ I \ll J}}{\simeq} \Lambda_{\text{Latt}}^{-2} K \left(\frac{\beta_0 g^2(a)}{16\pi^2} \right)^{-\beta_1/\beta_0} e^{-16\pi^2/\beta_0 g^2(a)}. \qquad (9.5.4)$$

If one evaluates for *small* Wilson loops, one obtains a result that is very different from (9.5.4); for example, and after a trivial calculation, we find

$$\chi(1, 1) = g^2(a)/3. \qquad (9.5.5)$$

Thus we expect that for finite I, J, $\chi(I, J)$ will deviate from the true value (9.5.4) to which, however, it will tend as we increase I, J. Moreover, if there is no phase transition, the regions where (9.5.3) and (9.5.4) hold will be joined by letting g vary, with the transition remaining smooth as I, J increase.

In Fig. 9.5.1 we have plotted typical results of calculations for finite I, J, as well as the theoretical expectations both at large and small g^2 (based on the review of Moriarty (1983) and work quoted there). We see that there indeed seems to exist an approach to the theoretical limits both at small and large coupling, and that the interpolation appears smooth. The best fit is obtained for

$$\Lambda_{\text{Latt}} = (6 \pm 1) \times 10^{-3} K^{1/2}$$

(Creutz and Moriarty, 1982; using a 6^4 lattice). This may be compared with the relation, obtained with an improved 12^4 lattice by Barkai, Creutz and Moriarty (cf. Moriarty's 1983 review),

$$\Lambda_{\text{Latt}} = (8 \pm 1) \times 10^{-3} K^{1/2}, \qquad (9.5.6a)$$

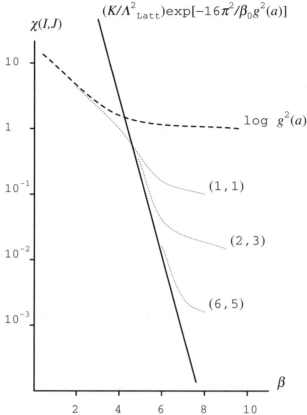

Fig. 9.5.1. Numerical and analytical results for $\chi(I, J)$. *Continuous line:* weak coupling. *Dashed line:* strong coupling. *Dotted lines:* eyeball interpolations of numerical results for various values of I, J. (Actually, this drawing is somewhat optimistic; larger values are necessary to obtain the correct asymptotic behaviour.)

while Fukugita, Kaneko and Ukawa (1983) found

$$\Lambda_{\text{Latt}} = (7.9 \pm 4) \times 10^{-3} K^{1/2}. \qquad (9.5.6b)$$

An up to date review, from which references to recent calculations may be retrieved, is that of Aoki et al. (1998).

The string tension K may be obtained from two sources: from Regge theory (Simonov, 1989a,b; Dubin, Kaidalov and Simonov, 1994) or from fits to the linear, long distance interquark potential, especially for heavy quarks; cf. Sect. 6.4. Both estimates agree to a value

$$K^{1/2} \simeq 420 \text{ MeV}. \qquad (9.5.7)$$

If we now use (9.5.6) we obtain a *prediction* for Λ_{Latt}, obtained from *long distance* phenomena, i.e., from the value of $K^{1/2}$ in (9.5.7). It is,

$$\Lambda_{\text{Latt}} \simeq 3.4 \text{ MeV} \qquad (9.5.8)$$

and we have *not* taken into account the errors in (9.5.6). After the long chain of reasoning involved, including unexpectedly large numbers (as in the connection between Λ_{Latt} and $\Lambda_{\overline{\text{MS}}}$, Eq. (9.3.10)) the consistency of (9.5.8) with the short distance value $\Lambda_{\text{Latt}} = 10 \pm 4$ MeV of Eq. (9.3.11) is remarkable.[9]

ii Hadronization of Jets

We will consider the simple case of two jets in e^+e^- annihilations. Thus, we start with the production of two quarks at a point 0, which then move to the spacetime points x, y. The corresponding amplitude will be connected with the expression

$$\mathcal{A}_\mu \sim \langle 0|T\bar{q}_j(x)q_j(y)J_\mu(0)|0\rangle,$$
$$J_\mu = \sum_k \bar{q}_k\gamma_\mu q_k, \qquad (9.5.9)$$

where j, k, are colour indices and we consider a single species of quark. (9.5.9) is *not* gauge invariant; to get a gauge invariant expression we should introduce a path ordered line integral,

$$\text{Pe}^{ig\int_y^x dz_\mu B^\mu(z)} = \lim_{a\to 0}\prod_y^x U(n, \mu_n),$$

between y and x. Writing the corresponding expression directly for Euclidean lattice QCD and recalling (9.2.24), we have that (9.5.9) becomes

$$\mathcal{A}_\mu = \int \mathcal{D}q\,\mathcal{D}\bar{q}\,\mathcal{D}U\; \bar{q}_{j,n_1} \prod_{l(n_2\to n_1)} U(n,\mu_n)q_{j,n_2} \sum_k \bar{q}_k(0)\gamma_\mu q_k(0)e^{-S},$$
$$S = S_G + S_{qG}.$$

$$(9.5.10)$$

Here $x = an_1$, $y = an_2$ and $l(n_2 \to n_1)$ is the straight line from n_2 to n_1.

We make now two remarks. First, as quarks get separated, we expect the coupling to increase; so we will work to leading order in the strong coupling limit. Secondly, the movement occurs in a region that is forbidden classically. In a potential language, the energy of the quarks is smaller than the potential,

---

[9] All the same, it is a fact that lattice evaluations tend to give *smallish* values for Λ – and for the light quark masses too; see Sects. 10.3, 4.

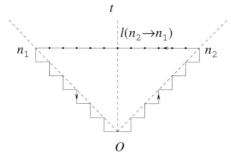

Fig. 9.5.2. The "triangle" $T = (0, n_1, n_2)$.

Kr, for large separation $|\mathbf{x} - \mathbf{y}| = r$. Therefore the Euclidean formalism is appropriate here, as it gives the *tunnelling* probability amplitude (Sect. 8.1).

It is easy to see that the first nonvanishing contribution to (9.5.10) is obtained by first bringing sufficiently many terms from the exponent containing S_{qG} to complete a loop around the "triangle" T with vertices 0, n_1, n_2 (Fig. 9.5.2). The leading order contribution to the WKB approximation will correspond to the quarks following the "classical" trajectory, i.e., straight lines from 0 to n_1 and n_2 (actually, the polygonal more closely approximating these lines). The introduced new links are sufficient to give the loop product

$$\prod_{n \in T} U(n, \mu_n)$$

so, with loose notation,

$$\mathcal{A}_\mu \sim \int \mathcal{D}U \prod_{n \in T} U(n, \mu_n) e^{-S_G} \sim W(T) \sim e^{-\frac{1}{2}Ktr} \sim e^{-Kt^2}, \qquad (9.5.11)$$

where $t = x_4 = y_4$ is the time elapsed since quarks were created, and we have taken $r/2 = t$ (quarks travelling with the speed of light; we neglect quark masses). As expected, (9.5.11) implies that the probability for finding two isolated quarks decreases exponentially with time and separation.

Consider then processes with *four* quarks[10] in the final state: for example, two created by a current, and the other two by materialization at short distance from O of a gluon radiated by any of the existing quarks. If, for simplicity, we assume the colours to be matched, we will get surfaces such as the A_i in Fig. 9.5.3, where we approximate polygonals by straight lines. For large t, r, the point where the two extra quarks were created is irrelevant. The probability amplitude is now

$$\mathcal{A} \sim e^{-KA_1 - KA_2},$$

---

[10]More accurately, two quarks and two antiquarks.

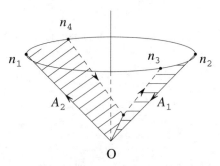

Fig. 9.5.3. World lines of four quarks.

where A_1, A_2 are the shaded surfaces in Fig. 9.5.3. For large t, we find that, for example, A_1 is

$$A_1 = \tfrac{1}{2}|n_3 - n_2|\,|n_2|.$$

Now,

$$|n_2| = \sqrt{t^2 + \tfrac{1}{4}r^2}, \quad |n_3 - n_2| = tv_2,$$

with v_2 the relative velocity of n_3 to n_2. In terms of energy, $v_2 = 2p_{2\perp}/s^{1/2}$, with s the total energy of the original $\bar{q}q$ pair and $p_{2\perp}$ the transverse momentum of n_3 with respect to n_2. Thus, $A_1 \sim 2t^{1/2}p_{2\perp}/s^{1/2}$, and

$$\mathcal{A} \sim \exp\frac{-Kt^2(p_{1\perp} + p_{2\perp})}{s^{1/2}} :$$

the probability decreases as the exponential of the transverse momentum.

We can iterate the short distance production of quarks, until all the transverse momenta are as small as the momenta of quarks in hadrons, at which time one may consider the hadronization process to be complete. This yields a picture qualitatively similar to that of the Lund model[11] which, as explained in Sects. 5.3, 4, provides a good description of hadronization of jets.

---

[11] Anderson, Gustafson and Peterson (1977, 1979).

iii Locality of the QCD Vacuum. Long Distance Behaviour of Invariant Propagators. Constituent Quark Mass

In this subsection we will consider a model for the quark propagator, in the nonperturbative vacuum. This model will incorporate the short distance behaviour of the propagator discussed in Sect. 3.9 and the long distance behaviour suggested by a strong coupling lattice calculation. As we will see, this propagator presents features which, particularly at long distances, correspond to particles with an effective invariant mass of the order of the constituent mass we discussed in Sect. 6.5; and this, even for quarks with zero "mechanical" mass.

The quark propagator,

$$S_{ij}(x) = \langle \text{vac} | \text{T} q_i(x) \bar{q}_j(0) | \text{vac} \rangle, \tag{9.5.12}$$

is a gauge dependent object. We define an invariant propagator by inserting a line integral. In matrix notation, but working in Minkowski space for now, we thus write an effective propagator as

$$S_{ij}^{\text{eff}}(x) = -\frac{\delta_{ij}}{N_c} \langle \text{vac} | \text{T} \bar{q}(0) \text{P} \exp^{-ig \int_x^0 dy^\mu \, B_\mu(y)} q(x) | \text{vac} \rangle. \tag{9.5.13}$$

We can interpret S_{ij}^{eff} as the propagator describing a quark as it moves in the gluonic soup inside a hadron. In p-space,

$$S_{ij}^{\text{eff}}(p) = \int d^4x \, e^{ip \cdot x} S_{ij}^{\text{eff}}(x).$$

At short distances we have the familiar expression

$$S_{ij}^{\text{eff}}(x) \underset{x \to 0}{\simeq} \delta_{ij} \left\{ \frac{-1}{4\pi^2} \not{\partial} \frac{1}{x^2 - i0} - \frac{1}{4N_c} \langle \bar{q}q \rangle \right\}; \tag{9.5.14}$$

we have taken the quark to be massless. At long distances we evaluate $S_{ij}^{\text{eff}}(x)$ as follows. First, we go to Euclidean space. Then, and because we expect confinement (and thus that the interaction grows at long distances), we calculate for large coupling, $g \to \infty$. Finally, the quenched approximation is used.

Under these circumstances, the evaluation of $S_{ij}^{\text{eff}}(x)$ is identical to that of the Wilson loop in Sect. 9.4. Underlining Euclidean quantities, we then find

$$\underline{S}_{ij}^{\text{eff}}(\underline{x}) \underset{\underline{x} \to \infty}{\sim} \delta_{ij} e^{-K^{1/2}|\underline{x}|};$$

K is the string tension. The corresponding Minkowski space expression,

$$S_{ij}^{\text{eff}}(x) \underset{\underline{x} \to \infty}{\sim} \delta_{ij} e^{-K^{1/2}\sqrt{-x^2}}, \tag{9.5.15}$$

is very appealing. According to it, the probability of a quark in the vacuum (inside a hadron) to propagate at a spacelike distance $r = \sqrt{-x^2}$ decreases exponentially when $r \gg K^{-1/2}$; but the quark may move freely along a timelike or lightlike trajectory, where $\sqrt{-x^2}$ is pure imaginary. This makes apparent the *local* character of the QCD vacuum.[12]

A simple ansatz incorporating both (9.5.14) and (9.5.15) is

$$S_{ij}^{\text{eff}}(x) = \delta_{ij} \left\{ \frac{-1}{4\pi^2} \partial\!\!\!/ \frac{1}{x^2 - i0} - \frac{1}{4N_c} \langle \bar{q}q \rangle \right\} e^{-K^{1/2}\sqrt{-x^2}}. \qquad (9.5.16)$$

The corresponding p-space expression is then easily evaluated to be

$$S_{ij}^{\text{eff}}(p) = \delta_{ij} \left\{ \frac{i}{p\!\!\!/} \left(1 - \frac{K^{1/2}}{(K - p^2 - i0)^{1/2}} \right) - \frac{3\pi^2 i K^{1/2} \langle \bar{q}q \rangle}{N_c(K - p^2 - i0)^{5/2}} \right\}. \qquad (9.5.17)$$

It is interesting to note that the piece originating in the quark condensate in (9.5.17), viz.,

$$-\delta_{ij} \frac{3\pi^2 i K^{1/2} \langle \bar{q}q \rangle}{N_c(K - p^2 - i0)^{5/2}},$$

provides a regularization for the nonperturbative quark propagator that we gave in Sect. 3.9ii, Eq. (3.9.12), with $m = 0$,

$$-(2\pi)^4 \frac{\langle \bar{q}q \rangle}{4N_c} \delta_4(p)\delta_{ij},$$

to which it indeed tends when letting $K \to 0$.

The expression (9.5.17) for the propagator fulfills the Bricmont–Fröhlich (1983) criterion for confinement and indeed exhibits many of the characteristics of the propagator for a particle with nonzero effective mass. Thus, $S_{ij}^{\text{eff}}(p)$ presents a cut starting at $p^2 = K$ and, what is more interesting, it behaves for $p \to 0$ like the propagator for a massive particle:

$$S_{ij}^{\text{eff}}(p) \underset{p \to 0}{\simeq} -\frac{ip\!\!\!/}{2K} + \frac{3\pi^2 i \langle \bar{q}q \rangle}{N_c K^2} \simeq -\frac{i}{\mu}, \qquad (9.5.18a)$$

where the effective mass μ is

$$\mu = \frac{N_c K^2}{-3\pi^2 \langle \bar{q}q \rangle}. \qquad (9.5.18b)$$

It is curious that in the last expression the quark condensate appears in the denominator. This indicates that $K^{1/2}$ and $(-\langle \bar{q}q \rangle)^{-1/3}$ are proportional and links, in ordinary QCD, quark confinement ($K \neq 0$) with spontaneous breaking of chiral invariance, $\langle \bar{q}q \rangle \neq 0$.

The numerology also works reasonably well. Taking the quark condensate renormalized at 1 GeV from the PCAC relation (7.3.4, 5) with the light quark masses of (7.4.9), and the value $\mu \simeq 320$ MeV obtained from phenomenological quark models, we can predict K using (9.5.18b) to get $K^{1/2} \simeq 470$ MeV. The excellent (and far from trivial) agreement with previous determinations for this quantity (e.g., Eq. (9.5.7) that gave 420 MeV) should, however, not

---

[12]As opposed, for example, to the Higgs vacuum, that extends over all spacetime.

make one forget the shortcomings of our calculation here; (9.5.17) is to be considered as no more than a phenomenological expression. In fact, not only is the interpolation (9.5.16) somewhat arbitrary, but, because the expression for the propagator only takes account of a certain class of gluon couplings, use of (9.5.17) *tel quel* into Feynman diagrams may lead to violations of gauge invariance.

A treatment of the gluon propagator along these lines is possible, but much less satisfactory.

iv Masses and Other Hadron Properties. Glueballs

As stated when we begun to discuss lattice QCD, in principle this tool allows us to calculate everything: the masses and decay constants of hadrons made of light quarks, form factors at finite, even small momentum transfer, structure functions, etc. Before these calculations make sense, however, a number of questions have to be understood. In particular, we have technical problems that we now must mention.

Consider first the integrals over the gauge fields, $\int \prod dU$. Because each group element depends on 8 parameters, we have an eight-dimensional integral for each lattice point, and hence 4×8 for each link. Now, putting a hadron in a lattice makes only sense if a is much smaller than the size of the hadron and, moreover, the hadron is much smaller than the lattice. Even if we understand "much smaller" to mean only half an order of magnitude, this implies $2N + 1 \sim 10$, so we have $4 \times 8 \times 10^4 > 10^5$ integrations: not a light task, even with Monte Carlo methods.

Secondly, we have integrals over the anticommuting fermion fields. The integral (cf. (9.2.23), (9.2.24))

$$
\int \mathcal{D}q \, \mathcal{D}\bar{q} \, \exp \sum_{nk} \bar{q}_n D_{nk} q_k,
$$

$$
D_{nk}(U) = \left(m + \frac{4r}{a} \right) \delta_{nk}
$$

$$
+ \frac{1}{2a} \sum_{\mu} \left[(r + \gamma_\mu) U(n + \hat{\mu}, -\mu) \delta_{k,n+\hat{\mu}} + (r - \gamma_\mu) U(n - \hat{\mu}, \mu) \delta_{k,n-\hat{\mu}} \right],
$$

$$
(9.5.19)
$$

can be carried out explicitly because

$$
\int \mathcal{D}q \, \mathcal{D}\bar{q} \, \exp \sum_{nk} \bar{q}_n D_{nk} q_k = \text{const.} \times \det D.
$$

So we should, to take dynamical quarks into account, evaluate this determinant for every calculation of the integral over gluon fields. This is a formidable task, and usually the *quenched approximation* is employed, in which we take $\det D \to 1$. Numerical calculations seem to justify this approximation, but only to a certain extent.

Besides this, there are problems of principle involving (among others) the restoration of the Lorentz invariance and chiral invariance in the continuum limit, or the dependence of the results, at finite a, on the action used for the gluons, etc.

All of this justifies the size of the errors present in lattice calculations, and the lack of realistic estimates of the *systematic* biases: it is with this in mind that one should appraise the results to be presented. Also, and as said before, almost everything has been calculated on the lattice; here we show only a few sets of results that can be considered representative.

The first set concerns the hadron masses. We have[13]

$$
\begin{aligned}
m_\rho &= 730 \pm 90 \text{ MeV} & (770) \\
m_{A_1} &= 1190 \pm 90 \text{ MeV} & (1260) \\
m_p &= 920 \pm 100 \text{ MeV} & (938) \\
m_{\pi'} &= 1100 \pm 150 \text{ MeV} & (1300)
\end{aligned}
$$

for $\Lambda_{\text{Latt}} = 2.4$ and $\beta = 5.7$. The errors are purely statistical; the numbers in brackets are the experimental values, in MeV. The agreement is certainly encouraging, albeit with a value of Λ_{Latt} too much on the small side.

As a complementary set of results, we consider the π, K decay constants. Then, with the same conventions as in the previous table,

$$
\begin{aligned}
f_\pi &= 99 \pm 14 \text{ MeV} & (93) \\
f_K &= 112 \pm 9 \text{ MeV} & (116) \\
f_K &= 122 \pm 59 \text{ MeV}.
\end{aligned}
$$

The results are due to Gavela et al. (1988), De Grand and Loft (1988) and Bernard et al. (1989). The size of the systematic errors can be inferred from the difference in the *statistical* errors given for both determinations of f_K.

We finish this section with a few comments on *glueballs*, somewhat special objects presumed to be mostly made of gluons.

There are in the Particle Data Group (1996) tables a number of hadronic resonances, without flavour quantum numbers, that cannot be fitted into a conventional constituent quark model; that is to say, they cannot be explained as $\bar{q}q$ states. Various interpretations have been advanced for these objects. Some can be viewed as states with more than two quarks, "molecules" $\bar{q}q\bar{q}q$; but there are a few for which the simplest structure consistent with experiment is to consider them to be *glueballs*. The experimental situation is far from clear. There are a number of scalar resonances,

$$f_0(975), \ f_0(1400), \ f_0(1590), \ f_0(1710),$$

all of which cannot be interpreted as excited $\bar{q}q$ states; but none of them is unambiguously a gluon state. There are also more tensor (spin two) resonances than one would expect on the basis of the $\bar{q}q$ classification, and the

---

[13]The results are due to J. P. Gilchrist, G. Schierholz and H. Schneider; see Schierholz's (1985) review, or, for more recent results, Aoki et al. (1998).

same thing happens for pseudoscalar resonances. For the latter we have the state

$$\eta(1440) \ (\text{formerly } \iota(1440)),$$

which is a strong candidate for a glueball. Indeed, it is copiously produced in decays

$$J/\psi \rightarrow \eta(1440) + \gamma,$$

which are easily interpreted as

$$J/\psi \rightarrow 2G + \gamma.$$

Moreover, it decays prominently in channels containing η, $\eta(980)$ which, because of the anomaly, are known to have a strong gluon component; or into states containing kaons, also favoured because the mixing of pseudoscalar glueballs with quarks is thought to be proportional to the mass of the latter. Even in this case, however, the situation is not totally clear. For example, it is still doubtful whether the $\eta(1440)$ is a single wide resonance or two narrow ones,

$$\eta(1410), \ \eta(1449).$$

From the point of view of the theory, it is not easy to treat glueballs either. As remarked in Sect. 6.1, constituent quark models for glueballs are mostly arbitrary. One could try to study these objects in the bag model, or using the SVZ sum rules. The drawbacks of the first method were discussed in Sect. 6.6. As for the SVZ sum rules, they are reasonably effective to get QCD parameters from experimentally known correlation functions; the opposite path is much more difficult to follow, as we would require extrapolation of high energy QCD expressions to yield low energy parameters, a notoriously unstable procedure.

There remain lattice calculations. In principle, glueballs are ideally suited for these: because the glueball operators can be constructed with only gluon fields, it would seem that we could dispense with quarks altogether. If we let N_{GG} be (composite) glueball operators, we can consider the correlator

$$\sum_{\mathbf{n}} \langle N_{GG}(\mathbf{n}, n_4), N_{GG}(0) \rangle_{\text{vac}}.$$

As x_4 grows, this quantity behaves as

$$\exp(-m_0 x_4),$$

with m_0 the mass of the lightest glueball state. Thus, even a rough calculation should provide a reasonable estimate of m_0. In this way one can get an idea of the spectrum of the lowest-lying glueball states corresponding to different spin-parity assignments. For example, one has

$$m(\text{scalar}) = 740 \pm 40 \text{ MeV}$$
$$m(\text{tensor}) = 1620 \pm 100 \text{ MeV}$$
$$m(\text{pseudoscalar}) = 1220 \pm 200 \text{ MeV}$$

(statistical errors only). The calculation is due to K. Ishikawa et al. in 1983; for a discussion of it, and more details and references, see the review of Halliday (1983). An idea of the systematic errors can be obtained by comparing with the values

$$m(\text{scalar}) = 1370 \pm 90 \text{ MeV}, \quad m(\text{tensor}) = 2115 \pm 125 \text{ MeV},$$

given in the Kronfeld (1989) compilation, as well as the recent estimates[14] that give

$$m(\text{scalar}) = 1.71 \pm 0.05 \text{ GeV}.$$

The main problem with these calculations relates to mixing. Consider, for example, pseudoscalar glueballs. Because of the existence of the axial anomaly, we expect that the quark and gluon operators

$$m\bar{q}\gamma_5 q, \quad \tilde{G}G,$$

should appear together as in fact they mix under renormalization. The situation for scalar states (using for example the energy–momentum tensor anomaly) is similar. Because of this mixing, it is clear that the quenched approximation cannot be as good as in other situations, and some evaluations indicate that it is not to be trusted to better than some 50%. Another problem is that there is no guarantee that gluon states will be lighter than multigluon ones. It remains, however, that, as explained above, lattice QCD is the only method with which to study these elusive entities with any reliability.

---

[14]Bali et al. (1997); Luo et al. (1997).

10 The Perturbative QCD Series. The Parameters of QCD

10.1 The Functions β, γ_m

A large number of the more reliable results in QCD come from perturbative expansions at large momenta, and are due to the asymptotic freedom property. This justifies devoting a section to presenting a summary of our knowledge of the basic functions β and γ_m. In *supersymmetric* extensions of QCD, both functions are related and, for some specific supersymmetric theories, they can be calculated *exactly*. Actually, and as proved by Mandelstam (1983), there are renormalization schemes in which both β, γ_m vanish identically, and in others they can be found to all orders, as remarked first by Shifman, Vainshtein and Zakharov (1983). We will not discuss these theories here. The interested reader may find information, and trace the relevant literature, from the monumental papers of Seiberg and Witten (1994); we turn now to ordinary QCD.

Since the pioneering papers of Gross and Wilczek (1973a) and Politzer (1973), much progress has been made in pushing the calculations of β, γ_m to higher orders. In both cases the coefficients in their expansions are known to *four* loops and, for γ_m (but not for β), the calculation has been checked by at least two independent groups. We now present the full set of results. It is convenient to organize the expansions in terms of the parameter $a_s \equiv g^2/16\pi^2$. Thus we will write the beta function, and its expansion, in the form

$$\beta(a_s) = -\sum_{n=0}^{\infty} \beta_n a_s^n,$$

$$a_s = \frac{\alpha_s}{4\pi} = \frac{g^2}{16\pi^2}; \ g = g(\mu^2).$$

(10.1.1)

Furthermore, we define the numbers $\zeta_n = \zeta(n)$, where

$$\zeta(z) = \sum_{j=1}^{\infty} \frac{1}{j^z}$$

is Riemann's zeta function. Numerically,

$$\zeta_2 = \frac{\pi^2}{6}, \quad \zeta_3 \simeq 1.20206, \quad \zeta_4 \simeq 1.08232, \quad \zeta_5 \simeq 1.03693.$$

We then have

$$\beta_0 = \tfrac{11}{3}C_A - \tfrac{4}{3}T_F n_f, \quad \beta_1 = \tfrac{34}{3}C_A^2 - 4C_F T_F n_f - \tfrac{20}{3}C_A T_F n_f,$$

$$\beta_2 = \tfrac{2857}{54}C_A^3 + 2C_F^2 T_F n_f - \tfrac{205}{9}C_F C_A T_F n_f$$
$$- \tfrac{1415}{27}C_A^2 T_F n_f + \tfrac{44}{9}C_F T_F^2 n_f^2 + \tfrac{158}{27}C_A T_F^2 n_f^2,$$

$$\beta_3 = C_A^4 \left(\tfrac{150653}{486} - \tfrac{44}{9}\zeta_3 \right) + C_A^3 T_F n_f \left(-\tfrac{39143}{81} + \tfrac{136}{3}\zeta_3 \right)$$
$$+ C_A^2 C_F T_F n_f \left(\tfrac{7073}{243} - \tfrac{656}{9}\zeta_3 \right) + C_A C_F^2 T_F n_f \left(-\tfrac{4204}{27} + \tfrac{352}{9}\zeta_3 \right)$$
$$+ 46C_F^3 T_F n_f + C_A^2 T_F^2 n_f^2 \left(\tfrac{7930}{81} + \tfrac{224}{9}\zeta_3 \right) + C_F^2 T_F^2 n_f^2 \left(\tfrac{1352}{27} - \tfrac{704}{9}\zeta_3 \right)$$
$$+ C_A C_F T_F^2 n_f^2 \left(\tfrac{17152}{243} + \tfrac{448}{9}\zeta_3 \right) + \tfrac{424}{243}C_A T_F^3 n_f^3 + \tfrac{1232}{243}C_F T_F^3 n_f^3$$
$$+ \frac{d_A^{abcd} d_A^{abcd}}{N_A} \left(-\tfrac{80}{9} + \tfrac{704}{3}\zeta_3 \right) + n_f \frac{d_F^{abcd} d_A^{abcd}}{N_A} \left(\tfrac{512}{9} - \tfrac{1664}{3}\zeta_3 \right)$$
$$+ n_f^2 \frac{d_F^{abcd} d_F^{abcd}}{N_A} \left(-\tfrac{704}{9} + \tfrac{512}{3}\zeta_3 \right).$$

$$(10.1.2a)$$

Here $[t^a t^a]_{ij} = C_F \delta_{ij}$ and $f^{acd} f^{bcd} = C_A \delta^{ab}$ are the familiar quadratic Casimir operators of the fundamental and the adjoint representation of the Lie algebra, and $\mathrm{Tr}(t^a t^b) = T_F \delta^{ab}$ is the trace normalization of the fundamental representation. N_A is the number of generators of the group (i.e., the number of gluons, $N_A = N^2 - 1 = 8$) and n_f is the number of quark flavours. At four loops there appear new, higher order group invariants that are expressed in terms of contractions between the following fully symmetrical tensors:

$$d_F^{abcd} = \tfrac{1}{6} \mathrm{Tr} \left[t^a t^b t^c t^d + t^a t^b t^d t^c + t^a t^c t^b t^d \right.$$
$$\left. + t^a t^c t^d t^b + t^a t^d t^b t^c + t^a t^d t^c t^b \right],$$

$$d_A^{abcd} = \tfrac{1}{6} \mathrm{Tr} \left[C^a C^b C^c C^d + C^a C^b C^d C^c + C^a C^c C^b C^d \right.$$
$$\left. + C^a C^c C^d C^b + C^a C^d C^b C^c + C^a C^d C^c C^b \right].$$

For QCD, with $N_c = 3$,

$$\beta_0 = 11 - \tfrac{2}{3}n_f \approx 11 - 0.66667n_f, \quad \beta_1 = 102 - \tfrac{38}{3}n_f \approx 102 - 12.6667n_f,$$
$$\beta_2 = \tfrac{2857}{2} - \tfrac{5033}{18}n_f + \tfrac{325}{54}n_f^2 \approx 1428.50 - 279.611n_f + 6.01852n_f^2,$$
$$\beta_3 = \left(\tfrac{149753}{6} + 3564\zeta_3 \right) - \left(\tfrac{1078361}{162} + \tfrac{6508}{27}\zeta_3 \right) n_f$$
$$+ \left(\tfrac{50065}{162} + \tfrac{6472}{81}\zeta_3 \right) n_f^2 + \tfrac{1093}{729}n_f^3$$
$$\approx 29243.0 - 6946.30n_f + 405.089n_f^2 + 1.49931n_f^3.$$

$$(10.1.2b)$$

For the anomalous dimension of the mass, we write

$$\gamma_m(a_s) = \sum_{n=0}^{\infty} \gamma_m^{(n)} a_s^n \tag{10.1.3a}$$

and then

$$\gamma_m^{(0)} = 3C_F, \quad \gamma_m^{(1)} = \tfrac{3}{2}C_F^2 + \tfrac{97}{6}C_F C_A - \tfrac{10}{3}C_F T_F n_f,$$

$$\gamma_m^{(2)} = \tfrac{129}{2}C_F^3 - \tfrac{129}{4}C_F^2 C_A + \tfrac{11413}{108}C_F C_A^2$$

$$+ C_F^2 T_F n_f(-46 + 48\zeta_3) + C_F C_A T_F n_f\left(-\tfrac{556}{27} - 48\zeta_3\right) - \tfrac{140}{27}C_F T_F^2 n_f^2,$$

$$\gamma_m^{(3)} = C_F^4\left(-\tfrac{1261}{8} - 336\zeta_3\right) + C_F^3 C_A\left(\tfrac{15349}{12} + 316\zeta_3\right)$$

$$+ C_F^2 C_A^2\left(-\tfrac{34045}{36} - 152\zeta_3 + 440\zeta_5\right) + C_F C_A^3\left(\tfrac{70055}{72} + \tfrac{1418}{9}\zeta_3 - 440\zeta_5\right)$$

$$+ C_F^3 T_F n_f\left(-\tfrac{280}{3} + 552\zeta_3 - 480\zeta_5\right)$$

$$+ C_F^2 C_A T_F n_f\left(-\tfrac{8819}{27} + 368\zeta_3 - 264\zeta_4 + 80\zeta_5\right)$$

$$+ C_F C_A^2 T_F n_f\left(-\tfrac{65459}{162} - \tfrac{2684}{3}\zeta_3 + 264\zeta_4 + 400\zeta_5\right)$$

$$+ C_F^2 T_F^2 n_f^2\left(\tfrac{304}{27} - 160\zeta_3 + 96\zeta_4\right)$$

$$+ C_F C_A T_F^2 n_f^2\left(\tfrac{1342}{81} + 160\zeta_3 - 96\zeta_4\right) + C_F T_F^3 n_f^3\left(-\tfrac{664}{81} + \tfrac{128}{9}\zeta_3\right)$$

$$+ \frac{d_F^{abcd} d_A^{abcd}}{N_F}(-32 + 240\zeta_3) + n_f \frac{d_F^{abcd} d_F^{abcd}}{N_F}(64 - 480\zeta_3). \tag{10.1.3b}$$

This produces the following expression for the running masses:

$$m_q(\mu^2) = \widetilde{m}_q a_s^{d_m}\left[1 + A_1 a_s + \left(A_1^2 + A_2\right)\frac{a_s^2}{2}\right.$$

$$\left. + \left(\tfrac{1}{2}A_1^3 + \tfrac{3}{2}A_1 A_2 + A_3\right)\frac{a_s^3}{3} + O(a^4)\right], \tag{10.1.4a}$$

where $\widetilde{m}_q = (2\beta_0)^{d_m}\hat{m}_q$ ($\hat{m}_q$ defined in Sect. 3.7) and

$$d_m = \gamma_m^{(0)}/\beta_0, \quad A_1 = -\frac{\beta_1 \gamma_m^{(0)}}{\beta_0^2} + \frac{\gamma_m^{(1)}}{\beta_0},$$

$$A_2 = \frac{\gamma_m^{(0)}}{\beta_0^2}\left(\frac{\beta_1^2}{\beta_0} - \beta_2\right) - \frac{\beta_1 \gamma_m^{(1)}}{\beta_0^2} + \frac{\gamma_m^{(2)}}{\beta_0},$$

$$A_3 = \frac{\gamma_m^{(0)}}{\beta_0^2}\left[\frac{\beta_1 \beta_2}{\beta_0} - \frac{\beta_1}{\beta_0}\left(\frac{\beta_1^2}{\beta_0} - \beta_2\right) - \beta_3\right]$$

$$+ \frac{\gamma_m^{(1)}}{\beta_0^2}\left(\frac{\beta_1^2}{\beta_0} - \beta_2\right) - \frac{\beta_1 \gamma_m^{(2)}}{\beta_0^2} + \frac{\gamma_m^{(3)}}{\beta_0}. \tag{10.1.4b}$$

We note that β_3 is positive for all positive values of n_f. The two loop coefficient β_1 was given by Caswell (1974), Jones (1974) and Egorian and Tarasov (1979); the three loop one, β_2, by Tarasov, Vladimirov and Zharkov

(1980). Finally, the coefficient β_3 was calculated by Larin, van Ritbergen and Vermaseren (1997a). The various terms in the expansion of γ_m have been calculated by Nanopoulos and Ross (1979) and by Tarrach (1981), who corrected a trivial error in the Nanopoulos–Ross evaluation (two loop, $\gamma_m^{(1)}$); Tarasov (1982) for $\gamma_m^{(2)}$; and Larin, van Ritbergen and Vermaseren (1997b) and Chetyrkin (1997) (four loop $\gamma_m^{(3)}$).

10.2 The Character of the QCD Perturbative Series. Renormalization Scheme Dependence of Calculations and Parameters. Renormalons. Saturation

i Truncation and Renormalization Effects

In QED there is a natural renormalization scheme: one renormalizes with photons and electrons on their mass shells. This is useful because of Thirring's (1950) theorem which states that, at zero photon energy, the Compton amplitude (and a number of other processes as well) is given exactly, i.e., to all orders in α, by the classical approximation: so we may use results from classical physics to determine the fundamental parameters α, m_e. In QCD there is no such preferred scheme, at least not based on physical grounds. Therefore, a discussion of what happens when we change the scheme is necessary. We will in the discussion neglect quark masses (but we will take into account the effective value of n_f) and gauge parameters; their introduction would not pose problems different from the ones we shall consider now.

Take a physical observable, P. Clearly, P must be independent of the renormalization scheme $\mathfrak{R}$ we use to calculate it. However, when we write a series expansion for P,

$$P = \sum_n C_n(\mathfrak{R})[\alpha_s(\mathfrak{R})]^n, \tag{10.2.1}$$

both C_n and α_s depend upon the scheme $\mathfrak{R}$ in which we are calculating. The relation with a new scheme $\mathfrak{R}'$ is found by writing

$$P = \sum_n C_n(\mathfrak{R}')[\alpha_s(\mathfrak{R}')]^n, \tag{10.2.2}$$

expanding $\alpha_s(\mathfrak{R}')$ in terms of $\alpha_s(\mathfrak{R})$ and equating. This expansion will be of the form

$$\alpha_s(\mathfrak{R}') = \alpha_s(\mathfrak{R})\{1 + a_1(\mathfrak{R}', \mathfrak{R})\alpha_s(\mathfrak{R}) + \cdots\}.$$

It is clear that the expansion must begin with unity because, to zero order, $\alpha_s = g_u^2/4\pi$, which is independent of the scheme. This also implies that $C_{0,1}(\mathfrak{R}') = C_{0,1}(\mathfrak{R})$. However, the other C_n are expected to vary:

$$C_2(\mathfrak{R}) = a_1(\mathfrak{R}', \mathfrak{R})C_2(\mathfrak{R}'), \quad \text{etc.}$$

As a simple example, consider the quantity R describing e^+e^- hadron annihilations.[1] We will consider two renormalization schemes: the minimal scheme, to be denoted by m.s. (in which, it will be remembered, one only cancels the $2/\epsilon$ divergences instead of the full $N_\epsilon = 2/\epsilon - \gamma_E + \log 4\pi$), and our familiar $\overline{\text{MS}}$, where the whole N_ϵ is subtracted. We will work only to order α_s^2.

In the m.s. scheme we would have obtained that Eq. (4.1.10) is replaced by

$$R(s) = 3 \sum_{f=1}^{n_f} Q_f^2 \left\{ 1 + \frac{\alpha_{s,\text{m.s.}}(s)}{\pi} + r_{2,\text{m.s.}} \left(\frac{\alpha_{s,\text{m.s.}}(s)}{\pi} \right)^2 \right\} + O(\alpha_s^3),$$

$$r_{2,\text{m.s.}} = r_2 + (\log 4\pi - \gamma_E) \frac{33 - 2n_f}{12}.$$
(10.2.3)

The expression for α_s also changes. We had, to two loops, and in the $\overline{\text{MS}}$ scheme (cf. Eqs. (3.7.4)),

$$\alpha_s(\mu^2) = \frac{12\pi}{(33 - 2n_f) \log \mu^2/\Lambda^2} \left\{ 1 - 3 \frac{153 - 19n_f}{(33 - 2n_f)^2} \frac{2 \log \log \mu^2/\Lambda^2}{\log \mu^2/\Lambda^2} \right\},$$

while in the m.s. we find

$$\alpha_{s,\text{m.s.}}(\mu^2) = \frac{12\pi}{(33 - 2n_f) \log \mu^2/\Lambda^2}$$
$$\times \left\{ 1 - 3 \frac{153 - 19n_f}{(33 - 2n_f)^2} \frac{2 \log \log \mu^2/\Lambda^2}{\log \mu^2/\Lambda^2} - \frac{\log 4\pi - \gamma_E}{\log \mu^2/\Lambda^2} \right\},$$
(10.2.4)

as could be expected. We can have the same functional form for both if we define a new parameter, $\Lambda_{\text{m.s.}}$,

$$\Lambda_{\text{m.s.}}^2 = e^{\gamma_E - \log 4\pi} \Lambda_{\overline{\text{MS}}}^2,$$
(10.2.5)

and then (10.2.4) becomes

$$\alpha_{s,\text{m.s.}}(\mu^2) = \frac{12\pi}{(33 - 2n_f) \log \mu^2/\Lambda_{\text{m.s.}}^2}$$
$$\times \left\{ 1 - 3 \frac{153 - 19n_f}{(33 - 2n_f)^2} \frac{2 \log \log \mu^2/\Lambda_{\text{m.s.}}^2}{\log \mu^2/\Lambda_{\text{m.s.}}^2} \right\},$$
(10.2.6)

up to terms of order α_s^3.

The simple point which this makes is that the parameters of the theory depend on the renormalization scheme used; this holds for Λ as well as for

---

[1] A discussion in the case of deep inelastic scattering may be found in Bardeen, Buras, Duke and Muta (1978).

the $\hat{m}$ masses.[2] The $\overline{MS}$ scheme is preferred in this book, for perturbative calculations, because of its simplicity; no unnecessary transcendentals (like the $-\gamma_E + \log 4\pi$ of Eqs. (10.1.3, 4)) appear. Generally speaking, it also gives reasonably small radiative corrections. For example, in the m.s.,

$$r_{2,\text{m.s.}} \simeq 7.4 - 0.44 n_f,$$

compared to the $\overline{MS}$ value $r_2 = 2.0 - 0.12 n_f$.

Besides the ambiguities caused by the different choices of renormalization scheme, we have ambiguities due to the fact that the perturbative series are in practice truncated. It is clear that, for example, the whole series (10.2.1) and (10.2.2) must be equal; but in general we will not have equality of

$$\sum_{n=0}^{N} C_n(\mathfrak{R})[\alpha_s(\mathfrak{R})]^n \tag{10.2.7a}$$

and

$$\sum_{n=0}^{N} C_n(\mathfrak{R}')[\alpha_s(\mathfrak{R}')]^n \tag{10.2.7b}$$

for finite N. What is more, for truncated series the expansion parameters are not well defined, nor is the truncated series unique. To see what this means in an example, consider that one has the ambiguity of using in (say) (10.2.7a) the expression of α_s to N loops in all the terms; or to N loops in the term $C_1\alpha_s$, to $N-1$ in the term $C_2\alpha_s^2,\ldots$, and to one loop in the term $C_N\alpha_s^N$: or anything in between. In all cases, the error is of higher order α_s^{N+1}. As a second example, consider the one loop coupling constant:

$$\alpha_s(\mu^2) = \frac{12\pi}{(33 - 2n_f)\log \mu^2/\Lambda^2}.$$

Now change $\Lambda^2 \to \Lambda'^2 = (1 + \delta)\Lambda^2$, δ small. We get,

$$\alpha_s(\mu^2) \to \alpha_s'(\mu^2) = \frac{12\pi}{(33 - 2n_f)[\log \mu^2/\Lambda^2 - \log(1 + \delta)]}$$

$$\simeq \alpha_s(\mu^2)\left\{1 + \delta\,\frac{33 - 2n_f}{12}\,\frac{\alpha_s(\mu^2)}{\pi}\right\}.$$

The modification of Λ induced by the alteration is only felt at the next order. Thus, more generally, we can see that in an expression such as (10.2.7a) one can modify $\Lambda^2 \to (1 + \delta)\Lambda^2$ provided only that $\delta \sim \alpha_s^N$.

---

[2] Another example of this has already been encountered when we discussed lattice QCD.

People have tried to get around these problems by devising "improved" series. In particular, one may try to optimize the parameter Λ, or the renormalization point μ^2; popular choices being to adjust them so that the coefficient of the last term (C_N in (10.2.7a)) vanishes, or that the derivative $\mathrm{d}/\mathrm{d}\mu$ is zero.

It is the author's opinion that what may be gained by these manipulations is offset by their disadvantages; notably that one has to use a different Λ for each process. Roughly speaking, we have two possibilities. First, it may happen in a given calculation that the value of α_s is small and the last coefficient is not large, so that

$$|C_N \alpha_s^N| \ll |C_k \alpha_s^k|, \quad k < N,$$

in a given scheme, say the $\overline{\mathrm{MS}}$ one. Then improvements are of little consequence. Or it may be that

$$|C_N \alpha_s^N| \sim |C_k \alpha_s^k|, \quad k < N.$$

In this case what happens is that we are not in the asymptotic regime (for the given process) and the inclusion, in particular, of the last term has only an indicative value. No amount of manipulation will alter this substantially. Actually, the situation is worse, because the QCD series is not expected to be convergent, nor to specify uniquely the theory, as will be discussed in the next subsection.

Note, however, that we object (but only mildly) to *formal* manipulations; in some cases *physical* considerations may suggest resummations (like for the K factor in Drell–Yan) or choices of scale (as in bound state problems) which may somewhat improve the calculation.

ii Renormalons

The running coupling constant, on which perturbation theory in QCD is based, was obtained in Chap. 3 using the renormalization group. The same equations may be derived summing leading logarithms (in this context, see also the discussion at the end of Sect. 4.8). Let us work in a lightlike gauge, so that the whole renormalization of the charge is connected to the gluon propagator, and consider a physical observable P which depends on the squared momenta q_i^2. The observable may be written, to a certain order in perturbation theory and after renormalization at the scale μ_0, as

$$P_R\left(q_i^2, \frac{g^2(\mu_0^2)}{4\pi}, \mu_0\right);$$

for simplicity we neglect quark masses and take the momenta to be spacelike, and we also assume P to be dimensionless. If all the q_i^2 are large, and we write $q_i^2 = Q^2 u_i^2$, $u_i^2 = -1$, then we find the renormalization group-improved expression

$$P = P_R \left(u_i^2, \frac{\bar{g}^2(Q^2)}{4\pi}, Q \right). \tag{10.2.8}$$

The running coupling constant $\bar{g}^2(Q^2)$ was obtained from the renormalization group before; now we will show how to get it by summing logarithms in the gluon propagator. What one does, for every gluon propagator that enters the expression for P, is to replace it by what may be called a *dressed* propagator in which one has included the sum to all orders of the quark and gluon bubble corrections. That is, we replace

$$
\frac{g^2(\mu_0^2)}{4\pi} D^{(0)\mu\nu}(k) = \frac{g^2(\mu_0^2)}{4\pi} \, \mathrm{i} \frac{-g^{\mu\nu} + (k^\mu n^\nu + k^\nu n^\mu)/(k \cdot n)}{k^2}
$$

$$
\rightarrow \frac{g^2(\mu_0^2)}{4\pi} D^{(0)\mu\nu}(k)
$$

$$
+ \frac{g^2(\mu_0^2)}{4\pi} D^{(0)\mu\alpha}(k) \frac{g^2(\mu_0^2)}{4\pi} \Pi^{(2)}_{\alpha\beta}(k) \frac{g^2(\mu_0^2)}{4\pi} D^{(0)\beta\nu}(k)
$$

$$
+ \frac{g^2(\mu_0^2)}{4\pi} D^{(0)\mu\alpha}(k) \frac{g^2(\mu_0^2)}{4\pi} \Pi^{(2)}_{\alpha\beta}(k) \frac{g^2(\mu_0^2)}{4\pi} D^{(0)\beta\rho}(k)
$$

$$
\times \frac{g^2(\mu_0^2)}{4\pi} \Pi^{(2)}_{\rho\sigma}(k) \frac{g^2(\mu_0^2)}{4\pi} D^{(0)\sigma\nu}(k) + \cdots ,
$$

$$\tag{10.2.9}$$

where we omit colour indices and $\Pi^{(2)}$ is the second order gluon vacuum polarization tensor. The procedure is shown graphically in Fig. 10.2.1.[3] If we replace the value of $\Pi^{(2)}$ (see Sect. 3.3) and keep only leading terms in $\log k^2$, we obtain the dressed propagator in the leading log approximation

$$
\frac{g^2(\mu_0^2)}{4\pi} D^{\mu\nu}_{\text{dressed}}(k) = \frac{\bar{g}^2(k^2)}{4\pi} \, \mathrm{i} \frac{-g^{\mu\nu} + (k^\mu n_\nu + k^\nu n_\mu)/(k \cdot n)}{k^2},
$$

and

$$
\alpha_s(k^2) \equiv \frac{\bar{g}^2(k^2)}{4\pi} = \frac{\alpha_s(\mu_0^2)}{1 + \alpha_s(\mu_0^2)\beta_0(\log k^2/\mu_0^2)/4\pi}
$$

$$
= \frac{4\pi}{\beta_0 \log k^2/\Lambda^2}. \tag{10.2.10}
$$

Equation (10.2.8) is obtained by rescaling the momenta k^2 by Q^2.

It is clear that the procedure can only be valid for $k^2 \gg \Lambda^2$ (here, and to avoid inessential sign complications, we work in Euclidean QCD). When $k^2 \sim \Lambda^2$, the quantity $\alpha_s(k^2)$ presents an unphysical pole. This pole, which is quite similar to the Landau pole in QED is, in QCD, connected to the breakdown of perturbation theory: a breakdown that must necessarily occur

---

[3] This is similar to the replacement of D by the expression (3.3.22b), with Π to second order.

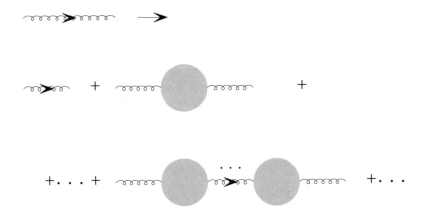

Fig. 10.2.1. Chain of iterations leading to a sum of leading logarithms. The dark blob represents a gluon or quark loop.

for phenomena such as the formation of condensates, spontaneous breaking of chiral symmetry or confinement to appear. The pole is called the *renormalon* pole and the ambiguities that its presence induces were first discussed in QED by Lautrup (1977) and in QCD by 't Hooft (1979), in particular in connection with the question of whether QCD is uniquely defined by the perturbative expansion. In fact, the renormalon singularities indicate that the answer to this last question is *negative*.

To see this, consider again the calculation of P that we discussed before; and assume to simplify that P depends on a single external momentum, and has, to the order in which we are working, a single internal gluon propagator, as occurs, for example, in the case of the second order correlators that enter the calculation of τ decay, cf. Sect. 4.2. We will then have

$$P(q^2) = \int d^4k\, F_{\mu\nu}(q,k) \frac{g^2(\mu_0^2)}{4\pi} D^{(0)\mu\nu}(k),$$

and F embodies the rest of the diagrams that contribute to P, properly renormalized and integrated. When replacing the gluon propagator by the dressed propagator, this becomes

$$P_{\text{dressed}}(q^2) = \int d^4k\, F_{\mu\nu}(q,k) D^{(0)\mu\nu}(k)\alpha_s(k^2)$$
$$= \frac{4\pi}{\beta_0} \int d^4k\, F_{\mu\nu}(q,k) D^{(0)\mu\nu}(k) \frac{1}{\log k^2/\Lambda^2}.$$

(10.2.11a)

The point is that the $\mathrm{d}^4 k$ integral runs through the pole of $\alpha_s(k^2)$ at $k^2 = \Lambda^2$. Now, from the general theory of singular integrals (see, e.g., Gel'fand and Shilov, 1962) it follows that we can write

$$\frac{1}{\log k^2/\Lambda^2} = \mathrm{P.P.}\,\frac{1}{\log k^2/\Lambda^2} + c\Lambda^2\delta(k^2 - \Lambda^2), \qquad (10.2.11\mathrm{b})$$

where P.P. means the principal part when crossing the singularity, and we have separated explicitly a factor of Λ^2 from the constant c for dimensional reasons. The constant c is arbitrary in that it is certainly *not* given by perturbation theory. It therefore follows that the value of P_{dressed} is also undetermined:

$$\begin{aligned}
P_{\mathrm{dressed}}(q^2) = {}& \mathrm{P.P.}\,\frac{4\pi}{\beta_0}\int \mathrm{d}^4 k\, F_{\mu\nu}(q,k)D^{(0)\mu\nu}(k)\frac{1}{\log k^2/\Lambda^2} \\
& + \frac{4\pi c\Lambda^2}{\beta_0}\int \mathrm{d}\Omega_k F_{\mu\nu}(q,k)\mathrm{i}\left[-g^{\mu\nu} + \frac{k^\mu n_\nu + k^\nu n_\mu}{k\cdot n}\right]
\end{aligned} \qquad (10.2.12\mathrm{a})$$

and the Ω_k are the angular variables of k. Suppose that the observable was dimensionless; then, on dimensional grounds, we have that

$$\frac{4\pi c\Lambda^2}{\beta_0}\int \mathrm{d}\Omega_k\, F_{\mu\nu}(q,k)\mathrm{i}\left[-g^{\mu\nu} + \frac{k^\mu n_\nu + k^\nu n_\mu}{k\cdot n}\right] \underset{q^2\to\infty}{\simeq} (\mathrm{constant})\frac{\Lambda^2}{q^2},$$
$$(10.2.12\mathrm{b})$$

and thus we find that perturbative QCD only defines the observable P within the ambiguity given by (10.2.12).

Clearly, this ambiguity disappears at large momenta where the error committed by neglecting terms $O(\Lambda^2/k^2)$ is of higher order than any perturbative correction, $O(1/\log^N k^2/\Lambda^2)$; but it persists for any finite k^2. Also, we have given here the proof for a simple case, but it is not difficult to see how it can be extended to a general situation (Muller, 1985). We consider an observable, P, to any order in perturbation theory and let k_i be the momenta of the internal *gluon* lines. We can split the integrals defining P,

$$P = \int \mathrm{d}^4 k_1 \ldots \int \mathrm{d}^4 k_n\, \Phi(q, k_1, \ldots, k_n),$$

where q now denotes the set of external momenta, into two pieces: the piece $k_i^2 < M_0^2$ with M_0 an arbitrary mass larger than Λ; and the region $k_i^2 > M_0^2$ (we still work in Euclidean QCD). With obvious notation, we write this as

$$P = P_{>M_0^2} + P_{<M_0^2}.$$

In the second piece the integrals in $\mathrm{d}^4 k_i$ run over a compact region and thus are convergent in the ultraviolet. It follows that, as $q \to \infty$, that piece vanishes relative to the first as powers of M_0^2/q^2. The first piece avoids the regions where one has renormalon singularities, and can therefore be evaluated without problems.

In some cases the renormalon ambiguity is of higher order (in Λ^2/q^2) than one would guess from (10.2.12), because of cancellations. This occurs, for example, in correlators like the ones entering the SVZ sum rules, or e^+e^- or τ, Z decays. Here the renormalon contribution is of relative order Λ^4/q^4; in other cases it is indeed of second order, as for example in deep inelastic scattering. In the first examples the renormalon ambiguity is related to the vacuum structure: one can fix it by requiring that it be absorbed into the $O(\langle \alpha_s G^2 \rangle/q^4)$ corrections. This, incidentally, gives a hint as to the nature of the ambiguities present in QCD: there appear to exist a number of different theories with the same perturbative expansion. Of these, one will have the vacuum with minimal energy, which will be the one that one finds in nature.

In deep inelastic scattering the renormalon ambiguities add to the higher twist corrections. A readable discussion of this case may be found in Martinelli and Sachrajda (1996). In jet physics renormalons also appear. In some favourable instances one can even prove a kind of approximate universality of these corrections, as in the evaluations considered by Akhoury and Zakharov (1996).

It is unclear to what extent one can use renormalons for phenomenological purposes. It is true that one can use them to guess nonperturbative properties from perturbation theory; but there are very few situations where *quantitative* results may be obtained from renormalon calculations.

As an example of a *qualitative* use of renormalons we will consider how a renormalon estimate suggests the short distance behaviour of the heavy quark potential that we obtained in Sect. 6.4 from the nonperturbative Dosch–Simonov model, as remarked by Aglietti and Ligeti (1995). The calculation is very simple. We consider the potential generated by the exchange of a renormalon chain. In momentum space this is

$$\widetilde{V}(k) = \frac{-4\pi C_F}{k^2} \frac{4\pi}{\beta_0 \log(k^2/\Lambda^2)},$$

and we have substituted the one-loop expression for $\alpha_s(k^2)$. This expression is undefined for *soft* gluons, with $k^2 \simeq \Lambda^2$. As follows from the general theory, and as we have already remarked, the ambiguity is of the form $c\delta(k^2 - \Lambda^2)$: upon Fourier transformation this produces an indetermination in the x-space potential of $\delta V(r) = c[\sin \Lambda r]/r$. At short distances we may expand this in powers of r and find

$$\delta V(r) \sim C_0 + C_1 r^2 + \dots,$$

which indeed coincides with the short distance behaviour of $U(r)$ as determined in Sect. 6.4, Eq. (6.4.16). This shows the uses, and also the limitations, of renormalon calculations: it is true that the short distance behaviour in the regime $|E_n^{(0)}| \ll \Lambda$ (cf. Sect. 6.4) is reproduced correctly; but neither the long distance behaviour, nor the short distance nonperturbative

Leutwyler–Voloshin corrections in the $|E_n^{(0)}| \gg \Lambda$ regime are given by these renormalons.[4]

Apart from the "ultraviolet" renormalons we have discussed up to now, other ambiguities exist which are connected with exchanges of *ladders* of dressed gluon propagators, and which are effective at low values of the momenta. We refer to the paper of Peris and de Rafael (1997), which contains a thorough discussion and references to earlier work.

iii Saturation

In the previous subsection we have seen that QCD, when defined by the perturbative series, has ambiguities. These ambiguities are associated with small momenta or, equivalently, with long distances. However, at least the *singularities* are clearly spurious. Indeed, not only the theory should be well defined but, because of confinement, long distances are never attained: the theory possesses an internal infrared cut-off of the order of the confinement radius, $R \sim \Lambda^{-1}$. To try and implement it we consider again the gluon propagator. To one loop it gets a correction involving the vacuum polarization tensor. Neglecting quarks this is, in x-space, given by an expression like

$$\Pi_{\alpha\beta}^{aa'}(x,0) \sim g^2 f_{abc} f_{a'de}$$

$$\times \langle 0| \int \mathrm{d}^4 y_1 \, \mathrm{d}^4 y_2 \, \mathrm{T} B_b^\alpha(y_1) \partial_\mu B_{c\alpha}(y_1) B_d^\beta(y_2) \partial_\nu B_{e\beta}(y_2) |0\rangle + \cdots.$$

We can take into account the *long distance* interactions by introducing a string between the field products at finite distances, i.e., in the matrix notation introduced in Sect. 8.3, by replacing

$$\mathcal{B}^\alpha(y_1)\mathcal{B}^\beta(y_2) \to \mathcal{B}^\alpha(y_1) \mathrm{P} \left(\exp \mathrm{i} \int_{y_2}^{y_1} \mathrm{d}z^\mu \, \mathcal{B}_\mu(z) \right) \mathcal{B}^\beta(y_2).$$

The process may be described as "filling the loop" (see Fig. 10.2.2) by introducing all exchanges between the gluonic lines there. If we, furthermore, replace the perturbative vacuum $|0\rangle$ by the nonperturbative one $|\mathrm{vac}\rangle$, then a calculation similar to that made for the long distance potential for heavy quarks in Sect. 6.4 yields a dressed propagator

$$D_{\mathrm{dressed}}^{\mu\nu}(k) = D^{(0)\mu\nu}(k) \frac{4\pi}{\beta_0 \log(M^2 + k^2)/\Lambda^2},$$

---

[4] Moreover, part of the renormalon singularity is spurious. If we consider an observable quantity, such as the mass of a quarkonium state, we have to add to V the rest energy, $2m$ with m the pole mass, which also has a renormalon ambiguity that partially cancels that in V; see Beneke (1998) and Hoang, Smith, Stelzer and Willenbrock (1998) for details.

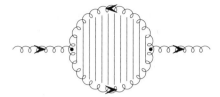

Fig. 10.2.2. "Filled in" gluon loop.

and M^2 is related to the gluon condensate at finite distances, $\langle G(x)G(0)\rangle_{\mathrm{vac}}$.

This may be described as a *saturation* property of the coupling constant; the calculation in fact suggests that, at small momenta, the expression for the running coupling constant should be modified according to

$$\alpha_s(k^2) = \frac{4\pi}{\beta_0 \log k^2/\Lambda^2} \rightarrow \alpha_s^{\mathrm{sat}}(k^2) = \frac{4\pi}{\beta_0 \log(k^2 + M^2)/\Lambda^2}. \qquad (10.2.13)$$

It is certain that an expression such as (10.2.13) incorporates, to some extent, long distance properties of the QCD interaction. For example, if we take (10.2.13) with $M = \Lambda$ in the tree level potential for heavy quarks, this becomes

$$\widetilde{V}^{(0)}(\mathbf{k}) = -\frac{4\pi C_F \alpha_s(\mathbf{k}^2)}{\mathbf{k}^2} \rightarrow \widetilde{V}^{(0),\mathrm{sat}}(\mathbf{k}) = -\frac{16\pi^2 C_F}{\beta_0 \mathbf{k}^2 \log(k^2 + \Lambda^2)/\Lambda^2}.$$

When one has $\mathbf{k}^2 \gg \Lambda^2$, the short distance Coulombic potential is, of course, recovered. For $\mathbf{k}^2 \ll \Lambda^2$, however,

$$\widetilde{V}^{(0),\mathrm{sat}}(\mathbf{k}) \underset{k^2 \ll \Lambda^2}{\simeq} \frac{16\pi^2 C_F \Lambda^2}{\beta_0 \mathbf{k}^4},$$

whose Fourier transform gives

$$V^{(0),\mathrm{sat}}(r) \underset{r \gg \Lambda^{-1}}{\simeq} (\text{constant}) \times r,$$

i.e., a linear potential. Indeed, a reasonably accurate description of spin-independent splittings in quarkonia states is obtained with such a potential (Richardson, 1979). Likewise, use of (10.2.13) with $M = \Lambda$ provides a surprisingly good description of small-x deep inelastic scattering down to $Q^2 \sim 0$, as we discussed in Sect. 4.12iii; and these two cases are not unique.

In spite of these successes, it should nevertheless be obvious that (10.2.13) can only be of limited applicability. For example, consider the correlator of two currents in the spacelike region, $\Pi(Q^2)$. We know that in some cases such as the correlators of vector or axial currents for massless quarks studied in Sects. 4.1, 2, or that of pseudoscalar ones in Sect. 7.4, one has

$$\Pi(Q^2) \underset{Q^2 \to \infty}{\simeq} \Pi_{\text{perturbative}} \left\{ 1 + O(\langle \alpha_s G^2 \rangle) Q^{-4} \right\},$$

whereas (10.2.13) would give a correction of order $M^2 Q^{-2}$. The Richardson potential is also a good example of the limitations of the uses of saturation, in particular in connection with the extent to which saturation really does (or does not) represent a real, physical improvement, or merely the addition of a somewhat arbitrary new parameter. Indeed, the linear potential induced by saturation in the Richardson model is the *fourth component* of a Lorentz vector, while we know that the Wilson linear potential, as obtained, e. g., in the stochastic vacuum model or in lattice calculations, should be a Lorentz four-scalar: it thus follows that the linear potential obtained from saturation can be only of phenomenological use in some specific situations.

iv The Limit of Accuracy of QCD Calculations

Let us consider a physical observable P which, for simplicity, we take to depend on a single momentum, Q^2; for example, P could be the quantity R in $e^+ e^-$ annihilations. We write a perturbative series for it, and assume that we have selected a definite renormalization scheme, say the $\overline{\text{MS}}$ one. We thus have

$$P(Q^2) = \sum_{n=0}^{\infty} c_n \alpha_s(Q^2).$$

There are a number of arguments, in particular the existence of renormalons, that indicate that the QCD series are *not* convergent. What one expects (and we will assume this for definiteness) is that the series are *asymptotic*. These kind of series have the property that they converge to the exact quantity for $\alpha_s \to 0$ in the following sense: the terms in the sum, $c_n \alpha_s^n$, decrease (in modulus) with increasing n up to a certain, α_s-dependent $N(\alpha_s)$ where one has

$$c_{N(\alpha_s)} \alpha_s^{N(\alpha_s)} \sim c_{N(\alpha_s) \pm 1} \alpha_s^{N(\alpha_s) \pm 1}$$

and then higher $|c_{N+k} \alpha_s^{N+k}|$ increase. The best one can do is to consider the approximation

$$P_{N(\alpha_s)} \equiv \sum_{n=0}^{N(\alpha_s)} c_n \alpha_s^n, \tag{10.2.14}$$

and the error committed by doing so is of the order of the last term included,

$$|\epsilon_{N(\alpha_s)}| = |P - P_{N(\alpha_s)}| \sim |c_{N(\alpha_s)} \alpha_s^{N(\alpha_s)}|. \tag{10.2.15}$$

For some divergent series it is possible to devise summation methods which produce unique answers for the exact quantity, P; popular ones being

Borel or Cesaro summation. However, and as the existence of renormalon ambiguities shows, this is not the case for the QCD perturbative series. It thus follows that, for each observable like P, there is, for each value of $\alpha_s(Q^2)$ (and hence for each Q^2) a maximum possible accuracy.

The statement is at times found in the literature that adding renormalon contributions one may improve on this. That is, one could imagine replacing (10.2.14) by

$$P^{\text{improv.}}_{N(\alpha_s)} = \sum_{n=0}^{N(\alpha_s)} c_n \alpha_s^n + \frac{M^2}{Q^2}; \qquad (10.2.16)$$

i.e., we add the renormalon that will cancel the singularity of the original, divergent series. One hopes that the truncated series expansion $\sum_{n=0}^{N(\alpha_s)} c_n \alpha_s^n$ will represent the quantity

$$P(Q^2) - \frac{M^2}{Q^2} = P(Q^2) - k e^{-4\pi/\beta_0 \alpha_s},$$

$M^2 = k\Lambda^2$, better than P itself. In this way one would improve the convergence at the cost of adding a new parameter, M, which could be obtained from other sources or at least be fitted to experiment.

Except in some favourable cases, this improvement is unlikely to occur. By hypothesis, the renormalon contribution is of the order of the error in the series,

$$\frac{M^2}{Q^2} \sim \epsilon_N \sim c_{N(\alpha_s)} \alpha_s^{N(\alpha_s)}.$$

However, the series starts to diverge when its terms begin to increase,

$$c_{N-1}\alpha_s^{N-1} \sim c_N \alpha_s^N \sim c_{N+1}\alpha_s^{N+1},$$

and thus the error could just as easily be ϵ_N as $2\epsilon_N$ or $3\epsilon_N$: in any event, of the order of the renormalon contribution. We have to admit that the perturbative QCD series have intrinsic limits to their accuracy. In the typical case of the observable $R(Q^2 = s)$, Eqs. (4.1.11) tell us that (with $n_f = 4$ for definiteness)

$$R(s) = 3 \sum_f Q_f^2 \left\{ 1 + 0.318\alpha_s(s) + 0.154\alpha_s^2(s) - 0.372\alpha_s^3(s) + \cdots \right\},$$

so the series diverges with the terms shown here for $\alpha_s(s) \sim 0.41$, i.e., when $s^{1/2} \leq 1.4$ GeV. The error here would be of order $0.372\alpha_s^3 \sim 0.026$, to be compared with the renormalon contribution, of order $(\alpha_s/\pi)\langle \alpha_s G^2 \rangle / s^2 \sim 0.0020$. For tau decay, the last term given in (4.2.13) shows that the series, for $s = m_\tau^2$, is at the limit of convergence.

Among other consequences, these considerations show that, in some cases, the calculations of higher orders of perturbation theory do not improve our knowledge of the quantity in question; and, in particular, this suggests that a truly precise determination of α_s requires high momenta (e.g., Z decay) where the smallness of α_s allows us effectively to use high terms in the perturbative expansion.

10.3 Coupling Constants: θ, α_s, Λ

i The Parameter θ

QCD is a theory with two coupling constants. We have the coupling $g^2/4\pi$ that we trade for the running coupling constant $\alpha_s(\mu^2)$ defined at the reference scale μ; and we have the coupling θ associated with the piece of the Lagrangian $(-g^2/32\pi^2)\theta\widetilde{G}G$ discussed in, among other places, Sect. 7.7. We will start with the second because its status is much simpler: we only have a bound, coming from the absence of a dipole moment of the neutron (see Sect. 7.7)

$$|\theta| < 10^{-9}. \tag{10.3.1}$$

A word on this: because θ is only defined modulo $U(1)$ rotations, the bound (10.3.1) should really be understood as a bound for the effective parameter θ_{eff}, defined by

$$\theta_{\text{eff}} \equiv \theta - \arg \det \underset{\sim}{M}$$

where $\underset{\sim}{M}$ is the quark mass matrix of Kobayashi and Maskawa, including the weak CP-violating phase.

ii Λ and α_s

In recent years, it has fortunately become customary to express fits to the strong interaction coupling directly in terms of $\alpha_s(\mu^2)$ defined at a convenient reference momentum, instead of giving it as values of the parameter Λ. To be sure, the two specifications are equivalent; but the latter suffers from more ambiguities than the former. Here we will give values for both. We will use the very recent compilation of Bethke (1998).[5] The values of μ we will choose will be M_Z and m_τ, because it is here that we have the best *direct* measurements of α_s; recall Sect. 4.2. It is a pity that the experimental value of $Z \to$ hadrons still presents large systematic errors; this process is the one that is potentially more appropriate for extracting an accurate value of α_s.

Let us turn to the results. We define α_s and Λ in the $\overline{\text{MS}}$ scheme, with an effective number of flavours; the relation between this and two other definitions, the μ-scheme (also called the *momentum renormalization scheme*)

---

[5] It is not easy to connect directly some of our results as given here with those presented in the last edition of the Particle Data Group (1996) tables. The reason is that these authors have chosen to employ definitions both for β_n and of $\Lambda(3 \text{ loop})$ which are at variance with the ones used here. Of course, we have verified that the figures given there for α_s agree with ours.

and the lattice renormalization are (Hasenfratz and Hasenfratz, 1980; Buras, 1981):

$$\Lambda_{\text{MOM}}(n_f = 4) \simeq (2.16)\Lambda_{\overline{\text{MS}}},$$

$$\Lambda_{\text{Latt}}(n_f = 0) \simeq \frac{1}{38.6}\Lambda_{\overline{\text{MS}}}(n_f = 0),$$

$$\Lambda_{\text{Latt}}(n_f = 4) \simeq \frac{1}{55}\Lambda_{\overline{\text{MS}}}(n_f = 4),$$

relations valid to one loop.

Considering then only the processes where the theoretical calculation has been pushed to the NNLO level and the experimental data are good enough to make the analysis meaningful (which excludes e^+e^- annihilations), we have the following table:

Process	Average Q [GeV]	$\alpha_s(M_Z^2)$
DIS; ν, Bj	1.58	0.122
DIS; ν, GLS	1.73	0.115
τ decays	1.777	0.119
$Z \to$ hadrons	91.2	0.124

Here DIS means deep inelastic scattering, Bj stands for the Bjorken, and GLS for the Gross–Llewellyn Smith sum rules. The average value, also taking into account NLO calculations, is

$$\alpha_s(M_Z^2) = 0.118 \pm 0.006, \tag{10.3.2a}$$

which corresponds to

$$\alpha_s(m_\tau^2) = 0.320^{+0.044}_{-0.055}, \tag{10.3.2b}$$

slightly below the value found directly from tau decay, where we have

$$\alpha_s(m_\tau^2) \simeq 0.330 \pm 0.030.$$

The values of the parameter Λ that reproduce (10.3.2) are, in MeV,

$$\Lambda(3 \text{ loop}, n_f = 5) = 208^{+80}_{-61},$$
$$\Lambda(3 \text{ loop}, n_f = 4) = 293^{+97}_{-79}, \tag{10.3.3}$$
$$\Lambda(3 \text{ loop}, n_f = 3) = 344^{+131}_{-107}.$$

We have made the matching between $n_f = 5$ and 4 at $\mu = 5$ GeV, and between $n_f = 4$ and 3 at $\mu = 1.777$ GeV. To two loops, the value of α_s given in (10.3.2) corresponds to

$$\Lambda(2 \text{ loop}, n_f = 4) = 314^{+122}_{-81}. \tag{10.3.4}$$

This is compatible (within errors) but slightly larger than the values used in the eighties, dominated by Υ decays which give $\alpha_s^{\Upsilon \text{ decay}}(M_Z^2) = 0.112$ or $\Lambda(2 \text{ loop}, n_f = 4) \simeq 230$. It is of some historical interest to compare with the earliest theoretical determinations, performed analyzing the old CDHS (De Groot et al., 1979) and SLAC (Bodek et al., 1979) data for deep inelastic scattering of neutrinos and electrons. In particular, the determination based on ep, the more precise of the two, gave a result[6] that, when translated to the definitions used now, produces the figure

$$\Lambda(2 \text{ loop}, n_f = 4) = 320 \pm 130. \tag{10.3.5}$$

Of course, the more recent values (10.3.4) are tested to NNLO, and also by agreement with a large variety of determinations so that the error given in, for example, (10.3.4) is much more reliable than the earlier ones; still, the coincidence between (10.3.4) and (10.3.5) is remarkable and a tribute to the quality of the fixed target experimental data.

We end this section with a comment. If we only keep, in the determination of α_s, processes with *spacelike* momenta, then, as noted by Bethke (1998), a slightly smaller value is obtained:

$$\alpha_s(M_Z^2) = 0.114 \pm 0.005 \text{ (spacelike momenta)}. \tag{10.3.6}$$

This may be a fluctuation, and in fact (10.3.6) is still compatible with the overall average, (10.3.2a); but one may also think that (10.3.6) is to be preferred, at least when comparing with processes involving spacelike momenta. This is because the effect of analytical continuation (cf. Eqs. (4.1.11) for e^+e^- annihilations) first appears at NNLO and, at least for the processes e^+e^- and $Z \to$ hadrons, partially cancels (and even reverses the sign) of the contribution one would have for spacelike momenta. So it would appear that it is the corresponding increase of α_s that should be considered a fluctuation. The matter will not be resolved until NNNLO calculations, and more precise measurements for $Z \to$ hadrons become available; the difference between (10.3.6) and (10.3.2a) should for the moment be classed as a systematic uncertainty in the theoretical treatment.

---

[6] González-Arroyo, López and Ynduráin (1979).

10.4 Quark Masses

Let us start by considering the light quarks. We will give the values of the running masses, in the $\overline{\text{MS}}$ scheme, renormalized at 1 GeV^2. The *errors* given for the values of the masses in certain calculations (Gasser and Leutwyler, 1982; Domínguez and de Rafael, 1987; Chetyrkin et al., 1995; Jamin and Müntz, 1995; Chetyrkin, Pirjol and Schilcher, 1997) have been shown to be somewhat optimistic, both for the values of the combination $m_u + m_d$ and of m_s from sum rules, and for the ratios from PCAC and chiral dynamics. It would seem that safe bounds and estimates (Bijnens, Prades and de Rafael, 1995; Ynduráin, 1998) would be as follows. First, one has the model independent ratio

$$\frac{\bar{m}_u}{\bar{m}_d} = 0.44 \pm 0.22, \tag{10.4.1a}$$

and positivity bounds

$$\bar{m}_d + \bar{m}_u \geq 9 \text{ MeV}, \quad \bar{m}_d - \bar{m}_u \geq 3 \text{ MeV}, \quad \bar{m}_s \geq 150 \text{ MeV}. \tag{10.4.1b}$$

Then, with reasonable models for the low energy discontinuity of the correlators, (Bijnens, Prades and de Rafael, 1995; Chetyrkin et al., 1995; Chetyrkin, Pirjol and Schilcher, 1997) one obtains the estimates

$$\bar{m}_d + \bar{m}_u = 13 \pm 4 \text{ MeV},$$
$$\bar{m}_s = 200 \pm 50 \text{ MeV}, \quad \bar{m}_d = 8.9 \pm 4.3 \text{ MeV}, \quad \bar{m}_u = 4.2 \pm 2 \text{ MeV}. \tag{10.4.1c}$$

Lattice calculations of the light quark masses are available[7] but, for the moment, they are not very accurate; in fact, some of them amply violate the positivity bounds given in Sect. 7.4. For the s quark one can obtain independent evaluations from τ decay into strange particles, or from e^+e^- annihilations, using experiment and current algebra relations. The results that one finds are, from the first method, $m_s(1 \text{ GeV}) = 235^{+35}_{-42}$, compatible with the sum rule–chiral dynamics ones presented in (10.4.1c), but are less precise in that *systematic* errors are not included. A recent review of this is given by Chen (1998). From the second method, Narison (1995b) gives $m_s(1 \text{ GeV}) = 197 \pm 29$ MeV, and systematic errors are only partially included.

Next, we consider the masses of the heavy quarks. The t quark mass is given by the experiments at Fermilab as

$$m_t \simeq 172 \pm 10 \text{ GeV}. \tag{10.4.2}$$

The analyses do not specify which definition one takes for the mass, so the error in (10.4.2) should be probably increased by a contribution $C_F \alpha_s(m_t^2)/\pi \sim$

---

[7] Cf. the reviews of Aoki et al. (1998) and especially Bhattacharya and Gupta (1998).

7 GeV. The value of m_t had been predicted from consistency of the radiative electroweak corrections. For example, in the previous (2nd) edition of this book, the value quoted (corresponding to the 1992 analyses) was

$$100 \text{ GeV} \leq m_t \leq 170 \text{ GeV},$$

the value of this mass and that of the c quark being remarkably successful predictions of the standard theory of electroweak (and strong) interactions.

For the c, b quarks we have two main sources of values: sum rules, and fits to quarkonium spectra. Besides this, the c quark mass can be obtained from GIM-violating decays, of historical interest because they gave the first estimate of the mass prior to discovery, but not very precise as they suggested

$$1.2 \text{ GeV} \leq m_c \leq 1.8 \text{ GeV}.$$

Lattice determinations are also not very accurate; one finds

$$\bar{m}_c(\bar{m}_c^2) = 1.5 \pm 0.3 \quad \text{and} \quad \bar{m}_c(\bar{m}_c^2) = 1.22 \pm 0.5$$

(Allton et al., 1994; Bochkarev and de Forcrand, 1997, respectively).

The b quark mass can also be found from Z decays (Rodrigo, Santamaría and Bilenkii, 1997), but this is as yet less precise than the two other methods. The Z-decay determination, however, is of interest in that it checks experimentally the running of the mass, since it yields $\bar{m}_b(M_Z^2)$ directly. The experimental analysis, as performed by the DELPHI group at LEP, gives

$$\bar{m}_b(M_Z^2) = 2.67 \pm 0.50 \text{ GeV},$$

which translates into

$$\bar{m}_b(\bar{m}_b^2) = 4.0 \pm 0.7 \text{ GeV}.$$

Sum rule and quarkonium spectroscopy determinations are barely compatible among themselves, even within errors. The sum rule values are systematically lower than the spectroscopic ones. There may be a number of reasons for this, of which we mention three. First of all, SVZ sum rules are based on the assumption of local duality. While it is true that local duality permits reasonable fits, it is also true that it cannot be exact, so sum rule determinations have a built-in source of systematic error. Secondly, sum rule determinations only go to an accuracy $O(\alpha_s^2)$, while quarkonium ones reach to $O(\alpha_s^3)$ and $O(\alpha_s^4)$ for the pole mass, and this is the most likely reason for the bulk of the discrepancy. A possible third reason could be due to the fact that in sum rule determinations one obtains the $\overline{\text{MS}}$ masses directly, while in quarkonium calculations one goes through the pole masses as an intermediate step. We then have the results given in the following table, where we present two determinations based on sum rules, two *ab initio* calculations from spectroscopy and a third based on spectroscopy but with a phenomenological potential to enforce confinement:

Reference	$m_b(\text{pole})$	$\bar{m}_b(\bar{m}_b^2)$	$m_c(\text{pole})$	$\bar{m}_c(\bar{m}_c^2)$
PY	5001^{+104}_{-66}	4440^{+43}_{-28}	1866^{+190}_{-154}	1531^{+132}_{-127}
TY	4906^{+70}_{-65}	4397^{+18}_{-32}	1570 ± 20 (*)	1306 ± 30 (*)
PTN	4860	—	1480	—
N; GL	—	4250 ± 100	—	1270 ± 50
JP	4604 ± 20	4133 ± 60	—	—

b and c quark masses.   (*) Systematic errors not included.

PY: Pineda and Ynduráin (1998). [Full $O(\alpha_s^4)$ for m^{pole}];
TY: Titard and Ynduráin (1994). [$O(\alpha_s^3)$ plus $O(\alpha_s^3)v$, $O(v^2)$];
PTN: Pantaleone, Tye and Ng (1986). [$O(\alpha_s^3)$, phenomenological Kr potential];
N: Narison (1995a); GL: Gasser and Leutwyler (1982);
JP: Jamin and Pich (1997). [$O(\alpha_s^2)$, duality assumption].

One can take (unweighted) averages as central values:

$$m_b(\text{pole}) = 4840 \pm 150, \quad \bar{m}_b(\bar{m}_b) = 4306 \pm 100,$$
$$m_c(\text{pole}) = 1640 \pm 200, \quad \bar{m}_c(\bar{m}_c) = 1370 \pm 150,$$

(10.4.3)

with rather generous errors, or consider that, at least in principle, the PY calculation for $m(\text{pole})$, being of order α_s^4, is the more accurate one; but this is not necessarily the case. Given the large size of the two loop corrections, it is not clear to what extent the PY result (for example) is an improvement over the TY one.

10.5 Condensates

Condensates are *derived* quantities which, at least in principle, can be obtained from the more fundamental parameters (masses and coupling constant). For this, however, we require a formulation of QCD that does not rely on perturbation theory: in practice, this means lattice QCD. The evaluations of the condensates in lattice QCD however, are not very precise,[8] so it is of interest to discuss their values as obtained from *experimental* information.

Four quark condensates and three gluon condensates are usually obtained from SVZ sum rules. The calculations are little more than order of magnitude guesses and indeed, as discussed in previous sections, these quantities are not even well defined. The two gluon condensate is better defined, and more reliable estimates exist for it; and the same occurs for $\bar{q}q$ condensates. These are the quantities whose values we will discuss in the present section.

---

[8] See, for example, Di Giacomo and Rossi (1981).

i Quark Condensates, $\langle \bar{q}q \rangle$

The PCAC relations (7.3.4, 5) plus flavour independence of light quark condensates,

$$\langle \bar{u}u \rangle \simeq \langle \bar{d}d \rangle \simeq \langle \bar{s}s \rangle \equiv \langle \bar{q}q \rangle,$$

relate these to light quark masses. Thus, the value of $\langle \bar{q}q \rangle$ can be read off from the estimates on light quark masses of the previous section: using (7.3.4), we have, for the average of u, d condensates,

$$\langle \bar{q}q \rangle (1 \text{ GeV}^2) = -\frac{f_\pi^2 m_\pi^2}{2[\bar{m}_d(1 \text{ GeV}^2) + \bar{m}_u(1 \text{ GeV}^2)]} \simeq (0.22 \text{ GeV})^3.$$

The *differences* among the three $\langle \bar{q}q \rangle$ are more difficult to find. According to Domínguez and de Rafael (1987), we have

$$5 \times 10^{-3} \leq 1 - \frac{\langle \bar{d}d \rangle}{\langle \bar{u}u \rangle} \leq 17 \times 10^{-3}, \tag{10.5.1}$$

but nothing comparable exists for $\langle \bar{s}s \rangle$.

Heavy quark condensates are not of much phenomenological interest, at least at the present time, but we will give them for completeness. Consider a heavy quark, q with $m_q \gg \Lambda$. One would assume that no such quarks are present "primordially" in the physical vacuum, $|\text{vac}\rangle$; but, since $|\text{vac}\rangle$ contains gluons, there is some probability that the gluon splits (virtually) into a pair $\bar{q}q$, so we expect $\langle \bar{q}q \rangle \neq 0$ at second order in the QCD coupling. The actual calculation is rather simple. We write

$$\langle : \bar{q}(0)q(0) : \rangle = -\sum_{ij} \delta_{ij} \operatorname{Tr}\langle : q_i(0)\bar{q}_j(0) : \rangle = -\sum_{ij} \delta_{ij} \operatorname{Tr} \int \frac{d^4p}{(2\pi)^4} \langle : q_i\bar{q}_j(p) : \rangle.$$

The ij are colour indices, and the trace refers to Dirac indices. The p-space expression is evaluated, to lowest order, with the help of the diagram of Fig. 10.5.1. The nonzero value of the result is due to $\langle : BB : \rangle$ being nonzero. We get,

$$\langle : \bar{q}(0)q(0) : \rangle = -(ig)^2 T_F$$

$$\times \sum_{ab} \delta_{ab} \operatorname{Tr} \int d^D\hat{k} \, D^{\mu\nu}_{NPab}(k) \int d^D\hat{p} \, \frac{i}{\not{p}-m} \gamma_\mu \frac{i}{\not{p}+\not{k}-m} \gamma_\nu \frac{i}{\not{p}-m};$$

substituting the expression (3.9.15) for D_{NP}, and setting $D = 4$ because the expression is convergent, this becomes

$$\langle : \bar{q}(0)q(0) : \rangle = -iT_F \frac{\pi\langle \alpha_s G^2 \rangle}{72}$$

$$\times \operatorname{Tr} \int \frac{d^4p}{(2\pi)^4} \gamma_\nu \frac{1}{\not{p}-m} \frac{1}{\not{p}-m} \gamma_\mu (5g^{\mu\nu}\partial^2 - 2\partial^\mu\partial^\nu) \frac{1}{\not{p}+\not{k}-m} \Big|_{k=0},$$

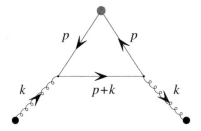

Fig. 10.5.1. Diagram involved in the calculation of the heavy quark condensate.

and the derivatives are with respect to k. The rest of the calculation is elementary. We find,

$$\langle : \bar{q}(0)q(0) : \rangle = -\frac{T_F\langle \alpha_s G^2\rangle}{12\pi m}. \tag{10.5.2a}$$

If we take $\langle \bar{q}q\rangle$ to be normalized at the momentum $\mu^2 = \bar{m}_q^2$, and we use the values of the gluon condensates and quark masses given below, this implies very small values for the heavy quark condensates:

$$\langle \bar{c}c\rangle \sim -5 \times 10^{-5} \text{ GeV}^3,$$
$$\langle \bar{b}b\rangle \sim -2 \times 10^{-5} \text{ GeV}^3. \tag{10.5.2b}$$

ii The Gluon Condensate, $\langle \alpha_s G^2\rangle$

The values of the gluon condensate are not directly related to an observable, so one has to find them by indirect methods, particularly SVZ sum rules. Because of this, the results are somewhat uncertain. Thus, the value favoured in the older papers (e.g., Shifman, Vainshtein and Zakharov, 1979a,b; Launer, Narison and Tarrach, 1984), of the order of

$$\langle \alpha_s G^2\rangle \sim 0.04 \text{ GeV}^4,$$

have been superseded by more recent evaluations. Here, and to some extent as a reflection of the fact that the favoured value of Λ has increased, one obtains *larger* values for the condensate, of the order of 0.06 GeV^4 or more; see, for example, Narison (1997).

Let us show in some detail the kind of calculation involved. We will present an evaluation of the gluon condensate based on spectroscopy and sum rules, following a method which improves an early calculation (Novikov et al., 1978). We consider the correlator for the vector current of heavy quarks,

$$\Pi_{\mu\nu} = (p^2 g_{\mu\nu} - p_\mu p_\nu)\Pi(p^2) = i \int d^4x\, e^{ip\cdot x}\langle TJ_\mu(x)J_\nu(0)\rangle, \quad J_\mu = \bar{q}\gamma_\mu q,$$

and sum over omitted colour indices is understood. This will give information on triplet, $l = 0$ states; information on states with other quantum numbers would be obtained with other correlators. For definiteness we will assume the quarks to be $\bar{b}b$; a similar, but less precise, evaluation would hold for $\bar{c}c$. The function $\Pi(t)$ satisfies a dispersion relation,

$$\Pi(t) = \frac{1}{\pi} \int ds \frac{\rho(s)}{s-t},$$

where $\rho(s) \equiv \text{Im}\,\Pi(s)$. Actually, this equation should have been written with one subtraction. We will not bother to do so, as its contribution drops out for the quantities of interest for us here.

Let us denote by $\Pi_{\text{p.t.}}, \rho_{\text{p.t.}}$ the corresponding quantities calculated in perturbation theory, albeit *to all orders*,[9] but nonperturbative effects are neglected in $\Pi_{\text{p.t.}}, \rho_{\text{p.t.}}$. In particular, the gluon condensate contribution is not included here.

At large t, both spacelike and timelike, the OPE is applicable to $\Pi(t)$, and we have the results of Sect. 4.11,

$$\Pi(t) \simeq \Pi_{\text{p.t.}}(t) + \frac{\langle \alpha_s G^2 \rangle}{12\pi t^2}, \tag{10.5.3a}$$

and of Sect. 5.6,

$$\rho(s) \simeq \rho_{\text{p.t.}}(s) - \frac{N_c C_F}{128} \frac{\langle \alpha_s G^2 \rangle}{s^2} \frac{(1+v^2)(1-v^2)^2}{v^5}, \tag{10.5.3b}$$

with $v = (1 - 4m^2/s)^{\frac{1}{2}}$ the velocity of the quarks. If we then define $\Pi_{\text{NP}}, \rho_{\text{NP}}$ as the results of subtracting the perturbative parts,

$$\Pi_{\text{NP}} \equiv \Pi - \Pi_{\text{p.t.}}; \quad \rho_{\text{NP}} \equiv \rho - \rho_{\text{p.t.}},$$

it follows that $\Pi_{\text{NP}}(t)$ decreases at infinity as t^{-2}. So it satisfies a superconvergent dispersion relation, and therefore we have the sum rule

$$\int ds\, \rho_{\text{NP}}(s) = 0. \tag{10.5.4}$$

---

[9] Of course, we shall not be able to evaluate *all* orders in perturbation theory; actually, we we will sum the one-gluon exchange to all orders (which can be done explicitly in the nonrelativistic regime) and add one loop corrections to this.

In fact, it would appear that one still has another sum rule because of the following argument. At large t, $\Pi_{\mathrm{NP}}(t)$ behaves like

$$\Pi_{\mathrm{NP}}(t) \simeq \frac{\langle \alpha_s G^2 \rangle}{12\pi t^2},$$

while the contribution from the bound states to the dispersion relation (see below),

$$\Pi_{\mathrm{NP;bound\ states}}(t) \sim \frac{\langle \alpha_s G^2 \rangle}{t^2 \alpha_s^3},$$

dominates over this. Therefore we have the extra relation

$$\int \mathrm{d}s \, s \rho_{\mathrm{NP}}(s) = 0.$$

It turns out that this is actually equivalent to (10.5.4), up to radiative corrections. This is because the region where any of the integrals in both sum rules are appreciably different from zero is for $s \simeq 4m^2(1 + \mathrm{O}(\alpha_s^2))$, so the integrals differ only by terms of order α_s^2, smaller than the radiative corrections which none of them take into account.

Let us return to the sum rule (10.5.4). The function $\rho(s)$ consists of a continuum part, for s above threshold for open bottom production, and a sum of bound states. Both can be calculated theoretically provided that s is larger than a certain critical $s(v_0)$, and n smaller or equal than a critical n_0. $s(v_0)$ and n_0 are defined as the points where the perturbation-theoretic contribution to ρ and the nonperturbative one are of equal magnitude, and form the limits of the regions where a full theoretical evaluation is possible.

To be precise, for the continuum we use (10.5.3b), so that above the critical $s(v_0)$,

$$\rho_{\mathrm{NP}}^{\mathrm{cont}}(s) = \frac{N_c C_F}{128} \frac{\langle \alpha_s G^2 \rangle}{s^2} \frac{(1+v^2)(1-v^2)^2}{v^5}, \quad s > s(v_0), \qquad (10.5.5\mathrm{a})$$

and v_0 is such that $\rho_{\mathrm{NP}}^{\mathrm{cont}}(s(v_0)) \simeq \rho_{\mathrm{p.t.}}^{\mathrm{cont}}(s(v_0))$; numerically, and for $\bar{b}b$, $v_0 \simeq 0.2$. For the bound states, ρ is proportional to the square of the wave function at the origin:

$$\rho(s) = \frac{N_c}{M_n} |R_n(0)|^2 \delta(s - M_n).$$

We may get $\rho_{\mathrm{p.t.}}^{\mathrm{b.s}}(s)$ and $\rho_{\mathrm{NP}}^{\mathrm{b.s.}}(s)$ by splitting $|R_n(0)|^2$ into a Coulombic piece,

$$|R_n^{\mathrm{Coul.}}(0)|^2 = \frac{m^3 C_F^3 \alpha_s^3}{2n^3},$$

and the (leading) nonperturbative correction

$$|R_n(0)|^2 \simeq |R_n^{\mathrm{Coul.}}(0)|^2 + |R_n^{\mathrm{NP}}(0)|^2,$$

the latter being given by the Leutwyler–Voloshin analysis of Sect. 6.3, Eqs. (6.3.2) ff. So we have

$$\rho_{\text{NP}}^{\text{b.s.}}(s) = \frac{3 N_c C_F^3 \pi m^3 \langle \alpha_s G^2 \rangle}{8 \alpha_s^3 m^4} \sum_{n=1}^{n_0} \frac{\lambda_n}{M_n} \delta(s - M_n^2), \quad n \leq n_0. \qquad (10.5.5b)$$

The numbers λ_n have been calculated by Leutwyler (1981) and Voloshin (1982). For $n = 1$, λ_1 was given in Sect. 6.3iii, Eq. (6.3.25). This is all we really need since, for bottomium, $n_0 = 1$.

The sum rule (10.5.4) can then be written schematically as

$$\int_{s(v_0)}^{\infty} \rho_{NP} + \sum_{n=1}^{n_0} \text{Residue of } \rho_{NP} = \left\{ \int_{\text{threshold}}^{s(v_0)} \rho_{NP} + \sum_{n=n_0+1} \text{Residue of } \rho_{NP} \right\}.$$

The left hand side is given in terms of $\langle \alpha_s G^2 \rangle$ by Eqs. (10.5.5); the right hand side can be connected with experiment using the following argument. The sum over higher bound states, "$\sum_{n \geq n_0+1}$ Residue of ρ_{NP}", may be identified as the difference between the sum over the *experimental* residues of the poles of the bound states, and what we would get by a Coulombic formula, for all $n \geq n_0 + 1$. Certainly, this Coulombic formula will not be valid for large n, since here the radiative corrections will become large; but, because the residues decrease as $1/n^3$, the contribution of these states will be negligible. We write this decomposition as

$$(\text{bound states with } n > n_0) = \rho_{\exp, \, n > n_0}^{\text{b.s.}}(s) - \rho_{\text{Coulombic}, \, n > n_0}^{\text{b.s.}}(s).$$

As for the continuum piece below $s(v_0)$, we may likewise interpret it as the difference between experiment and a perturbative evaluation, which we write as

$$\rho_{\text{NP}}^{\text{cont}}(s) = \rho_{\exp}^{\text{cont}}(s) - \rho_{\text{p.t.}}^{\text{cont}}(s), \quad s < s(v_0),$$

and, because we are close to threshold, we have (Sect. 5.4, Eq. (5.4.5))

$$\rho_{\text{p.t.}}^{\text{cont}}(s) = \frac{N_c C_F \alpha_s}{8} \frac{1}{1 - e^{-\pi C_F \alpha_s / v}} \simeq \frac{N_c C_F \alpha_s}{8}.$$

Taking everything into account, the sum rule (10.5.4) becomes

$$\sum_{n=n_0+1} \frac{1}{m^2 M_n} |R_n^{\exp}(0)|^2 + f_{\text{back}}(v_0)$$

$$= 2 C_F^3 \left\{ \alpha_s^3 \sum_{n=n_0+1}^{\infty} \frac{1}{n^3} - \frac{\pi \langle \alpha_s G^2 \rangle}{m^4 \alpha_s^3} \sum_{n=1}^{n_0} \lambda_n n^5 \right\} \qquad (10.5.6)$$

$$+ \frac{2}{3} \left\{ 8 \epsilon^2 \alpha_s^3 + \frac{\langle \alpha_s G^2 \rangle}{48 \epsilon^3 \alpha_s^3 m^4} \right\}.$$

We have defined $v_0 \equiv \epsilon \alpha_s$ and the expression is valid up to corrections of relative order α_s. The function $f_{\text{back}}(v_0)$ is the contribution of the background which, when added to the resonances above threshold (included in the sum in the left hand side of (10.5.6)), give the experimental value of $\int_{\text{threshold}}^{s(v_0)} \rho_{\text{NP}}$. The function f_{back} would be obtained by integrating the cross sections for production of $\Upsilon + GG$ and $B\bar{B}$, where by GG we mean a "glueball" decaying into 2π, and B is any of the states $B^0, B^\pm, B^*$. Because we may assume that the structure is provided by the resonances, we can take f_{back} to be given by phase space only. So we have

$$f_{\text{back}}(v_0) = f_1 v_0^{5/2} + f_2 v_0^3,$$

where the first term refers to the channel $\Upsilon + GG$, and the second to $B\bar{B}$. We have in this expression neglected m_{GG}, α_s^2.

In principle, the procedure would appear to be straightforward. One would fit the resonance and bound state residues and f to the data, and then, after substituting into (10.5.6), obtain a determination of $\langle \alpha_s G^2 \rangle$. In practice, however, things do not work out so nicely. The quality of the experimental data does not allow any precise determination of the constants $f_{1,2}$; any values in the range $f_{1,2} \sim 0.03$ to 0.1 would do the job. Secondly, the effective dependence of $\langle \alpha_s G^2 \rangle$ in Eq. (10.5.6) on experiment is proportional to α_s^{-6}: so the result will depend very strongly on the value of α_s that we choose. This is particularly true because radiative corrections to the nonperturbative contribution to the bound states have not been calculated, so there is not even a "natural" renormalization point.

These two difficulties may be partially overcome with the following tricks. First, since we are assuming that the $n = 1$ bound state is described with the bound state analysis as discussed in Sect. 6.3, we may fix the value of α_s that produces such agreement. This means that we will take $0.35 \leq \alpha_s \leq 0.4$. Secondly, we may alter the treatment of the continuum in the following manner. We split *not* from v_0, but from v_1, arbitrary provided only that $v_1 \geq v_0$. Thus, for $s \leq s(v_1)$, we use $\rho_{\text{NP}}^{\text{cont}}(s) = \rho_{\text{exp}}^{\text{cont}}(s) - \rho_{\text{p.t.}}^{\text{cont}}(s)$, and for $s \geq s(v_1)$ we take the theoretical expression

$$\rho_{\text{NP}}^{\text{cont}}(s) = \frac{N_c C_F}{128} \frac{\langle \alpha_s G^2 \rangle}{s^2} \frac{(1+v^2)(1-v^2)^2}{v^5}.$$

The sum rule is thus written as

$$\sum_{n=2} \frac{1}{m^2 M_n} |R_n^{\text{exp}}(0)|^2 + f_{\text{back}}(v_1) = 2C_F^3 \left\{ [\zeta(3) - 1]\alpha_s^3 - \frac{4.9 \langle \alpha_s G^2 \rangle}{m^4 \alpha_s^3} \right\}$$

$$+ \frac{2}{3} \left\{ 8\epsilon_1^2 \alpha_s^3 + \frac{\langle \alpha_s G^2 \rangle}{48\epsilon_1^3 \alpha_s^3 m^4} \right\}, \quad \epsilon_1 \alpha_s = v_1.$$

Then we may profit from the fact that the sum rule should be valid for *all* values of $v_1 \geq v_0$ to fix $f_{1,2}$ requiring this independence, at least in the mean.

That is to say, that when we increase v_1 past a particle threshold from $\Upsilon(2)$ to $\Upsilon(6)$, the variation of the corresponding determinations of $\langle \alpha_s G^2 \rangle$ around their average should be a minimum. The calculation may be further simplified by replacing

$$f_{\text{back}}(v_1) \to 2 f_0 v_1^{2.75}.$$

The results of the analysis are summarized in the following tables, where the column "Res" indicates at which resonance the cut in v_1 occurs. We have taken two rather extreme values of f_0:

Res.	v_1	$\langle \alpha_s G^2 \rangle$		Res.	v_1	$\langle \alpha_s G^2 \rangle$
$\Upsilon(2)$	0.21	0.014		$\Upsilon(2)$	0.21	0.037
$\Upsilon(3)$	0.34	0.034		$\Upsilon(3)$	0.34	0.057
$\Upsilon(4)$	0.40	0.048		$\Upsilon(4)$	0.40	0.067
$\Upsilon(5)$	0.43	0.039		$\Upsilon(5)$	0.43	0.048
$\Upsilon(6)$	0.46	0.046		$\Upsilon(6)$	0.46	0.052

For $\alpha_s = 0.35$, $f_0 = 0.04$ For $\alpha_s = 0.40$, $f_0 = 0.09$

This derivation shows very clearly the kind of errors one encounters. To the variations that may be called "statistical", apparent in the different values found in the tables above,

$$0.014 \le \langle \alpha_s G^2 \rangle \le 0.067,$$

we have to add "systematic" ones, e.g., the influence of the not calculated radiative corrections, easily of some 30%: not to mention our including the Coulombic wave functions at the origin for large values of n. Given all these uncertainties, which do even make it dubious that one can really define with precision the condensate in terms of experimental observables, it is not surprising that one cannot pin down the gluon condensate with any accuracy. We would suggest an estimate, taking into account the above figures as well as other determinations, of

$$\langle \alpha_s G^2 \rangle \sim 0.055 \pm 0.03 \text{ GeV}^4. \tag{10.5.7}$$

Appendices

Appendix A: γ-Algebra in Dimension D

The gamma matrices are taken to be of dimension 4. We have D matrices γ^μ,

$$\gamma^0, \gamma^1, \ldots, \gamma^{D-1},$$

and the matrix[1] γ_5. They verify the anticommutation relations,

$$\{\gamma^\mu, \gamma^\nu\} = 2g^{\mu\nu}, \quad \gamma_5^2 = 1,$$

with

$$g^{\mu\nu} = 0, \ \mu \neq \nu; \quad g^{00} = 1, \quad g^{ii} = -1, \text{ for } i = 1, \ldots, D-1$$

and $g^{\mu\nu} = g_{\mu\nu}$.

A few useful relations are

$$\text{Tr}\,\gamma^\mu\gamma^\nu = 4g^{\mu\nu}, \quad \text{Tr}\,\gamma_5\gamma^\mu\gamma^\nu = 0, \quad \text{Tr}\,\overbrace{\gamma^\mu \ldots \gamma^\tau}^{\text{odd}} = 0, \quad \text{Tr}\,\gamma_5\,\overbrace{\gamma^\mu \ldots \gamma^\tau}^{\text{odd}} = 0;$$

$$\text{Tr}\,\gamma^\mu\gamma^\nu\gamma^\alpha\gamma^\beta = 4S^{\mu\nu\alpha\beta}, \quad S^{\mu\nu\alpha\beta} = g^{\mu\nu}g^{\alpha\beta} + g^{\mu\beta}g^{\alpha\nu} - g^{\mu\alpha}g^{\beta\nu};$$

$$S^{\mu\nu\rho\alpha}S_{\mu\nu\rho\beta} = (3D-2)g^\alpha_\beta;$$

$$\slashed{a}\slashed{a} = a^2, \quad \slashed{a}\slashed{b}\slashed{a} = -a^2\slashed{b} + 2(a \cdot b)\slashed{a};$$

$$\gamma^\mu\gamma_\mu = D, \quad \gamma^\mu\gamma^\alpha\gamma_\mu = (2-D)\gamma^\alpha, \quad \gamma^\mu\gamma_5\gamma_\mu = (D-8)\gamma_5;$$

$$\gamma^\mu\gamma^\alpha\gamma^\beta\gamma_\mu = 4g^{\alpha\beta} + (D-4)\gamma^\alpha\gamma^\beta;$$

$$\gamma^\mu\gamma^\alpha\gamma^\beta\gamma^\delta\gamma_\mu = -2\gamma^\delta\gamma^\beta\gamma^\alpha + (4-D)\gamma^\alpha\gamma^\beta\gamma^\delta.$$

For $D = 4$, $\gamma_5 = i\gamma^0\gamma^1\gamma^2\gamma^3$ and we define the totally antisymmetric tensor,

$$\epsilon^{\mu\nu\rho\sigma} = \begin{cases} 0, \text{ if two indices are equal} \\ -1, \quad \text{if } \mu\nu\rho\sigma = 0123 \\ 1, \quad \text{if } \mu\nu\rho\sigma = 1230 \end{cases}$$

---

[1] More about γ_5 may be found in the main text in Sects. 3.1 and 7.5.

and cyclically. For $D = 4$ then,

$$\gamma^\mu\gamma^\alpha\gamma^\nu = S^{\mu\alpha\nu\beta}\gamma_\beta - i\epsilon^{\mu\alpha\nu\beta}\gamma_\beta\gamma_5, \quad \gamma_5\gamma^\mu\gamma^\nu = \gamma_5 g^{\mu\nu} + \frac{1}{2i}\epsilon^{\mu\nu\alpha\beta}\gamma_\alpha\gamma_\beta;$$

$$\gamma^\alpha\gamma^{\mu_1}\ldots\gamma^{\mu_{2n+1}}\gamma_\alpha = -2\gamma^{\mu_{2n+1}}\ldots\gamma^{\mu_1};$$

$$\operatorname{Tr}\gamma_5\gamma^\mu\gamma^\nu\gamma^\alpha\gamma^\beta = 4i\epsilon^{\mu\nu\alpha\beta}.$$

We also have

$$g_{\alpha\beta}\epsilon^{\alpha\mu\rho\sigma}\epsilon^{\beta\nu\tau\lambda} = -g^{\mu\nu}(g^{\rho\tau}g^{\sigma\lambda} - g^{\rho\lambda}g^{\sigma\tau})$$
$$-g^{\mu\lambda}(g^{\rho\nu}g^{\sigma\tau} - g^{\rho\tau}g^{\sigma\nu})$$
$$+g^{\mu\tau}(g^{\rho\nu}g^{\sigma\lambda} - g^{\rho\lambda}g^{\sigma\nu});$$

$$g_{\alpha\alpha}g_{\beta\beta}\epsilon^{\mu\nu\alpha\beta}\epsilon^{\rho\sigma\alpha\beta} = 2(g^{\mu\sigma}g^{\rho\nu} - g^{\mu\rho}g^{\sigma\nu}).$$

Moreover, $\{\gamma_5, \gamma^\mu\} = 0$. In the Pauli or Weyl realizations, $\gamma_2\gamma_\mu\gamma_2 = -\gamma_\mu^*$, $\gamma_0\gamma_\mu\gamma_0 = \gamma_\mu^\dagger$. Finally, if w_1, w_2 are spinors and $\Gamma_1, \ldots, \Gamma_n$ any of the matrices γ_μ, $i\gamma_5$, then

$$(\bar{w}_1\Gamma_1\ldots\Gamma_n w_2)^* = \bar{w}_2\Gamma_n\ldots\Gamma_1 w_1.$$

Appendix B: Some Useful Integrals

Integration in Feynman Amplitudes

In D dimensions,

$$\int \frac{d^D k}{(2\pi)^D}\frac{(k^2)^r}{(k^2 - R^2 + i0)^m} = i\frac{(-1)^{r-m}}{(16\pi^2)^{D/4}}\frac{\Gamma(r+D/2)\Gamma(m-r-D/2)}{\Gamma(D/2)\Gamma(m)(R^2)^{m-r-D/2}};$$

$$\int d^D k\frac{1}{k^2 + i0} = 0.$$

The last formula may be obtained from the general one by replacing $i0$ by $i\delta$, calculating for finite δ and taking the limit $\delta \to 0$ at the end.

In Euclidean space,

$$\int d^D\underline{k}\,\delta(1 - |\underline{k}|) = \frac{2\pi^{D/2}}{\Gamma(D/2)}.$$

Symmetric integration:

$$\int d^D k\,k^\mu k^\nu f(k^2) = \frac{g^{\mu\nu}}{D}\int d^D k\,k^2 f(k^2);$$

$$\int d^D k\,k^\mu k^\nu k^\rho k^\sigma f(k^2) = \frac{g^{\mu\nu}g^{\rho\sigma} + g^{\mu\rho}g^{\nu\sigma} + g^{\mu\sigma}g^{\nu\rho}}{D^2 + 2D}\int d^D k\,k^4 f(k^2);$$

$$\int d^D k\,k^{\mu_1}\ldots k^{\mu_{2n+1}} f(k^2) = 0.$$

Taylor expansion of the logarithm of Euler's gamma function (no simple formula exists for the expansion of Γ itself):

$$\log \Gamma(1 + \epsilon) = -\gamma_E \epsilon + \sum_{n=2}^{\infty} \frac{(-\epsilon)^n}{n} \zeta(n).$$

ζ is Riemann's function and $\gamma_E \simeq 0.5772$ is the Euler–Mascheroni constant.

Feynman parameters:

$$\frac{1}{A^\alpha B^\beta} = \frac{\Gamma(\alpha + \beta)}{\Gamma(\alpha)\Gamma(\beta)} \int_0^1 dx \, \frac{x^{\alpha-1}(1-x)^{\beta-1}}{[xA + (1-x)B]^{\alpha+\beta}};$$

$$\frac{1}{A^\alpha B^\beta C^\gamma} = \frac{\Gamma(\alpha + \beta + \gamma)}{\Gamma(\alpha)\Gamma(\beta)\Gamma(\gamma)} \int_0^1 dx \, x \int_0^1 dy \, \frac{u_1^{\alpha-1} u_2^{\beta-1} u_3^{\gamma-1}}{[u_1 A + u_2 B + u_3 C]^{\alpha+\beta+\gamma}},$$

$$u_1 = xy, \quad u_2 = x(1-y), \quad u_3 = 1 - x;$$

$$\frac{1}{A^\alpha B^\beta C^\gamma D^\delta} = \frac{\Gamma(\alpha + \beta + \gamma + \delta)}{\Gamma(\alpha)\Gamma(\beta)\Gamma(\gamma)\Gamma(\delta)}$$

$$\times \int_0^1 dx \, x^2 \int_0^1 dy \, y \int_0^1 dz \, \frac{u_1^{\alpha-1} u_2^{\beta-1} u_3^{\gamma-1} u_4^{\delta-1}}{[u_1 A + u_2 B + u_3 C + u_4 D]^{\alpha+\beta+\gamma+\delta}},$$

$$u_1 = 1 - x, \; u_2 = xyz, \; u_3 = x(1-y), \; u_4 = xy(1-z), \text{ etc.}$$

In general,

$$\frac{1}{A_1 \ldots A_n} =$$

$$(n-1)! \int_0^1 dx_1 \ldots \int_0^1 dx_n \, \delta\left(\sum_1^n x_i - 1\right) \frac{1}{(x_1 A_1 + \cdots + x_n A_n)^n}.$$

More formulas may be found in Narison (1989).

Feynman amplitudes are evaluated with the help of the preceding formulas as follows. An amplitude with external momenta $p_1, \ldots, p_n$ is given by an integral

$$F(p_1, \ldots, p_n) = \int d^D k_1 \ldots \int d^D k_J \, \frac{N(p,k)}{[q_1(p,k)^2 - m_1^2] \cdots [(q_L(p,k)^2 - m_L^2]}.$$

Here $N(p,k)$ is a polynomial in the p_i, k_j; the latter are the loop variables and the momenta $q_l(p,k)$ are lineal combinations (given by energy–momentum conservation at the vertices) of the p_i, k_j.

Combining the denominators we may write this as, for example,

$$F(p_1, \ldots, p_n) = (L-1)! \int_0^1 dx_1 \ldots \int_0^1 dx_L \, \delta\left(\sum_1^L x_i - 1\right)$$

$$\times \int d^D k_1 \ldots \int d^D k_J \, \frac{N(p,k)}{\{x_1[q_1(p,k)^2 - m_1^2] + \cdots + x_L[(q_L(p,k)^2 - m_L^2]\}^L}.$$

The term in curly brackets in the last denominator, which we will call Δ, can be cast, by replacing the $q(p,k)$ by their explicit expressions in terms of ps and ks, in the form

$$\Delta \equiv x_1[q_1(p,k)^2 - m_1^2] + \cdots + x_L[(q_L(p,k)^2 - m_L^2]$$
$$= \sum_{i,j} a_{ij}(x,p)k_i \cdot k_j + \sum_i b_i(x,p) \cdot k_i + A(x,p),$$

and we will not write explicitly the dependence on the masses, m. This quadratic form may be diagonalized by an orthogonal transformation S and a translation. If we add a dilatation, so that we set

$$k_i = \lambda_i(x,p) \sum_j S_{ij}(x,p)l_j + c_i(x,p),$$

then Δ becomes

$$\Delta = \sum_i l_i^2 + R^2(x,p).$$

Likewise, the numerator can be written, in the new variables, as

$$N(p,k) = N_0(x,p) + \sum N_{2ij}^{\mu\nu}(x,p)l_{i\mu}l_{j\nu} + \sum N_{4ijst}^{\mu\nu\rho\sigma}(x,p)l_{i\mu}l_{j\nu}l_{s\rho}l_{t\sigma} + \cdots;$$

we do not write terms that are odd in the l, as they vanish by symmetric integration. Thus, we have

$$F(p_1,\ldots,p_n) = (L-1)! \int_0^1 \mathrm{d}x_1 \ldots \int_0^1 \mathrm{d}x_L \, \delta\left(\sum_1^L x_i - 1\right)$$
$$\times \int \mathrm{d}^D l_1 \ldots \int \mathrm{d}^D l_J \frac{N_0(x,p) + \sum N_{2ij}^{\mu\nu}(x,p)l_{i\mu}l_{j\nu} + \cdots}{[\sum_i l_i^2 + R^2(x,p)]^L}.$$

Finally, this may be written in terms of a single, (DJ)-dimensional integral by defining the (DJ)-dimensional vector l with components l_i; in the full space we define of course $l^2 = l_1^2 + \cdots + l_L^2$. The individual $l_{i\mu}$ in the numerator above can be interpreted as the μth component in the ith subspace. We may further use symmetric integration, which is generalized in (DJ)-dimensional space to

$$\int \mathrm{d}^{DJ}l \, l_{i\mu}l_{j\nu}f(l^2) = \frac{\delta_{ij}g_{\mu\nu}}{DJ} \int \mathrm{d}^{DJ}l \, l^2 f(l^2), \text{ etc.,}$$

and so we get

$$F(p_1,\ldots,p_n) = (L-1)! \int_0^1 \mathrm{d}x_1 \ldots \int_0^1 \mathrm{d}x_L \, \delta\left(\sum_1^L x_i - 1\right)$$
$$\times \int \mathrm{d}^{DJ}l \, \frac{\bar{N}_0(x,p) + \bar{N}_2(x,p)l^2 + \bar{N}_4(x,p)l^4 + \cdots}{[l^2 + R^2(x,p)]^L}.$$

The integral over $\mathrm{d}^{DJ}l$ is carried out with the help of the first formula of this appendix: thus reducing the evaluation of the Feynman amplitude F to that of the finite range integrals over the Feynman parameters $\mathrm{d}x_1,\ldots, \mathrm{d}x_L$.

Some Numerical Integrals

$$\int_0^1 dx \, \log(1+x) = 2\log 2 - 1; \qquad \int_0^1 dx \, \frac{\log(1+x)}{x} = \frac{\pi^2}{12}.$$

Many useful integrals can be obtained from Euler's formula,

$$\int_0^1 dx \, x^\alpha (1-x)^\beta = \frac{\Gamma(1+\alpha)\Gamma(1+\beta)}{\Gamma(2+\alpha+\beta)}.$$

For example, by differentiation we obtain,

$$\int_0^1 dx \, x^\alpha \log x = -\frac{1}{(\alpha+1)^2};$$

$$\int_0^1 dx \, x^\alpha (1-x)^\beta \log x = [S_1(\alpha) - S_1(1+\alpha+\beta)] \frac{\Gamma(1+\alpha)\Gamma(1+\beta)}{\Gamma(2+\alpha+\beta)};$$

$$\int_0^1 dx \, \frac{x^\alpha - 1}{1-x} = -S_1(\alpha);$$

$$\int_0^1 dx \, x^\alpha \log x \log(1-x) = \frac{S_1(1+\alpha)}{(1+\alpha)^2} + \frac{S_2(1+\alpha)}{1+\alpha} - \frac{\pi^2}{6(1+\alpha)};$$

$$\int_0^1 dx \, x^\alpha \frac{\log^2 x}{1-x} = 2\zeta(3) - 2S_3(\alpha),$$

$$\int_0^1 dx \, \frac{x^\alpha}{1-x} \log x \log(1-x) = \frac{\pi^2}{6} S_1(\alpha) - S_1(\alpha)S_2(\alpha) - S_3(\alpha) + \zeta(3),$$

$$\int_0^1 dx \, x^\alpha (1-x)^\beta \log x \log(1-x) = \frac{\Gamma(1+\alpha)\Gamma(1+\beta)}{\Gamma(2+\alpha+\beta)}$$
$$\times \left\{ S_2(1+\alpha+\beta) - \frac{\pi^2}{6} + [S_1(\alpha) - S_1(1+\alpha+\beta)][S_1(\beta) - S_1(1+\alpha+\beta)] \right\};$$

$$\int_0^1 dx \, x^\alpha (1-x)^\beta \log^2 x = \frac{\Gamma(1+\alpha)\Gamma(1+\beta)}{\Gamma(2+\alpha+\beta)}$$
$$\times \left\{ [S_1(\alpha) - S_1(1+\alpha+\beta)]^2 + S_2(1+\alpha+\beta) - S_2(\alpha) \right\},$$

etc. Here,

$$S_l(\alpha) = \sum_{k=1}^\infty \left[\frac{1}{k^l} - \frac{1}{(k+\alpha)^l} \right];$$

$$S_l(\alpha) = \sum_{j=1}^\alpha \frac{1}{j^l} \quad \text{for } \alpha = \text{positive integer}.$$

Also, $S_1(\alpha) = \psi(1+\alpha) + \gamma_E$. For the special functions ψ, Γ, ζ, see Abramowicz and Stegun (1965).

Appendix C: Group-theoretic Quantities. Group Integration

Lie Algebra. Invariants

We let σ^j be the ordinary Pauli matrices:

$$\sigma^1 = \begin{pmatrix} 0 & 1 \\ 1 & 0 \end{pmatrix}, \quad \sigma^2 = \begin{pmatrix} 0 & -i \\ i & 0 \end{pmatrix}, \quad \sigma^3 = \begin{pmatrix} 1 & 0 \\ 0 & -1 \end{pmatrix}.$$

For $SU(2)$, $t^a = \sigma^a/2$; for $SU(3)$, $t^a = \lambda^a/2$ with

$$\lambda^j = \begin{pmatrix} \sigma^j & 0 \\ 0 & 0 \end{pmatrix}, \quad \lambda^4 = \begin{pmatrix} 0 & 0 & 1 \\ 0 & 0 & 0 \\ 1 & 0 & 0 \end{pmatrix}, \quad \lambda^5 = \begin{pmatrix} 0 & 0 & -i \\ 0 & 0 & 0 \\ i & 0 & 0 \end{pmatrix},$$

$$\lambda^6 = \begin{pmatrix} 0 & 0 & 0 \\ 0 & 0 & 1 \\ 0 & 1 & 0 \end{pmatrix}, \quad \lambda^7 = \begin{pmatrix} 0 & 0 & 0 \\ 0 & 0 & -i \\ 0 & i & 0 \end{pmatrix}, \quad \lambda^8 = \frac{1}{\sqrt{3}}\begin{pmatrix} 1 & 0 & 0 \\ 0 & 1 & 0 \\ 0 & 0 & -2 \end{pmatrix}.$$

Still for $SU(3)$, we can introduce also the matrices C^a with elements $C^a_{bc} = -if_{abc} \equiv -if^{abc}$. The commutation relations of the t, C are

$$[t^a, t^b] = i\sum f^{abc}t^c, \quad [C^a, C^b] = i\sum f^{abc}C^c,$$

and the anticommutation relations of the t are

$$\{t^a, t^b\} = \sum d^{abc}t^c + \tfrac{1}{3}\delta^{ab}.$$

The f are totally antisymmetric, and the $d_{abc} \equiv d^{abc}$ are totally symmetric. The only nonzero elements, up to permutations, are as follows:

$$1 = f_{123} = 2f_{147} = 2f_{246} = 2f_{257} = 2f_{345}$$

$$= -2f_{156} = -2f_{367} = \frac{2}{\sqrt{3}}f_{458} = \frac{2}{\sqrt{3}}f_{678};$$

$$\frac{1}{\sqrt{3}} = d_{118} = d_{228} = d_{338} = -d_{888}, \quad -\frac{1}{2\sqrt{3}} = d_{448} = d_{558} = d_{668} = d_{778},$$

$$\tfrac{1}{2} = d_{146} = d_{157} = d_{247} = d_{256} = d_{344} = d_{355} = -d_{366} = -d_{377}.$$

For an arbitrary group $SU(N)$ we define the invariants C_A, C_F, T_F by

$$\delta_{ab}C_A = \operatorname{Tr} C^a C^b = \sum_{cc'} f^{acc'} f^{bcc'},$$

$$\delta_{ik}C_F = \left(\sum_a t^a t^a\right)_{ik} = \sum_{a,l} t^a_{il} t^a_{lk},$$

$$\delta_{ab}T_F = \operatorname{Tr} t^a t^b = \sum_{k,i} t^a_{ik} t^b_{ki}.$$

Then,

$$C_A = N, \quad C_F = \frac{N^2 - 1}{2N}, \quad T_F = \tfrac{1}{2}.$$

Other useful relations are

$$\mathrm{Tr}\, t^a t^b t^c = \tfrac{1}{4}\left(\mathrm{i} f^{abc} + d^{abc}\right), \quad \sum_a t^a_{ij} t^a_{lk} = -\frac{1}{2N}\delta_{il}\delta_{jk} + \tfrac{1}{2}\delta_{ik}\delta_{lj};$$

$$\sum_{abc} d^2_{abc} = \tfrac{40}{3}, \quad \sum_{abc} f^2_{abc} = 24, \quad \sum_{rka} \epsilon_{irk} t^a_{jr} t^a_{kl} = -\tfrac{1}{6}\epsilon_{ijl}.$$

Here the last relations are valid for $SU(3)$, and ϵ_{ijk} is the three-dimensional antisymmetric symbol, $\epsilon_{123} = 1, \epsilon_{132} = -1$ and cyclically.

A large set of group relations may be found in the paper by Cvitanović (1976).

Group Integration

Let us consider a group, G. We will assume that the group is topological, which is the case in the particular instances where G is a discrete group or a Lie group. Then, there exists a measure, called the *Haar measure*, positive definite and left invariant: if we denote by $\mathrm{d}\mu_L(g)$ the measure

$$\int_G \mathrm{d}\mu_L(g)\, f(g) = \int_G \mathrm{d}\mu_L(g) f(hg),$$

for any smooth function f that decreases sufficiently fast at infinity of the group, and for any group element h. A right invariant measure also exists:

$$\int_G \mathrm{d}\mu_R(g)\, f(g) = \int_G \mathrm{d}\mu_R(g) f(gh).$$

For Abelian groups, $\mathrm{d}\mu_R$ obviously coincides with $\mathrm{d}\mu_L$; and the same can be proved to be the case for discrete and compact groups (Abelian or not). Moreover, $\mathrm{d}\mu_L = \mathrm{d}\mu_R = \mathrm{d}\mu$ is unique up to a multiplicative constant. For discrete groups, the Haar measure is merely the sum over group elements:

$$\int \mathrm{d}\mu(g)\, f(g) \to \sum_{\text{all } g} f(g).$$

For Lie groups, that we take as being compact for simplification, one can reduce the Haar measure to ordinary integration as follows. Let $\alpha_1, \ldots, \alpha_n$ be the parameters determining the group elements, $g = g(\alpha_1, \ldots, \alpha_n)$. If we assume that $g(\alpha_1, \ldots, \alpha_n)$ is the product of $g(\beta_1, \ldots, \beta_n)$ and $g(\gamma_1, \ldots, \gamma_n)$,

$$g(\alpha_1, \ldots, \alpha_n) = g(\beta_1, \ldots, \beta_n)g(\gamma_1, \ldots, \gamma_n),$$

then obviously the α are functions of the β and the γ,

$$\alpha_i = \varphi_i(\beta_1, \ldots, \beta_n; \gamma_1, \ldots, \gamma_n).$$

We will assume that the unit of the group has parameters $0, \ldots, 0$, and define the Jacobian

$$J^{-1}(\gamma_1, \ldots, \gamma_n) = \det\left(\frac{\partial\varphi_i(\beta_1, \ldots, \beta_n; \gamma_1, \ldots, \gamma_n)}{\partial\beta_j}\right)\Bigg|_{\beta=0}.$$

The expression for the invariant measure is then

$$\int d\mu(g)\, f(g) \equiv \int dg\, f(g) = \int d\alpha_1 \ldots d\alpha_n\, J(\alpha_1, \ldots, \alpha_n) f(g(\alpha_1, \ldots, \alpha_n)).$$

An elementary proof may be found in the book of Creutz (1983). We will adjust the arbitrary constant in the definition of dg so that $\int_G dg = 1$.

A basic theorem in group integration is the following:

THEOREM (PETER–WEYL). *Let $D^{(a)}(g)$ be a unitary, irreducible representation of a compact group, and $D^{(b)}(g)$ another one inequivalent to the first when $b \neq a$. Then*

$$\int_G dg D_{ij}^{(a)}(g) D_{kl}^{(b)}(g)^* = \delta_{ab}\delta_{ik}\delta_{jl}.$$

Moreover, the set of functions $D_{ij}^{(a)}(g)$, where a runs over all the irreducible representations, is complete. That is to say, if the function $f(g)$ is L^2-integrable on the group, it may be expanded on the basis formed by the D:

$$f(g) = \sum_{aij} c_{a,ij} D_{ij}^{(a)}(g).$$

All integrals over representations of the group can be deduced from the Peter–Weyl theorem. For example, consider the integral of an arbitrary product of representations

$$\int_G dg\, D_{i_1 j_1}^{(1)}(g) \ldots D_{i_\nu j_\nu}^{(\nu)}(g).$$

One may decompose this as a sum of irreducible representations using the appropriate Clebsch–Gordan coefficients:

$$D_{i_1 j_1}^{(1)}(g) \ldots D_{i_\nu j_\nu}^{(\nu)}(g) = C^{(0)}(i_1, j_1, \ldots, i_\nu, j_\nu) D^{(0)}$$
$$+ \sum C^{(I)}(i_1, j_1, \ldots, i_\nu, j_\nu | kl) D_{kl}^{(I)}(g),$$

and $D^{(0)} \equiv 1$ is the identity representation. Because of the Peter–Weyl theorem, all terms in the integral save the first give zero, so

$$\int_G dg\, D_{i_1 j_1}^{(1)}(g) \ldots D_{i_\nu j_\nu}^{(\nu)}(g) = C^{(0)}(i_1, j_1, \ldots, i_\nu, j_\nu).$$

Useful algorithms for calculating these Clebsch–Gordan coefficients may be found in the quoted text of Creutz (1983).

Appendix D: Feynman Rules for QCD

Ordinary Formalism

The Feynman rules for the S-matrix are as follows: we have to include an overall factor $(2\pi)^4\delta(P_i - P_f)$ for energy–momentum conservation, and a factor (-1) for each closed quark or ghost loop. Each loop integration requires a factor

$$\nu_0^{4-D} \int \frac{\mathrm{d}^D k}{(2\pi)^D} \equiv \int \mathrm{d}^D \hat{k}\,.$$

Diagrams with disconnected bubbles are excluded. A diagram is to be read *against* the direction of the oriented lines. External and internal lines produce the following factors:

p *ingoing quark:* $(2\pi)^{-3/2}u(p,\lambda)$

p *ingoing antiquark:* $(2\pi)^{-3/2}\overline{v}(p,\lambda)$

p *outgoing quark :* $(2\pi)^{-3/2}\overline{u}(p,\lambda)$

p *outgoing antiquark :* $(2\pi)^{-3/2}v(p,\lambda)$

k *ingoing gluon :* $(2\pi)^{-3/2}\varepsilon_\mu(k,\eta)$

k *outgoing gluon :* $(2\pi)^{-3/2}\varepsilon*_\mu(k,\eta)$

p *quark propagator:* $\delta_{lj}\,\mathrm{i}/(p\cdot\gamma - m + \mathrm{i}0)$
j $\quad$ l

k *gluon propagator:* $\delta_{ba}\,\mathrm{i}\,\dfrac{-g_{\mu\nu}+\xi k^\mu k^\mu/(k^2+\mathrm{i}0)}{k^2+\mathrm{i}0}$
a,μ $\quad$ b,ν *(Lorentz gauges)*

p *ghost propagator:* $\delta_{ba}\,\mathrm{i}/(p^2+\mathrm{i}0)$
a $\quad$ b

For the vertices:

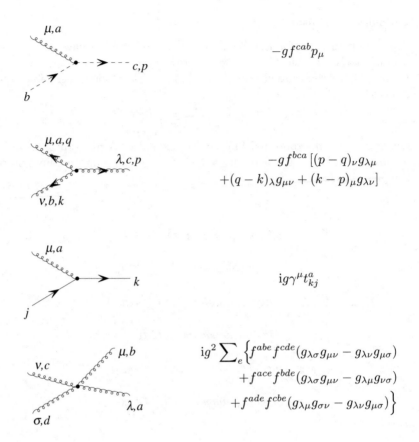

$$-gf^{cab}p_\mu$$

$$-gf^{bca}\left[(p-q)_\nu g_{\lambda\mu}\right.$$
$$+(q-k)_\lambda g_{\mu\nu} + (k-p)_\mu g_{\lambda\nu}\right]$$

$$ig\gamma^\mu t^a_{kj}$$

$$ig^2\sum_e \left\{ f^{abe}f^{cde}(g_{\lambda\sigma}g_{\mu\nu} - g_{\lambda\nu}g_{\mu\sigma})\right.$$
$$+ f^{ace}f^{bde}(g_{\lambda\sigma}g_{\mu\nu} - g_{\lambda\mu}g_{\nu\sigma})$$
$$\left.+ f^{ade}f^{cbe}(g_{\lambda\mu}g_{\sigma\nu} - g_{\lambda\nu}g_{\mu\sigma})\right\}$$

Combinatorial factors, due to the identity of the gluons, are to be included:

The spinors and polarization vectors are normalized to

$$\sum_\sigma u(p,\sigma)\bar{u}(p,\sigma) = \slashed{p} + m, \quad \sum_\lambda \epsilon^\mu(k,\lambda)^*\epsilon^\nu(k,\lambda) = -g^{\mu\nu} \quad \text{(Feynman gauge)}.$$

Our rules differ from the ones in e.g. Bjorken and Drell (1965) by the normalization of the spinors, $\sum_\sigma u_{\mathrm{BD}}(p,\sigma)\bar{u}_{\mathrm{BD}}(p,\sigma) = (\not{p}+m)/2m$, and the factors $(2\pi)^{-3/2}$ due to our normalization of $\mathcal{T}$, which differs from $\mathcal{T}_{\mathrm{BD}}$ by precisely these factors.

Background Field Formalism

We have only to give the trilinear and quadrilinear couplings; external lines are identical for quarks and gluons to those already presented, and do not exist for the background field. Likewise, quark, gluon and ghost propagators are identical to the ordinary ones. For the couplings, we have, defining $1 - a = \xi$ and representing the background field (as in the main text) by a gray blob,

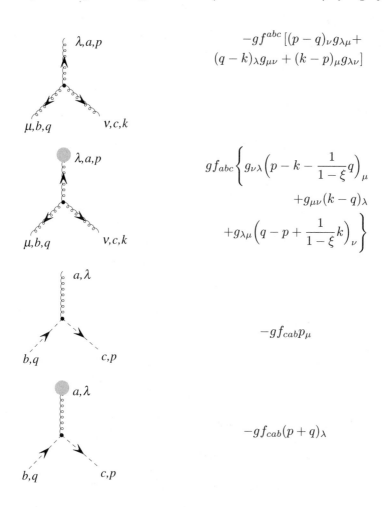

$$-gf^{abc}\left[(p-q)_\nu g_{\lambda\mu} + (q-k)_\lambda g_{\mu\nu} + (k-p)_\mu g_{\lambda\nu}\right]$$

$$gf_{abc}\left\{g_{\nu\lambda}\left(p - k - \frac{1}{1-\xi}q\right)_\mu + g_{\mu\nu}(k-q)_\lambda + g_{\lambda\mu}\left(q - p + \frac{1}{1-\xi}k\right)_\nu\right\}$$

$$-gf_{cab}p_\mu$$

$$-gf_{cab}(p+q)_\lambda$$

and

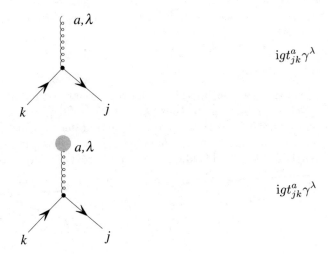

$$igt^a_{jk}\gamma^\lambda$$

$$igt^a_{jk}\gamma^\lambda$$

for the trilinear couplings.

Quadrilinear couplings:

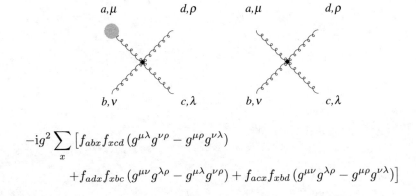

$$-ig^2 \sum_x \left[f_{abx}f_{xcd}\left(g^{\mu\lambda}g^{\nu\rho} - g^{\mu\rho}g^{\nu\lambda}\right) \right.$$
$$\left. + f_{adx}f_{xbc}\left(g^{\mu\nu}g^{\lambda\rho} - g^{\mu\lambda}g^{\nu\rho}\right) + f_{acx}f_{xbd}\left(g^{\mu\nu}g^{\lambda\rho} - g^{\mu\rho}g^{\nu\lambda}\right) \right]$$

(both graphs yield the same factor).

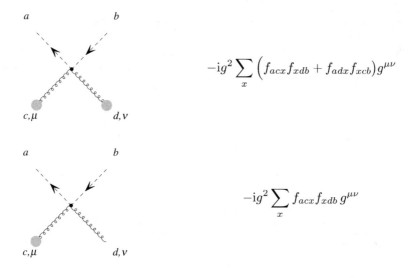

$$-\mathrm{i}g^2 \sum_x \left(f_{acx} f_{xdb} + f_{adx} f_{xcb} \right) g^{\mu\nu}$$

$$-\mathrm{i}g^2 \sum_x f_{acx} f_{xdb}\, g^{\mu\nu}$$

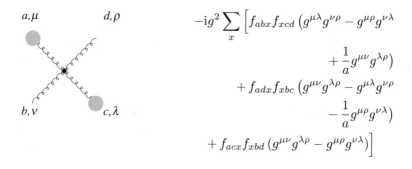

$$-\mathrm{i}g^2 \sum_x \left[f_{abx} f_{xcd} \left(g^{\mu\lambda} g^{\nu\rho} - g^{\mu\rho} g^{\nu\lambda} \right. \right.$$
$$\left. + \frac{1}{a} g^{\mu\nu} g^{\lambda\rho} \right)$$
$$+ f_{adx} f_{xbc} \left(g^{\mu\nu} g^{\lambda\rho} - g^{\mu\lambda} g^{\nu\rho} \right.$$
$$\left. - \frac{1}{a} g^{\mu\rho} g^{\nu\lambda} \right)$$
$$\left. + f_{acx} f_{xbd} \left(g^{\mu\nu} g^{\lambda\rho} - g^{\mu\rho} g^{\nu\lambda} \right) \right]$$

The combinations of gauge fields and background fields not shown here vanish. For example, there is no quadrilinear vertex with three or four background fields.

Appendix E: Feynman Rules for Composite Operators

We let $\gamma_+ = 1$, $\gamma_- = \gamma_5$, and let also Δ be an arbitrary four-vector with $\Delta^2 = 0$. Then we have the Feynman rules, for the operators that intervene in deep inelastic scattering at leading twist:

$: \bar{q}\gamma^{\mu_1}\ldots\partial^{\mu_n}\gamma_{\pm}q :$

$$\not{\Delta}(\Delta \cdot k)^{n-1}\gamma_{\pm}$$

$: G^{\mu\mu_1}\partial^{\mu_2}\ldots\partial^{\nu}G :$

$$g_{\mu\nu}(\Delta \cdot k)^n + k^2\Delta_\mu\Delta_\nu(\Delta \cdot k)^{n-2}$$
$$- (k_\mu\Delta_\nu + k_\nu\Delta_\mu)(\Delta \cdot k)^{n-1}$$

$: \bar{q}_j\gamma^{\mu_1}\ldots gB_a^\mu t_{lj}^a \ldots \gamma^{\mu_n}\gamma_{\pm}q_l :$

$$gt_{lj}^a\Delta^\mu\not{\Delta}\sum_{j=0}^{n-2}(\Delta \cdot p_1)^j(\Delta \cdot p_2)^{n-j-2}\gamma_{\pm}$$

$: G^{\mu\nu_1}\partial^{\mu_2}\ldots gB^{\mu_i}\ldots G :$

$$\frac{ig}{3!}f_{abc}$$

$$\times\Big\{\Delta_\nu\big[\Delta_\lambda k_\mu(\Delta \cdot p) + p_\lambda\Delta_\mu(\Delta \cdot k)$$
$$- g_{\mu\lambda}(\Delta \cdot p)(\Delta \cdot k) - \Delta_\mu\Delta_\lambda(k \cdot p)\big]$$

$$+ \sum_{j=1}^{n-2}(-1)^j(\Delta \cdot p)^{j-1}(\Delta \cdot k)^{n-j-2}$$

$$+ \big[(g_{\mu\lambda}\Delta_\nu - g_{\nu\lambda}\Delta_\mu)(\Delta \cdot k)$$
$$+ \Delta_\lambda(\Delta_\mu k_\nu - k_\nu\Delta_\mu)\big](\Delta \cdot k)^{n-2}\Big\}$$

$$+ \text{ permutations}$$

The operators are taken at $x = 0$ so, for example, $: \bar{q}\gamma^{\mu_1} \ldots \partial^{\mu_n} \gamma_{\pm} q :$ means $: \bar{q}\gamma^{\mu_1} \ldots \partial^{\mu_n} \gamma_{\pm} q : |_{x=0}$. See also Floratos, Ross and Sachrajda (1977, 1979).

Appendix F: Some Singular Functions

We define x-space free field causal functions by

$$\Delta(x; m^2) = i \int \frac{d^4 k}{(2\pi)^4} e^{-ik \cdot x} \frac{1}{k^2 - m^2 + i0},$$

$$D_\xi^{\mu\nu}(x) = i \int \frac{d^4 k}{(2\pi)^4} e^{-ik \cdot x} \frac{-g^{\mu\nu} + \xi k^\mu k^\nu / (k^2 + i0)}{k^2 + i0},$$

$$S(x; m) = i \int \frac{d^4 k}{(2\pi)^4} e^{-ik \cdot x} \frac{\not{k} + m}{k^2 - m^2 + i0}.$$

We will at times omit the variable m. In terms of free field VEVs,

$$\langle T\phi(x)\phi(0)\rangle_0 = \Delta(x; m^2), \quad \langle TB_a^\mu(x)B_b^\nu(0)\rangle_0 = \delta_{ab} D_\xi^{\mu\nu}(x),$$

$$\langle Tq^j(x)\bar{q}^i(0)\rangle_0 = \delta_{ij} S(x, m).$$

The character of Green's functions of the propagators is exhibited clearly by the equations $(\Box_x + m^2) i\Delta(x - y) = \delta(x - y)$, etc. We also have

$$S(x, m) = (i\not{\partial} + m)\Delta(x; m^2), \quad D_\xi^{\mu\nu}(x) = (-g^{\mu\nu} + \xi\partial^\mu\partial^\nu/\Box)\Delta(x; m^2).$$

The translation invariance of the VEVs follow from the invariance of the vacuum, and the transformation properties of the fields: for a generic field,

$$U(a, \Lambda)\Phi_\alpha(x)U(a, \Lambda)^{-1} = \sum P_{\alpha\beta}(\Lambda)\Phi_\beta(\Lambda x + a).$$

On the light cone,

$$\Delta(x; m^2) \underset{x^2 \to 0}{\sim} -\frac{1}{4\pi^2} \frac{1}{x^2 - i0} + \frac{im^2 \theta(x^2)}{16\pi} + \frac{m^2}{8\pi^2} \log \frac{m|x^2|^{\frac{1}{2}}}{2} + \cdots$$

$$S(x; m) \underset{x^2 \to 0}{\sim} \frac{1}{4\pi^2} \frac{2ix_\mu \gamma^\mu}{(x^2 - i0)^2} + \cdots \text{ etc.}$$

Additional relations may be found in Bjorken and Drell (1965).[2] Fourier transforms of distributions, including in particular propagator-type ones, are

---

[2] Our causal functions differ from those of Bjorken and Drell (1965) by an i: $S = iS_{BD}$, $D = iD_{BD}, \ldots$.

given in Gel'fand and Shilov (1962), pp. 277ff, 316ff. The ones used in the text are

$$\int d^4x\, e^{-ik\cdot x}\frac{1}{x^2 \pm i0} = 4\pi^2\frac{i}{k^2 \mp i0},$$

$$\int d^4x\, e^{-ik\cdot x}\frac{1}{(x^2 \pm i0)^2} = \pi^2 i \log(k^2 - +i0) + \text{constant}.$$

Equal-time and light-cone commutation relations for fermions:

$$\{q^i_\alpha(x), q^j_\beta(y)\} = 0, \quad \delta(x_0 - y_0)\{q^i_\alpha(x), q^j_\beta(y)^\dagger\} = \delta_{\alpha\beta}\delta_{ij}\delta(x - y),$$

$$\{q^i_\alpha(x), \bar{q}^j_\beta(0)\} \underset{x^2 \to 0}{\simeq} (i\not\partial - im)_{\alpha\beta}\left\{\frac{1}{2\pi}\epsilon(x^0)\delta(x^2) - \frac{m}{4\pi\sqrt{x^2}}\theta(x^2)\epsilon(x^0) + \cdots\right\}.$$

Appendix G: Kinematics, Cross Sections, Decay Rates. Units

The states of a particle with spin (or helicity) λ and momentum p are normalized according to

$$\langle p', \lambda'|p, \lambda\rangle = 2p^0\delta_{\lambda\lambda'}\delta(\mathbf{p} - \mathbf{p}'), \quad P^\mu|p, \lambda\rangle = p^\mu|p, \lambda\rangle.$$

This corresponds to a density of particles per unit volume of

$$\rho(p) = \frac{2p^0}{(2\pi)^3}.$$

We define the scattering amplitude in terms of the S matrix by

$$S = 1 + i\mathcal{T}, \quad \langle f|\mathcal{T}|i\rangle = \delta(P_f - P_i)F(i \to f).$$

For $|i\rangle$ a state of two particles A, B with masses m_A, m_B the cross section is then

$$d\sigma(i \to f) = \frac{2\pi^2}{\lambda^{1/2}(s, m_A^2, m_B^2)}|F(i \to f)|^2\delta(P_i - P_f)\frac{d^3\mathbf{p}'_1}{2p'_{10}}\cdots\frac{d^3\mathbf{p}'_n}{2p'_{n0}}$$

where $p'_1, \ldots, p'_n$ are the momenta of the particles in the final state, $P_i = p_A + p_B$, $P_f = p'_1 + \cdots + p'_n$ and

$$\lambda(a, b, c) = a^2 + b^2 + c^2 - 2ab - 2ac - 2bc.$$

The differential cross-section in the centre of mass (c.m.) system of reference, at given solid angle $\cos\theta, \varphi$ for quasi-elastic scattering,

$$A + B \to A' + B',$$

with A', B' different or equal to A, B, is

$$\frac{d\sigma}{d\Omega}\bigg|_{c.m.} = \frac{\pi^2}{4s} \frac{\lambda^{1/2}(s, m_{A'}^2, m_{B'}^2)}{\lambda^{1/2}(s, m_A^2, m_B^2)} |F(i \to f)|^2,$$

$$d\Omega = d\cos\theta d\varphi,$$

and the moduli of initial and final three-momenta are

$$|\mathbf{p}_{1,cm}| = \frac{\lambda^{1/2}(s, m_A^2, m_B^2)}{2s^{1/2}}, \quad |\mathbf{p}'_{1,cm}| = \frac{\lambda^{1/2}(s, m_{A'}^2, m_{B'}^2)}{2s^{1/2}}.$$

For fixed momentum transfer, $t = (p_A - p'_{A'})^2$,

$$\frac{d\sigma}{dt} = \frac{\pi^3}{\lambda(s, m_A^2, m_B^2)} |F(i \to f)|^2.$$

Optical theorem:

$$\sigma_{total} = \frac{4\pi^2}{\lambda^{1/2}(s, m_A^2, m_B^2)} \operatorname{Im} F(i \to i).$$

For *decay rates*, we have the *differential decay rate* of a particle $|i\rangle = |p, \lambda\rangle$ into a final state[3] $|f\rangle = |p'_1, \lambda'; \ldots; p'_n, \lambda'_n\rangle$, in the rest system of the decaying particle:

$$d\Gamma(i \to f) = \frac{1}{4\pi m} |F(i \to f)|^2 \delta(m - P_f^0)\delta(\mathbf{P}_f) \frac{d^3\mathbf{p}'_1}{2p'_{10}} \cdots \frac{d^3\mathbf{p}'_n}{2p'_{n0}};$$

m is the mass of the decaying particle.

A useful and simple formula is that of the phase space integral for massless particles:

$$\int \frac{d^3\mathbf{p}_1}{2p_1^0} \cdots \int \frac{d^3\mathbf{p}_n}{2p_n^0} \delta(P - \sum p_i) = (P^2)^{n-2} \frac{(\pi/2)^{n-1}}{(n-1)!(n-2)!}.$$

Our units are such that $\hbar = c = 1$. Some useful formulas are the following:

1 MeV^{-1} = 1.973×10^{-11} cm = 6.582×10^{-22} s.

1 GeV^{-2} = 3.894×10^{-4} barn.

Classical electron radius: $r_e = \alpha/m_e = 2.817938 \times 10^{-15}$ m.

Rydberg (energy): Ry = $\frac{1}{2}m_e\alpha^2 = 13.6058$ eV.

1 J = 6.241×10^{18} eV.

1 eV = 1.60219×10^{-19} J = 2.418×10^{14} cycles s^{-1}.

1 fermi (or femtometer) = 10^{-15} m.

1 barn = 10^{-28} m^2.

Fine-structure constant: $\alpha^{-1} = 137.0360$.

---

[3] All the formulas are valid for distinguishable or indistinguishable particles. When calculating integrated rates, or cross sections, we have to divide by the number of redundant permutations. For example, if we integrate over the momenta of j gluons, then divide by $j!$.

Appendix H: Functional Derivatives

A functional F is an application of the space of sufficiently smooth functions, $\{f(x)\}$, into the complex numbers:

$$F : f \to F[f].$$

Note that F will in general not be linear. We treat functionals of several variables $F[f, g, \ldots]$ in the same way. A functional may be considered a generalization of an ordinary function in the following sense: divide the space of the x variable in N cells,[4] and let x_j lie one on each cell. Then $F[f]$ is the limit for vanishing cell size of the ordinary function of N variables $F_N(f_1, \ldots, f_j, \ldots, f_N)$, $f_j \equiv f(x_j)$. The derivative $\partial F_N / \partial f_j$ is obtained by shifting, $f_i \to f_i + \epsilon \delta_{ij}$:

$$\frac{\partial F_N(f_1, \ldots, f_N)}{\partial f_j} = \lim_{\epsilon \to 0} \frac{F_N(\ldots, f_i + \epsilon \delta_{ij}, \ldots) - F_N(f_1, \ldots, f_N)}{\epsilon},$$

so, in the continuum limit, we define the functional derivative

$$\frac{\delta F[f]}{\delta f(y)} = \lim_{\epsilon \to 0} \frac{F[f + \epsilon \delta_y] - F[f]}{\epsilon},$$

where δ_y is the delta function at y; $\delta_y(x) = \delta(x - y)$.

An important type of functional consists of those given by integrals:

$$F[f] = \int dx\, K_F(x) f(x).$$

In this case, an elementary evaluation yields the formula

$$\frac{\delta F[f]}{\delta f(y)} = K_F(y).$$

Taylor series may be generalized to functional series. If the kernel functions $K_n(x_1, \ldots, x_n)$ are symmetric (antisymmetric for fermionic f), the functional defined by

$$F[f] = \sum_{n=0}^{\infty} \frac{1}{n!} \int dx_1 \ldots dx_n\, K_n(x_1, \ldots, x_n) f(x_1) \ldots f(x_n)$$

is such that

$$K_n(x_1, \ldots, x_n) = \frac{\delta^n F[f]}{\delta f(x_n) \ldots \delta f(x_1)}.$$

---

[4] We assume here the space of finite size, L. Otherwise, an extra limit, $L \to \infty$, has to be taken.

Functional derivatives of expressions that do not involve integrals may often be evaluated by re-expressing them as integrals. For example, the derivative entering Eq. (2.5.9) in the main text is so evaluated:

$$\frac{\delta\left(\partial \cdot B_a(x)\right)}{\delta B_b^\rho(y)} = \frac{\delta}{\delta B_b^\rho(y)} \frac{\partial}{\partial x^\mu} \sum_c \int d^4z \, \delta(z-x)\delta_{ac} B_c^\mu(z) = \delta_{ab}\frac{\partial}{\partial x^\rho}\delta(x-y).$$

A concept related to that of functional derivatives is that of functional integration. We define

$$\int \mathcal{D}f \, F[f] \equiv \int \prod_x df(x) F[f] \equiv \lim_{N\to\infty} \int df_1 \dots df_N \, F_N(f_1, \dots, f_N).$$

Functional integration obeys rules similar to those of ordinary integration. Both for functional differentiation and integration, some modifications are necessary to accommodate anti-commuting numbers; they are described in Sect. 1.3.

Appendix I: Gauge Invariant Operator Product

It is intuitively obvious that, in a gauge theory, an expression such as those appearing in the OPE,

$$\bar{q}(0)q(x) = \sum \frac{x^{\mu_1} \dots x^{\mu_n}}{n!} \bar{q}(0)\partial_{\mu_1} \dots \partial_{\mu_n} q(0),$$

should be replaced by another with derivatives substituted by covariant derivatives, $\partial_\mu \to D_\mu$. Here we sketch a formal proof of how this comes about.

When the fields are interacting, the propagators are not free propagators. For example, for a fermion in the presence of the gluon field, the propagator satisfies the equation, derived directly from the Lagrangian,

$$(i\slashed{D} - m)S_{\text{int}}(x, y) = i\delta(x-y).$$

Retaining only the more singular (lower twist) terms, the solution to this is

$$S_{\text{int}}(x, y) \approx \left\{ \mathrm{P} \exp i \int_y^x dz_\mu \sum t^a B_a^\mu(z) \right\} S(x-y),$$

where S is the (ordinary) free propagator, and P indicates ordering along the path from y to x. If we now repeat the OPE taking this into account, we find that operator products $\bar{q}(x)q(y)$ are replaced by the gauge invariant combination

$$\bar{q}(x) \left\{ \mathrm{P} \exp i \int_y^x dz_\mu \sum t^a B_a^\mu(z) \right\} q(y),$$

whose expansion for $x \simeq y$ is precisely that with covariant derivatives. The same is, of course, true for operators built from gluon fields. Additional details may be found in the paper of Wilson (1975) and the review of Efremov and Radyushkin (1980b).

References

Abad, J., and Humpert, B. (1978). *Phys. Lett.* **B77**, 105.

Abarbanel, H. D., and Bronzan, J. B. (1974). *Phys. Rev.* **D8**, 2397.

Abarbanel, H. D., Goldberger, M. L., and Treiman, S. B. (1969). *Phys. Rev. Lett.* **22**, 500.

Abbott, L. F. (1981). *Nucl. Phys.* **B185**, 189.

Abbott, L. F., Atwood, W. B., and Barnett, R. M. (1980). *Phys. Rev.* **D22**, 582.

Abramowicz, M., and Stegun, I. E. (1965). *Handbook of Mathematical Functions*, Dover, New York.

Adel, K., Barreiro, F., and Ynduráin, F. J. (1997). *Nucl. Phys.* **B495**, 221.

Adel, K., and Ynduráin, F. J. (1995). *Phys. Rev.* **D52**, 6577.

Adler, S. L. (1966). *Phys. Rev.* **143**, 1144.

Adler, S. L. (1969). *Phys. Rev.* **177**, 2426.

Adler, S. L. (1971). In *Lectures in Elementary Particles and Quantum Field Theory* (Deser, Grisaru and Pendleton, eds.), MIT Press.

Adler, S. L. and Bardeen, W. A. (1969). *Phys. Rev.* **182**, 1517.

Aglietti, U., and Ligeti, Z. (1995). *Phys. Lett.* **B364**, 75.

Akhiezer, A., and Berestetskii, V. B. (1963). *Quantum Electrodynamics*, Wiley.

Akhoury, A., and Zakharov, V. I. (1996). *Nucl. Phys.* **B465**, 295.

Ali, A. (1981). *Phys. Lett.* **110B**, 67.

Ali, A., and Barreiro, F. (1982). *Phys. Lett.* **118B**, 155.

Ali, A., et al. (1980). *Nucl. Phys.* **110B**, 67.

Allton, C., et al. (1994). *Nucl. Phys.* **B431**, 667.

Alphonse X "The Wise" (1221 - 1284). Quoted in *The Harvest of the Quick Eye: a Selection of Scientific Quotations*, A. L. Mackay, The Institute of Physics, London, 1977.

Altarelli, G. (1982). *Phys. Rep.* **81C**, 1.

Altarelli, G. (1983). In Proc. 11th Winter Meeting on Fundamental Physics (Ferrando, ed.), I.E.N.

Altarelli, G. and Maiani, L. (1974). *Phys. Lett.* **52B**, 351.

Altarelli, G., and Parisi, G. (1977). *Nucl. Phys.* **B126**, 298.

Altarelli, G., Ellis, R. K., and Martinelli, G. (1978). *Nucl. Phys.* **B143**, 521, and (E) **B146**, 544.

Altarelli, G., Ellis, R. K., and Martinelli, G. (1979). *Nucl. Phys.* **B157**, 461.

Altarelli, G., Ellis, R. K., and Martinelli, G. (1985). *Phys. Lett.* **151B**, 457.

Altarelli, G., Parisi, G., and Petronzio, R. (1978). *Phys. Lett.* **76B**, 351 and 356.

Altarelli, G., Diemoz, M. Martinelli, G., and Nason, P. (1988). *Nucl. Phys.* **B308**, 724.

Altarelli, G., Ellis, R. K., Greco, M., and Martinelli, G. (1984). *Nucl. Phys.* **B246**, 12.

Altarelli, G., et al. (1982). *Nucl. Phys.* **B208**, 365.

Alvarez-Estrada, R. F., Fernández, F., Sánchez-Gómez, J. L., and Vento, V. (1986). *Models of Hadron Structure Based on Quantum Chromodynamics*, Springer, Berlin.

Amati, D., et al. (1980). *Nucl. Phys.* **B173**, 429.

Anaximander, (-546). Quoted in Aristotle's *Physics*. [From the German translation by W. Capelle, *Die Vorsokratiker*, Krönen-Verlag, Stuttgart, 1963.]

Anderson, B., Gustafson, G., and Peterson, C. (1977). *Phys. Lett.* **71B**, 337

Anderson, B., Gustafson, G., and Peterson, C. (1979). *Z. Phys.* **C1**, 105

Angelis, A. L. S., et al. (1979). *Phys. Lett.* **87B**, 398.

Aoki, S., et al. (1998). *Nucl. Phys. Proc. Suppl.* **60A**, 14.

Appelquist, T., and Carrazzone, J. (1975). *Phys. Rev.* **D11**, 2865.

Appelquist, T., and Georgi, H. (1973). *Phys. Rev.* **D8**, 4000.

Appelquist, T., and Politzer, H. D. (1975). *Phys. Rev. Lett.* **34**, 43.

Armstrong, T. A., et al. (1992). *Phys. Rev. Lett.* **69**, 2337.

Aubert, J. J., et al. (1981). *Phys. Lett.* **105B**, 315.

Badalian, A. M., and Yurov, V. P. (1990). *Phys. Rev.* **D42**, 3138.

Bali, G. S., et al. (1997). *Nucl. Phys. B (Proc. Suppl.)* **53**, 239.

Balitskii, Ya. Ya., and Lipatov, L. N. (1978). *Sov. J. Nucl. Phys.* **28**, 822.

Ball, R. D., and Forte, S. (1995). *Phys. Lett.* **358B**, 365.

Baluni, V. (1979). *Phys. Rev.* **D19**, 2227.

Barbieri, R., Curci, G., d'Emilio, E., and Remiddi, E. (1979). *Nucl. Phys.* **B154**, 535.

Barbieri, R., Ellis, J., Gaillard, M. K., and Ross, G. G. (1976). *Nucl. Phys.* **B117**, 50.

Barbieri, R., Gatto, R., Kögerler, R., and Kunszt, Z. (1975). *Phys. Lett.* **57B**, 455.

Bardeen, W. A. (1974). *Nucl. Phys.* **B75**, 246.

Bardeen, W. A., and Buras, A. J. (1979). *Phys. Rev.* **D20**, 166.

Bardeen, W. A., Fritzsch, H., and Gell-Mann, M. (1972). Report CERN-TH.1538.

Bardeen, W. A., Buras, A. J., Duke, D. W., and Muta, T. (1978). *Phys. Rev.* **D18**, 3998.

Barger, V. D., and Cline, D. B. (1969). *Phenomenological Theories of High Energy Scattering*, Benjamin, New York.

Barnett, R. M., Dine, M., and McLerran, L. (1980). *Phys. Rev.* **D22**, 594.

Barreiro, F. (1986). *Fortsch. Phys.* **34**, No.8, 503.

Basham, C. L., Brown, L. S., Ellis, S. D., and Love, S. T. (1978). *Phys. Rev. Lett.* **41**, 1585.

Baulieu, L., Ellis, J., Gaillard, M. K., and Zakrewski, W. J. (1979). *Phys. Lett.* **81B**, 224.

Becchi, C., Rouet, A., and Stora, R. (1974). *Phys. Lett.* **52B**, 344.

Becchi, C., Rouet, A., and Stora, R. (1975). *Commun. Math. Phys.* **42**, 127.

Becchi, C., Narison, S., de Rafael, E., and Ynduráin, F. J. (1981). *Z. Phys.* **C8**, 335.

Beenakker, W., et al. (1991). *Nucl. Phys.* **B351**, 507.

Belavin, A., Polyakov, A., Schwartz, A., and Tyupkin, Y. (1975). *Phys. Lett.* **59B**, 85.

Bell, J. S., and Jackiw, R. (1969). *Nuovo Cimento* **60A**, 47.

Beneke, M. (1998). *Phys. Lett.* **B434**, 115.

Beneke, M., Signer, A., and Smirnov, V. A. (1998). *Phys. Rev. Lett.* **80**, 2535.

Berge, P., et al. (1991). *Z. Phys.* **C49**, 187.

Bernard, C., et al. (1989). *Nucl. Phys. B (Proc. Suppl.)* **9**, 155.

Berstetskii, V. B., Lifshitz, E. M., and Pitaevskii, L. P. (1971). *Relativistic Quantum Theory*, Pergamon, London.

Bertmann, R. A., Dosch, H. G., and Krämer, A. (1989). *Phys. Lett.* **B223** (1989), 105.

Berzin, F. A. (1966). *Methods of Second Quantization*, Academic Press, New York.

Bethke, S. (1998). *Nucl. Phys. Proc. Suppl.* **B64**, 54.

Bhattacharya, T., and Gupta, R. (1998). *Nucl. Phys. B (Proc. Suppl.)* **63**, 95.

Bijnens, J. (1993). *Phys. Lett.* **B306**, 343.

Bijnens, J., Prades, J., and de Rafael, E. (1995). *Phys. Lett.* **348**, 226.

Billoire, A. (1980). *Phys. Lett.* **92B**, 343.

Bjorken, J. D. (1969). *Phys. Rev.* **179**, 1547.

Bjorken, J. D., and Drell, S. D. (1965). *Relativistic Quantum Fields*, McGraw-Hill, London.

Bjorken, J. D., and Paschos, E. A. (1969). *Phys. Rev.* **185**, 1975.

Blatt, J. M., and Weisskopf, V. F. (1952). *Theoretical Nuclear Physics*, Chapman.

Bochkarev, A., and de Forcrand, Ph. (1997). *Nucl. Phys. B (Proc. Suppl.)* **53**, 305.

Bodek, A., et al. (1979). *Phys. Rev.* **D20**, 1471.

Bogoliubov, N. N. (1967). *Ann. Inst. H. Poincaré* **8**, 163.

Bogoliubov, N. N., and Shirkov, D. V. (1959). *Introduction to the Theory of Quantized Fields*, Interscience, London.

Bogoliubov, N. N., Logunov, A. A., and Todorov, I. T. (1975). *Axiomatic Field Theory*, Benjamin, New York.

Bohr, N., and Rosenfeld, L. (1933). *K. Dansk. Vid. Selsk. Matt.-Fys. Medd.* **12**, No. 8.

Bohr, N., and Rosenfeld, L. (1950). *Phys. Rev.* **78**, 794.

Bollini, C. G., and Gianbiagi, J. J. (1972). *Phys. Lett.* **40B**, 566.

Bollini, C. G., Gianbiagi, J. J., and González-Domínguez, A. (1964). *Nuovo Cimento* **31**, 550.

Brambilla, N., Consoli, P., and Prosperi, G. M. (1994). *Phys. Rev.* **D50**, 5878.

Brandt, R., and Preparata, G. (1970). *Ann. Phys.* (N.Y.) **61**, 119.

Brandt, R., and Preparata, G. (1971). *Nucl. Phys.* **B24**, 541.

Bricmont, J., and Fröhlich, J. (1983). *Phys. Lett.* **122B**, 73.

Broadhurst, D. J. (1981). *Phys. Lett.* **B101**, 423.

Brodsky, S. J., and Farrar, G. (1973). *Phys. Rev. Lett.* **31**, 1153.

Brodsky, S. J., and Lepage, G. P. (1980). *Phys. Rev.* **D22**, 2157.

Brodsky, S. J., Frishman, Y., Lepage, G. P., and Sachrajda, C. T. (1973). *Phys. Lett.* **91B**, 239.

Brown, G. E., Rho, M., and Vento, V. (1979). *Phys. Lett.* **84B**, 383.

Buchmüller, W. (1982). *Phys. Lett.* **112B**, 479.

Buchmüller, W., Ng, Y. J., and Tye, S. H.-H. (1981). *Phys. Rev.* **D24**, 3003.

Buras, A. J. (1980). *Rev. Mod. Phys.* **52**, 199.

Buras, A. J. (1981). In *Topical Questions in QCD*, *Phys. Scripta* **23**, No. 5.

Buras, A. J., and Gaemers, K. J. F. (1978). *Nucl. Phys.* **B132**, 249.

Buskulic, D., et al. (1997). *Z. Phys.* **C73**, 409.

Cabibbo, N., Parisi, G., and Testa, M. (1970). *Nuovo Cimento Lett.* **4**, 35.
Camici, G., and Ciafaloni, M. (1998). *Phys. Lett.* **B430**, 349.
Calahan, C. G., Geer, K. A., Kogut, J., and Susskind, L. (1975). *Phys. Rev.* **D11**, 1199.
Callan, C. G. (1970). *Phys. Rev.* **D2**, 1541.
Callan, C. G., and Gross, D. J. (1969). *Phys. Rev. Lett.* **22**, 156.
Callan, C. G., Coleman, S., and Jackiw, R. (1970). *Ann. Phys.* (N.Y.) **59**, 42.
Callan, C. G., Dashen, R., and Gross, D. J. (1976). *Phys. Lett.* **66B**, 379.
Calvo, M. M. (1977). *Phys. Rev.* **D15**, 730.
Campostrini, M, Di Giacomo, A., and Olejnik, S. (1986). *Z. Phys.* **C31**, 577.
Carroll, L. (1986). *Through the Looking Glass.* [Reprinted in *The Complete Works of Lewis Carroll*, Random House, New York.]
Carroll, L. (1987(?)). *The Hunting of the Snark.* [Reprinted in *The Complete Works of Lewis Carroll*, Random House, New York.]
Caswell, W. E. (1974). *Phys. Rev. Lett.* **33**, 224.
Caswell, W. E., and Lepage, G. P. (1986). *Nucl. Phys.* **B167**, 437.
Caswell, W. E., and Wilczek, F. (1974). *Phys. Lett.* **49B**, 291.
Catani, S., and Trentadue, L. (1989). *Nucl. Phys.* **B327**, 323.
Catani, S., Fiorini, F., and Marchesini, G. (1990). *Nucl. Phys.* **B336**, 18.
Cerveró, J., Jacobs, L., and Nohl, C. (1977). *Phys. Lett.* **69B**, 351.
Chen, S. (1998). *Nucl. Phys. B (Proc. Suppl.)* **64**, 265.
Chetyrkin, K. G. (1997). *Phys. Lett.* **B404**, 161.
Chetyrkin K. G., Groshny, S. G., and Tkachov, F. V. (1982). *Phys. Lett.* **B119**, 407.
Chetyrkin, K. G., Kataev, A. L., and Tkachov, F. V. (1979). *Phys. Lett.* **85B**, 277.
Chetyrkin, K. G., Kuhn, J. H., and Kwiatowski, A. (1996). *Phys. Rep.* **C277**, 189.
Chetyrkin, K. G., Kuhn, J. H., and Steinhauser, M. (1997). *Nucl. Phys.* **B505**, 40.
Chetyrkin, K. G., Pirjol, D., and Schilcher, K. (1997). *Phys. Lett.* **B404**, 337.
Chetyrkin, K. G., Domínguez, C. A., Pirjol, D., and Schilcher, K. (1995). *Phys. Rev.* **D51**, 5090.
Chetyrkin, K. G., Harlander, R., Kuhn J.H., and Steinhauser, M. (1997). *Nucl. Phys.* **B503**, 339.
Chodos, A., Jaffe, R. L., Johnson, K., and Thorn, C. B. (1974). *Phys. Rev.* **D10**, 2599.
Chodos, A., et al. (1974). *Phys. Rev.* **D9**, 3471.
Christ, N., and Lee, T. D. (1980). *Phys. Rev.* **D22**, 939.
Christ, N., Hasslacher, B., and Muller, A. (1972). *Phys. Rev.* **D6**, 3543.
Chung, Y., Dosch, H. G., Kremer, M., and Schall, D. (1984). *Z. Phys.* **C25**, 151.
Chyla, J., and Kataev, A. L. (1992). *Phys. Lett.* **B297**, 385.
Ciafaloni, M. (1988). *Nucl. Phys.* **B296**, 49.
Ciafaloni, M., and Curci, G. (1981). *Phys. Lett.* **102B**, 352.
Close, F., and Dalitz, R. H. (1981). In *Low and Intermediate Energy Kaon–Nucleon Physics* (Ferrari and Violini, eds.), Reidel, Dordrecht.
Coleman, S. (1966). *J. Math. Phys.* **7**, 787.
Coleman, S., and Gross, D. J. (1973). *Phys. Rev. Lett.* **31**, 851.
Collins, J. C., and Soper, D. E. (1982). *Nucl. Phys.* **B197**, 446.

Collins, J. C., Duncan, A., and Joglekar, S. D. (1977). *Phys. Rev.* **D16**, 438.

Combridge, B. L. (1979). *Nucl. Phys.* **B151**, 429.

Combridge, B. L., Kripfganz, J., and Ranft, J. (1978). *Phys. Lett.* **70B**, 234.

Cooper-Sarkar, A. M., Devenish, R. C. E., and De Roeck, Λ. (1998). *J. Mod. Phys.* **A13**, 3385 .

Coquereaux, R. (1980). *Ann. Phys.* (N.Y.) **125**, 401.

Coquereaux, R. (1981). *Phys. Rev.* **D23**, 1365.

Cornwall, J. M., and Norton, R. E. (1969). *Phys. Rev.* **177**, 2584.

Corrigan, E., and Fairlie, D. B. (1977). *Phys. Lett.* 67B, 69.

Creutz, M. (1980). *Phys. Rev. Lett.* **45**, 313.

Creutz, M. (1983). *Quarks, Gluons and Lattices*, Cambridge University Press, Cambridge.

Creutz, M., and Moriarty, K. J. M. (1982). *Phys. Rev.* **D26**, 2166.

Crewther, R. J. (1972). *Phys. Rev. Lett.* **28**, 1421.

Crewther, R. J. (1979a). *Riv. Nuovo Cimento* **2**, No. 7.

Crewther, R. J. (1979b). In *Field Theoretical Methods in Elementary Particle Physics*, Proc. Kaiserslautern School.

Crewther, R. J., Di Vecchia, P., Veneziano, G., and Witten, E. (1980). *Phys. Lett.* **88B**, 123 and (E), **91B**, 487.

Curci, G., and Greco, M. (1980). *Phys. Lett* **92B**, 175.

Curci, G., Furmanski, W., and Petronzio, R. (1980). *Nucl. Phys.* **B175**, 27.

Curci, G., Greco, M., and Srivastava, Y. (1979). *Nucl. Phys.* **B159**, 451.

Cutler, R., and Sivers, D. J. (1977). *Phys. Rev.* **D16**, 679.

Cvitanović, P. (1976). *Phys. Rev.* **D14**, 1536.

Danckaert, D., et al. (1982). *Phys. Lett.* **114B**, 203.

Dashen, R., and Gross, D. J. (1981). *Phys. Rev.* **D23**, 2340.

Davies, C. T. H., and Stirling, W. J. (1984). *Nucl Phys.* **B244**, 337.

Dawson, S., Ellis, R. K., and Nason, P. (1989). *Nucl. Phys.* **B327**, 49, and (E) **B335**, 260 (1990).

De Alfaro, V., Fubini, S., and Furlan, G. (1976). *Phys. Lett.* **65B**, 163.

De Alfaro, V., Fubini, S., and Furlan, G. (1977). *Phys. Lett.* **72B**, 203.

De Grand, T. A., and Loft, R. D. (1988). *Phys. Rev.* **D38**, 954.

De Groot, J. G., et al. (1979). *Z. Phys.* **C1**, 143.

De Rújula, and Glashow, S. L. (1975). *Phys. Rev. Lett.* **34**, 46.

De Rújula, A., Georgi, H., and Glashow, S. L. (1975). *Phys. Rev.* **D12**, 147.

De Rújula, A., Georgi, H., and Politzer, H. D. (1977a). *Ann. Phys.* (N.Y.) **103**, 315.

De Rújula, A., Georgi, H., and Politzer, H. D. (1977b). *Phys. Rev.* **D15**, 2495.

De Rújula, A., Ellis, J. Floratos, E. G., and Gaillard, M. K. (1978). *Nucl. Phys.* **B138**, 387.

De Rújula, et al. (1974). *Phys. Rev.* **D10**, 1649.

DeWitt, B. (1964). *Relativity, Groups and Topology*, p. 587 ff., Blackie & Son, London.

DeWitt, B. (1967). *Phys. Rev.* **162**, 1195 and 1239.

Di Giacomo, A., and Rossi, G. C. (1981). *Phys. Lett.* **100B**, 481.

Dine, M., and Sapiristein, J. (1979). *Phys. Rev. Lett.* **43**, 668.

Dixon, J. A., and Taylor, J. C. (1974). *Nucl. Phys.* **B78**, 552.

Dokshitzer, Yu. L. (1977). *Sov. Phys. JETP* **46**, 641.

Dokshitzer, Yu. L., Dyakonov, D. I., and Troyan, S. I. (1980). *Phys. Rep.* **C58**, 269.

Domínguez, C. A. (1978). *Phys. Rev. Lett.* **41**, 605.

Domínguez, C. A., and de Rafael, E. (1987). *Ann. Phys.* (N.Y.) **174**, 372.

Donoghue, J. F., Holstein, B. R., and Wyler, D. (1993). *Phys. Rev.* **D47**, 2089.

Dosch, H. G. (1987). *Phys. Lett.* **B190**, 177.

Dosch, H. G., and Simonov, Yu. A. (1988). *Phys. Lett.* **B205**, 339.

Doyle, A. C. (1892). "Silver Blaze", Strand Magazine, London. [Reprinted in *The Memoirs of Sherlock Holmes*, Penguin Books, 1970.]

Drell, S. D., and Yan, T. M. (1971). *Ann. Phys.* (N.Y.) **66**, 578.

Dubin, A. Yu., Kaidalov, A. B., and Simonov, Yu. A. (1994). *Phys. Lett.* **B323**, 41.

Duncan, A. (1981). In *Topical Questions in QCD, Phys. Scripta*, **23**, No. 5.

Duncan, A., and Muller, A. (1980a). *Phys. Lett.* **93B**, 119.

Duncan, A., and Muller, A. (1980b). *Phys. Rev.* **D21**, 1636.

Ecker, G. (1995). *Prog. Part. Nucl. Phys.* **35**, 71.

Eden, R. J., Landshoff, P. V., Olive, D. I., and Polkinghorne, J. C. (1966). *The Analytic S-Matrix*, Cambridge University Press, Cambridge.

Efremov, A. V., and Radyushin, A. V. (1980a). *Phys. Lett.* **94B**, 245.

Efremov, A. V., and Radyushin, A. V. (1980b). *Riv. Nouvo Cimento* **3**, No. 2.

Egorian, E. S., and Tarasov, O. V. (1979). *Theor. Mat. Phys.* **41**, 863.

Eichten, E., and Feinberg, F. (1981). *Phys. Rev.* **D23**, 2724.

Ellis, J. (1976). In *Weak and Electromagnetic Interactions at High Energy*, North Holland, Amsterdam.

Ellis, J. and Gaillard, M. K. (1979). *Nucl. Phys.* **B150**, 141.

Ellis, R. K., Ross, D. A., and Terrano, A. E. (1981). *Nucl. Phys.* **B178**, 421.

Ellis, S. D., Richards, D., and Stirling, W. J. (1982). *Phys. Lett.* **119B**, 193.

Epstein, H., Glaser, V., and Martin, A. (1969). *Commun. Math. Phys.* **13**, 257.

Espriu, D., and Ynduráin, F. J. (1983). *Phys. Lett.* **132B**, 187.

Espriu, D., Pascual, P., and Tarrach, R. (1983). *Nucl. Phys.* **B214**, 285.

Fadeyev, L. D. (1976). In *Methods in Field Theory* (Balian and Zinn-Justin, eds.) North Holland, Amsterdam.

Fadeyev, L. D., and Popov, Y. N. (1967). *Phys. Lett.* **25B**, 29.

Fadeyev, L. D., and Slavnov, A. A. (1980). *Gauge Fields*, Benjamin, New York.

Fadin, V. S., and Lipatov, L. N. (1998). *Phys. Lett.* **B429**, 127.

Fadin, V. S., Khoze, V., and Sjöstrand, T. (1990). *Z. Phys.* **C48**, 613.

Farhi, E. (1977). *Phys. Rev. Lett.* **39**, 1587.

Farrar, G., and Jackson, D. R. (1979). *Phys. Rev. Lett.* **43**, 246.

Fermi, E. (1934). *Z. Phys.* **88**, 161.

Ferrara, S., Gatto, R., and Grillo, A. F. (1972). *Phys. Rev.* **D5**, 5102.

Feynman, R. P. (1963). *Acta Phys. Polonica* **24**, 697.

Feynman, R. P. (1969). *Phys. Rev. Lett.* **23**, 1415.

Feynman, R. P. (1972). *Photon Hadron Interactions*, Benjamin, New York.

Feynman, R. P., and Field, R. D. (1977). *Phys. Rev.* **D15**, 2590.

Feynman, R. P., and Hibbs, A. R. (1965). *Quantum Mechanics and Path Integrals*, McGraw-Hill, New York.

Fischler, W. (1977). *Nucl. Phys.* **B129**, 157.

Flamm, D., and Schoberl, F. (1981). *Introduction to the Quark Model of Elementary Particles*, Gordon and Breach, New York.

Floratos, E. G., Ross, D. A., and Sachrajda, C. T. (1977). *Nucl. Phys.* **B129**, 66; and (E) **B139**, 545 (1978).

Floratos, E. G., Ross, D. A., and Sachrajda, C. T. (1979). *Nucl. Phys.* **B152**, 493.

Forgács, P., Horváth, Z., and Palla, L. (1981). *Phys. Rev. Lett.* **46**, 392.

Fritzsch, H., and Gell-Mann, M. (1971). In *Broken Scale Invariance and the Light Cone* (Dal Cin, Iverson and Perlmutter, eds.), Gordon and Breach, New York.

Fritzsch, H., and Gell-Mann, M. (1972). Proc. XVI Int. Conf. on High Energy Physics, Vol. 2, p.135, Chicago.

Fritzsch, H., Gell-Mann, M., and Leutwyler, H. (1973). *Phys. Lett.* **B47**, 365.

Fujikawa, K. (1980). *Phys. Rev.* **D21**, 2848 and (E) **D22**, 1499.

Fujikawa, K. (1984). *Phys. Rev.* **D29**, 285.

Fujikawa, K. (1985). *Phys. Rev.* **D31**, 341.

Fujikawa, K., Kaneko, T., and Ukawa, A. (1983). *Phys. Rev.* **D28**, 2696.

Furmanski, W., and Petronzio, R. (1980). *Phys. Lett.* **97B**, 437.

Gaemers, K. J. F., Oldham, S. J., and Vermaseren, J. A. M. (1981). *Nucl. Phys.* **B187**, 301.

Gaillard, M. K. (1978). In Proc. SLAC Summer Institute on Particle Physics.

Gaillard, M. K., and Lee, B. W. (1974a). *Phys. Rev. Lett.* **33**, 108.

Gaillard, M. K., and Lee, B. W. (1974b). *Phys. Rev.* **D10**, 897.

Gasser, J., and Leutwyler, H. (1982). *Phys. Rep.* **C87**, 77.

Gasser, J., and Leutwyler, H. (1984). *Ann. Phys.* (N.Y.) **158**, 142.

Gastmans, R., and Meuldermans, M. (1973). *Nucl. Phys.* **B63**, 277.

Gavela, M. B., et al. (1988). *Nucl. Phys.* **B306**, 677.

Gel'fand (Guelfand), I. M., and Shilov (Chilov), G. E. (1962). *Les Distributions*, Vol. I, Dunod, Paris.

Gell-Mann, M. (1961). Caltech preprint CTSL-20, unpublished.

Gell-Mann, M. (1962). *Phys. Rev.* **125**, 1067.

Gell-Mann, M. (1964a). *Phys. Lett.* **8**, 214.

Gell-Mann, M. (1964b). *Physics* **1**, 63.

Gell-Mann, M., and Low, F. (1954). *Phys. Rev.* **95**, 1300.

Gell-Mann, M., Oakes, R. L., and Renner, B. (1968). *Phys. Rev.* **175**, 2195.

Generalis, S. C. (1990). *J. Phys.* **G16**, 785.

Georgi, H. (1984). *Weak Interactions and Modern Particle Theory*, Benjamin, New York.

Georgi, H., and Politzer, H. D. (1974). *Phys. Rev.* **D9**, 416.

Georgi, H., and Politzer, H. D. (1976). *Phys. Rev.* **D14**, 1829.

Glashow, S. L., and Weinberg, S. (1968). *Phys. Rev. Lett.* **20**, 224.

Glashow, S. L. (1968). In *Hadrons and their Interactions*, p. 83, Academic Press, New York.

Glashow, S. L., Iliopoulos, J., and Maiani, L. (1970). *Phys. Rev.* **D2**, 1285.

Glück, M., Owens, J. F., and Reya, E. (1978). *Phys. Rev.* **D18**, 1501

Goldstone, J. (1961). *Nuovo Cimento* **19**, 154.

González-Arroyo, A., and Korthals Altes, C. P. (1982). *Nucl. Phys.* **B205**, 46.

González-Arroyo, A., and López, C. (1980). *Nucl. Phys.* **B166**, 429.

González-Arroyo, A., López, C., and Ynduráin, F. J. (1979). *Nucl. Phys.* **B153**, 161 and **B159**, 512.

González-Arroyo, A., Martinelli, G., and Ynduráin, F. J. (1982). *Phys. Lett.* **117B**, 437 and (E) **122B**, 486 (1983).

Gorishny, S. G., Kataev, A. L., and Larin, S. A. (1991). *Phys. Lett.* **B259**, 144.

Gorishny, S. G., Kataev, A. L., Larin, S. A., and Sugurladze, L. R. (1991). *Phys. Rev.* **D43**, 1633.

Gray, N., Broadhurst, D. J., Grafe, W., and Schilcher, K. (1990). *Z. Phys.* **C48**, 673.

Greenberg, O. W. (1964). *Phys. Rev. Lett.* **13**, 598.

Greiner, W., Müller, B., and Rafelski, J. (1985). *Quantum Mechanics of Strong Fields*, Springer, Berlin.

Gribov, V. N., and Lipatov, L. N. (1972). *Sov. J. Nucl. Phys.* **15**, 438 and 675.

Grinstein, B. (1991). In *Proc. Workshop on High Energy Phenomenology* (Pérez and Huerta, eds.), World Scientific, Singapore.

Gross, D. J. (1974). *Phys. Rev. Lett.* **32**, 1071.

Gross, D. J. (1976). In *Methods in Field Theory* (Balian and Zinn-Justin, eds.) , North Holland, Amsterdam.

Gross, D. J., and Llewellyn Smith, C. H. (1969). *Nucl. Phys.* **B14**, 337.

Gross, D. J., and Wilczek, F. (1973a). *Phys. Rev. Lett.* **30**, 1323.

Gross, D. J., and Wilczek, F. (1973b). *Phys. Rev.* **D8**, 3635.

Gross, D. J., and Wilczek, F. (1974). *Phys. Rev.* **D9**, 980.

Gross, D. J., Pisarski, R. D., and Yaffe, L. G. (1981). *Rev. Mod. Phys.* **53**, 43.

Gupta, S. N., and Radford, S. F. (1981). *Phys. Rev.* **D24**, 2309 and **D25**, 3430 (1982).

Gupta, S. N., Radford, S. F., and Repko, W. W. (1982). *Phys. Rev.* **D26**, 3305.

Gürsey, F., and Radicati, L. A. (1964). *Phys. Rev. Lett.* **13**, 173.

Halliday, I. G. (1983). In Proc. Europhysics Conf. High Energy Physics (Guy and Constantin, eds.), Brighton.

Halzen, F., Olson, C., Olsson, M. G., and Stong, M. L. (1993). *Phys. Rev.* **D47**, 3013.

Han, M., and Nambu, Y. (1965). *Phys. Rev.* **139**, 1006.

Harada, K. and Muta, T. (1980). *Phys. Rev.* **D22**, 663.

Hasenfratz, A., and Hasenfratz, P. (1980). *Phys. Lett.* **93B**, 165.

Hasenfratz, P., and Kuti, J. (1978). *Phys. Rep.* **40C**, 75.

Hey, A. J. G., and Kelly, R. L. (1983). *Phys. Rep.* **96C**, 72.

Hinchliffe, I., and Llewellyn Smith, C. H. (1977). *Nucl. Phys.* **B128**, 93.

Hoang, A. H., Smith, M. C., Stelzer, T., and Willenbrok, S. (1998). UCSD/PTH 98-13 (hep-ph/9804227).

Hubschmid, W., and Mallik, S. (1981). *Nucl. Phys.* **B193**, 368.

Humpert, B., and van Neerven, W. L. (1981). *Nucl. Phys.* **B184**, 225.

Iofa, M. Z., and Tyutin, I. V. (1976). *Theor. Math. Phys.* **27**, 316.

Ioffe, B. L. (1981). *Nucl. Phys.* **B188**, 317.

Isgur, N., and Karl, G. (1979). *Phys. Rev.* **D20**, 1191 and **D21**, 3175.

Isgur, N., and Llewellyn Smith, C. H. (1989). *Nucl. Phys.* **B317**, 526.

Ito, A. S., et al. (1981). *Phys. Rev.* **D23**, 604.

Itzykson, C., and Zuber, J. B. (1980). *Quantum Field Theory*, McGraw-Hill, New York.

Izuka, J., Okada, K., and Shito, D. (1966). *Progr. Theor. Phys.* **35**, 1061.

Jackiw, R., Nohl, C., and Rebbi, C. (1977). *Phys. Rev.* **D15**, 1642.

Jacob, M., and Landshoff, P. (1978). *Phys. Rep.* **C48**, 285.

Jaffe, R., and Ross, G. G. (1980). *Phys. Lett.* **93B**, 313.

Jaffe, R., and Soldate, M. (1981). *Phys. Lett.* **105B**, 467.

Jamin, M. and Müntz, M. (1995). *Z. Phys.* **C66**, 633.

Jamin, M., and Pich, A. (1997). *Nucl. Phys.* **B507**, 334.

Johnson, K. (1975). *Acta Phys. Polonica* **B6**, 865.

Jones, D. T. R. (1974). *Nucl. Phys.* **B75**, 730.

Jost, R., and Luttinger, J. (1950). *Helvetica Phys. Acta* **23**, 201.

Kaplan, D. B., and Manohar, A. V. (1986). *Phys. Rev. Lett.* **56**, 2004.

Karsten, L. H., and Smit, J. (1981). *Nucl. Phys.* **B183**, 103.

Kataev, A. L., Krasnikov, N. V., and Pivovarov, A. A. (1983). *Phys. Lett.* **123B**, 93.

Kawai, H., Nakayama, R., and Seo, K. (1981). *Nucl. Phys.* **B189**, 40.

Khriplovich, I. B. (1969). *Yad. Fiz.* **10**, 409.

Kingsley, R. L. (1973). *Nucl. Phys.* **B60**, 45.

Kinoshita, T. (1962). *J. Math. Phys.* **3**, 650.

Kluberg-Stern, H., and Zuber, J. B. (1975). *Phys. Rev.* **D12**, 467.

Kobel, M. et al. (1992). *Z. Phys.* **C53**, 193.

Kodaira, J., and Trentadue, L. (1983). *Phys. Lett.* **123B**, 335.

Korthals Altes, C. P., and de Rafael, E. (1977). *Nucl. Phys.* **B125**, 275.

Kourkoumelis, C., et al. (1980). *Phys. Lett.* **91B**, 475.

Krasnikov, N. V., and Pivovarov, A. K. (1982). *Phys. Lett.* **116B**, 168.

Kronfeld, A. S. (1989). *Nucl. Phys. B (Proc. Suppl.)* **9**, 227.

Kubar-André, J., and Paige, F. E. (1979). *Phys. Rev.* **D19**, 221.

Kummer, W. (1975). *Acta Phys. Austriaca* **41**, 315.

Kuraev, E. A., Lipatov, L. N., and Fadin, V. S. (1976). *Sov. Phys. JETP* **44**, 443.

Laenen, E., Smith, J., and van Neerven, W. L. (1992). *Nucl. Phys.* **B369**, 543.

Laenen, E., Riemersma, S., Smith, J., and van Neerven, W. L. (1993). *Nucl. Phys.* **B392**, 162 and 229.

Laenen, E., Riemersma, S., Smith, J., and van Neerven, W. L. (1994). *Phys. Rev.* **D49**, 5753.

Landau, L. D., and Lifshitz, E. M. (1958). *Quantum Mechanics*, Pergamon, London.

Larin, S. A., and Vermaseren, J. A. M. (1991). *Phys. Lett.* **B259**, 345.

Larin, S. A., and Vermaseren, J. A. M. (1993). *Z. Phys. C* **57**, 93.

Larin, S. A., van Ritbergen, T., and Vermaseren, J. A. M. (1997a). *Phys. Lett.* **B400**, 379.

Larin, S. A., van Ritbergen, T., and Vermaseren, J. A. M. (1997b). *Phys. Lett.* **B405**, 327.

Larin, S. A., Nogueira, P., van Ritbergen, T., and Vermaseren, J. A. M. (1997). *Nucl. Phys.* **B492**, 338.

Launer, G., Narison, S., and Tarrach, R. (1984). *Z. Phys.* **C26**, 433.
Lautrup, B. (1977). *Phys. Lett.* **B69**, 109.
Le Diberder, F., and Pich, A. (1992). *Phys. Lett.* **289B**, 165.
Lee, B. W. (1976). In *Methods in Field Theory* (Balian and Zinn-Justin, eds.), North Holland, Amsterdam.
Lee, B. W., and Zinn-Justin, J. (1972). *Phys. Rev.* **D5**, 3121.
Lee, T. D., and Nauenberg, M. (1964). *Phys. Rev.* **133**, B1549.
Lepage, G. P., and Thacker, B. A. (1988). *Nucl. Phys. Proc. Suppl.* **4**, 199.
Leutwyler, H. (1974). *Nucl. Phys.* **B76**, 413.
Leutwyler, H. (1981). *Phys. Lett.* **98B**, 447.
Lipatov, L. N. (1975). *Sov. J. Nucl. Phys.* **20**, 94.
Lipatov, L. N. (1976). *Sov. J. Nucl. Phys.* **23**, 338.
Llewellyn Smith, C. H. (1972). *Phys. Rep.* **C3**, 261.
López, C., and Ynduráin, F. J. (1981). *Nucl. Phys.* **B187**, 157.
Luo, X.-Q., et al. (1997). *Nucl. Phys. B (Proc. Suppl.)* **53**, 243.

Mackenzie, P. B., and Lepage, G. P. (1981). *Phys. Rev. Lett.* **47**, 1244.
Mandelstam, S. (1983). *Nucl. Phys.* **B213**, 149.
Marchesini, G., and Webber, B. R. (1984). *Nucl. Phys.* **B238**, 1.
Marciano, W. J. (1984). *Phys. Rev.* **D29**, 580.
Marciano, W. J., and Sirlin, A. (1988). *Phys. Rev. Lett.* **61**, 1815.
Marshak, R. E., Riazzuddin, and Ryan, C. P. (1969). *Theory of Weak Interactions in Particle Physics*, Wiley, New York.
Marshall, R. (1989). *Z. Phys.* **C43**, 595.
Martin, F. (1979). *Phys. Rev.* **D19**, 1382.
Martinelli, G., and Sachrajda, C. T. (1996). *Nucl. Phys.* **B478**, 660.
Méndez, A. (1978). *Nucl. Phys.* **B145**, 199.
Migdal, A. A., Polyakov, A. M., and Ter-Martirosian, K. A. (1974). *Phys. Lett.* **B48**, 239.
Moriarty, K. J. M. (1983). In Proc. Europhysics Conf. High Energy Physics (Guy and Constantin, eds.), Brighton.
Moshe, M. (1978). *Physics Rep.* **37**, 255.
Muller, A. H. (1978). *Phys. Rev.* **D18**, 3705.
Muller, A. H. (1985). *Nucl. Phys.* **B250**, 327.

Nachtmann, O. (1973). *Nucl. Phys.* **B63**, 237.
Nambu, Y. (1960). *Phys. Rev. Lett.* **4**, 380.
Nambu, Y., and Jona-Lasinio, G. (1961a). *Phys. Rev.* **122**, 345.
Nambu, Y., and Jona-Lasinio, G. (1961b). *Phys. Rev.* **124**, 246.
Nanopoulos, D. V. (1973). *Nuovo Cimento Lett.* **8**, 873.
Nanopoulos, D. V., and Ross, D. (1979). *Nucl. Phys.* **B157**, 273.
Narison, S. (1989). *QCD Spectral Sum Rules*, World Scientific, Singapore.
Narison, S. (1995a). *Acta Phys. Pol.* **B26**, 687.
Narison, S. (1995b). *Phys. Lett.* **B358**, 112.
Narison, S. (1997). *Nucl. Phys. B (Proc. Suppl.)* **54**, 238.
Narison, S., and de Rafael, E. (1981). *Phys. Lett.* **103B**, 57.
Ne'eman, Y. (1961). *Nucl. Phys.* **26**, 222.
Nielsen, H. B., and Ninomiya, M. (1981). *Nucl. Phys.* **B185**, 20 and **B193**, 173.

Novikov, V. A., et al. (1978). *Phys. Rep.* **C41**, 1.

Okubo, S. (1963). *Phys. Lett.* **5**, 165.
Okubo, S. (1969). *Phys. Rev.* **188**, 2295 and 2300.

Pagels, H. (1975). *Phys. Rep.*, **C16**, 219.
Pagels, H., and Zepeda, A. (1972). *Phys. Rev.* **D5**, 3262.
Pais, A. (1964). *Phys. Rev. Lett.* **13**, 175.
Pantaleone, J., Tye, S.-H. H., and Ng, Y. J. (1986). *Phys. Rev.* **D33**, 777.
Parisi, G. (1980). *Phys. Lett.* **90B**, 295.
Parisi, G., and Petronzio, R. (1979). *Nucl. Phys.* **B154**, 427.
Particle Data Group (1996). *Phys. Rev.* **D54**, 1.
Pascual, P., and de Rafael, E. (1982). *Z. Phys.* **C12**, 127.
Pascual, P., and Tarrach, R. (1984). *QCD: Renormalization for the Practitioner*, Springer, Berlin.
Peccei, R. D., and Quinn, H. R. (1977). *Phys. Rev.* **D16**, 1751.
Pennington, M. R., Roberts, R. G., and Ross, G G. (1984). *Nucl. Phys.* **B242**, 69.
Peris, S., and. de Rafael, E. (1997). *Nucl. Phys.* **B500**, 325.
Peter, M. (1997). *Phys. Rev. Lett.* **78**, 602.
Pich, A. (1995). *Rep. Progr. Phys.* **58**, 563.
Pich, A. (1997). *Tau Physics*, In *Heavy Flavours, II* (Buras and Lidner, eds.) World Scientific, Singapore.
Pich, A., and de Rafael, E. (1991). *Nucl. Phys.* **B358**, 311.
Pineda, A. (1997a). *Phys. Rev.* **D55**, 407.
Pineda, A. (1997b). *Nucl. Phys.* **B494**, 213.
Pineda, A., and Ynduráin, F. J. (1998). *Phys. Rev.* **D58**, 094022, and CERN–TH/98-402 (hep-ph/9812371).
Politzer, H. D. (1973). *Phys. Rev. Lett.* **30**, 1346.
Polyakov, A. M. (1977). *Nucl. Phys.* **B120**, 429.

de Rafael, E. (1977). *Lectures on Quantum Electrodynamics*, Universidad Autónoma de Barcelona, UAB-FT-D1.
de Rafael, E. (1979). In *Quantum Chromodynamics* (Alonso and Tarrach, eds.), Springer, Berlin.
de Rafael, E. (1995). *Chiral Lagrangians and Kaon CP-Violation*, in *CP Violation and the Limits of the Standard Model*, Proc. TASI'94 (J. F. Donoghue, ed.), World Scientific, Singapore.
Reinders, L., Rubinstein, H., and Yazaki, S. (1981). *Nucl. Phys.* **B186**, 109.
Richardson, L. (1979). *Phys. Lett.* **B82**, 272.
Rodrigo, G., Santamaría, A., and Bilenkii, M. (1997). *Phys. Rev. Lett.* **79**, 193.
Rose, M. E. (1961). *Relativistic Electron Theory*, Wiley.

Sachrajda, C. T. (1978). *Phys. Lett.* **76B**, 100.
Sachrajda, C. T. (1979). In *Quantum Chromodynamics* (Alonso and Tarrach, eds.), Springer, Berlin.
Sánchez-Guillén, J., et al. (1991). *Nucl. Phys.* **B353**, 337.
Schierholz, G. (1985). CERN-TH-4139/95, Proc. Scottish Summer School 1984.

Schiff, L. I. (1968). *Quantum Mechanics*, McGraw-Hill, New York.

Schröder, Y. (1998). DESY 98-191 (hep-ph/9812205).

Sciuto, S. (1979). *Riv. Nuovo Cimento* **2**, No. 8.

Seiberg, N., and E. Witten, E. (1994). *Nucl.Phys.* **B426**, 15, and (E) **B430**, 485 and **B431**, 484.

Shifman, M. A., Vainshtein, A. I., and Zakharov, V. I. (1977a). *Nucl. Phys.* **B120**, 316.

Shifman, M. A., Vainshtein, A. I., and Zakharov, V. I. (1977b). *JETP* **45**, 670.

Shifman, M. A., Vainshtein, A. I., and Zakharov, V. I. (1979a). *Nucl. Phys.* **B147**, 385.

Shifman, M. A., Vainshtein, A. I., and Zakharov, V. I. (1979b). *Nucl. Phys.* **B147**, 448.

Shifman, M. A., Vainshtein, A. I., and Zakharov, V. I. (1983). *Nucl. Phys.* **B229**, 381.

Siegel, W. (1979). *Phys. Lett.* **84B**, 193.

Simonov, Yu. A. (1988). *Nucl. Phys.* **B307**, 512.

Simonov, Yu. A. (1989a). *Phys. Lett.* **B226**, 151; ibid. **B228**, 413.

Simonov, Yu. A. (1989b). *Nucl. Phys.* **B324**, 67.

Simonov, Yu. A., Titard, S., and Ynduráin, F. J. (1995). *Phys. Lett.* **B354**, 435.

Slavnov, A. A. (1975). *Sov. J. Particles Nuclei* **5**, 303.

Smith, K. F., et al. (1990). *Phys. Lett.* **234B**, 191.

Söding, P. (1983). In Proc. Europhysics Conf. High Energy Physics (Guy and Constantin, eds.), Brighton.

Speer, E. R. (1968). *J. Math. Phys.* **9**, 1404.

Speer, E. R. (1975). In *Renormalization Theory* (Velo and Wightman, eds.), Reidel.

Steinberger, J. (1949). *Phys. Rev.* **76**, 1180.

Sterman, G. (1987). *Nucl. Phys.* **B281**, 310.

Sterman, G., and Libby, S. (1978). *Phys. Rev.* **D18**, 3252 and 4737.

Sterman, G., and Weinberg, S. (1977). *Phys. Rev. Lett.* **39**,1436.

Stückelberg, E. C. G., and Peterman, A. (1953). *Helvetica Phys. Acta* **26**, 499.

Sudakov, V. V. (1956). *Sov. Phys. JETP* **30**, 87.

Sugurladze, L. R., and Samuel, M. A. (1991). *Phys. Rev. Lett.* **66**, 560.

Sugurladze, L. R., and Tkachov, F. V. (1990). *Nucl. Phys.* **B331**, 35.

Sutherland, D. G., (1967). *Nucl. Phys.* **B2**, 433.

Symanzik, K. (1970). *Commun. Math. Phys.* **18**, 227.

Symanzik, K. (1973). *Commun. Math. Phys.* **34**, 7.

Tarasov, O. V. (1982). Dubna Report JINR P2-82-900 (unpublished).

Tarasov, O. V., Vladimirov, A. A., and Zharkov, A. Yu. (1980). *Phys. Lett.* **B93**, 429.

Tarrach, R. (1981). *Nucl. Phys.* **B183**, 384.

Tarrach, R. (1982). *Nucl. Phys.* **B196**, 45.

Taylor, J. C. (1971). *Nucl. Phys.* **B33**, 436.

Taylor, J. C. (1976). *Gauge Theories of Weak Interactions*, Cambridge.

Terentiev, M. V., and Vanyashin, V. S. (1965). *Zh. Eksp. Teor. Fiz.* **48**, 565.

Thirring, W. (1950). *Phil. Mag.* **41**, 113.

't Hooft, G. (1971). *Nucl. Phys.* **B33**, 173.

't Hooft, G. (1973). *Nucl. Phys.* **B61**, 455.

't Hooft, G. (1974a). *Nucl. Phys.* **B72**, 461.

't Hooft, G. (1974b). *Nucl. Phys.* **B75**, 461.

't Hooft, G. (1976a). *Phys. Rev. Lett.* **37**, 8.

't Hooft, G. (1976b). *Phys. Rev.* **D14**, 3432 and (E) **D18**, 2199 (1978).

't Hooft, G. (1979). In *The Whys of Subnuclear Physics*, Proc. Int. School, Erice, 1977 (A. Zichichi, ed.) Plenum, New York.

't Hooft, G., and Veltman, M. (1972). *Nucl. Phys.* **B44**, 189.

Titard, S., and Ynduráin, F. J. (1994). *Phys. Rev.* **D49**, 6007 and **D51**, 6348 (1995).

Titchmarsh, E. C. (1939). *Theory of Functions*, Oxford.

Tomboulis, E. (1973). *Phys. Rev.* **D8**, 2736.

Tyutin, I. V. (1974) Lebedev Institute preprint (unpublished).

Vainshtein, A. I., et al. (1978). *Sov. J. Nucl. Phys.* **27**, 274.

van Neerven, W. L., and Zijlstra, E. B. (1991a). *Phys. Lett.* **B272**, 127.

van Neerven, W. L., and Zijlstra, E. B. (1991b). *Phys. Lett.* **B272**, 476.

van Neerven, W. L., and Zijlstra, E. B. (1991c). *Phys. Lett.* **B273**, 476.

van Neerven, W. L., and Zijlstra, E. B. (1992a). *Phys. Lett.* **B297**, 377.

van Neerven, W. L., and Zijlstra, E. B. (1992b). *Nucl. Phys.* **B382**, 11.

van Neerven, W. L., and Zijlstra, E. B. (1992c). *Nucl. Phys.* **B383**, 525.

Vento, V., et al. (1980). *Nucl. Phys.* **A345**, 413

Veltman, M. (1967). *Proc. Roy. Soc.* (London) **A301**, 107.

Voloshin, M. B. (1979). *Nucl. Phys.* **B154**, 365.

Voloshin, M. B. (1982). *Sov. J. Nucl. Phys.* **36**, 143.

Walsh, T. (1983). In Proc. Europhysics Conf. High Energy Physics (Guy and Constantin, eds.), Brighton.

Walsh, T., and Zerwas, P. (1973). *Phys. Lett.* **B44**, 195.

Webber, B. R. (1984). *Nucl. Phys.* **B238**, 492.

Weeks, B. J. (1979). *Phys. Lett.* **81B**, 377.

Weinberg, S. (1973a). *Phys. Rev. Lett.* **31**, 494.

Weinberg, S. (1973b). *Phys. Rev.* **D8**, 3497.

Weinberg, S. (1975). *Phys. Rev.* **D11**, 3583.

Weinberg, S. (1978a). In a *Festschrift for I. I. Rabi*, New York Academy of Sciences, New York.

Weinberg, S. (1978b). *Phys. Rev. Lett.* **40**, 223.

Weinberg, S. (1980). *Phys. Lett.* **91B**, 51.

Wess, J., and Zumino, B. (1971). *Phys. Lett.* **37B**, 95.

Wiener, N. (1923). *J. Math. and Phys.* **2**, 131.

Wiik, B., and Wolf, G. (1979). *Electron-Positron Interactions*, Springer, Berlin.

Wilczek, F. (1977). In *Quark Confinement and Field Theory* (Stump and Weingartner, eds.), Plenum, New York.

Wilczek, F. (1978). *Phys. Rev. Lett.* **40**, 279.

Wilson, R. (1969). *Phys. Rev.* **179**, 1499.

Wilson, R. (1975). *Phys. Rev.* **D10**, 2445.

Wilson, R. (1977). In *New Phenomena in Subnuclear Physics* (Zichichi, ed.), Plenum, New York.

Wilson, R., and Zimmermann, W. (1972). *Comunn. Math. Phys.* **24**, 87.

Witten, E. (1976). *Nucl. Phys.* **B104**, 445.

Witten, E. (1977). *Nucl. Phys.* **B120**, 189.
Witten, E. (1979a). *Nucl. Phys.* **B156**, 269.
Witten, E. (1979b). *Nucl. Phys.* **B160**, 57.
Witten, E. (1980). *Phys. Today*, July, p. 38.

Yang, C. N., and Mills, R. L. (1954). *Phys. Rev.* **96**, 191.
Ynduráin, F. J. (1996). *Relativistic Quantum Mechanics*, Springer.
Ynduráin, F. J. (1998). *Nucl. Phys.* **B517**, 324.

Zachariasen, F. (1980). In *Hadronic Matter at Extreme Density* (Cabibbo and Ser-
    torio, eds.), p. 313. Plenum, New York.
Zee, A. (1973). *Phys. Rev.* **D8**, 4038.
Zee, A., Wilczek, F., and Treiman, S. B. (1974). *Phys. Rev.* **D10**, 2881.
Zepeda, A. (1978). *Phys. Rev. Lett.* **41**, 139.
Zimmermann, W. (1970). In *Lectures in Elementary Particles and Field Theory*,
    M.I.T. Press.
Zweig, G. (1964). CERN preprints Th. 401 and 412 (unpublished).

Subject Index

Computer to plate: Mercedes Druck, Berlin
Binding: Buchbinderei Lüderitz & Bauer, Berlin